Lecture Notes in Physics

Springer-Verlag Berlin Heidelberg GmbH

The Editorial Policy for Proceedings

The series Lecture Notes in Physics reports new developments in physical research and teaching – quickly, informally, and at a high level. The proceedings to be considered for publication in this series should be limited to only a few areas of research, and these should be closely related to each other. The contributions should be of a high standard and should avoid lengthy redraftings of papers already published or about to be published elsewhere. As a whole, the proceedings should aim for a balanced presentation of the theme of the conference including a description of the techniques used and enough motivation for a broad readership. It should not be assumed that the published proceedings must reflect the conference in its entirety. (A listing or abstracts of papers presented at the meeting but not included in the proceedings could be added as an appendix.)
When applying for publication in the series Lecture Notes in Physics the volume's editor(s) should submit sufficient material to enable the series editors and their referees to make a fairly accurate evaluation (e.g. a complete list of speakers and titles of papers to be presented and abstracts). If, based on this information, the proceedings are (tentatively) accepted, the volume's editor(s), whose name(s) will appear on the title pages, should select the papers suitable for publication and have them refereed (as for a journal) when appropriate. As a rule discussions will not be accepted. The series editors and Springer-Verlag will normally not interfere with the detailed editing except in fairly obvious cases or on technical matters.
Final acceptance is expressed by the series editor in charge, in consultation with Springer-Verlag only after receiving the complete manuscript. It might help to send a copy of the authors' manuscripts in advance to the editor in charge to discuss possible revisions with him. As a general rule, the series editor will confirm his tentative acceptance if the final manuscript corresponds to the original concept discussed, if the quality of the contribution meets the requirements of the series, and if the final size of the manuscript does not greatly exceed the number of pages originally agreed upon. The manuscript should be forwarded to Springer-Verlag shortly after the meeting. In cases of extreme delay (more than six months after the conference) the series editors will check once more the timeliness of the papers. Therefore, the volume's editor(s) should establish strict deadlines, or collect the articles during the conference and have them revised on the spot. If a delay is unavoidable, one should encourage the authors to update their contributions if appropriate. The editors of proceedings are strongly advised to inform contributors about these points at an early stage.
The final manuscript should contain a table of contents and an informative introduction accessible also to readers not particularly familiar with the topic of the conference. The contributions should be in English. The volume's editor(s) should check the contributions for the correct use of language. At Springer-Verlag only the prefaces will be checked by a copy-editor for language and style. Grave linguistic or technical shortcomings may lead to the rejection of contributions by the series editors. A conference report should not exceed a total of 500 pages. Keeping the size within this bound should be achieved by a stricter selection of articles and not by imposing an upper limit to the length of the individual papers. Editors receive jointly 30 complimentary copies of their book. They are entitled to purchase further copies of their book at a reduced rate. As a rule no reprints of individual contributions can be supplied. No royalty is paid on Lecture Notes in Physics volumes. Commitment to publish is made by letter of interest rather than by signing a formal contract. Springer-Verlag secures the copyright for each volume.

The Production Process

The books are hardbound, and the publisher will select quality paper appropriate to the needs of the author(s). Publication time is about ten weeks. More than twenty years of experience guarantee authors the best possible service. To reach the goal of rapid publication at a low price the technique of photographic reproduction from a camera-ready manuscript was chosen. This process shifts the main responsibility for the technical quality considerably from the publisher to the authors. We therefore urge all authors and editors of proceedings to observe very carefully the essentials for the preparation of camera-ready manuscripts, which we will supply on request. This applies especially to the quality of figures and halftones submitted for publication. In addition, it might be useful to look at some of the volumes already published. As a special service, we offer free of charge LaTeX and TeX macro packages to format the text according to Springer-Verlag's quality requirements. We strongly recommend that you make use of this offer, since the result will be a book of considerably improved technical quality. To avoid mistakes and time-consuming correspondence during the production period the conference editors should request special instructions from the publisher well before the beginning of the conference. Manuscripts not meeting the technical standard of the series will have to be returned for improvement.

For further information please contact Springer-Verlag, Physics Editorial Department II, Tiergartenstrasse 17, D-69121 Heidelberg, Germany

Zygmunt Petru Jerzy Przystawa
Krzysztof Rapcewicz (Eds.)

From Quantum Mechanics to Technology

Proceedings of the XXXIInd Winter School of Theoretical Physics, Held in Karpacz, Poland 19–29 February 1996

Springer

Editors

Zygmunt Petru
Jerzy Przystawa
Instytut Fizyki Teoretycznej, Uniwersytet Wrocławski
Pl. M. Borna 9, PL-50-205 Wrocław, Poland

Krzysztof Rapcewicz
Department of Physics, North Carolina State University
Raleigh, NC 27645-8202, USA

Cataloging-in-Publication Data applied for.

Die Deutsche Bibliothek - CIP-Einheitsaufnahme

From quantum mechanics to technology : proceedings of the XXXIInd Winter School of Theoretical Physics, held in Karpacz Poland, 19 - 29 February 1996 / Zygmunt Petru ... (ed.).

(Lecture notes in physics ; Vol. 477)
ISBN 978-3-662-14088-8 ISBN 978-3-540-70724-0 (eBook)
DOI 10.1007/978-3-540-70724-0
NE: Petru, Zygmunt [Hrsg.]; Zimowa Szkoła Fizyki Teoretycznej <32, 1996, Karpacz>; GT

ISSN 0075-8450
ISBN 978-3-662-14088-8

Originally published by Springer-Verlag Berlin Heidelberg New York in 1996
Softcover reprint of the hardcover 1st edition 1996

Typesetting: Camera-ready by the authors
Cover design: *design & production* GmbH, Heidelberg
SPIN: 10550463 55/3144-543210 - Printed on acid-free paper

PREFACE

It is with great satisfaction that we present the Proceedings of the 32nd Karpacz Winter School of Theoretical Physics. It was our intention that the programme of the school should cover topics in condensed matter physics ranging from quantum mechanics to technology – a broad field, indeed. In practice semiconductors, superconductors and spin systems were emphasized, reflecting the scientific interests of the Wrocław physical community; other topics connected with the subject of the school were also treated, but to a lesser degree. We were aware of the risks contained in such a general formula; the usual Karpacz meeting is something between a school and a conference and it was possible that some of the lectures might be too general for specialists or too involved for other participants. However, it seemed to us that there was a need for a school that would provide a forum of discussion for several aspects of condensed matter physics. The success of the 32nd Karpacz School confirmed that the right formula was chosen for the meeting.

These proceedings contain contributions that are outstanding both in their importance and for their timeliness. Therefore, we must express, first of all, our gratitude to the authors of these reports for their collaboration. We also thank the invited speakers who regretfully did not send their manuscripts in time and J. Villain who was not able to attend the school but nonetheless submitted his lecture notes, which are included in the proceedings.

The 32nd Karpacz School was organized jointly by three Wrocław Institutes: the Institute of Theoretical Physics of University of Wrocław (**Jerzy Przystawa**), the Institute of Physics of the Wrocław Technical University (**Lucjan Jacak**) and the Institute of Low Temperature and Structure Research of the Polish Academy of Sciences (**Józef Sznajd**).

The school was attended by 120 participants from 15 countries and the programme comprised 30 invited lectures and poster contributions.

It is a pleasure to thank all of the participants who in spite of the sometimes heated discussions and disputes about physics helped to create the warm and informal atmosphere characteristic of the Karpacz Schools.

We are most obliged to Dr. Czesław Oleksy who took upon himself the burden of finacial matters and to members of the Organizing Committee: Wojciech Gańcza, Bogdan Grabiec, Marek Mozrzymas and Mirosław Pruchnik for their assistance.

Further, we should like to express our deep gratitude to Mrs. Anna Jadczyk for her skillful work in preparing the camera-ready manuscript of the book.

It would have been impossible to organize the school without financial support from our sponsors: the German Stiftung für Deutsch-Polnische Zusammenarbeit, the German Physical Society via WE-Heraeus-Stiftung, the Polish Ministry of National Education (MEN), the International Science Foundation, the State Committee for Scientific Research (KBN), and the Physics Committee of the Polish Academy of Sciences. In this respect, the efforts of Professor Werner Rühl of Kaiserslautern and Professor Jan Stankowski of Poznań should especially be acknowledged.

Wrocław, August 1996 *Jerzy Przystawa and Józef Sznajd*

CONTENTS*

---

* Lecturers names are underline

Theory of Dense Hydrogen: Proton Pairing

N. W. Ashcroft

Laboratory of Atomic and Solid State Physics, Clark Hall,
Cornell University, Ithaca
NY 14853-2501, U. S. A.
and
New Zealand Institute for Industrial Research
Lower Hutt
NZ

Abstract: Dense hydrogen, a dual Fermion system, possesses a Hamiltonian of high symmetry and especial simplicity. As a consequence of the latter its ground state energy satisfies general scaling conditions, independent of phase. Electron exchange is an important contributor, and its role in proton pairing (so evident at low densities) can be argued as a persistent feature. In the single particle description instabilities associated with band-gap closure can be seen as incipient charge density waves but in pair coordinates. This gives rise to a notion of higher pairing within which there can be an associated broken symmetry in electron density (consistent with the observed infrared activity). The persistence of exchange driven pairing under conditions where temperatures approach characteristic vibron energies is discussed in the context of recent reports of the metallization of hydrogen by dynamic methods.

1 Introduction

In first quantization the non-relativistic problem of N protons and N electrons is described by the spin-independent Hamiltonian

$$\hat{H} = \sum_{\alpha=1,2} \Big\{ \sum_{i=1}^{N} (-\hbar^2/2m_\alpha) c\nabla^2_{\alpha i} + \frac{1}{2} \sum_{\alpha'} (-1)^{\alpha+\alpha'} \int_V d\boldsymbol{r} \int_V d\boldsymbol{r}' \phi_c(\boldsymbol{r}-\boldsymbol{r}') \hat{\rho}^{(2)}_{\alpha\alpha'}(\boldsymbol{r},\boldsymbol{r}') \Big\}. \tag{1}$$

The index α differentiates between electrons (e, α=1) and protons (p, α=2), whose corresponding one-particle density operators are

$$\hat{\rho}^{(1)}_\alpha = \sum_{i=1}^{N} \delta(\boldsymbol{r} - \boldsymbol{r}_{\alpha i}). \tag{2}$$

The two-particle density operators appearing in (1) are defined by

$$\hat{\rho}^{(2)}_{\alpha\alpha'} = \hat{\rho}^{(1)}_{\alpha}(\boldsymbol{r})\hat{\rho}^{(1)}_{\alpha'}(\boldsymbol{r}') - \delta_{\alpha\alpha'}\delta(\boldsymbol{r},\boldsymbol{r}')\hat{\rho}^{(1)}_{\alpha}(\boldsymbol{r}), \tag{3}$$

and are obviously linked to pairwise Coulomb interactions ($\phi_c = e^2/|\boldsymbol{r} - \boldsymbol{r}'|$). Though (1) holds for any choice of N, in the approach to the thermodynamic limit it is convenient to regroup the terms as

$$\begin{aligned}\hat{H} &= \sum_{i=1}^{N}(-\hbar^2/2m_e)\boldsymbol{\nabla}^2_{ei}\\ &+\frac{1}{2}\int_V dr \int_V dr' \phi_c(\boldsymbol{r} - \boldsymbol{r}')\left\{\hat{\rho}^{(2)}_{ee}(\boldsymbol{r},\boldsymbol{r}') - 2\hat{\rho}^{(1)}_{e}(\boldsymbol{r})\bar{\rho} + \bar{\rho}^2\right\} &(4a)\\ &+\sum_{j=1}^{N}(-\hbar^2/2m_p)\boldsymbol{\nabla}^2_{pj}\\ &+\frac{1}{2}\int_V dr \int_V dr' \phi_c(\boldsymbol{r} - \boldsymbol{r}')\left\{\hat{\rho}^{(2)}_{pp}(\boldsymbol{r},\boldsymbol{r}') - 2\hat{\rho}^{(1)}_{p}(\boldsymbol{r})\bar{\rho} + \bar{\rho}^2\right\} &(4b)\\ &-\int_V dr \int_V dr' \phi_c(\boldsymbol{r} - \boldsymbol{r}')\left\{\hat{\rho}^{(1)}_{e} - \bar{\rho}\right\}\left\{\hat{\rho}^{(1)}_{p} - \bar{\rho}\right\}. &(4c)\end{aligned}$$

This has been achieved by supplying for the electron system a rigid uniform compensating background charge density $e(N/V) = e\bar{\rho}$, and one of opposite sign for the protons. Then (4a) and (4b) constitute a pair of well-defined and separate problems, but in very different limits of mass, and, for given densities, entirely different forms for their ground state densities, ($\rho^{(1)}_e(\boldsymbol{r}) = \langle\hat{\rho}^{(1)}_e(\boldsymbol{r})\rangle$ and ($\rho^{(1)}_p(\boldsymbol{r}) = \langle\hat{\rho}^{(1)}_p(\boldsymbol{r})\rangle$)). If the average density $\bar{\rho}$ is recorded by the standard parameter r_s ($\frac{4\pi}{3}r_s^3a_o^3 = \bar{\rho}^{-1}$; $a_o = \hbar^2/m_e e^2$), then for $r_s \sim 1$, the ground-state of the uniform interacting electron gas is usually taken as a normal paramagnetic Fermi liquid, though states with off-diagonal long range order have also been proposed [1,2]. The proton problem, represented by (4b) and taken at comparable densities is well into the Coulomb (Wigner) crystal regime, and possibly antiferromagnetic in its spin character.

Since the interactions in (1) or (4) are entirely Coulombic, the ground state energy $E = \langle\hat{H}\rangle_o$ must conform to the general constraints of the virial theorem. In addition, however, some general properties for the ground state are implied by the simplicity of (1) (or (4)). First, independent of phase the ground state energy has a scaling form, consequences of the Hellman–Feynman theorem and the virial theorem together. As shown by Moulopoulos and Ashcroft [3] the general form for E is

$$E = Nf(m_e r_s; m_p r_s)/r_s. \tag{5}$$

Second, if the protons are held fixed (or viewed in the adiabatic approximation), the electronic density $\rho^{(1)}_e(\boldsymbol{r})$ evaluated at any proton (designated here by

o) satisfies the cusp-condition [4,5], namely

$$\partial \rho_e^{(1)}(\boldsymbol{r})/\partial r|_o = (-2/a_o)\rho_e^{(1)}(\boldsymbol{r})|_o \tag{6}$$

and is valid for any N. Third, for a *one-component* Fermion system, with an assumed pairwise interaction v, the canonical partition function $Z = Tr\, exp(-\beta \hat{H})$ can be written as a coherent-state functional integral where, for chemical potential μ, the corresponding action S takes the form [6]

$$\begin{aligned} S = \int_0^{\beta\hbar} d\tau \Big\{ &\int dx \Psi^*(x\tau)\Big(\hbar\frac{\partial}{\partial\tau} - \mu\Big)\Psi(x\tau) \\ &+ \int dx \Psi^*(x\tau)\epsilon(-i\hbar\boldsymbol{\nabla})\Psi(x\tau) \\ &+ \frac{1}{2}\int dx \int dy \Psi^*(x\tau)\Psi^*(y\tau)v(x-y)\Psi(y\tau)\Psi(x\tau)\Big\} \end{aligned}$$

where $\int dx = \sum_s \int d\boldsymbol{r}$. A Hubbard–Stratonovich transformation then introduces new (collective) fields $\{\Delta, \Delta^*\}$ and correspondingly a new action $\tilde{S}$, quadratic in the Ψ's. A Gaussian integration can therefore be carried out leading to

$$Z = (const)\int D[\Delta^*]D[\Delta]exp(\overline{S}/\hbar)$$

where

$$\overline{S} = Tr\ln(\hat{A}/2) + \frac{1}{2}\int d1d2|\Delta(12)|^2/v(1-2) \tag{7}$$

and where $1 \equiv sx\tau$. In (7) the operator $\hat{A}$ is

$$\hat{A}(12) = \begin{pmatrix} \delta(1-2)[\hbar\frac{\partial}{\partial\tau} + \epsilon(-i\hbar\boldsymbol{\nabla}) - \mu] & \Delta(12) \\ \Delta^*(12) & \delta(1-2)[\hbar\frac{\partial}{\partial\tau} + \epsilon(-i\hbar\boldsymbol{\nabla}) - \mu] \end{pmatrix}.$$

These procedures, which have a straightforward parallel for the two-component electron-proton problem, are exact. Extremization of $\overline{S}$ with respect to the choice of the Δ's, is imposed after a stationary phase approximation; it leads to

$$\Delta(12) = -v(12)Tr[\hat{A}^{-1}(12)\begin{pmatrix} 0 & 0 \\ 1 & 0 \end{pmatrix}],$$

and immediately to

$$\Delta(\boldsymbol{k}) = -\sum_{\boldsymbol{k}'} v(\boldsymbol{k}-\boldsymbol{k}')(\Delta(\boldsymbol{k}')/2E(\boldsymbol{k}'))tanh(\beta E(\boldsymbol{k}')/2), \tag{8}$$

after Fourier transformation and a summation over Matsubara frequencies. (Here $E^2(\boldsymbol{k}) = \Delta^2(\boldsymbol{k}) + (\epsilon(\boldsymbol{k}) - \mu)^2$.)

Equation (8), the BCS gap equation for pairing states, is therefore obtained by a compact route which as shown by Moulopoulos and Ashcroft [7] has an immediate and direct parallel for the *two − component* Fermion problem represented by Hamiltonian (1). The corresponding structure of S (and then $\overline{S}$)

admits, as might be expected, of 4 gap equations ($p-p, p-e, e-p$, and $e-e$) in which the effective interactions between like-particles are *always* less repulsive than the direct Coulomb interactions. The important, and intuitive point, is that in a dense phase of hydrogen, a breaking of the symmetry of (1) (or (4)) is expected in a *persistence* of proton pairing [7]. It is the generalization of the Heitler London molecular states to be discussed below, and the extraordinary lack of variation of the principal vibron, (only $\sim 2\%$ change in over 10 fold compression) is ample testimony to it. However, the same arguments suggest that electron-order is also anticipated at high density and low temperature. The expected channels are charge-density waves (diagonal long-range order) and superconductivity (off-diagonal long-range order), the latter following for itinerant electron states from the two-component extension of the steps just outlined.

2 Paired Proton States

The general characteristics just described result from the symmetry and simplicity of the many-body Hamiltonian ((1) or (4)). By action of pressure the average density ($\overline{\rho}$ in (4)) can be very considerably altered, either dynamically [8] or statically [9]. Yet, although static compressions $\overline{\rho}/\overline{\rho}_o$ approaching 12 are now reported, the pairing of protons remains robustly evident. It is therefore useful to consider the physics of the paired proton state both in the molecular limit, and the limit of a pair of protons in an otherwise uniform interacting electron gas at a chosen density r_s ($\hat{H}_{eg} = \hat{H}_{eg}(r_s)$ as in (4a)).

The well known four-particle problem constituting the hydrogen molecule is described by (1) with the choice $N = 2$. For fixed protons the solution for the ground state of the two electrons, a spin antisymmetric state, is known from the definitive work of Kolos and Wolniewicz [10] to a high level of accuracy. Figure 1 shows the one electron density [11]

$$\rho_e^{(1)}(\boldsymbol{r}) = \langle \hat{\rho}_1^{(1)}(\boldsymbol{r})\rangle$$

and it is compared there with densities corresponding to uniform, but dense electron systems. (For the sake of comparison, $\overline{\rho}$ for the valence bands of Be and Al, corresponds respectively to $r_s = 1.87$ and 2.07.) The evident inhomogeneity is also typical of the emerging solid-hydrogen problem and is an important issue in both ordering and energetics.

Within the Heitler-London approximation the one-particle density for inter-proton separation R is

$$\rho_e^{(1)}(\boldsymbol{r}) = \{\rho_a^{(1)}(\boldsymbol{r}) + \rho_a^{(1)}(|\boldsymbol{R}-\boldsymbol{r}|) + 2S_R[\rho_a^{(1)}(\boldsymbol{r})\rho_a^{(1)}(|\boldsymbol{R}-\boldsymbol{r}|)]^{\frac{1}{2}}\}/(1+S_R^2) \quad (9)$$

where
$\rho_a^{(1)}(r) = (\lambda^3/\pi a_0^3)\exp(-2\lambda r/a_0)$, $S_R = (1+\lambda R/a_o + \frac{1}{3}\lambda^2 R^2/a_o^2)\exp(-\lambda R/a_o)$, and λ is a variational parameter. From this it is apparent that the contribution to the one-particle density originating with exchange is significant (it integrates

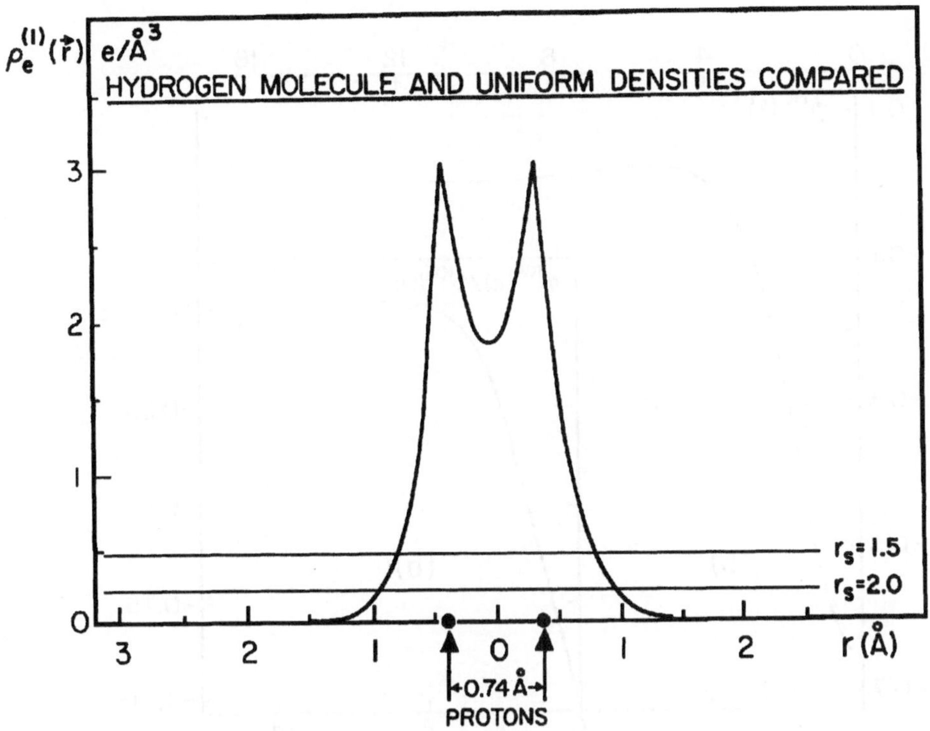

Fig. 1. One-electron density for the hydrogen molecule [11], compared with uniform state electron densities in highly metallic states of Hamiltonian (4b). Note that the valence state average electron density of trivalent aluminum corresponds to r_s=2.07.

to a fraction $S_R^2/(1+S_R^2)$). As is very well known it contributes in an essential way to the overall strength of bonding of the hydrogen molecule. If we describe this bonding by the pair-function $\Phi^{(2)}(R)$, then for later purposes the nature of the bonding can be equally well elucidated by examining the Fourier transform $\Phi^{(2)}(\boldsymbol{k}) = \int d\boldsymbol{R} e^{-i\boldsymbol{k}\cdot\boldsymbol{R}}\Phi^{(2)}(R)$. Figure 2a shows the Fourier transform of the Kolos-Wolniewicz [10] interaction, and its interesting form is readily understood by examination of the basic Heitler-London result. In this the dominant terms, apart from normalization, are

(a) the direct electrostatic contribution

$$\int d\boldsymbol{r} \int d\boldsymbol{r}'(\delta(\boldsymbol{r}) - \rho_a^{(1)}(\boldsymbol{r}))(\delta(\boldsymbol{r}' - \boldsymbol{R}) - \rho_a^{(1)}(\boldsymbol{r}' - \boldsymbol{R}))\phi_c(\boldsymbol{R} - \boldsymbol{r} - \boldsymbol{r}') \quad (10)$$

whose Fourier transform is

$$\frac{4\pi e^2}{k^2}\left\{1 - \frac{(2\lambda/a_o)^4}{(k^2+(2\lambda/a_o)^2)^2}\right\}^2 \quad (11)$$

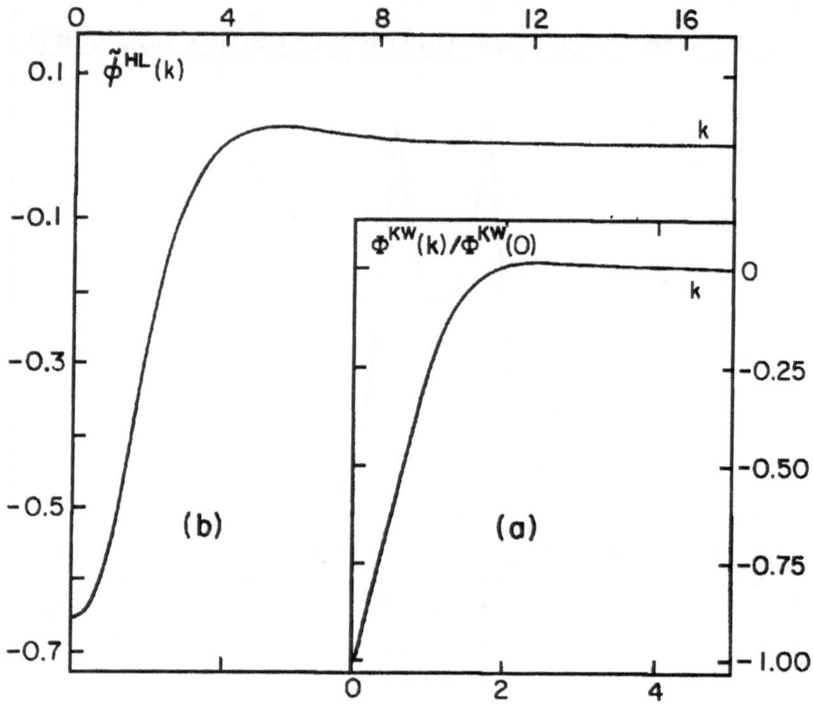

Fig. 2. (a) Fourier transform of the Kolos-Wolniewicz [10] effective intra-molecular interaction.
The strongly negative region at small k is attributable to electron exchange as can be seen from corresponding contributions within the Heitler-London approximation. (b) Fourier transform of Heitler-London effective interaction (except for normalizing denominators - see text).

and
(b), the electrostatic contribution [12] originating with exchange

$$-4(e^2/2a_0)\lambda S_R e^{-\lambda R/a_o}(1+\lambda R/a_o) \tag{12}$$

whose Fourier transform is

$$-8\pi e^2(2\lambda/a_o)^3 \frac{(k^4+6k^2(2\lambda/a_o)^2+21(2\lambda/a_o)^4)}{(k^2+(2\lambda/a_o)^2)^5}. \tag{13}$$

It is the combination of (8) and (10) (shown in Figure 2b) that mainly account for the characteristics of the Kolos-Wolniewicz interaction in reciprocal space. The exchange term is crucial to this behavior, and this will remain the case below (especially in the comparison to linear screening) when two protons are introduced into an interacting electron gas whose density, though high, is small in comparison to local densities about the protons and is typified by the horizontal lines of Figure 1. In assessing the likelihood of *ionic* character in the electronic density, it may be observed that possible admixtures in the two-electron wave function can be inferred by noting that hydrogen molecule can be constructed by (i) bringing together two neutral hydrogen atoms, (ii) bringing up a lone proton to the stable two-electron ion H^-, and (iii) bringing up a lone electron to the stable H_2^+ molecule. The internal electronic structure of the ensuing molecule leads to a considerable average linear polarizability $\alpha = \langle \alpha_\perp + \alpha_\parallel \rangle$. In units of a_o^3 it has the value 5.6; separately $\alpha_\perp = 3.72a_o^3$ and $\alpha_\parallel = 6.49a_o^3$.

3 Protons Adrift in a Fermi Sea

The problem of a single proton (fixed at the origin) in an interacting electron gas is described by the Hamiltonian

$$\hat{H} = \hat{H}_{eg} - \int_V d\boldsymbol{r}\hat{\rho}_e^{(1)}(\boldsymbol{r})\phi_c(\boldsymbol{r}) \tag{14}$$

where $\hat{H}_{eg}$ is given in equation (4a). Studies of (11) using density-functional theory indicate that a bound state sets in for $r_s \gtrsim 1.9$ [13]. At all densities the cusp theorem ensures that this density will have the ground-state atomic form for values of r close to the origin ($r \lesssim a_0$). To describe such a feature, repeated scattering of electrons from the proton must be included; it is therefore essential to proceed beyond linear response. In what follows we will assume that densities are high enough that the states are unbound and if so, that for small enough wave vectors the induced electronic density) will take the form

$$\rho_e^{(1)}(k) = k_o^2/(k_o^2 + k^2)$$

where k_o can be fixed by the compressibility sum rule for the interacting electron gas. On the other hand, for large wave-vector $\rho_e^{(1)}(k)$ must reflect the cusp in $\rho_e^{(1)}(r)$ (*i.e.*, $\rho_e^{(1)}(k) \sim (4\lambda/a_o)/(k^2 + (2\lambda/a_o)^2))^2$). Accordingly the simplest form which incorporates these limits is [14]

$$\rho_e^{(1)}(k) = \frac{(2\lambda/a_o)^4}{(k^4 + 2k^2(2\lambda/a_o)^2\alpha^2 + (2\lambda/a_o)^4)}$$

where $\alpha^2 = (\lambda/k_F)^2(2k_F/k_o)^2/2$. The two important length scales implied by the Hamiltonian (1), namely a_o, and k_F^{-1}, are both embodied in this form. If conditions are chosen so that $\alpha < 1$, then Fourier inversion gives

$$\rho_e^{(1)}(r) = (\frac{\lambda^3}{\pi a_o^3})\frac{exp(-2\lambda r\sqrt{(1+\alpha^2)/2}/a_o)}{\sqrt{(1+\alpha^2)/2}}\left\{\frac{\sin 2\lambda r\sqrt{(1-\alpha^2/2}/a_o}{2\lambda r\sqrt{(1-\alpha^2)/2}/a_o}\right\}. \tag{15}$$

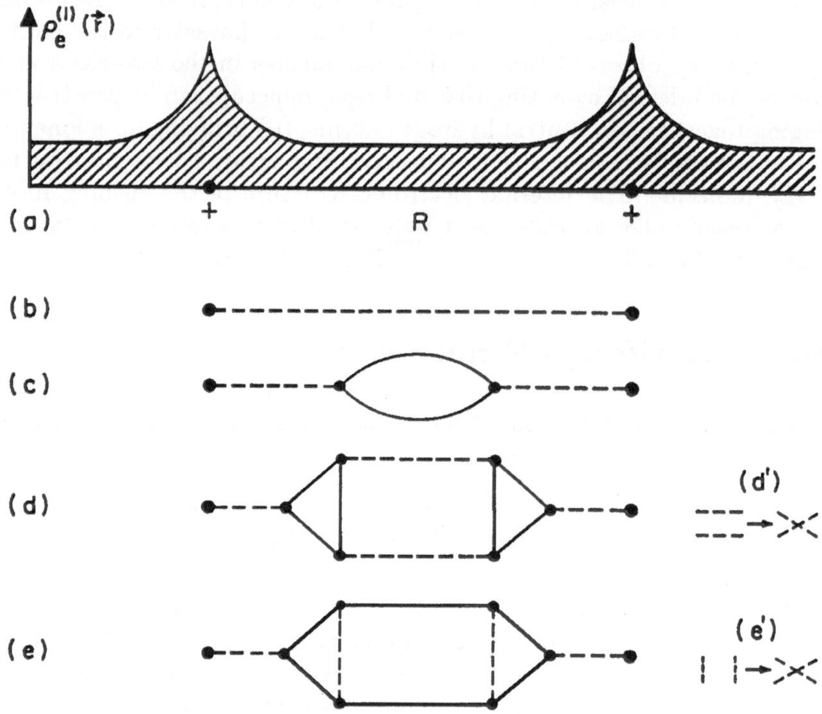

Fig. 3. (a) Form of electronic charge density expected for a pair of protons (separation R) embedded in an interacting electron gas.
(b) and (c) Respectively the direct and linear response proton - proton interactions for Hamiltonian (16). (d') and (e') respectively the lowest order ladder and exchanges diagrams the former leading to Van der Waals type attraction [15].

If this is added to a background density (say $r_s \gtrsim 1$), and the cusp condition applied, then values of λ close to unity emerge. This follows because in an average spin picture the electron density can be approximated by

$$\rho_e(r) = (\lambda^3 e^{-2\lambda r/a_o} + 3/4r_s^3)/\pi a_o^3$$

so that the cusp condition then reads

$$\lambda^4 = \lambda^3 + 3/4r_s^3$$

which has an approximate solution

$$\lambda = 1 + 3/2r_s^3(1 + \sqrt{(1 + 9/r_s^3)}).$$

whose values are close to unity for typical metallic densities. The local value of r_s at the proton in a single hydrogen atom is 0.909, and even at this density λ

rises only to a value 1.38. For background densities typical of ordinary metals the *rate of decline* of the electron density in the vicinity of the proton is very little affected. In the above the assumption $\alpha^2 < 1$ corresponds to $r_s \lesssim 1.9$, and for background densities lower than this the solution for $\rho_e^{(1)}(r)$ passes at long range from oscillatory to exponential (confirming with this model what is expected for a bound state transition). In either case the form of $\rho_e^{(1)}(r)$ for small r is completely dominated by atomic-like behavior. Note, however, that the discussion has been given within a paramagnetic view of the electron gas.

Turning now to a pair of protons fixed at separation $\boldsymbol{R}$ (see Figure 3), Equation (14) becomes

$$\hat{H} = \hat{H}_{eg} - \int_V d\boldsymbol{r}\hat{\rho}^{(1)}(\boldsymbol{r})\phi_c(\boldsymbol{r}) - \int_V d\boldsymbol{r}\hat{\rho}^{(1)}(r)\phi_c(\boldsymbol{r} - \boldsymbol{R}) + e^2/R. \qquad (16)$$

If $E(R)$ is the energy of the pair in the ground state of the electron gas, then from the Hellman-Feynman theorem we have

$$\partial E/\partial R = -e^2/R + \int d\boldsymbol{r}\langle\hat{\rho}^{(1)}(\boldsymbol{r})\rangle_R(-\partial/\partial R)\phi_c(\boldsymbol{r} - \boldsymbol{R}),$$

and the possibility of stabilization ($\partial E/\partial R = 0$) at all such electron densities (all densities leading to unbound states) is immediately raised, and is anticipated on general grounds as noted above. Once again, the situation is very different from the predictions of linear response, where the effective interaction between a proton pair is given by $4\pi e^2/k^2\epsilon(k)$, with $\epsilon(k)$ being the static dielectric function of $\hat{H}_{eg}$. Figure 4 shows the Fourier transform $(1/8\pi^3)\int d\boldsymbol{k}e^{-i\boldsymbol{k}\cdot\boldsymbol{R}}[4\pi e^2/k^2\epsilon(k)]$ for $r_s = 1.5$ and for a choice of $\epsilon(k)$ satisfying the compressibility sum rule of the interacting electron gas. The location and depth of the first minimum are actually both on the scale of the corresponding minimum predicted by the Fourier transform of (8) (i.e. just the Hartree view of the hydrogen molecule); both properties are quite far from what is expected when both repeated scattering and exchange are included. (The evident oscillatory behavior is a reflection of the Friedel structure.)

Long-range *attractive* behavior is, however, expected on far more general grounds. For example, as shown by Maggs and Ashcroft [15], the lowest order ladder diagrams already lead at long range to modified Van der Waals *attraction* [16]. These diagrams are given in Figure 3 along with two of the dominant exchange diagrams which are important at short range. It is worth noting again that repeated scattering of electrons from protons is important in this problem (terms that are omitted, for example, in the treatment given by Ferraz and March [17]).

The linear response view of the problem of dense hydrogen proceeds by associating a coupling constant λ $(0 \leq \lambda \leq 1)$ with electron–proton and proton–proton interactions so that

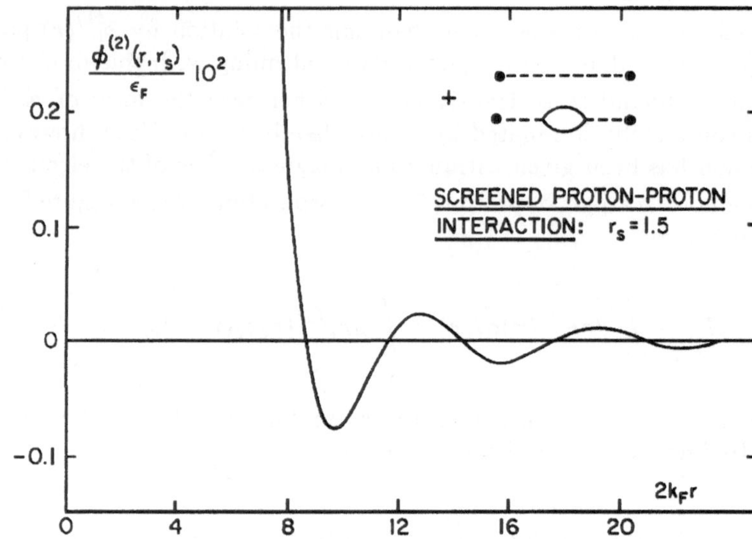

Fig. 4. Typical form of screened (linear response) proton-proton interaction at high densities ($\varepsilon_F = 50.1/r_s^2 eV$). The scale of attraction is very much less than is typical of Heitler-London bonding where exchange (of Figure 3e and 3e′) plays such an important role.

$$\hat{H}(\lambda) = \sum_{i=1}^{N} (-\hbar^2/2m_e)\nabla_{ei}^2 + \frac{1}{2}\int_V d\boldsymbol{r} \int_V d\boldsymbol{r}' \phi_c(\boldsymbol{r}-\boldsymbol{r}')\left\{\hat{\rho}_{ee}^{(2)}(\boldsymbol{r},\boldsymbol{r}') - 2\rho_e^{(1)}(\boldsymbol{r})\bar{\rho} + \bar{\rho}^2\right\} \quad (17a)$$

$$+\sum_{j=1}^{N} (-\hbar^2/2m_p(\lambda))\nabla_{pj}^2 + \frac{\lambda}{2}\int_V d\boldsymbol{r} \int_V d\boldsymbol{r}' \phi_c(\boldsymbol{r}-\boldsymbol{r}')\left\{\hat{\rho}_{pp}^{(2)}(\boldsymbol{r},\boldsymbol{r}') - 2\rho_p^{(1)}(\boldsymbol{r})\bar{\rho} + \bar{\rho}^2\right\} \quad (17b)$$

$$-\lambda \int_V d\boldsymbol{r} \int_V d\boldsymbol{r}' \phi_c(\boldsymbol{r}-\boldsymbol{r}')\left\{\hat{\rho}_e^{(1)}(\boldsymbol{r}') - \bar{\rho}\right\}\left\{\hat{\rho}_p^{(1)}(\boldsymbol{r}) - \bar{\rho}\right\} \quad (17c)$$

where $m_p(\lambda) = m_p/\lambda$.

To illustrate the remarkably different states encountered in the progression $0 \leq \lambda \leq 1$, we may start with (17a) and an r_s value of, say 1.72, corresponding as it happens to average densities actually somewhat higher than are found in the valence bands of the elemental metal Be. This situation is depicted in Figure 5; it is associated with $\lambda = 0$, and the ground state is of deeply metallic character. Keeping the average electron density fixed, now imagine aggregating the background charge into units of the fundamental charge, endowing each with mass m_p, and spin 1/2, i.e., just the progression envisioned in (17b) and (17c).

The result of this elementary construction is dense hydrogen, (Fig. 5), and it is not at all necessary to guess at its states. For the chosen density actually corresponds to near 6-fold compression, and under these conditions experiment appears to have accurately revealed the state of hydrogen. In the first place it is not a metal; it is an insulating crystal [18], though one that is far from static. Next, the protons are strongly paired, tenaciously retaining the 4-particle molecular problem referred to above, a grouping that very much reflects the importance of intra-molecular exchange (with near adiabatic adjustment of the electrons in response to the motion of the protons). Finally, the pairing implies an associated proton dynamics, the primary manifestations being vibrononic motion and orientational motion, the latter including the possibility of complete rotational excitations [19]. Though the structure is not known at higher densities, the pairing character (via the measured vibron excitation) appears established to a solid compression of about a factor of 11. Even at these compressions (corresponding to $r_s \sim 1.41$) there seems little evidence for any ground state metallic character even though band-theory has indicated otherwise. The role of exchange in stabilizing the pairing of protons which self-consistently underlies the evident preference for the insulating state can be seen as starting with the Heitler-London description. Per proton the contribution to the overall energy from exchange related one-body terms (i.e., the difference arising from the choice of Hartree and Heitler-London two-electron states) is just

$$-2S_x e^{-x}\left\{(1+x)-(1+x+x^{2/3})(1-(1+x)e^{-2x})/x\right\}/(1+S_x^2)$$

where $x = \lambda R$ with λ a variational parameter. For small interproton spacing R this difference vanishes quadratically, and at large separation it also declines rapidly. Maximum exchange benefit to the energy of pairing occurs at $x = 1.45$ and is at the level of $0.13Ry$ per proton for $\lambda = 1$. In a condensed phase, with Figure 1 as a guide, the presence of an overall background density will lead through the consequences of the cusp theorem to values of $\lambda > 1$. When due account of the direct interaction is also included, a corresponding decline in the stabilizing pair separation is therefore indicated by the above form of the exchange benefit. It is worth repeating that in hydrogen this benefit is very much tied to the rapidity with which $\rho_e(\boldsymbol{r})$ declines in the vicinity of each proton.

But dense hydrogen appears to be a system where effects going beyond the single-particle description seem to be unusually important. It is useful to note that the energy scale associated with the free-molecular vibron is $\hbar\omega_v \sim 1/2eV$, and this will eventually be carried into the condensed state as an optical band. If at this point the electron energy bands are wide (say ε_F), as they are predicted

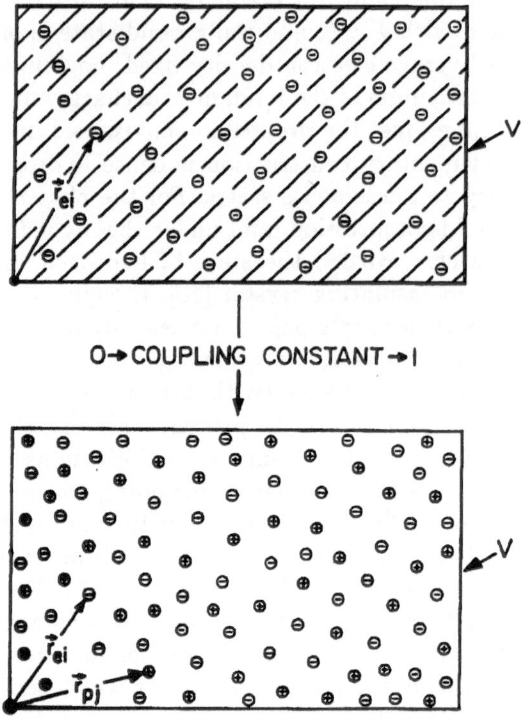

Fig. 5. Coupling constant development of the Hamiltonian representing dense hydrogen (equations 4 and 17) starting with the interacting electron gas (upper, $\lambda=0$) in a uniform background and progressing to a two component system of electrons and protons (lower, $\lambda=1$).

to be at high pressure, then an electron-phonon coupling is expected to be scaled by $\{\hbar\omega_v\varepsilon_F\}^{\frac{1}{2}}$ which, at $r_s \sim 1.5$ is already $\sim 4eV$. This result was important in the early suggestion [20] that metallic hydrogen might be a high temperature superconductor for monatomic states, but also for states based on H_2^+ which is also quite strongly bound and which, on account of its spin, might also be paired. However, as will be seen, a range of different orderings (including those involving spin) is open to hydrogen, each owing its possible origin to strong many-body effects.

4 Bands, Band-Overlap, and the Metallic Transition

A quite general approach to the expected density driven metallic transition in hydrogen can be formulated by augmenting (1) to include the presence of a static but spatially independent vector potential $\boldsymbol{A}$. It is to be chosen to always satisfy

$\nabla \times \boldsymbol{A} = 0$, and if the macroscopic volume V is now taken as a torus, then $\boldsymbol{A}$ will be associated with a magnetic flux permeating the exterior region, these regions being inaccessible to protons and electrons. The Hamiltonian corresponding to (1) now becomes

$$H = \sum_{\alpha=1,2} \Big\{ \sum_{i=1}^{N} (\hat{p}_{\alpha i} + (-1)^{\alpha} e\boldsymbol{A}/c)^2/2m_\alpha + \frac{1}{2}\sum_{\alpha'}(-1)^{\alpha+\alpha'} \int_V d\boldsymbol{r} \int_V d\boldsymbol{r}' \phi_c(\boldsymbol{r}-\boldsymbol{r}')\hat{\rho}^{(2)}_{\alpha\alpha'}(\boldsymbol{r},\boldsymbol{r}') \Big\}$$

where $\hat{p}_{\alpha i} = (\hbar/i)\nabla_{\alpha i}$. It was noted earlier that independent of phase, the ground state energy E of (1) satisfies scaling relations; the same is true here [21], and in addition the dependence of E on any total current $\boldsymbol{J} = \langle \hat{J} \rangle$ that might be established is given by

$$-r_s d/dr_s(E/N) = (\langle \hat{T} \rangle + E)/N - (V/N)\boldsymbol{J}.\boldsymbol{A}/c,$$

where

$$\boldsymbol{J} = \langle \hat{J} \rangle = (e/V)\langle \sum_{\alpha i} (\hat{p}_{\alpha i} + (-1)^{\alpha} e\boldsymbol{A}/c) \rangle.$$

Then so far as the energy is concerned, it follows that for insulating states ($\boldsymbol{J} = 0$) there can be no dependence on $\boldsymbol{A}$ with the specified form. Following Kohn [22], this can be taken as general criterion for the persistence of the insulating phase of hydrogen. A non-vanishing dependence on $\boldsymbol{A}$ will then signify a transition to the metallic state. for the dual system of protons and electrons, the equivalents of (5) for such a state are [21]

$$E = Nf(m_e r_s; m_p r_s; A r_s)/r_s,$$

and

$$E = Ng(e^2 r_s; A^2 r_s)/r_s^2.$$

Though perturbative treatments may be possible [21], general solutions for the hydrogen problem have not yet been established. However, some qualitative arguments can be advanced to suggest that the metallic transion will occur at values of r_s lower than 1.5 (but leaving open still the possibility of further ordering preceding the progression to a metallic state). They begin by first appealing to the dimensional implications of electronic structure of an isolated hydrogen molecule. The Coulomb interactions in (1) are then manifested, for example, in the static dipole polarizability it is a tensor with components $\alpha_{\parallel} = 6.49a_o^3, \alpha_{\perp} = 3.72a_o^3$, and equivalent as stated above to a spherically symmetric object with average polarizability $\alpha = 5.6a_o^3$. If a static charge -e is brought to within r of such an object then the mutual energy is $-(e^2/2a_o)(\alpha/a_o^3)/(r/a_o)^4$. A second measure of the internal structure of the molecule is the ionization energy $E_I(e^2/2a_o) = 1.1351(e^2/2a_o)$. Imagine therefore a possible balance of energies involved in removing an electron from one molecule and then inserting it in a crystal of hydrogen at a distant location. (This may be likened to the process of electron detachment and solvation that accompanies solution of, say, Li, in anhydrous ammonia.) Since the free radical H_2^- is actually unbound, the symmetric choice for localizing this electron will be an interstitial site. If the crystal has high symmetry (for example, face centered cubic with lattice constant a) the energy gain arising from polarization is, in a simple semi-classical approximation

$$-\left(\frac{e^2}{2a_o}\right)\frac{(\alpha/a_o^3)}{(a/2a_o)^4}C$$

where C $=\{6+2\pi/\sqrt{3}\}$. Here the first term in C originates with immediate near neighbors, and the second from all others, but treated as a continuum. Note the conversion $(a/2a_o) = (4\pi/3)^{\frac{1}{3}} r_s$ where r_s is still taken as a convenient measure of the density. It follows that an approximation to the polarization energy is just $-\gamma/r_s^4 Ry$, with γ taking the value $\sim$ 7.9. Because an electron has been localized in an interstitial region there is a kinetic energy δ/r_s^2 Ry, where δ is easily seen to be of order unity. The assumption being made is that electronic timescales internal to the molecule are shorter than those associated with the localized electron. It now appears that a density can be found where an electron can be detached from a molecular unit and located without energy penalty at a distant interstitial site. The condition for this is approximate equality of E_I and $\gamma/r_s^4 - \delta/r_s^2$. The solution is $r_s \sim 1.45$ corresponding to about 9-fold compression.

This estimate is not greatly altered by inclusion of the effects of what may be considered a distant surrounding dielectric on the ionization energy of the atom concerned. To take the case of a hydrogen atom it can be imagined that within the coordination shell of neighbors the potential seen by an electron is $-e^2/r$, but beyond is $-e^2/\epsilon r$ where ϵ is the dielectric constant of what is now taken as a continuum. Perturbation theory then asserts that the ionization energy becomes

$$E_I - 2\left(\frac{\epsilon-1}{\epsilon}\right)e^{-2r_o/a_o}\left(1+\frac{2r_o}{a_o}\right).$$

But ϵ itself is determined by the level structure of the atoms as expressed by the average linear polarizability α of the distant distribution. For example, at the level of the Clausius-Mossotti approximation, $\epsilon = 1 + (4\pi/3)\overline{\rho}\alpha/(1 - 4\pi\overline{\rho}\alpha/3)$. Again, for the hydrogen atom the excited levels will penetrate far more into the surrounding continuum dielectric than will the ground state wave function. Merely to obtain an estimate we can take all states to penetrate equally; then we have a scaling $\alpha = \epsilon^3\alpha_H$, which can be used as a first correction to the constant polarizability models of dielectric divergence. These corrections do not significantly affect the prediction that at compression corresponding to a reduction by about 50% in the separation of near neighbor molecular centers, an electronic instability is anticipated. The same prediction emerges from the more traditional application of the Hertzfeld criterion [23].

In the argument just given it is important to note that H_2^+ has a spin, (and in a suitable background is therefore itself susceptible to exchange driven pairing), and also internal dynamics very different from H_2 (the principal vibron has an energy ~ 0.25 eV). Further, as noted the linear polarizability of H_2 is very anisotropic. From this it can be concluded that in the presence of transferred charges it is aligned states that may take preference as indicated above. Moreover, transferred charges in significant numbers will be subject to electrostatic redistribution (assured by energies of a Madelung character), and also of course to band-broadening. The expectation therefore is of an eventual state of a charge density wave type, that is, of a self-sustaining electronic distortion as envisaged by, for example, Onsager [24].

With this hint of impending electronic instability, dense hydrogen can now be considered from the pragmatic viewpoint of a perfect crystalline state, and therefore from the standard mean-field approach to the electronic problem. As is well known this leads to paramagnetic Bloch waves and energy-bands $\varepsilon_{n\boldsymbol{k}}$, the solutions to

$$\left\{-\frac{\hbar^2}{2me}\boldsymbol{\nabla}^2 + V(\boldsymbol{r})\right\}\psi_{n\boldsymbol{k}} = \varepsilon_{n\boldsymbol{k}}\psi_{n\boldsymbol{k}}$$

with $V(\boldsymbol{r})$ as the best self-consistent approximation to the one- and two-body interactions and to the required overall antisymmetry of the original many-electron problem. For paired proton states an integral number of bands can be exactly filled providing an overall band-gap exists. The magnitude of this gap depends on density, on structure (particularly the orientation of proton pairs) and on the quality of approximation in the construction of $V(\boldsymbol{r})$.

The proposition that dense hydrogen could achieve a metallic state with pairing of protons retained was put forward in 1968 by the author [20], and by Ramaker *et al.* in terms of zone occupancy [25] (and also by Friedli and Ashcroft [26] for a quite specific band structure). The underlying physics rests with the observation that the pairing of protons is very much driven by exchange arising with a density reflecting the cusp structure, and this changes little when electrons are driven from localized to itinerant with r_s in the range $r_s \sim 1$. The reduction of the gap in this process (the gap is generally indirect) is confirmed by most band-structure calculations, those being largely carried out nowadays within the local approximation of density functional theory [27-31]. Since it is precisely the closing of a gap which is the issue here, it is actually necessary to go beyond this local approximation to decide the question. The study by Chacam and Louie [31] achieves this within hexagonal structures; they confirm for hydrogen the general rule that the local approximation under-estimates gaps in a one-electron band-structure.

Residual many-particle effects are also particularly important to this matter of projected band gap closure. On general grounds it is known that a gap cannot close continuously [32]; residual many-electron effects lead to the possibility of instabilities, for example the condensation of excitons, or the formation of charge density waves. For the latter it may be anticipated that if the valence band maximum and conduction band minimum are separated in reciprocal-space by $\boldsymbol{k}_o$ (Figure 6), then a change density-wave accompanying the closure of the associated gap will reflect this basic dimension. As has already been noted [33], distortions in *molecular* coordinates can now arise as a consequence, and given the importance of exchange it can be seen that a relative shift in intermolecular separation can be compensated by an exchange energy lowering. All that is required for this is an associated asymmetry in the charge within a molecular pair (Figure 7). The inference is clear; at a density close to where a single-particle gap is predicted to close, broken electronic symmetry in electron charge should develop. This is observed, for $r_s \leq 1.45$, in recent total energy calculations of progressively densified hydrogen (Edwards and Ashcroft [34]). Locally a molecule possesses a multipole structure now including a dipole. It is known that structures of a filamentary kind can often occur with assemblies of dipoles.

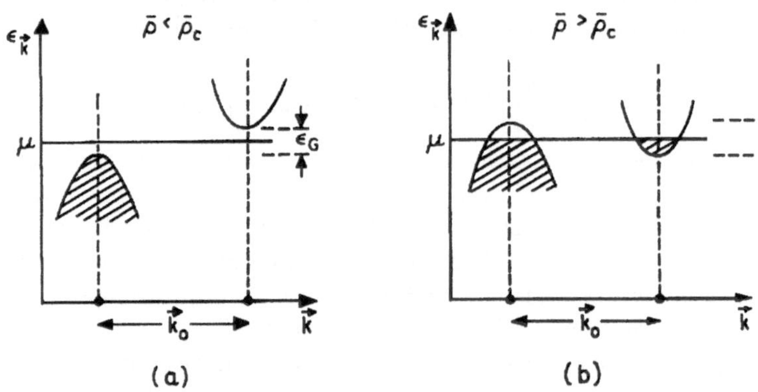

Fig. 6. (a) Typical single-particle band structure of dense hydrogen for average densities $\overline{\rho}$ less than a characteristic density $\overline{\rho}_c$ at which an indirect energy gap (E_G) vanishes. (b) For situations where structure is maintgained increase in pressure to $\rho < \rho_c$ results in a band overlap metallic state, with hole and electron pockets separated by a characteristic wave-vector $\boldsymbol{k}_o$. In the presence of many electron effects gap closure is expected to result in a molecular charge density wave.

5 Liquid Metallic Hydrogen

The tenacity of (electron) exchange driven pairing of protons in dense hydrogen may have ramifications in conditions that begin to approach classical. In a recent experimental advance utilizing dynamic compression techniques, Weir *et al.* [8] report the observation of metallic conductivity in a liquid phase of dense hydrogen at a relative compression $\overline{\rho}/\overline{\rho}_o \sim 9$ and at a temperature T ~3000K (note that this temperature has not reached the equivalent energy of the intramolecular vibron). The experiment measures conductivity directly and its values at these densities is typical of other liquid metals if projected to the temperatures characteristic of the experiment. From analysis of the progression of temperature under dynamic loading it is reported that very little translational energy has been expended in, for example, significant rupturing of proton pairs. If this is accepted then it follows that a metal has been fleetingly formed in hydrogen that remains largely paired, evidently a legacy of exchange under these very extreme conditions.

The detailed nature of this state is not completely clear but a possibility based on the expected high average coordination of the presumed fluid phase presents itself (as is well known, in simple systems the average change in near neighbor coordination on melting is small). On the assumption that the system remains above the melting curve, we have a situation where it is reasonable to assume the diffusional time scale to be far longer than rotational. An isotropic solid phase which, in the sense of orientational disposition, best mimics both the high coordination and rotational averaging characteristics of the liquid is the Pa3 structure. Both the total energy and band-structure of hydrogen in this arrangement can be determined by the methods outlined above. Moreover it is straightforward to show [33] that the lower bands in such a structure can be represented by an equivalent face-centered cubic description where

$$V_{111} = (3/4\pi r_s^3 a_o^3) S(111) v(111) \tag{18}$$

and

$$V_{200} = (3/4\pi r_s^3 a_o^3) S(200) v(200). \tag{19}$$

Here S_o is the structure factor per proton, and is given by

$$\begin{aligned} S_o(l,m,n) = \frac{1}{4}\Big\{ & \cos 2\pi\alpha(l+m+n) + (-l)^{l+m} \cos 2\pi\alpha(-l+m+n) \\ & + (-l)^{m+n} \cos 2\pi\alpha(l-m+n) \\ & + (-l)^{n+l} \cos 2\pi\alpha(l+m-n) \Big\} \end{aligned} \tag{20}$$

with $\alpha = d(r_s)/\sqrt{3}a$ (2a being the separation, on average, between protons). In (18) and (19), the quantity $v(K)$ is the Fourier transform of potential presented by a single proton and associated electron response. Accounting for effective mass rescaling (originating with the higher bands) an elementary condition signalling the onset of band gap closure can be determined in this picture (33), and once again it corresponds to $r_s \sim 1.5$.

This picture can be used to examine, in an approximate way, the role of temperature as it is expected to influence the disposition of hydrogen in a fluid phase. A common expectation is that kinetic effects associated with large temperature rise will lead to dissociation of the proton pairs. Indeed this matter is discussed at length, for example, by Saumon and Chabrier [35] whose main contention is that at high densities H_2, H, H^+ etc. can be given a well defined meaning as separate physical entities entering a fully statistical description. But at $r_s = 1.5$, for example, the one-electron bands are already very wide (the free electron estimate is $50/r_s{}^2 eV$) which means that there must be significant overlap between the orbitals assigned to these entities. Now it has already been observed [33] that beyond densities corresponding to $r_s = 1.76$, the long-range contributions to the interaction potentials associated with neutral entities cannot be defined. Further, as an alternative to complete dissociation there will be processes where a given pair is simply promoted to a state of high vibron excitation resulting in large amplitude spatial fluctuation, but not in overall diffusion. Accordingly, on the issue of the origin of electronic charge at high density and temperature, it may be more profitable to pursue the energy-band route, where it is well known that for low levels of excitation a standard Debye-Waller factor reduction of energy gaps is expected.

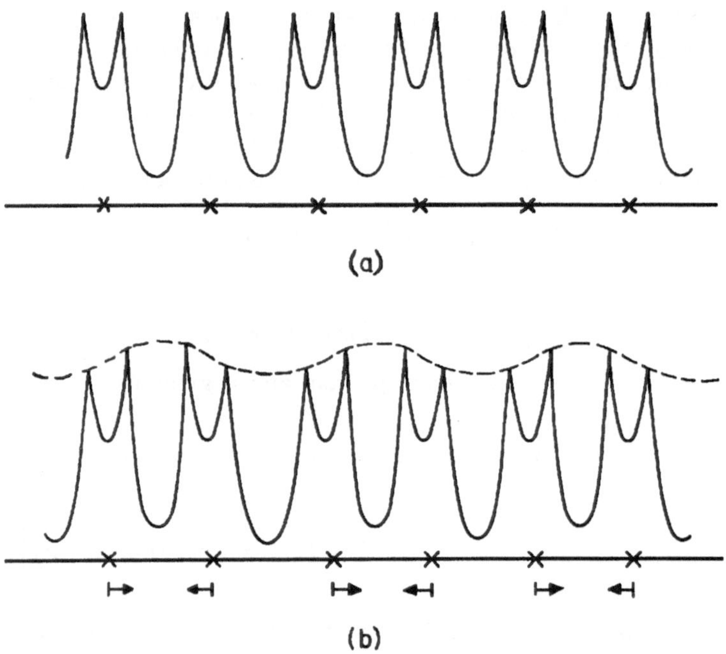

Fig. 7. (a) One-dimensional representation of the normal pairing state in hydrogen, itself originating with a Peierls type distortion from a monatomic arrangement. (b) Higher pairing that results [36] from the charge-density wave associated with the gap closure in Figure 6(b); note the possible charge asymmetry within pairs.

To gauge the general effect of fractional pair extension, we may take the Pa3 structure as a guide, but simply extend one of the four pairs, that is to assign it a separation $2d'(2d' > 2d)$ different from the three remaining basis pairs which are kept at their values for $r_s = 1.5$. Protons are therefore being driven towards what was previously an interstitial region. Figure 8 shows the consequent change in the energy per proton (obtained by density functional theory within the local density approximation for exchange and correlation [36]). On the same scale, the reported temperatures at which metallization has been achieved correspond to $k_BT \simeq 0.019$ (Ry). From Fig. 8 this is seen to associate with a significant possible excursion of the separation of a pair (a tendency towards dissociation). However, this will be in competition with kinetic processes in which energy is accepted into rotational degrees of freedom, and even vibrational. Further with respect to the origin of charge carriers in the reported metallic state [8], there is a further consequence of increasing the separation between protons of a fraction of the pairs; this is an additional lowering of the gap.

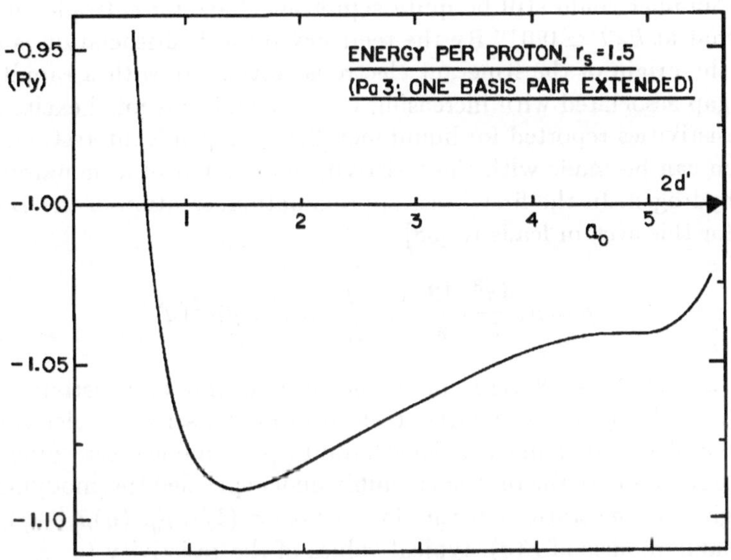

Fig. 8. Energy per proton as a function of separation $(2d')$ of a single proton-pair, the others held fixed in a Pa3 structure at their minimising values (when $d' = d$). Here the average density corresponds to $r_s = 1.5$. The recent shock experiment of Weir et al [8] reports metallization for a dense fluid state at a temperature $k_BT \simeq 0.019$ Ry.

To see this, suppose in a Pa3 structure that three of the pairs maintain their separations at 2d, but the fourth has its associated separation increased to $2d'$, as in the exercise underlying Fig. 8. Then from (20) we see that the structure factor per proton becomes

$$S(l,m,n) = S_o(l,m,n) - \frac{1}{2} sin\pi(\alpha' - \alpha)(l+m+n) sin\pi(\alpha' + \alpha)(l+m+n) \quad (21)$$

and is therefore reduced when a pair is extended. In consequence (see (18) and (19)) the tendency towards band-gap closure will be accelerated by any kinetically induced extensions in bond-lengths. These processes will make it more propitious for electrons to be promoted into conducting states and whether dissociation occurs in significant amounts remains therefore a matter directly connected with the persistence of the exchange driven bonding. For, at a given density this bonding leads within an adiabatic framework to the small oscillations problem associated with the vibron spectrum. There are some 13 levels of excitation for the free molecule; this number is expected to decrease upon compression. The acceptance of energy exceeding the highest vibron (or the highest part of an associated band) could be taken as the signature of the onset of dissociation. Little is known about the progressive decline with pressure of the total number of vibron levels (or bands), but estimates have indicated [37] that at r_s = 1.5 the number could still be quite significant. Save for entropic effects, this suggests that at $k_B T \simeq 0.019$ Ry the tendency towards dissociation is not large and that the origin of the itinerant electrons rests more with a rapidly declining band gap associated with increasing vibronic and rotational excitation. The scale of resistivities reported for liquid metallic hydrogen is interesting. A useful comparison can be made with the resistivity ρ expected of a monatomic phase of liquid hydrogen. In the first Born approximation, solution of the Boltzmann equation for this system leads to [38]

$$\rho = \rho_a \frac{4\pi^3}{3} (\frac{12}{\pi})^{\frac{1}{3}} r_s \int_0^1 dy y^3 S(y) v^2(y) \quad (22)$$

where $\rho_a = a_o \hbar / e^2 \equiv 21.7 \mu\Omega cm$ is the atomic unit of resistivity. In (22), $y = q/2k_F$, and $v(q)$ is the Fourier transform of the screened electron-proton interaction scaled to its long-wave length limit $(\frac{2}{3}\epsilon_F)$ in linear screening. Further the average structure of the proton assembly under specified thermodynamic conditions is given in the static structure factor $S(q) = (1/N)\langle \hat{\rho}_p(\mathbf{q}) \hat{\rho}_p(-\mathbf{q}) \rangle - N\delta_{qo}$. For reasonable models of $S(q)$, typical values of the resistivity (at $r_s \sim 1.5$, for example) are about $15 \mu\Omega cm$ [37]. But as reported by Weir, *et al.*, the state of liquid metallic hydrogen is *far* from monatomic.

The estimate just given corresponds to a static structure for the fluid state which is typical of conditions just above melting. The experiment of Weir *et al.* is likely to correspond to conditions well above melting where $S(q)$ begins to lose the features normally associated with a highly correlated fluid. Since kinetic energy is now playing such a prominent role, it is useful to consider the opposite limit, i.e., $S(q) = 1$. Then with Thomas-Fermi screening, the resistivity in this limit becomes

$$\rho = \rho_a 8 r_s^3 r_o \{ log(r_s + r_o)/r_s - 1 \} / 9(r_s + r_o)$$

where $r_o = (9\pi^4/4)^{\frac{1}{3}} = 6.03$. This leads to resistivities in the vicinity of $40\mu\Omega cm$, notably far below the reported $200 - 500\mu\Omega cm$. However, these values can be partially accounted for by a picture in which the persistence of exchange driven pairing, which plays such a prominent role in the above, actually carries into the high temperatures states (temperatures which still only begin to reach the vibron excitations). The states, now extended, retain the character of the band-overlap phase, in which the high electron density close to the protons persists and continues to favor proton pairing. On the other hand a mobility edge has been crossed, and the phase is conducting. As shown by Louis and Ashcroft [39] the remnants of the band overlap situation typical of the crystal leads to a situation quite different from the monatomic fluid state, and to conductivities that are also very much lower.

The notion of a possible state of dense hydrogen in which electrons are cooperatively but incompletely bound to proton (or deuteron) pairs has been introduced earlier [40], and its consequences with respect to a paired low temperature liquid phase have also been partly investigated. The experiments of Weir *et al.* indicate that this concept of the persistence of proton pairing endures in temperature regimes that appear high, but are still modest on the scale of proton zero point energies. They also indicate that further experimental and theoretical pursuit of the paired metallic phase, and its competition with ionic distortion (including spin ordering [41]) remains now of considerable interest.

Acknowledgments

This work has been supported by the National Science Foundation under Grant DMR-9319864. I am grateful to Prof. M. Teter, Mr. Byard Edwards and Mr. Ard Lewis for discussions on various aspects of the dense hydrogen problem.

References

1. W. Kohn and J. M. Luttinger, Phys. Rev. Lett **15**, 524 (1965).
2. Y. Takada, Phys. Rev. B **47**, 5202 (1993).
3. K. Moulopoulos and N.W. Ashcroft, Phys. Rev. B **41**, 6500 (1990).
4. T. Kato, Commun. Pure Appl. Math **10**, 151 (1957); J.C. Kimball, Phys. Rev. **7**, 1648 (1973).
5. A.E. Carlsson and N.W. Ashcroft, Phys. Rev. **25**, 3474 (1982).
6. H. Kleinert, Fort. Phys. **26**, 565 (1978); V.N. Popov, in *Functional Integrals and Collective Excitations* (Cambridge University Press, Cambridge, 1987).
7. K. Moulopoulos and N.W. Ashcroft, Phys. Rev. Lett. **66**, 2915 (1991).
8. S.T. Weir, A.C. Mitchell, and W. Nellis, Phys. Rev. Lett. **76**, 1860 (1996).
9. See, for example, M. Hanfland, R.J. Hemley, and H.K. Mao, Phys. Rev. Lett. **70**, 3760 (1993); L. Cui, N. Chen, and I.F. Silvera, Phys. Rev. B **51**, 14987 (1995).
10. W. Kolos, and L. Wolniewicz, J. Chem Phys. **49**, 404 (1968).
11. W. Kolos, and C.J.J. Roothan, Rev. Mod. Phys. **32**, 219 (1960).
12. J. C. Slater, in *Quantum Theory of Molecules and Solids* (McGraw Hill, New York, 1963).
13. C.O. Almbladh, U. von Barth, Z.D. Popovic, and M.J. Stott, Phys. Rev. B **14**, 2250 (1976).

14. K. Moulopoulos and N.W. Ashcroft (to be published)
15. A.C. Maggs and N.W. Ashcroft, Phys. Rev. Lett. **59**, 113 (1987).
16. C. Rapcewicz and N.W. Ashcroft, Phys. Rev. B **44**, 4032 (1991).
17. A. Ferraz and N.H. March, J. Phys. Chem. Solids **45**, 627 (1984).
18. M. Li, R. J. Hemley, and H. K. Mao, Bull. Am. Phys. Soc. **39**, 336 (1994).
19. L. Pauling, Phys. Rev. **36**, 430 (1930).
20. N. W. Ashcroft, Phys. Rev. Lett. **21**, 1748 (1968).
21. K. Moulopoulos and N.W. Ashcroft, Phys. Rev. B **45**, 1151B(1992).
22. W. Kohn, Phys. Rev. Lett. **133**A, 171(1964).
23. B. Edwards and N.W. Ashcroft (to be published).
24. L. Onsager, J. Phys. Chem., **43**, 189 (1939).
25. D.E. Ramaker, L. Kumar, and F.E. Harris, Phys. Rev. Lett. **34**, 812 (1975).
26. C. Friedli and N.W. Ashcroft, Phys. Rev. B **16**, 662 (1977).
27. T.W. Barbee, A. Garcia, M.L. Cohen, and J.L. Martins, Phys. Rev. Lett. **62**, 1150 (1989).
28. H. Nagara and T. Nakamura, Phys. Rev. Lett. **68**, 2915 (1991).
29. E. Kaxiras and Z. Guo, Phys. Rev. B **49**, 11822 (1994).
30. E. Kaxiras, J. Broughton, and R.J. Hemley, Phys. Rev. Lett. **67**, 1138 (1991).
31. H. Chacham and S.G. Louie, Phys. Rev. Lett. **66**, 64 (1991).
32. N.F. Mott, Phil Mag **6**, 287 (1961).
33. N. W. Ashcroft, "Elementary Processes in Dense Plasmas"; S. Ichimaru and S. Ogata, Eds. (Addison Wesley, 1995), p. 251.
34. B. Edwards and N.W. Ashcroft, Europhysics Letters, to appear (1996).
35. D. Saumon and G. Chabrier, Phys. Rev. A **44**, 5122 (1991); Phys. Rev. A **46**, 2084 (1992).
36. B. Edwards and N.W. Ashcroft (to be published).
37. N.W. Ashcroft, Phys. Rev. B. **41**, 10983 (1990).
38. D.J. Stevenson and N.W. Ashcroft, Phys. Rev. A **9**, 782 (1974).
39. A. Louis and N.W. Ashcroft (to be published).
40. N.W. Ashcroft, Zeits. für Phys. Chemie. **156**, 41 (1988).
41. N.W. Ashcroft, Physics World **8**, 45 (1995).

Quantum Dots

P.A. Maksym

Department of Physics and Astronomy,
University of Leicester, University Road
Leicester LE1 7RH
UK

Abstract: Quantum dots are reviewed, with emphasis on the theory of electron correlation. A brief survey of dot fabrication and experimental results is also given. The theoretical model of a dot as a 2D system with parabolic confinement and Coulomb interaction is explained and numerical results for spin polarised dots are presented to illustrate the physics in a typical case. When a dot is placed in a magnetic field a series of transitions occurs in which the ground state angular momentum increases with field. The angular momentum values that are selected form a sequence of magic numbers that is characteristic of the electron number and total spin. The ground state transitions affect most physical properties of dots and generally cause magnetic field dependent oscillations. The sequence of magic numbers is related to the symmetry of the classical minimum energy configuration and an approximation technique, based on a harmonic expansion about the classical minimum, is shown to give a good account of dot physics in the high angular momentum limit. Future prospects for studies of coupled dots are briefly discussed.

1 Introduction

Quantum dots are semiconductor nanostructures that are capable of confining very small numbers of electrons. One of the original motivations for studying them was the possible application to novel electro-optic devices [1]. Rapid advances in fabrication technology have subsequently led to a large number of experimental and theoretical studies of dot physics and uncovered some quite novel effects [2, 3] such as the Coulomb blockade and magic numbers. Very recently there has been renewed interest in applications and a quantum dot memory device that operates at room temperature has been demonstrated [4]. The present work is a tutorial review of recent progress with emphasis on the theory of correlation in dots that contain less than about 10 electrons.

The general plan of this work is first to describe some basic properties of dots and then to discuss some of the underlying principles. In keeping with the tutorial aim, the discussion centres on the simplest system, a dot containing just two interacting electrons, which is treated in detail. The general properties of dots are detailed in Section 2 which covers dot fabrication, experimental results and theoretical models. Following this the two electron system is treated in Section 3. This begins with an examination of the classical two electron system and shows how some features of dots have classical origins, while others are definitely not classical and can be traced back to the Pauli exclusion principle. After the basic ideas have been developed for the two electron system they are applied to more electrons in Section 4 and finally some future prospects are discussed in Section 5. This work is solely concerned with quantum dots. A discussion of electron correlation in related nanostructures such as quantum rings and quantum wires can be found in the literature [5, 6, 7].

2 General Overview

Most dots are fabricated by applying a lateral confining potential to a two dimensional electron system. Typically, the two dimensional system is in the form of a heterojunction [8] or a quantum well [9] and the dot is made by applying a confining potential in the plane of the two dimensional layer. One way of doing this is to etch away part of the system but a more common alternative is to use a modulated gate electrode [10]. A small part of the gate is in the form of a cap which is further away from the two dimensional electron layer that the rest. When a negative voltage is applied to the gate the electrons in the regions closest to the gate are fully depleted but a few electrons remain in the region under the cap, which forms the quantum dot. Typically, the region that confines the electrons is about 50-100nm in diameter and the number of electrons can be increased from zero upwards by changing the gate voltage. Many experiments have been done with dots of this kind. A third and more recent fabrication technique is self-organised growth of InAs dots on GaAs substrates [11].

Experimental studies of dots include work on transport through dots, charging of dots and spectroscopy. Early work showed that the far infra-red (FIR) absorption spectrum of dots containing more than one electron is very similar to the absorption spectrum of a single electron in a dot [10, 12]. The similarity of the observed spectra to the single electron spectrum provides strong evidence that the confining potential is nearly parabolic. If it was *exactly* parabolic the centre of mass motion (CM) would separate from the relative motion (RM). Because the dipole matrix element for optical absorption only depends on the centre of mass co-ordinate this means that FIR absorption only depends on the centre of mass motion and hence is unaffected by electron-electron interactions - a result known as the generalised Kohn theorem [13, 14]. The observed form of the experimental spectra is evidence that the confining potential of the real dots is nearly parabolic. However the experimental spectra do contain some interesting splittings which are attributed to residual mixing of CM and RM states by a small non-parabolic component of the confining potential. There have been a number of detailed studies of this effect which is still only partly understood [15, 16]. Optical studies of dots continue to advance and recently it has become possible to do optical experiments on single dots [17].

Studies of transport through dots and charging of dots are closely related. Both make use of the Coulomb blockade to control tunnelling of electrons into and out of a dot [3]. In each case the dot is constructed so that electrons can tunnel into and out of it from a reservoir with electrochemical potential μ. The dot is defined by a system of gates whose potentials V_i can be adjusted. The electrochemical potential of the dot $\mu(N+1;\{V_i\}) = E(N+1,0;\{V_i\}) - E(N,0;\{V_i\})$ is the energy required to add the $(N+1)$th electron to the dot and depends on the gate voltages, V_i, as well as the details of the confinement and interaction potentials. Here $E(N,0;\{V_i\})$ is the ground state energy of the N-electron dot and it is assumed that electrons tunnel into and out of the ground state. When the dot is at low temperature the number of electrons is fixed at N if $\mu(N+1,\{V_i\}) > \mu > \mu(N,\{V_i\})$. However an electron can tunnel into the dot when $\mu(N+1,\{V_i\}) = \mu$. This resonance condition can be set for different values of N by changing the voltage of a gate and leads to a series of conductance peaks in transport of electrons through the quantum dot as a function of the gate voltage [18, 19, 20, 21, 22, 23, 24]. A recent development is the study of photon assisted single electron tunnelling [25, 26]. An elegant and alternative way of probing the resonance condition is to measure the charge in the dot directly. This is done by integrating the dot and a field effect transistor on the same substrate so that the whole system functions as a sensitive electrometer which enables the charge in the dot to be measured [9, 27].

The current theoretical model of a dot is that of a system in which electrons are constrained to move in two dimensions, are confined by a parabolic potential and interact via the Coulomb interaction. This is sufficient to describe the physics of dots semi-quantatively but it is becoming apparent that a more detailed model is needed to treat real dots accurately [28]. For example, it has already been mentioned that the confining potential is not exactly parabolic [15, 16]. In addition, the electron motion is not exactly two dimensional and there are corrections to the Coulomb interaction [29]. A further complication is that the ionised donor density is spatially non-uniform in systems with modulated gate electrodes [30]. At present, the combined effect of all these corrections is not understood and most theoretical studies have been done with the 2D parabolic confinement model.

One of the most useful tools for theoretical work on electron correlation in dots is exact diagonalization of the Hamiltonian. This was pioneered by Bryant [31] who studied interacting electrons in small quantum boxes. Subsequently, Maksym and Chakraborty [14] considered dots with 2D parabolic confinement in a perpendicular magnetic field, B. In this case the Hamiltonian of an N-electron dot is

$$\begin{aligned} H &= \sum_{i=1}^{N}\left[\frac{1}{2m^*}\left(\boldsymbol{p}_i + e\boldsymbol{A}(\boldsymbol{r}_i)\right)^2 + \frac{1}{2}m^*\omega_0^2 r_i^2\right] \\ &+ \frac{1}{2}\left(\frac{e^2}{4\pi\epsilon\epsilon_0}\right)\sum_{i=1}^{N}\sum_{\substack{j=1\\ j\neq i}}^{N}\frac{1}{|\boldsymbol{r}_i - \boldsymbol{r}_j|} + g^*\mu_B B S_z, \end{aligned} \tag{1}$$

where the first term is the one electron term, the second term is the Coulomb interaction term and the last term is the Zeeman energy. The confinement energy is $\hbar\omega_0$, m^* is the effective electron mass, g^* is the effective g-factor and ϵ is the dielectric constant. The circular gauge has been used so the magnetic vector potential, $\boldsymbol{A} = (B/2)(\hat{\boldsymbol{k}} \times \boldsymbol{r})$. The eigenstates of non-interacting electrons with 2D parabolic confinement are the Fock-Darwin states [14]. They have the form

$$\psi_{nl}(\boldsymbol{r}) = \frac{1}{\sqrt{2\pi\lambda^2}} \left[\frac{n!}{(n+|l|)!}\right]^{\frac{1}{2}} \left(\frac{r}{\sqrt{2}\lambda}\right)^{|l|} L_n^{|l|}(r^2/2\lambda^2)\exp(-r^2/4\lambda^2)\exp(-il\phi), \tag{2}$$

with energies given by

$$E_{nl} = (2n+1+|l|)\hbar\Omega - l\hbar\omega_c/2, \tag{3}$$

where l and n respectively are angular momentum and radial quantum numbers, the $L_n^{|l|}$ are associated Laguerre polynomials, $\lambda^2 = \hbar/(2m^*\Omega)$, $\Omega^2 = \omega_0^2 + \omega_c^2/4$ and ω_c is the cyclotron frequency, eB/m^*. The parameter λ is a measure of the length scale of the system and it is important to note that it *decreases* with magnetic field. In addition, the length scale is also affected by the angular momentum and the typical radius of the Fock-Darwin states is given by

$$R^2 \sim 2\lambda^2(2n+|l|+1). \tag{4}$$

The basis for exact diagonalization of the Hamiltonian in Eq. 1 is constructed from Slater determinants of Fock-Darwin states. After the Hamiltonian has been expressed in this basis standard numerical procedures such as the QR algorithm or the Lanczos method can be used to diagonalize it. Calculations for upto about 6 interacting electrons are relatively easy to do on modern workstations but beyond that the size of the basis grows rapidly and method becomes prohibitively expensive.

A particularly interesting aspect of correlation in quantum dots is the way the magnetic field affects the ground state. The ground state energy as a function of magnetic field [28] is shown in Fig. 1 for the case of 3 spin polarised electrons in a GaAs dot with $\hbar\omega_0 = 2$meV (solid line, upper frame).

The contributions of the one electron energy, the Coulomb energy and the Zeeman energy have been calculated by taking the ground state expectation value of the Hamiltonian and are also shown in the figure. It can be seen that the ground state energy is a fairly smooth function of field but the one electron and Coulomb terms have discontinuities. The one electron energy first decreases with increasing field, then increases. The initial decrease is a consequence of the decrease in the length parameter, λ. Physically, it occurs because shrinkage of the wave function puts the electrons closer to the centre of the dot where the confinement potential has a minimum. The increase of one electron energy at higher fields is due to the increasing zero point energy of the cyclotron motion. The shrinkage of the wave function also affects the Coulomb energy which initially increases with field as the electron separation shrinks, but this process does not continue indefinitely. Because the length scale increases with angular

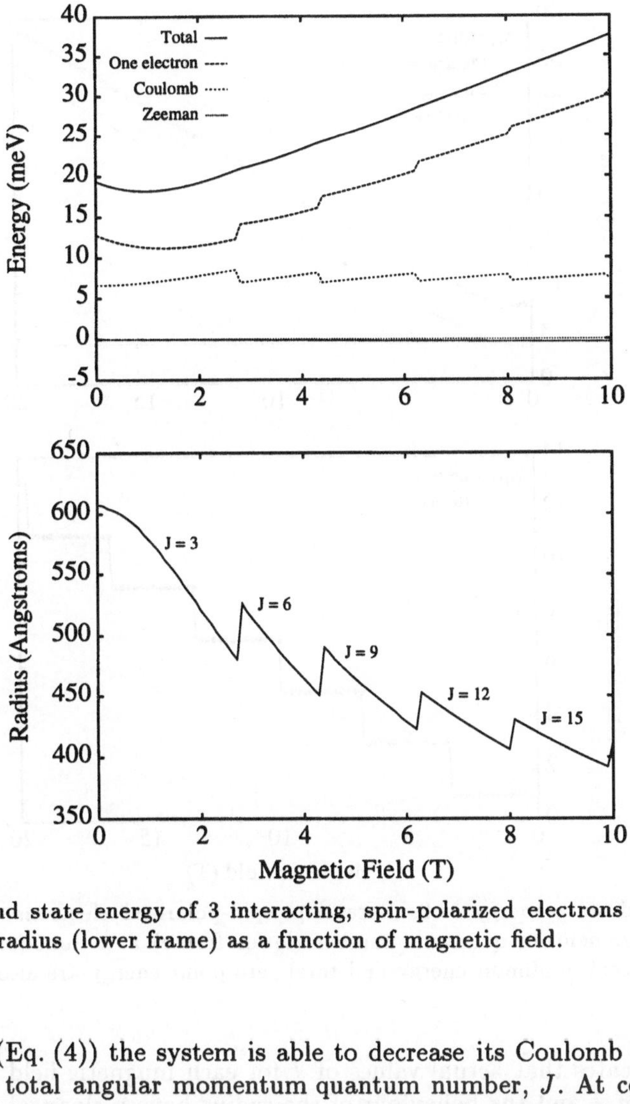

Fig. 1. Ground state energy of 3 interacting, spin-polarized electrons (upper frame) and effective radius (lower frame) as a function of magnetic field.

momentum (Eq. (4)) the system is able to decrease its Coulomb energy by increasing the total angular momentum quantum number, J. At certain critical magnetic fields the increase of one electron energy that results from an increase in J becomes less than the corresponding decrease in the Coulomb energy. The system then undergoes a transition to a new ground state which has a higher value of J. The cycle then repeats with increasing B and this leads to the oscillations in the Coulomb energy and the step-like increase of one electron energy which can be seen in Fig. 1.

The ground state transitions are accompanied by abrupt expansions of the effective dot radius, R_{eff}. This can be estimated by calculating the radius of the circle that contains 95% of the charge, that is, R_{eff} is defined by $2\pi \int_0^{R_{eff}} n_e(r) r dr = 0.95N$, where $n_e(r)$ is the electron density. The lower frame of Fig. 1 shows the abrupt expansions superimposed on top of a decreasing ra-

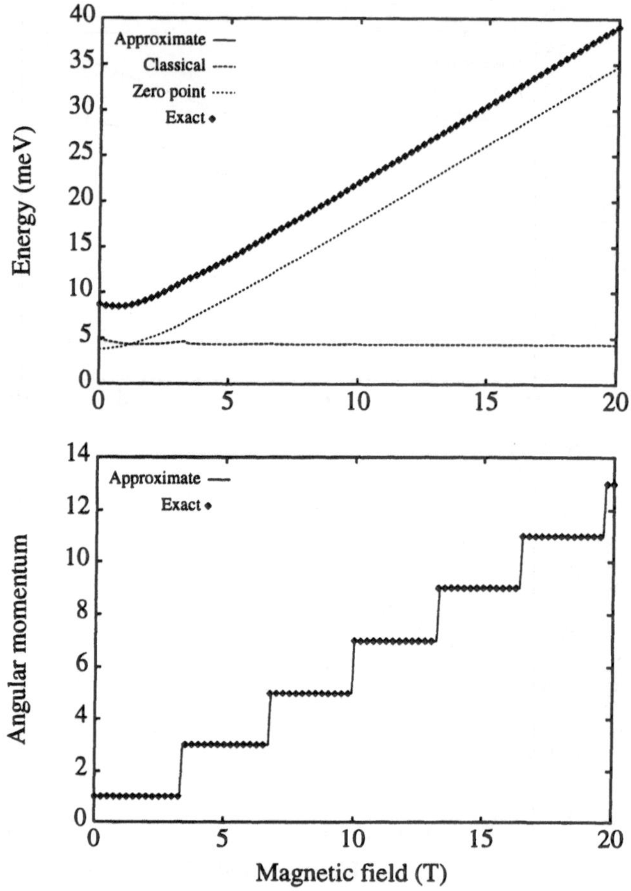

Fig. 2. Ground state energy of 2 interacting, spin-polarized electrons (upper frame) and angular momentum quantum number (lower frame) as a function of magnetic field. The classical minimum energy and total zero point energy are also shown.

dius and indicates that actual values of J for each magnetic field region. Both the increase of J and the behaviour of the radius have a classical origin and it is shown in the next section that the classical orbits of two interacting electrons have similar properties. The real quantum behaviour lies in the actual ground state J values. The ground states only have certain J values, which are characteristic both of the number of electrons and the total spin. These values are known as the magic numbers. For example, the ground state magic numbers for 3 spin polarised electrons are multiples of 3 as indicated in Fig. 1. Roughly speaking, the magic numbers are a consequence of the Pauli exclusion principle: without the exclusion principle, some spatial symmetry is associated with the state that minimises the total energy and the magic J values are the only ones where this spatial symmetry is compatible with the anti-symmetry imposed by the Pauli principle. However, as explained in Section 4, this is an oversimplification because it is necessary to transform to a moving frame to find the spatial symmetry.

Almost all physical properties of quantum dots are affected by the ground state transitions. For instance, they are predicted to cause oscillations in thermodynamic properties such as the electronic heat capacity [14] and magnetisation [32, 33]. In addition, calculations have shown that they affect transport [34, 35], luminescence [36], optical properties [15, 37, 38] and the chemical potential [39, 40, 41]. There is evidence that some of the transitions have been observed [27, 39].

3 Two Interacting Electrons

3.1 Classical Treatment

In the minimum energy configuration of two classical electrons in a 2D parabolic dot the electrons move in circular orbits and remain diametrically opposite each other. The centre of mass is at rest and the Hamiltonian for the relative motion is

$$H_{cl} = \frac{L^2}{4m^* a^2} + \frac{e^2}{8\pi\epsilon\epsilon_0 a} + m^* a^2 \Omega^2 + \frac{\omega_c}{2} L, \tag{5}$$

where a is the orbit radius and L is the total angular momentum. Minimising H_{cl} leads to the following equation which determines a as a function of L and B:

$$\left(\frac{L}{2m^* a^2}\right)^2 + \frac{e^2}{8\pi\epsilon\epsilon_0} \frac{1}{2m^* a^3} - \Omega^2 = 0. \tag{6}$$

This can be solved numerically to find a for any arbitrary value of L but to explore the physics it is sufficient to look at the large L limit. In this case a is approximately given by

$$a = \sqrt{\frac{|L|}{2m^* \Omega}} + O\left(\frac{1}{|L|}\right), \tag{7}$$

and the minimum energy in the same approximation is

$$E_0 \simeq \left(|L|\Omega + L\frac{\omega_c}{2}\right) + \frac{e^2}{8\pi\epsilon\epsilon_0}\sqrt{\frac{2m^*\Omega}{|L|}} + O\left(\frac{1}{|L|^2}\right). \tag{8}$$

The first term is similar to the energy of a Fock-Darwin state, while the second is a Coulomb correction. Clearly this term increases with $|L|$ while the Coulomb energy decreases and this behaviour is the same as that of the quantum one electron and Coulomb terms described in the previous section. The optimal L value minimises the energy. To compare with the quantum case it is convenient to put $L = -\hbar J$ (the J quantum number used in Fig. 1 is defined such that the total angular momentum is $-\hbar J$). The J value, J^*, that minimises E_0 is given by

$$J^{*3/2} = \frac{1}{16\pi\epsilon\epsilon_0} \frac{\sqrt{2m^*\Omega}}{\hbar^{3/2}(\Omega - \omega_c/2)}. \tag{9}$$

This is a monotonically increasing function of B so the classical J value increases in the same way as the quantum one. The limiting behaviour of the orbit radius can be found from the high field limit of Eq. 9, which can be written in the form:

$$B = \frac{m^*}{e}\left[\frac{e^2}{16\pi\epsilon\epsilon_0}\frac{\sqrt{m^*}}{\hbar^{3/2}\omega_0^2}\right]^{-2/3} J^* . \tag{10}$$

This shows that J^* is linear in B and hence a approaches a constant as B becomes large. However in the quantum case the values of J are restricted to integers and the optimal J value is either $[J^*]$ or $[J^*]+1$ where $[J^*]$ is the integer part of J^*, depending on which of these integers gives the lowest value of E_0. With this restriction, a series of transitions occurs in which the J value increases in a step like way and a has the oscillatory behaviour shown in Fig. 1. It is clear, then, that the ground state transitions have a classical origin while quantum mechanics is required to explain the J values that are selected.

The classical treatment also gives some insight into the physics of the transitions. In the classical picture a transition occurs when two orbits with different angular momenta have the same energy, that is when $E_0(J,B) = E_0(J+1,B)$, or approximately when $\partial E_0/\partial J = 0$. This is just the condition that defines J^* so the transition fields in the high field limit can be found from Eq. 10. This shows that the transition field is proportional to J and therefore the transitions are regularly spaced in the high field limit. This behaviour is quite peculiar. It only occurs for special types of potential and the combination of r^2 confinement and $1/r$ interaction is one of them. It is easy to repeat the classical treatment for a general confinement potential of the form r^q and an interaction of the form r^{-p}. In this case the approximate transition fields in the high field regime are given by

$$B = \frac{m^*}{e}\left[\frac{e^2}{16\pi\epsilon\epsilon_0}\frac{\sqrt{m^*}}{\hbar^{3/2}\omega_0^2}\right]^{-2/3} (J^*)^{\frac{p+q}{3}} . \tag{11}$$

and it is clear that the linear J dependence only occurs when $p+q = 3$. A possible use of this result is to probe the form of the interaction potential in real dots. Because the confinement is nearly parabolic, measurements of the transition fields in the high field regime could be used to determine the exponent in the interaction potential and hence to probe screening effects [29].

3.2 Quantum Treatment

In the quantum treatment CM and RM co-ordinates are defined by $\boldsymbol{R} = (\boldsymbol{r}_1 + \boldsymbol{r}_2)/2$ and $\boldsymbol{r} = (\boldsymbol{r}_1 - \boldsymbol{r}_2)/2$ respectively, where $\boldsymbol{r}_1$ and $\boldsymbol{r}_2$ are the individual electron co-ordinates. The Hamiltonian then separates into

$$H = H_{CM} + H_{RM} , \tag{12}$$

where

$$H_{CM} = \frac{1}{4m^*}\Big(\boldsymbol{p}_{CM} + 2e\boldsymbol{A}(\boldsymbol{R})\Big)^2 + m^*\omega_0^2 R^2, \tag{13}$$

$$H_{RM} = \frac{1}{4m^*}\Big(\boldsymbol{p}_{RM} + 2e\boldsymbol{A}(\boldsymbol{r})\Big)^2 + m^*\omega_0^2 r^2 + \frac{e^2}{8\pi\epsilon\epsilon_0 r}. \tag{14}$$

These equations describe a system which is sometimes called quantum dot helium and serves as the simplest example of electron correlation in quantum dots. It has been studied extensively by exact diagonalization, Hartree-Fock and analytic approximation methods [33, 42, 43, 44, 45]. The quantum states of H_{CM} are the Fock-Darwin states given by Eq. 2. The states of H_{RM} are eigenstates of the RM angular momentum and have the form $\exp(-iJ_{RM}\chi)\phi(r)$ where J_{RM} is the RM angular momentum quantum number, χ is a polar angle and $\phi(r)$ is the radial function for the RM motion. This function and the energy of the RM motion, E, are determined by the equation:

$$\frac{-\hbar^2}{4m^*}\frac{1}{r}\frac{d}{dr}\left(r\frac{d\phi}{dr}\right) + \left[\frac{L^2}{4m^*r^2} + m^*\Omega^2 r^2 + \frac{e^2}{8\pi\epsilon\epsilon_0 r} + \frac{\omega_c L}{2}\right]\phi(r) = E\phi(r), \tag{15}$$

where, $L = -\hbar J_{RM}$. The derivative in this equation can be simplified by putting $u(r) = \sqrt{r}\phi(r)$ which leads to

$$\frac{-\hbar^2}{4m^*}\frac{d^2u}{dr^2} + \left[\frac{L^2 - \hbar^2/4}{4m^*r^2} + m^*\Omega^2 r^2 + \frac{e^2}{8\pi\epsilon\epsilon_0 r} + \frac{\omega_c L}{2}\right]u(r) = Eu(r). \tag{16}$$

Except for the $-\hbar^2/(16m^*r^2)$ correction to the centrifugal potential, the potential energy in this equation is identical to H_{cl}. One method of solving the equation approximately is to expand the potential about the classical minimum. The zeroth order term is exactly the classical energy (Eq. 5) and is unaffected by the correction to the centrifugal potential which is of order $1/|L|$. The harmonic term gives the energy of vibrations about the classical minimum. The higher order terms are of order $1/\sqrt{|L|}$ or smaller so the harmonic approximation should describe the physics reasonably well in the large angular momentum limit. The approximate ground state energy is

$$E_{GS} = E_{CM} + E_{cl} + E_0, \tag{17}$$

where $E_{CM} = \hbar\Omega$ is the CM ground state energy, E_{cl} is the classical energy of the potential minimum and E_0 is the RM ground state energy relative to E_{cl}. In this approximation, the total ground state energy is just the sum of the classical energy and quantum zero point energy associated with the RM and CM motion.

When the spatial quantum states are combined with spin states the possible values of J are restricted. The total J value is the sum of J_{CM} and J_{RM} but J_{CM} is always zero for the ground state of the two electron system. The electron state is required to be anti-symmetric and this forces spatial asymmetry when the system is spin polarised. Because the RM wave co-ordinate is antisymmetric this restricts the J values of the spin polarised system to odd integers. Similarly, even J values occur when the system is spin unpolarised. These are the magic numbers for the two electron system.

Fig. 2 (upper frame) shows the approximate and exact ground state energies for two spin polarised electrons, with parameters as used for the calculations leading to Fig. 1. The solid line shows the results of the harmonic approximation and the diamonds give the results of the exact numerical diagonalization. The remaining lines show the classical energy and the sum of the classical and quantum zero point energy. It is clear that the exact and approximate energies agree well even in the small angular momentum regime. The lower frame of Fig. 2 shows the ground state J value. Again, it is clear that the classical and quantum calculations agree well. In addition, the spacing of the transition fields is nearly regular, in agreement with the general discussion after Eq. 10. Using this equation and allowing for the fact that the J step height is 2 gives 3.22T while the numerically calculated spacing is 3.2 ± 0.1T.

4 More Electrons and More Magic Numbers

For systems with more electrons it is not possible to identify the magic numbers from the Pauli principle alone. The two electron system is exceptional in this respect and a simple connection between spin and orbital angular momentum does not occur for larger numbers of electrons. For example, in the 3-electron system spin polarised states are allowed to occur at every angular momentum yet the magic numbers are multiples of 3. To understand why these particular numbers occur it is necessary to identify a class of states that are energetically favourable and then to ask whether these states are compatible with anti-symmetry.

From the 2-electron example, it is reasonable to suppose that the ground state in the high angular momentum limit is localised about the classical minimum. However the classical minimum is not well defined in the laboratory frame because the electrons move in circular orbits. To see a time independent minimum one has to transform to a rotating frame, for example, with suitable choice of frame the optimal positions of two electrons would be diametrically opposite each other on the x axis. This raises the question of a suitable frame for the general N-electron case. Whenever a moving frame is used Coriolis forces appear and this leads to a Hamiltonian in which co-ordinates and momenta are coupled. It is desirable to have a frame in which this coupling is minimised and a suitable choice is the Eckardt frame which has long been used to study the vibrational states of molecules [46, 47]. The Eckardt frame is chosen so that the angular momentum associated with vibrations about the classical minimum vanishes to first order in displacements. This leads to the Eckardt condition,

$$\sum_i \boldsymbol{a}_i \times \boldsymbol{r}'_i = \boldsymbol{0}, \tag{18}$$

where the $\boldsymbol{a}_i$ are the equilibrium positions that minimise the classical energy, the $\boldsymbol{r}'_i$ are positions relative to the CM and all vectors are in the Eckardt frame.

With normal co-ordinates used as dynamical variables, the classical Eckardt frame Hamiltonian is

$$H_{RM} = \frac{1}{2}\mu(L_{RM} - L_v)^2 + \frac{1}{2m^*}\sum_{k=1}^{2N-3} P_k^2 + V + \frac{\omega_c}{2}L_{RM}, \tag{19}$$

where L_{RM} is the RM angular momentum, $L_v = \sum_k \mathcal{Z}_k P_k$, $\mathcal{Z}_k = \sum_{ij}(\boldsymbol{Q}_{ij} \times \boldsymbol{Q}_{ik}) \cdot \hat{\boldsymbol{k}} Q_j$ and $\mu = I_0/(I_0 + m^* \sum_{ij} \boldsymbol{a}_i \cdot \boldsymbol{Q}_{ij} Q_j)^2$. Here P_j is the momentum conjugate to Q_j, I_0 is the equilibrium moment of inertia, the $\boldsymbol{Q}_{ij}$ are elements of the matrix that relates cartesian and normal co-ordinates and V is the total potential (including confinement and interaction terms together with a term quadratic in the magnetic vector potential). The quantity L_v is an angular momentum associated with vibrational motion and it involves products of co-ordinates and momenta. Physically, the coupling occurs because of Coriolis forces and the special feature of the Eckardt frame is that the coupling term is of second order in the displacements from the equilibrium positions. In quantum mechanics the Eckardt frame Hamiltonian is similar to the classical one except for a small extra term which is called the Watson term and is analogous to the correction to the centrifugal potential that appears in Eq. 16. The RM Hamiltonian defined by Eq. 19 depends only on $2N - 3$ co-ordinates although $2N - 2$ co-ordinates are needed to describe the relative motion. The final RM co-ordinate is an Euler angle, χ, that gives the orientation of the Eckardt frame. The transformation to the Eckardt frame is an *exact* transformation of the RM Hamiltonian but it is usually followed by a harmonic expansion of the potential to find the classical vibrational frequencies.

Before applying the harmonic expansion of the Eckardt frame Hamiltonian to study the magic numbers it is necessary to consider the possibility of quantum tunnelling. There are actually $N!$ equilibrium configurations, each corresponding to a different permutation of the electrons. Some of these configurations are connected by rotations but others are not. For example, equilateral triangular configurations of 3 electrons with vertices labelled (123) and (312) can be rotated into each other but it is not possible to rotate (123) into (213). The configurations that cannot be rotated into each other are called symmetrically equivalent. A harmonic expansion can be done about each symmetrically equivalent configurations and the resulting quantum states are degenerate. In reality tunnelling breaks the degeneracy but this effect is small in the large angular momentum limit and the magic numbers can be identified by considering a state localised about any one of the symmetrically equivalent minima. In the harmonic approximation these states take the general form

$$\Psi = \psi_{CM} \exp(-iJ_{RM}\chi) f_{J_{RM}, n_1 \ldots n_{2N-3}}(Q_1 \ldots, Q_{2N-3}) \psi_{spin}(S_z), \tag{20}$$

where ψ_{CM} is the CM wave function, f is a vibrational wave function, $n_1, \ldots n_{2N-3}$ are the numbers of quanta in each vibrational mode, ψ_{spin} is the spin function and S_z is the z component of the total spin, S. The key question now is whether these states can be anti-symmetrized. Applying the anti-symmetrization operator to Ψ either results in an anti-symmetric state or zero. The important point is that the anti-symmetrization operator operates on the laboratory frame co-ordinates. Because of the Eckardt condition, Eq. 18, there are some permutations that are equivalent to spatial rotations. Their effect is to rotate the Euler angle χ and to rotate and permute the Eckardt frame displacements. The classical minimum has definite point symmetry and the corresponding symmetry group determines how the vibrational states transform under the permutations that are equivalent to spatial rotations. The question of whether Ψ can be anti-symmetrized can be decided by considering the effect of these special permutations and because of their equivalence to rotations it is possible to relate the magic numbers to the symmetry of the classical minimum. This requires some group theory but the physics can be illustrated with the special case of the 3-electron, spin-polarised ground state which can be treated without resort to the general procedure. In this case the classical minimum energy configuration is equilateral triangular and the special permutations are equivalent to three-fold rotations, for example (123). The vibrational ground state is symmetric under these rotations while the CM ground state and spin state are symmetric under all permutations. The symmetry properties are therefore determined solely by the angular momentum factor $\exp(-iJ_{RM}\chi)$. The permutations that are equivalent to three-fold rotations have even parity so the angular momentum factor must be invariant under them if the ground state is anti-symmetric. Because $J_{CM} = 0$ in the ground state, this immediately restricts the J_{RM} values to multiples of 3 as shown in Fig. 1. A detailed group theoretical analysis [47, 48, 49, 50] gives all the magic numbers that are found in numerical calculations for electron numbers up to 7 and it is remarkable that exactly the same magic numbers emerge from a composite fermion treatment [51].

When the number of electrons becomes large there is the possibility that the ground state can be influenced by tunnelling between inequivalent minima. That is, there can be competing classical minima which have different symmetry but very similar energies. Approximate quantum states can be found that are localised about each of these minima. However if the states localised on minima of different symmetry have the same quantum numbers tunnelling between the minima is allowed and the ground state is best described as a state of mixed symmetry. The intuitive picture of the system when tunnelling between inequivalent minima is forbidden is that of a "molecule" while the mixed symmetry state might be called a "liquid". Systems of parabolically confined classical electrons have been studied in detail [52, 53, 54] and it is known that competing ground states first occur when the electron number is 6 and the two competing symmetry types in this case are a six-fold ring and a five-fold ring with an electron at the centre. The global minimum energy occurs in the five-fold case. The symmetry analysis for spin polarised electrons shows that five-fold rings occur when the $J_{RM} = 5k$ and six-fold rings occur when $J_{RM} = 6k + 3$, where k is an integer. Mixing can take place when $J_{RM} = 15, 45, 75...$ When these J values are converted to filling factors [55] with the aid of the expression $\nu = N(N-1)/2J$ the resulting sequence of filling factors is $\nu = 1, 1/3, 1/5...$ In other words the "liquid" states of the 6-electron system occur at exactly the same odd denominator filling factors as fractional quantum Hall liquids. An open question is whether tunnelling between symmetrically inequivalent states only takes place at odd denominator filling for an arbitrary number of electrons.

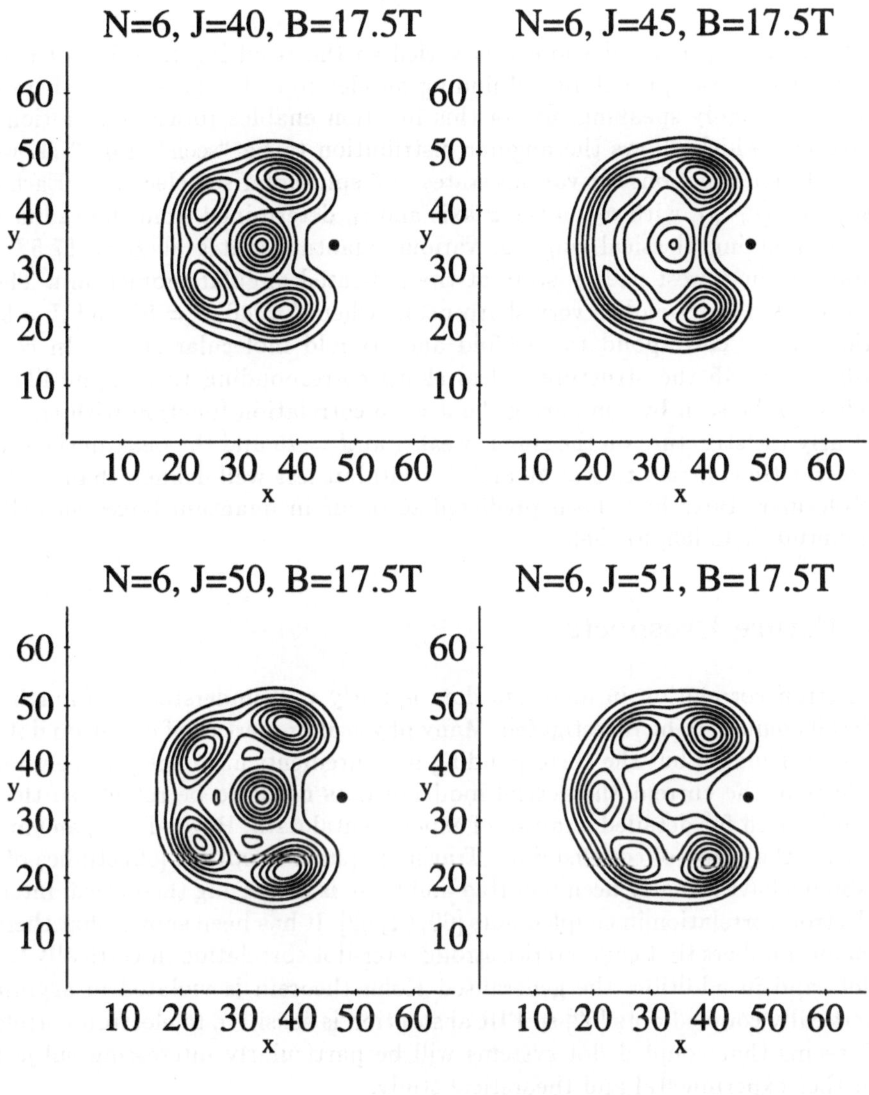

Fig. 3. Pair correlation function $P(\boldsymbol{r}, \boldsymbol{r}_0)$ for 6 interacting, spin-polarized electrons at $B = 17.5$T. Pair correlation functions for the lowest energy state at the indicated angular momenta are shown. The black spots denote $\boldsymbol{r}_0$. The x and y unit is 1.89nm.

The intuitive picture of "molecular" and "liquid" states is consistent with numerical calculations of the pair correlation function [47]. This function has proved extremely useful in characterising quantum dots. For spin polarised electrons it is defined as the ground state expectation value:

$$P(\boldsymbol{r},\boldsymbol{r}_0)=\frac{(2\pi\lambda^2)^2}{N(N-1)}\langle\sum_{i\neq j}\delta(\boldsymbol{r}_i-\boldsymbol{r})\delta(\boldsymbol{r}_j-\boldsymbol{r}_0)\rangle. \tag{21}$$

The vector $\boldsymbol{r}_0$ is fixed while $\boldsymbol{r}$ is varied so the resulting function of $\boldsymbol{r}$ is proportional to the probability of finding an electron at $\boldsymbol{r}$ given that there is one at $\boldsymbol{r}_0$. Roughly speaking, use of this function enables rotational motion to be "frozen" which allows the angular distribution to be "seen". Fig. 3 shows pair correlation functions for various states of 6 spin-polarised electrons. Each frame shows $P(\boldsymbol{r},\boldsymbol{r}_0)$, with $\boldsymbol{r}_0$ on the x axis and r_0 determined from the radius of the corresponding classical ring, for various quantum states at $B=17.5T$. Each state is the lowest energy state at the indicated angular momentum. The pair correlation function has very sharp peaks when $J=40$, $J=50$ and $J=51$ and these cases correspond to five-fold and six-fold molecular states. In contrast, when $J=45$ the structure is less sharp corresponding to a liquid-like state. This can be seen by comparing the $J=45$ correlation function with the others: clearly the structure on the ring is weaker at $J=45$ and the peak in the centre is present (in contrast to $J=40$ and $J=50$) but less well defined than at $J=51$. Molecular states have been predicted to occur in quantum boxes as well as in quantum dots [56, 57, 58].

5 Future Prospects

Electron correlation in quantum dots is fairly well understood although many details remain to be investigated. Many physical properties of quantum dots have been calculated but the corresponding measurements have not yet been done. In addition, the current theoretical model of dots needs to be refined so that it is can be used for detailed analysis of experimental data. Beyond this, an emerging area is the study of coupled dots. Transport [26] and optical [59] studies of these systems have recently been reported and there is increasing theoretical interest in electron correlation in coupled dots [60, 61, 62]. It has been shown that there new magic numbers that characterise strong inter-dot correlation in vertically coupled dots and in addition, the generalised Kohn theorem is violated in asymmetric vertically coupled dots [62] so FIR absorption is sensitive to electron correlation. It seems that coupled dot systems will be particularly interesting subjects for further experimental and theoretical study.

Acknowledgments

It is a pleasure to acknowledge contributions from T. Chakraborty, D. Pfannkuche, V. Gudmundsson, L. D. Hallam, J. Weis, N. Bruce, H. Aoki and H. Imamura. This work was supported by the UK Engineering and Physical Sciences Research Council.

References

1. S. Schmitt-Rink, D.A.B. Miller and D.S. Chemla, Phys. Rev. B **35**, 8113 (1987).
2. T. Chakraborty, Comments in Condensed Matter Physics **16** 35 (1992).
3. M. Kastner, Physics Today, **46**, 17 (1993).
4. K. Imamura, Y. Sugiyama, Y. Nakata, S. Muto and N. Yokoyama, Japan J. Appl. Phys, **34** L1445 (1995).
5. P. Pietiläinen, V. Halonen and T. Chakraborty, Physica B **212**, 256 (1995).
6. D. Yoshioka, Physica B **184**, 86 (1993).
7. T. Kimura, K. Kuroki and H. Aoki, Phys. Rev. B **51**, 13860 (1995).
8. B. Meurer, D. Heitmann and K. Ploog, Phys. Rev. Lett. **68**, 1371 (1992).
9. R.C. Ashoori, H.L. Stormer, J.S. Weiner, L.N. Pfeiffer, S.J. Pearton, K.W. Baldwin and K.W. West, Phys. Rev. Lett. **68**, 3088 (1992).
10. D. Heitmann and J.P. Kotthaus, Physics Today, **46**, 56 (1993).
11. J.M. Marzin, J.M. Gérard, A. Izraël, D. Barrier and G. Bastard, Phys. Rev. Lett. **73**, 716 (1994).
12. C. Sikorski and U. Merkt, Phys. Rev. Lett., **62**, 2164 (1989).
13. L. Brey, N.F. Johnson and B.I. Halperin, Phys. Rev. B **40**, 10647 (1989).
14. P.A. Maksym and T. Chakraborty, Phys. Rev. Lett. **59**, 1140 (1990).
15. D. Pfannkuche and R.R. Gerhardts, Phys. Rev. B **44**, 13132 (1991).
16. V. Gudmundsson, A. Brataas, P. Grambow, B. Meurer, T. Knurth and D. Heitmann, Phys. Rev. B **51**, 17744 (1995).
17. A. Zrenner, Surface Science, in press.
18. P.L. McEuen, E.B. Foxman, U. Meirav, M.A. Kastner, Y. Meir, N.S. Wingreen, and S.J. Wind, Phys. Rev. Lett. **66**, 1926 (1991).
19. Bo Su and V.J. Goldman, Phys. Rev. **B 46**, 7644 (1992).
20. M. Tewordt, L. Martin-Moreno, V.J. Law, M.J. Kelly R. Newbury, M. Pepper, D.A. Ritchie, J.E.F. Frost, G.A.C. Jones Phys. Rev. **B 46**, 3948 (1992).
21. A.T. Johnson, L.P. Kouwenhoven, W. de Jong, N.C. van der Vaart, C.J.P.M. Harmans, and C.T. Foxon, Phys. Rev. Lett. **69**, 1592 (1992).
22. J. Weis, R.J. Haug, K. v. Klitzing and K. Ploog, Phys. Rev. Lett. **71**, 4022 (1993).
23. T. Heinzel, D.A. Wharam, J.P. Kotthaus G. Böhm, W. Klein, G. Tränkle and G. Weimann, Phys. Rev. **B 50**, 15113 (1994).
24. N.C. van der Vaart, M.P. Ruyter van Stevenick, C.J.P.M. Harmans and C.T. Foxon, Physica **B 194**, 1251 (1994).
25. L. Kouwenhoven, S. Jauhar, J. Orenstein, P.L. McEuen, Y. Nagamune, I. Motohisa and H. Sakaki, Phys. Rev. Lett. **73**, 3443 (1994).
26. R.H. Blick, R.J. Haug, K. v. Klitzing and K. Eberl, Surface Science, in press.
27. R.C. Ashoori, H.L. Störmer, J.S. Weiner, L.N. Pfeiffer, K.W. Baldwin and K.W. West, Phys. Rev. Lett. **71**, 613 (1993).
28. P.A. Maksym, L.D. Hallam and J. Weis, Physica B **212**, 213 (1995).
29. L.D. Hallam, J. Weis and P.A. Maksym, Phys. Rev. B, in press.
30. K. Lier and R.R. Gerhardts, Phys. Rev. B **48**, 14416 (1993).
31. G.W. Bryant, Phys. Rev. Lett. **59**, 1140 (1987).
32. P.A. Maksym and T. Chakraborty, Phys. Rev. B **45**, 1947 (1992).
33. M. Wagner, U. Merkt and A.V. Chaplik, Phys. Rev. B **45**, 1951 (1992).
34. N.F. Johnson and M.C. Payne, Phys. Rev. Lett. **70**, 1513 (1993).
35. D. Pfannkuche and S. Ulloa, Phys. Rev. Lett., **74**, 1194 (1995).
36. P. Hawrylak and D. Pfannkuche, Phys. Rev. Lett. **70**, 485 (1993).
37. P. Hawrylak, Solid State Commun. **93**, 915 (1994).

38. V. Halonen, P. Pietiläinen and T. Chakraborty, submitted to Europhys. Lett.
39. P. Hawrylak, Phys. Rev. Lett. **71**, 3347 (1993).
40. S.R. Yang, A.H. MacDonald and M.D. Johnson, Phys. Rev. Lett. **71**, 3194 (1993).
41. J.J. Palacios, L. Martín-Moreno, G. Chiappe, E. Louis and C. Tejedor, Phys. Rev. B **50**, 5760 (1994).
42. D. Pfannkuche, V. Gudmundsson and P.A. Maksym, Phys. Rev. **B47**, 2244, (1993).
43. D. Pfannkuche, R.R. Gerhardts, P.A. Maksym and V. Gudmundsson, Physica B **189**, 6 (1993).
44. M. Taut, J. Phys. A **28**, 2081 (1995).
45. M. El Said, Semiconductor Sci. Tech. **10**, 1310 (1995).
46. E.B. Wilson Jr, J.C. Decius and P. C. Cross, Molecular Vibrations, McGraw-Hill, New York (1955).
47. P.A. Maksym, Phys. Rev. B, in press.
48. P.A. Maksym, Physica B **184**, 385 (1993).
49. W.Y. Ruan, Y.Y. Liu, C.G. Bao and Z.O. Zhang, Phys. Rev. B **51**, 7942 (1995).
50. P.A. Maksym, Europhys. Lett., **31** 405 (1995).
51. J.K. Jain and T. Kawamura, Europhys. Lett., **29**, 321 (1995).
52. F. Bolton and U. Rössler, Superlattices and Microstructures, **13**, 139 (1993).
53. V.M. Bedanov and F.M. Peeters, Phys. Rev. B **49**, 2667 (1994).
54. V.A. Schweigert and F.M. Peeters, Phys. Rev. B **51**, 7700 (1995).
55. S.M. Girvin and T. Jacah, Phys. Rev. B **28**, 4506 (1984).
56. W. Häusler, B. Kramer and J. Mǎsek, Z. Phys. B, **85** 435 (1991).
57. W. Häusler and B. Kramer, Phys. Rev. B **47**, 16353 (1993).
58. K. Jauregui, W. Häusler and B. Kramer, Europhys. Lett. **24**, 581 (1993).
59. K. Bollweg, T. Kurth, D. Heitmann, E. Vasiliadou, P. Grambow and K. Eberl, Surface Science, in press.
60. T. Chakraborty, V. Halonen and P. Pietiläinen, Phys. Rev. B **43**, 14289 (1991).
61. S.C. Benjamin and N.F. Johnson, Phys. Rev. B **51**, 14733 (1995).
62. H. Imamura, P.A. Maksym and H. Aoki, Phys. Rev. B, in press.

On the Electronic Properties of Quantum Dots

L. Jacak[1], *J. Krasnyj*[2], *P. Hawrylak*[3], *A. Wójs*[1] *and W. Nawrocka*[4]

[1] Institute of Physics, Technical University of Wrocław
Wybrzeże Wyspiańskiego 27
50-370 Wrocław, Poland
[2] Institute of Physics, University of Odessa
Petra Velikogo 2, Odessa 270100, Ukraina
[3] Institute for Microstructural Sciences, National Research Council of Canada
Ottawa, Canada K1A 0R6
[4] Institute of Theoretical Physics, University of Wrocław
Pl. M. Borna 9, 50-205 Wrocław, Poland

Abstract:

The electronic states of a parabolic quantum dot in a magnetic field are studied with the spin-orbit interaction included. The analytical formulae for the ground-state energy of the interacting system are derived. The spin-orbit interaction introduces a new feature into the far infra-red absorption spectrum, namely a splitting of the two principal modes. This conclusion is compared with the charging experiments of Ashoori *et al.* and the far infra-red absorption measurements of Demel *et al.*

A model of an exciton confined in a quantum dot is also analyzed. At a critical dot-size, an additional strong line in the photo-luminescence spectrum appears, a consequence of the occurrence of a meta-stable weakly excited state.

1 Spin-orbit interaction in the quantum dot

The spin-orbit coupling in a quasi two-dimensional quantum dot is included in an analogous way to the many-electron atom, i.e. via the single-particle potential

$$V_{LS} = \alpha \boldsymbol{l} \cdot \boldsymbol{\sigma}, \tag{1}$$

where $\boldsymbol{l}$ and $\boldsymbol{\sigma}$ are the orbital and spin angular-momenta of an electron, respectively. The coupling constant α is related to the average self-consistent field $< \delta U >$ acting on the electron via the relation

$$\alpha = \beta < \delta U >, \tag{2}$$

where the dimensionless parameter β is

$$\beta = \left(\frac{Ze^2}{\hbar c}\right)^2 \approx \left(\frac{Z}{137}\right)^2 \tag{3}$$

for a Z-electron atom. In the case of a quantum dot, β will be treated as a fitting parameter, while the magnitude of the field $< \delta U >$ will be estimated from the electronic structure of the dot.

The complete many-electron hamiltonian in the effective-mass approximation, including the kinetic energy in the perpendicular magnetic field B, parabolic confinement of strength ω_0, spin-orbit coupling as above, Zeeman splitting for the effective g-factor and electron-electron Coulomb interaction screened by the dielectric constant ϵ is

$$\begin{aligned}\mathcal{H} &= \sum_i \left[\frac{1}{2m^*}\left(\frac{\hbar}{i}\nabla_i + \frac{e}{c}\boldsymbol{A}_i\right)^2 + \frac{1}{2}m^*\omega_0^2\boldsymbol{r}_i^2 + \alpha l_i\sigma_i - g\mu_B\sigma_i B\right] \\ &+ \frac{1}{2}\sum_{i\neq j}\frac{e^2}{\epsilon|\boldsymbol{r}_i - \boldsymbol{r}_j|} \\ &\equiv \sum_i (H_B)_i + \frac{1}{2}\sum_{i\neq j}\frac{e^2}{\epsilon|\boldsymbol{r}_i - \boldsymbol{r}_j|}. \end{aligned} \tag{4}$$

In the above $\hbar$ is the Planck constant, m^* is the effective mass, $\boldsymbol{r}_i$ are the positions, $\boldsymbol{A}_i = \frac{1}{2}B(y_i, -x_i, 0)$ are the vector potentials in the symmetric gauge, l and σ are the projections of the orbital angular-momentum $\boldsymbol{l}$ and spin $\boldsymbol{\sigma}$ on to the plane of motion.

In the Hartree-Fock (HF) approximation, the equation for the HF wavefunctions ψ is

$$[H_B + V_i]\psi_i(\boldsymbol{r}\sigma) + \sum_{\sigma'}\int d\boldsymbol{r}'\, \Delta_i(\boldsymbol{r}\sigma, \boldsymbol{r}'\sigma')\psi_i(\boldsymbol{r}'\sigma') = \varepsilon_i\psi_i(\boldsymbol{r}\sigma), \tag{5}$$

where H_B is the hamiltonian of a single (non-interacting) electron in the field B defined in Eq. (4), V_i denotes the Hartree potential

$$V_i = \frac{e^2}{\epsilon}\int d\boldsymbol{r}'\, \frac{n_i(\boldsymbol{r}')}{|\boldsymbol{r}' - \boldsymbol{r}|}, \tag{6}$$

with

$$n_i(\boldsymbol{r}) = \sum_\sigma {\sum_j}' |\psi_j(\boldsymbol{r}\sigma)|^2, \tag{7}$$

and Δ_i is the Fock correction, viz.

$$\Delta_i(\boldsymbol{r}\sigma, \boldsymbol{r}'\sigma') = -\frac{e^2}{\epsilon}\delta_{\sigma'\sigma}{\sum_j}' \frac{\psi_j^*(\boldsymbol{r}'\sigma')\psi_j(\boldsymbol{r}\sigma)}{|\boldsymbol{r}' - \boldsymbol{r}|}. \tag{8}$$

Introducing the exchange operator G,

$$G_i\psi_i(\boldsymbol{r}\sigma) = \sum_{\sigma'}\int d\boldsymbol{r}'\, \Delta_i(\boldsymbol{r}\sigma, \boldsymbol{r}'\sigma')\psi_i(\boldsymbol{r}'\sigma'), \tag{9}$$

the HF equations, Eq. (5), can be written in compact form, namely

$$[H_B + V_i + G_i]\,\psi_i(\boldsymbol{r}\sigma) = \varepsilon_i \psi_i(\boldsymbol{r}\sigma). \tag{10}$$

Following the work of Shikin *et al.* [1] for the charge density within the parabolic dot (of confining frequency ω_0), we use the approximate form valid in the classical regime and therefore applicable to case of a large number of electrons N, i.e.

$$n(r) = \begin{cases} n(0)\frac{1}{R}\sqrt{R^2 - r^2} \text{ for } r \le R \\ 0 \text{ for } r > R \end{cases}, \tag{11}$$

where the charge density in the centre is $n(0) = 3N/2\pi R^2$. The dot radius R for the classical system [1] will be calculated from the minimum energy condition in order to take into account the quantum corrections. Using Eq. (11), the Hartree potential is

$$V_R(r) = \frac{3\pi N e^2}{4\epsilon R}\left(1 - \frac{r^2}{2R^2}\right) \tag{12}$$

(we also take this form for $r > R$). Neglecting for the moment the exchange term, which will be included later as a perturbation, we obtain the set of Hartree equations,

$$[H_B + V_R]\,\psi_i(\boldsymbol{r}\sigma) = \varepsilon_i \psi_i(\boldsymbol{r}\sigma), \tag{13}$$

parametrized by the dot radius R.

1.1 In the absence of a magnetic field

Consider first the case of zero magnetic field. The explicit form of Eq. (13) is

$$\left(-\frac{\hbar^2}{2m^*}\Delta + \frac{1}{2}m^* \Omega_0'^2 r^2 + \alpha l\sigma\right)\psi_i(\boldsymbol{r}\sigma) = \left(\varepsilon_i - \frac{3\pi N e^2}{4\epsilon R}\right)\psi_i(\boldsymbol{r}\sigma), \tag{14}$$

where the effective frequency, renormalized by the Hartree term Eq. (12), is

$$\Omega_0'^2 = \omega_0^2 - \frac{3\pi N e^2}{4\epsilon m^* R^3}. \tag{15}$$

Eq. (14) can be solved analytically to obtain the eigenstates

$$\psi_i(\boldsymbol{r}\sigma) = \psi_{nm\sigma}(\boldsymbol{r}\sigma) = \phi_m(\theta) R_{nm}(r)\chi_\sigma, \tag{16}$$

where the spin eigenfunction χ_σ (with eigenvalue $\sigma = \pm\hbar/2$), the angular wavefunction of the angular-momentum eigenvalue m, is

$$\phi_m(\theta) = \frac{1}{\sqrt{2\pi}} e^{im\theta} \tag{17}$$

and the orbital wavefunction is

$$R_{nm}(r) = \frac{\sqrt{2}}{l_0'}\sqrt{\frac{n_r!}{(n_r + |m|)!}}\left(\frac{r}{l_0'}\right)^{|m|} e^{-r^2/2l_0'^2} L_{n_r}^{|m|}\left(\frac{r^2}{l_0'^2}\right). \tag{18}$$

In the above, $L_{n_r}^{|m|}$ are Laguerre polynomials, i.e.

$$L_{n_r}^{|m|}(z) = \frac{1}{m!} z^{-|m|} e^z \frac{d^{n_r}}{dz^{n_r}} (z^{n_r+|m|} e^{-z}), \tag{19}$$

and $l'_0 = \sqrt{\hbar/m\Omega'_0}$ is a characteristic length; $n = 0, 1, 2, \ldots$, the principal quantum number, m, the azimuthal quantum number ($|m| \leq n$ and the parity of m and n is the same), and $n_r = \frac{n-|m|}{2}$ is the radial quantum number.

The eigenenergies of the eigenfunctions $\psi_{nm\sigma}$ are

$$\varepsilon_{nm\sigma} = \hbar\Omega'_0(n+1) + \alpha m\sigma + \frac{3\pi N e^2}{4\epsilon R}. \tag{20}$$

In the absence of a spin-orbit interaction ($\alpha = 0$), they form degenerate shells labelled by n. A non-vanishing α splits these shells into doubly degenerate sublevels.

The complementary Fock-Darwin representation,

$$\psi_i(\boldsymbol{r}\sigma) = \psi_{n_+n_-\sigma}(\boldsymbol{r}\sigma) = \phi_{n_+n_-}(\boldsymbol{r})\chi_\sigma, \tag{21}$$

with $n_\pm = 0, 1, 2, \ldots$, is often used. The two sets of quantum numbers $[n, m]$ and $[n_+, n_-]$ are connected by the simple relations $n = n_+ + n_-$ and $m = n_+ - n_-$. The orbital part of ψ_i is

$$\psi_{n_+n_-}(\boldsymbol{r}) = \frac{1}{\sqrt{2\pi} l'_0} \frac{(a^+)^{n_+}(b^+)^{n_-}}{\sqrt{n_+! n_-!}} e^{-r^2/2l_0'^2}, \tag{22}$$

where the raising operators a^+ and b^+ are

$$\begin{aligned} a^+ &= \frac{1}{2i}\left[\frac{x+iy}{l'_0} - l'_0\left(\frac{\partial}{\partial x} + i\frac{\partial}{\partial y}\right)\right] \\ b^+ &= \frac{1}{2}\left[\frac{x-iy}{l'_0} - l'_0\left(\frac{\partial}{\partial x} - i\frac{\partial}{\partial y}\right)\right]. \end{aligned} \tag{23}$$

The eigenenergies labelled by $n_\pm$ and σ are

$$\varepsilon_{n_+n_-\sigma} = \varepsilon_+(n_+ + \frac{1}{2}) + \varepsilon_-(n_- + \frac{1}{2}) + \frac{3\pi N e^2}{4\epsilon R}, \tag{24}$$

where $\varepsilon_\pm = \Omega'_0 \pm \alpha\sigma$.

The ground state energy of the system written in terms of the N lowest Hartree eigenenergies ε_i, where i stands for the composite index $[n, m, \sigma]$, is

$$\mathcal{E}_0 = \sum_{i=1}^{N} \varepsilon_i - \frac{e^2}{2\epsilon} \int d\boldsymbol{r} \int d\boldsymbol{r}' \frac{n(r)n(r')}{|\boldsymbol{r} - \boldsymbol{r}|'}; \tag{25}$$

the second term removes the double counting of the direct Coulomb energy from the summation of the Hartree energies ε_i. We introduce the Fermi energy

ε_F which separates the occupied and unoccupied Hartree energy-levels in the ground state and calculate the self-interaction

$$\mathcal{E}_0 = \sum_i \Theta(\varepsilon_F - \varepsilon_i)\,\varepsilon_i - \frac{3\pi N^2 e^2}{10\epsilon R}, \tag{26}$$

where Θ is the Heaviside function. The Fermi energy is determined by a condition on the number of electrons N, i.e.

$$N = \sum_i \Theta(\varepsilon_F - \varepsilon_i). \tag{27}$$

We now determine the magnitude of the spin-orbit coupling constant α. It follows from Eq. (2) that we need to estimate the average self-consistent field $< \delta U >$ experienced by an electron. The energy of a classical particle moving on an orbit of radius r in the two-dimensional potential $\frac{1}{2}m^* \Omega_0'^2 r^2$ is

$$\delta U = \frac{1}{2}m^* \Omega_0'^2 r^2 + \frac{1}{2}\Omega_0' l. \tag{28}$$

Replacing the classical variables by their respective operators and averaging δU first over the quantum state $[n, m, \sigma]$,

$$< nm\sigma|\delta U|nm\sigma > = \frac{1}{2}\hbar\Omega_0'(n+m+1), \tag{29}$$

and then over all occupied states,

$$< \delta U > = \frac{1}{2N}\hbar\Omega_0' \sum_{nm\sigma} \Theta(\varepsilon_F - \varepsilon_{nm\sigma})\,(n+m+1), \tag{30}$$

we find that [2]

$$\alpha = \frac{1}{3}\beta\hbar\Omega_0'^2\sqrt{N}. \tag{31}$$

The details of the calculation of the Hartree energy $\mathcal{E}_0$, defined by Eq. (26), have been given elsewhere [2] and we shall only present the final result, viz.

$$\mathcal{E}_0 = \frac{9\pi N^2 e^2}{20\epsilon R} + \frac{2}{3}N^{3/2}\hbar\Omega_0'\sqrt{1 - \frac{\beta^2 N}{36}}. \tag{32}$$

The radius of the dot R can now be determined from the minimum condition $\partial\mathcal{E}_0/\partial R = 0$ which is equivalent to

$$\omega_0^2 = \frac{3\pi N e^2}{4\epsilon R^3 m^*}\left[1 - \frac{100 a_B}{27\pi R}\left(1 - \frac{\beta^2 N}{36}\right)\right], \tag{33}$$

where $a_B = \epsilon\hbar^2/m^* e^2$ is the effective Bohr radius. Restricting our considerations to the case of a large number of electrons, we can solve this equation perturbatively with respect to the small parameter $a_B/R \ll 1$. The zeroth order approximation R_0 can be written as

$$R_0^3 = \frac{3\pi N e^2}{4\epsilon m^* \omega_0^2} \tag{34}$$

and coincides with the classical result [1]. Assuming the first-order approximation is of the form $R = R_0(1+\delta)$, the correction, δ, is

$$\delta = \frac{100 a_B}{81\pi R_0}\left(1 - \frac{\beta^2 N}{36}\right). \tag{35}$$

Neglecting higher-order corrections (non-linear in δ), we calculate the effective confining frequency Ω_0' from its definition Eq. (15), i.e.

$$\Omega_0'^2 = \Omega_0^2\left(1 - \frac{\beta^2 N}{36}\right), \tag{36}$$

where

$$\Omega_0^2 = \frac{100 a_B}{27\pi R_0}\omega_0^2. \tag{37}$$

Finally, we have an expression for the ground-state energy in the Hartree approximation, viz.

$$\mathcal{E}_0 = \frac{9\pi N^2 e^2}{20\epsilon R_0} + \frac{1}{3}\hbar\Omega_0\left(1 - \frac{\beta^2 N}{36}\right) N^{3/2}. \tag{38}$$

The first term in this equation [1] is the classical energy of N interacting electrons confined to a parabolic well, i.e.

$$\frac{9\pi N^2 e^2}{20\epsilon R_0} = \int d\boldsymbol{r}\, n(r)\cdot\frac{1}{2}m^*\omega_0^2 r^2 + \frac{e^2}{2\epsilon}\int d\boldsymbol{r}\int d\boldsymbol{r}'\,\frac{n(r)n(r')}{|\boldsymbol{r}-\boldsymbol{r}'|}. \tag{39}$$

The second term is the quantum correction and decomposes into the energy of an oscillator with frequency, Ω_0, given by

$$\frac{1}{3}\hbar\Omega_0 N^{3/2} = \sum_i \Theta(\varepsilon_F - \varepsilon_i)\,\frac{1}{2}m^*\Omega_0^2 < i|r^2|i >, \tag{40}$$

and the spin-orbit interaction.

We now include perturbatively the exchange interaction, which has been hitherto neglected in the Hartree approximation. We shall calculate the first-order correction $\Delta\mathcal{E}$ as the average value of the exchange operator G, Eq. (9), in the Hartree ground state, neglecting, however, the spin-orbit interaction

$$\Delta\mathcal{E} = \sum_i \Theta(\varepsilon_F - \varepsilon_i) < i|G_i|i > |_{\beta=0}\,. \tag{41}$$

This correction [2] is

$$\Delta\mathcal{E} = -\frac{4\sqrt{5}}{9\sqrt{3}}\left(1 - \frac{3}{4\sqrt{N}}\right)\frac{N^{7/4}e^2}{\epsilon R_0}(1-\delta_0), \tag{42}$$

where $\delta_0 = \delta(\beta = 0)$. Thus we have obtained the total ground state energy $\mathcal{E}$, including the kinetic energy, the direct and exchange Coulomb interaction, and the spin-orbit coupling,

$$\mathcal{E} = \mathcal{E}_0 + \Delta\mathcal{E}, \tag{43}$$

of a system of N electrons confined to a parabolic well. In Fig. 1 we show the average ground state energy per electron $\varepsilon = \mathcal{E}/N$ as a function of the number of electrons N. The two curves corresponding to the parameter β equal 0.3 and 0.6 are a reasonable interpolation between the classical result of Shikin *et al.* [1] and the experimental data of Ashoori *et al.* [3]. We now consider the

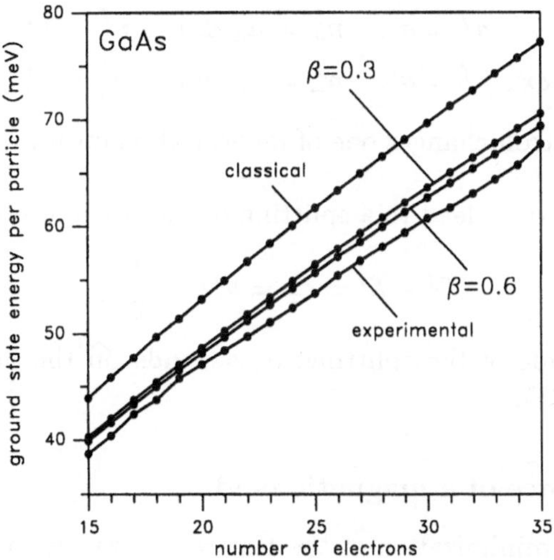

Fig. 1. The average ground state energy per electron as a function of the number of electrons in the dot. The classical result is taken from Ref. [1], the experimental result, from Ref. [4], and the two curves in between are obtained within our model for two values of the spin-orbit coupling constant β (GaAs, $\hbar\omega_0 = 5.4$ meV).

selection rules for optical transitions of the system in the presence of far infrared (FIR) radiation. The absorption of FIR light, which results in an excitation of the electron droplet, has been a powerful tool in the experimental studies of quantum dots [4, 5].

Since the wavelength in the FIR region is much larger than the radius of the dot, we can use the dipole approximation to describe the interaction between light and the electrons. The probability of an optical transition between the initial, (i), and final, (f), state is proportional to the squared matrix element of this interaction, i.e.

$$d_{fi}^2 \sim |< f|e\boldsymbol{E} \cdot \sum_i \Theta(\varepsilon_F - \varepsilon_i)\,\boldsymbol{r}_i|i > |^2, \tag{44}$$

where $\boldsymbol{E}$, the electric field, is uniform over the volume of the dot. The dipole matrix element, d_{fi}, vanishes unless there is a pair of HF states, one in the initial and the other in the final many-electron state, with equal spins, each of the orbital quantum numbers $[n, m]$ differing by unity, i.e.

$$\sigma^f = \sigma^i, \quad |n^f - n^i| = 1, \quad |m^f - m^i| = 1, \tag{45}$$

and with all the other HF states identical in the initial and final state. In other words, the absorption of a FIR photon leads to the excitation of a single electron from its (initial) HF state to another (final) HF state with the same spin σ and the orbital quantum-numbers changed according to Eq. (45). Translating these selection rules to the Fock-Darwin representation, we have

$$\begin{aligned} &\sigma^f = \sigma^i, \quad n_+^f = n_+^i \pm 1, \quad n_-^f = n_-^i, \\ \textit{or} \quad &\sigma^f = \sigma^i, \quad n_-^f = n_-^i \pm 1, \quad n_+^f = n_+^i, \end{aligned} \tag{46}$$

i.e. the excited electron changes one of its orbital quantum-numbers $[n_+, n_-]$ by unity.

These selection rules lead to a splitting of the resonance energy

$$\mathcal{E}^f - \mathcal{E}^i = \varepsilon_\pm = \hbar\Omega_0' \pm \frac{1}{2}\alpha, \tag{47}$$

where the magnitude of the splitting, α, depends on the number of electrons according to Eq. (31).

1.2 In the presence of a magnetic field

The procedure for minimization of the Hartree energy with respect to the dot radius R with the later perturbative inclusion of the exchange interaction, as sketched in the previous section for the case of zero magnetic field, has been also carried out for nonzero fields. The explicit form of the Hartree equations with a nonzero perpendicular magnetic field present, Eq. (13), is

$$\begin{aligned} &\left(-\frac{\hbar^2}{2m^*}\Delta + \frac{1}{2}m^*\left(\Omega_B'^2 + \frac{1}{4}\omega_c^2\right)r^2 - \frac{1}{2}\hbar\omega_c l - g\mu_B\sigma B + \alpha l\sigma\right)\psi_i(\boldsymbol{r}\sigma) \\ &\qquad = \left(\varepsilon_i - \frac{3\pi N e^2}{4\epsilon R(B)}\right)\psi_i(\boldsymbol{r}\sigma), \end{aligned} \tag{48}$$

where $\omega_c = eB/m^* c$ is the effective cyclotron frequency and the zero-field radius R appearing in the definition of Ω_B' (15) is now replaced by $R(B)$. We shall also denote the total confining frequency by $\Omega'^2 = \Omega_B'^2 + \omega_c^2/4$, and its corresponding characteristic length by $l = \sqrt{\hbar/m^*\Omega'}$.

The eigenfunctions of Eq. (48) are of the same form as Eqs. (16) and (21), with the characteristic length replaced by l. The corresponding eigenenergies are

$$\varepsilon_{nm\sigma} = \hbar\Omega'(n+1) - \frac{1}{2}\hbar\omega_c m - g\mu_B\sigma B + \alpha m\sigma + \frac{3\pi N e^2}{4\epsilon R(B)} \tag{49}$$

or, in the other representation,

$$\varepsilon_{n_+n_-\sigma} = \varepsilon_+(n_+ + \frac{1}{2}) + \varepsilon_-(n_- + \frac{1}{2}) - g\mu_B\sigma B + \frac{3\pi Ne^2}{4\epsilon R}, \tag{50}$$

with $\varepsilon_\pm = \Omega_0' \pm (\hbar\omega_c/2 + \alpha\sigma)$. Since the Zeeman splitting is rather small for GaAs ($g \sim \frac{1}{2}$ yielding $g\mu_B \sim 0.05$ meV/T), the most common material used for quantum dots, we shall neglect it in our further considerations.

The inclusion of a magnetic field leads to the possibility of crossings between different energy levels ε_i. Whether the two close levels ε_1 and ε_2 actually cross, or whether such a crossing is forbidden, depends upon the vanishing of the off-diagonal matrix element of the operator describing the change of the hamiltonian due to a small change of the field. Thus the condition for the permitted level-crossing is

$$< 1|\frac{\partial H}{\partial B}|2 > \equiv 0. \tag{51}$$

The operator $\partial H/\partial B$ commutes with the spin and inversion ($\boldsymbol{r} \to -\boldsymbol{r}$) operators. Commutation with the angular momentum operator requires the already assumed circular symmetry of the confining potential. Thus for states with unequal quantum numbers n and σ, we have condition Eq. (51) guaranteed; for a pair of states differing only in m, it is, in general, no longer true. This leads to the anti-crossing of levels that can be taken into account by changing the formulae for the eigenenergies to

$$\varepsilon_{nm\sigma} = \hbar\Omega'(n+1) + m\left|\frac{1}{2}\hbar\omega_c + \alpha\sigma\right| + \frac{3\pi Ne^2}{4\epsilon R(B)}. \tag{52}$$

We have to modify analogously the definition of the energies ε_+ and ε_-, appearing in Eq. (50), to $\varepsilon_\pm = \Omega_0' \pm |\hbar\omega_c/2 + \alpha\sigma|$.

Following a similar procedure to the zero magnetic-field case, we can estimate the spin-orbit coupling constant, now a function of the field,

$$\alpha(B) = \frac{1}{3}\beta f_B \hbar\Omega_B'\sqrt{N}, \tag{53}$$

with the renormalizing function

$$f_B = \sqrt{1 + z^2/N}\left(1 - \frac{z}{\sqrt{1+z^2}}\right)\Bigg|_{z=\omega_c/2\Omega_B'}. \tag{54}$$

At low fields, the function f_B tends to unity; for $B \to \infty$, it decays to zero $f_B \sim 1/\sqrt{N}$. To a good approximation, we can therefore replace Ω_B' by Ω_0, defined by Eq. (37), in the definition of f_B.

The Hartree energy of the system can be now expressed as [2]

$$\mathcal{E}_0(B) = \frac{9\pi N^2e^2}{20\epsilon R_0} + \frac{2}{3}N^{3/2}\hbar\left(\omega_0^2 u_B - \frac{3\pi Ne^2}{4\epsilon m^* R(B)^3}\right)^{1/2}\sqrt{1 - \frac{\beta^2 f_B^2 N}{36}}, \tag{55}$$

where

$$u_B = 1 + (\omega_c/2\Omega_0)^2 \frac{1}{1-z}\left(\frac{1}{N} - 4\frac{z}{1-z}\right)\bigg|_{z=\beta^2 f_B^2 N/36} . \tag{56}$$

Analogously to the previous section, the ground state radius of the dot $R(B)$ can be found from the minimum energy condition

$$\frac{\partial \mathcal{E}_0(B)}{\partial R(B)} = 0 \tag{57}$$

which resolves into the equation

$$\omega_0^2 u_B = \frac{3\pi N e^2}{4\epsilon m^* R(B)^3}\left[1 + \frac{100 a_B}{27\pi R(B)}\left(1 - \frac{\beta^2 f_B^2 N}{36}\right)\right] . \tag{58}$$

In the zeroth order approximation, we obtain

$$R_0(B)^3 = \frac{3\pi N e^2}{4\epsilon m^* \omega_0^2 u_B} = \frac{1}{u_B} R_0^3, \tag{59}$$

where R_0 is given by Eq. (34). The first order correction $\delta(B)$, defined as $R(B) = R_0(B)(1+\delta(B))$, is

$$\delta(B) = \frac{100 a_B}{81\pi R_0(B)}\left(1 - \frac{\beta^2 f_B^2 N}{36}\right) . \tag{60}$$

Finally, the ground state energy in the Hartree approximation can be written in the form

$$\mathcal{E}_0(B) = \frac{9\pi N^2 e^2}{20\epsilon R_0(B)} + \frac{1}{3} u_B^{2/3} \hbar\Omega_0 \left(1 - \frac{\beta^2 f_B^2 N}{36}\right) N^{3/2}. \tag{61}$$

Calculating the correction due to the exchange energy in the Hartree ground-state is far more complicated for the case of a non-zero magnetic field. Therefore we assume that dependence of the exchange energy on the number of particles is not affected by the presence of the field [6] and obtain

$$\Delta\mathcal{E}(B) = -\frac{4\sqrt{5}}{9\sqrt{3}}\left(1 - \frac{3}{4\sqrt{N}}\right)\frac{N^{7/4} e^2}{\epsilon R_0(B)}(1 - \delta_0(B)), \tag{62}$$

for the exchange energy, where $\delta_0(B) = \delta\ (B; \beta = 0)$. The total ground state energy within our approach is

$$\mathcal{E}(B) = \mathcal{E}_0(B) + \Delta\mathcal{E}(B). \tag{63}$$

In Fig. 2 we show the magnetic field evolution of the average ground-state energy per electron $\varepsilon(B) = \mathcal{E}(B)/N$. The three frames correspond to the parameter β equal 0.0 (no spin-orbit coupling), 0.3 and 0.6. We find qualitative agreement between our curves and the data reported by Ashoori *et al.*, obtained in a single-electron capacitance spectroscopy (SECS) experiment [3]. Comparing

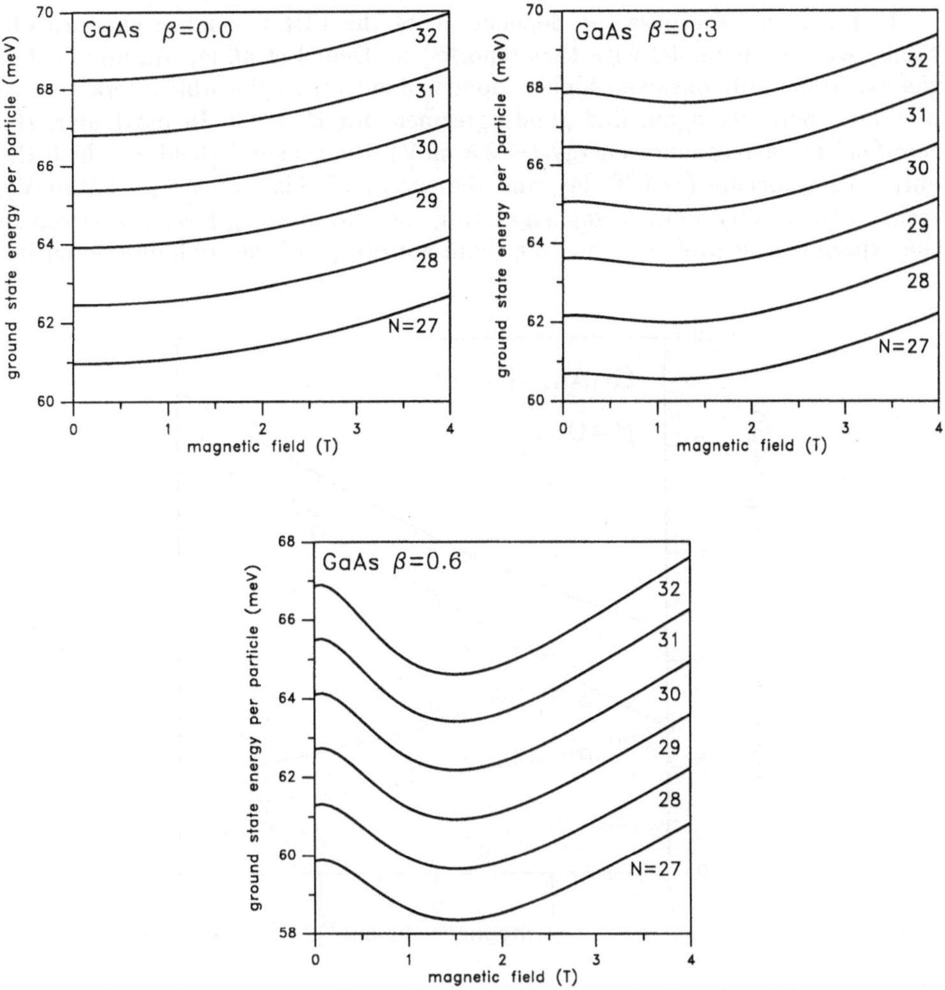

Fig. 2. The average ground-state energy per electron as a function of the magnetic field and number of electrons. The three frames correspond to a spin-orbit coupling constant β equal to 0.0, 0.3 and 0.6 (GaAs, $\hbar\omega_0 = 5.4$ meV).

the curves in the three frames, one can conclude that the inclusion of the spin-orbit interaction brings the model curves fairly close to the measured behaviour (the agreement is particularly good for $\beta = 0.3$).

We now discuss FIR absorption in the presence of a magnetic field. Since the magnetic field does not affect the structure of the HF wavefunctions, the selection rules, Eq. (45), remain unchanged and the transition energies are

$$\mathcal{E}^f(B) - \mathcal{E}^i(B) = \varepsilon_\pm = \hbar\Omega' \pm \left|\frac{1}{2}\hbar\omega_c \pm \frac{1}{2}\alpha(B)\right| \tag{64}$$

and we have four resonance branches.

In Fig. 3, we compare the dependence of the FIR resonance energies obtained within our model with that reported by Demel *et al.* [4]. Assuming that the experimentally observed higher mode is due to the spin-orbit interaction as presented here, we again find good agreement for $\beta = 0.3$. In particular, the zero-field lower resonance energy ($\sim$ 2.8 meV), the magnetic field at which the anti-crossing occurs ($\sim$ 1 Tesla), and the energy of this crossing ($\sim$ 3.9 meV), seem to be fit very well. A gap separating the anti-crossing levels, observed in the experiment, is probably due to a slight anisotropy of the confining potential.

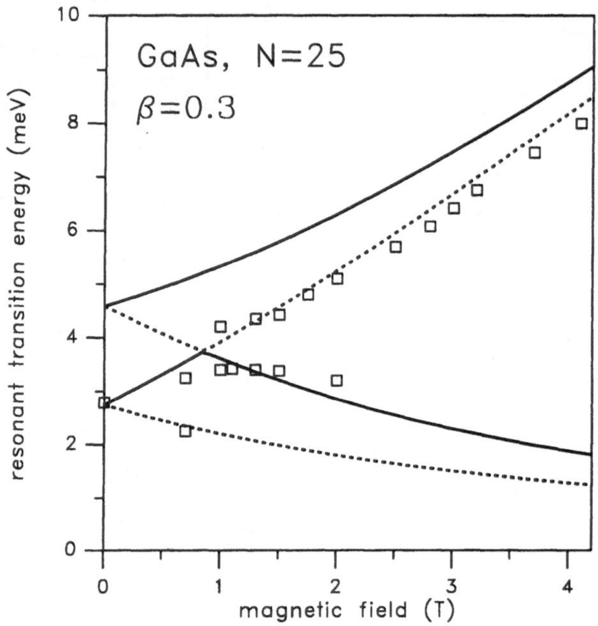

Fig. 3. FIR absorption spectra of a quantum dot containing 25 electrons. Squares – experiment of Demel *et al.* [4], lines – model (GaAs, $\beta = 0.3$, $\hbar\omega_0 = 7.5$ meV).

2 Electron-hole pair in a quantum dot

We consider an electron and a valence-band hole (an exciton) confined to a quantum dot and described within the effective-mass approximation. Since the electron is assumed to be confined, the hole is repelled by the bare potential of the dot as a particle of the opposite charge. Modelling the lateral potentials by Gaussians, the single-exciton hamiltonian is of the form

$$H = -\frac{\hbar^2}{2m_e}\Delta_e - V_0 e^{-r_e^2/L^2} - \frac{\hbar^2}{2m_h}\Delta_h + V_0 e^{-r_h^2/L^2} - \frac{e^2}{\epsilon|\boldsymbol{r}_e - \boldsymbol{r}_h|} + E_g, \quad (65)$$

where $\boldsymbol{r}_e$ and $\boldsymbol{r}_h$ are the electron and hole positions, m_e and m_h their effective masses, V_0 and L define the bare lateral potentials, and E_g is the energy gap of the underlying semiconductor.

Within the single-particle approach, we postulate the form of an exciton wavefunction to be the product of electron and hole wavefunctions, viz.

$$\psi(\boldsymbol{r}_e, \boldsymbol{r}_h) = \phi_e(\boldsymbol{r}_e)\phi_h(\boldsymbol{r}_h). \tag{66}$$

Using this form of the wavefunction ψ, the eigenequation of H resolves into the pair of equations

$$\left(-\frac{\hbar^2}{2m_{e,h}}\Delta_{e,h} + V_{e,h}\right)\phi_{e,h}(\boldsymbol{r}_{e,h}) = \varepsilon_{e,h}\phi_{e,h}(\boldsymbol{r}_{e,h}), \tag{67}$$

where the two self-consistent fields are

$$V_{e,h}(\boldsymbol{r}_{e,h}) = \mp V_0 e^{-r_{e,h}^2/L^2} - \frac{e^2}{\epsilon}\int d\boldsymbol{r}_{h,e}\,\frac{|\phi_{h,e}(\boldsymbol{r}_{h,e})|^2}{|\boldsymbol{r}_{e,h} - \boldsymbol{r}_{h,e}|}, \tag{68}$$

and the total energy of the pair is

$$E = E_g + \varepsilon_{e,h} + \int d\boldsymbol{r}_{h,e}\,\phi_{h,e}^*(\boldsymbol{r}_{h,e})\left(-\frac{\hbar^2}{2m_{h,e}}\Delta_{h,e} \pm V_0 e^{-r_{h,e}^2/L^2}\right)\phi_{h,e}(\boldsymbol{r}_{h,e}). \tag{69}$$

The system of equations Eq. (67) is treated perturbatively. In the zeroth-order approximation the electron motion is solved in the absence of the hole attraction and with the bare lateral-potential simplified to its expansion truncated at second order, i.e.

$$V_e^{(0)}(r) = -V_0 + \frac{1}{2}m_e\omega_e^2 r^2, \tag{70}$$

where ω_e is $V_0 = \frac{1}{2}m_e\omega_e^2 L^2$. Introducing the length scale $\lambda_e^2 = \hbar/m_e\omega_e$ and the dimensionless parameter $\kappa = \hbar\omega_e/V_0$ $(0 < \kappa < 1)$, the zeroth-order electron (bound) ground-state energy and corresponding wavefunction are

$$\varepsilon_e^{(0)} = \hbar\omega_e(1 - \kappa^{-1}), \quad \phi_e^{(0)}(r) = \frac{1}{\sqrt{\pi}\lambda_e}e^{-r^2/2\lambda_e^2}. \tag{71}$$

For a fixed curvature of the lateral potential ω_e (and, consequently, also λ_e), i.e. a fixed V_0/L^2, the parameter κ determines the potential depth V_0 and, hence, the radius L of the whole structure.

Using the charge distribution of the electron and solving Eq. (67), the hole state in the zeroth-order approximation can be calculated. After integrating the interaction term, we have

$$\left(-\frac{\hbar^2}{2m_h}\Delta + V_0 e^{-r^2/L^2} - \frac{\sqrt{\pi}e^2}{\epsilon\lambda_e}e^{-r^2/2\lambda_e^2}I_0(r^2/2\lambda_e^2)\right)\phi_h^{(0)}(\boldsymbol{r})$$

$$\equiv \left(-\frac{\hbar^2}{2m_h}\Delta + V_h^{(0)}\right)\phi_h(\boldsymbol{r}) = \varepsilon_h^{(0)}\phi_h^{(0)}(\boldsymbol{r}), \tag{72}$$

where I_0 is a Bessel function. Since this equation is axially symmetric, the angular part separates and the hole wavefunction can be written as

$$\phi_h^{(0)}(\boldsymbol{r}) = \frac{1}{\sqrt{2\pi}} e^{im\theta} R(r), \tag{73}$$

where m is the orbital angular-momentum. The radial function satisfies

$$\left(\frac{d^2}{dr^2} + \frac{1}{r}\frac{d}{dr} - \frac{m^2}{r^2} + \frac{2m_h}{\hbar}(\varepsilon_h^{(0)} - V_h^{(0)}) \right) R(r) = 0. \tag{74}$$

Before we solve this equation, it is helpful to note that the potential $V_h^{(0)}$, whose curvature defines the frequency ω_h, i.e.

$$m_h \omega_h^2 = \left. \frac{d^2 V_h^{(0)}}{dr^2} \right|_{r=r_0}, \tag{75}$$

has a smooth minimum at r_0. Since most of the hole charge resides near the minimum, the parabolic approximation holds, i.e.

$$V_h^{(0)}(r) \approx U_0 + \frac{1}{2} m_h \omega_h^2 (r - r_0)^2, \tag{76}$$

and we can assume a radial function R of the form

$$R(r) = A e^{-\gamma (r - r_0)^2 / 2\lambda_h^2}, \tag{77}$$

where $\lambda_h^2 = \hbar / m_h \omega_h$ gives the length scale for the hole, γ is a variational parameter, and A is a normalization constant. Minimization of the hole energy with respect to the parameter γ gives $\gamma \approx 1$ [7] and leads to the zeroth-order hole wavefunction

$$\phi_h^{(0)}(\boldsymbol{r}) = \frac{A}{\sqrt{2\pi}} e^{-(r - r_0)^2 / 2\lambda_h^2}, \tag{78}$$

with $A \approx (\sqrt{\pi} r_0 \lambda_h)^{-1/2}$.

We now calculate the next correction to the electron state. Taking into account the interaction with a hole in the state $\phi_h^{(0)}$, the electron potential is

$$V_e^{(1)}(\boldsymbol{r}) = -V_0 e^{-r^2/L^2} - \frac{A^2 e^2}{2\pi\epsilon} \int d\boldsymbol{r}' \, \frac{e^{-(r' - r_0)^2/\lambda_h^2}}{|\boldsymbol{r} - \boldsymbol{r}'|}. \tag{79}$$

For $|r - r_0| \gg \lambda_h$, the above is very well approximated by the intuitive expression [7] $V_e^{(1)}(r) \approx -V_0 \exp[-r^2/L^2] - e^2/\epsilon |r - r_0|$. The characteristic shape of the potential $V_e^{(1)}$ is presented in Fig. 4 for three values of the parameter κ. The potential is axially symmetric with two minima; one in the centre of the dot, originating at the minimum of the confining potential V, and the other one at a finite radius (forming a ring), due to the attractive potential of the hole. The attraction of the hole to the centre of the dot and, hence, the disappearance of the ring-like minimum in the effective electron potential $V_e^{(1)}$ is forbidden by the repulsive core $V_0 \exp[-r^2/2L^2]$ acting on the hole. The relative depths of the two minima in $V_e^{(1)}(r)$ depend on the strength of this core, parametrized here by κ. As shown in Fig. 4, there are situations in which either the two minima are equally deep or one of the minima is significantly deeper than the other.

In order to solve for the two lowest eigenstates of the electron motion in the two-well potential $V_e^{(1)}$, the wavefunction is assumed to be of the form

$$\phi_e^{(1)}(\boldsymbol{r}) = c_L \phi_L(\boldsymbol{r}) + c_R \phi_R(\boldsymbol{r}), \tag{80}$$

where the indices L and R correspond to the left (central) and right (outer) well. The two states ϕ_L and ϕ_R are the ground states calculated for the simplified single-well potentials

$$V_L(r) = v_L + \frac{1}{2} m_e \omega_L^2 r^2 \tag{81}$$

$$V_R(r) = v_R + \frac{1}{2} m_e \omega_R^2 (r - r_R)^2, \tag{82}$$

where v_L and v_R are the two minima of $V_e^{(1)}$ at $r_L = 0$ and r_R, respectively, and ω_L and ω_R are the curvatures of $V_e^{(1)}$ at these minima. Defining

$$\varrho = \int d\boldsymbol{r}\, \phi_L^*(\boldsymbol{r}) \phi_R(\boldsymbol{r}), \tag{83}$$

and $c_1 = c_L + \varrho c_R$, $c_2 = \varrho c_L + c_R$, we can write the equations for the coefficients c_L and c_R and the two eigenenergies ε_e as

$$\begin{aligned} c_1(E_L - \varepsilon_e^{(1)}) + c_2 I_{LR} &= 0, \\ c_1 I_{RL} + c_2(E_R - \varepsilon_e^{(1)}) &= 0. \end{aligned} \tag{84}$$

The symbols E_L and I_{LR} stand for the integrals

$$\begin{aligned} E_L &= \int d\boldsymbol{r}\, \phi_L^*(\boldsymbol{r}) \left(-\frac{\hbar^2 \Delta}{2m_e} + V_e^{(1)} \right) \frac{\phi_L(\boldsymbol{r}) - \varrho \phi_R(\boldsymbol{r})}{1 - \varrho^2}, \\ I_{LR} &= \int d\boldsymbol{r}\, \phi_L^*(\boldsymbol{r}) \left(-\frac{\hbar^2 \Delta}{2m_e} + V_e^{(1)} \right) \frac{\phi_R(\boldsymbol{r}) - \varrho \phi_L(\boldsymbol{r})}{1 - \varrho^2}, \end{aligned} \tag{85}$$

while the second pair, E_R and I_{RL}, can be obtained upon the interchange of the indices $L \leftrightarrow R$.

Using the condition for the vanishing of the determinant in Eq. (84), we can express the two lowest electron eigenenergies as

$$\varepsilon_e^{(1)} = \frac{E_L + E_R}{2} \pm \sqrt{\frac{(E_L - E_R)^2}{4} + I_{LR} I_{RL}}. \tag{86}$$

The corresponding electron wave-functions $\phi_{e1}^{(1)}$ and $\phi_{e2}^{(1)}$, multiplied by the hole state $\phi_h^{(0)}$, are the two lowest electron-hole pair states in our approximation

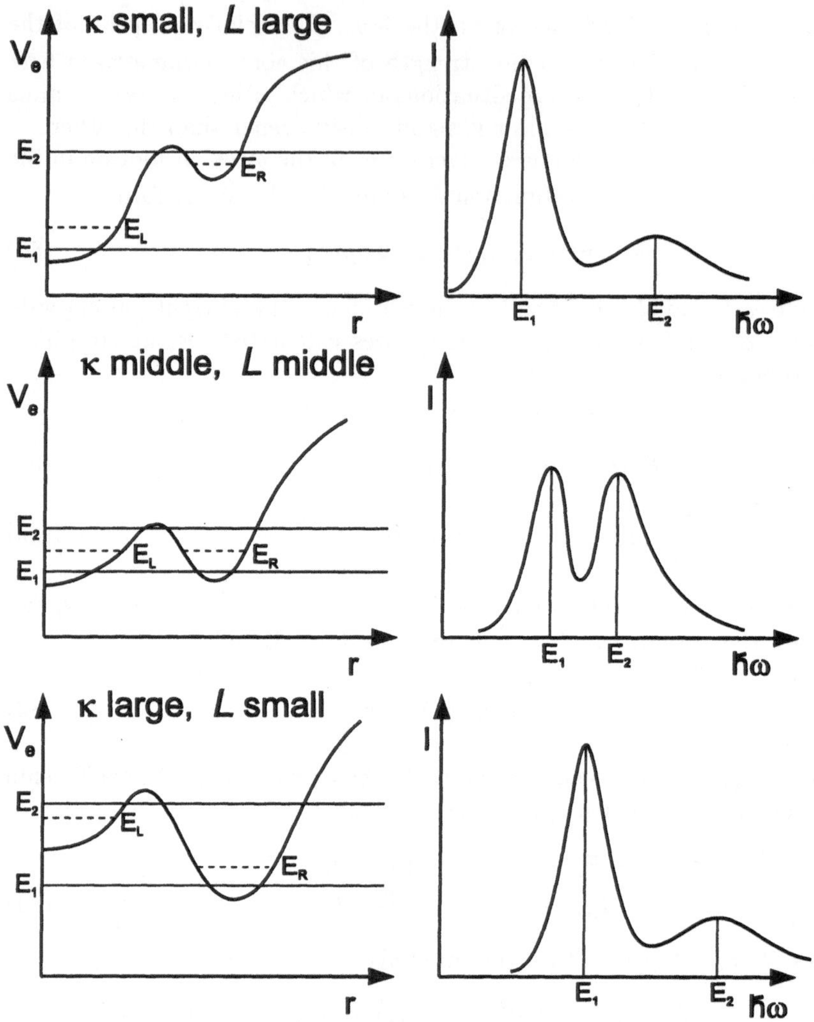

Fig. 4. The effective self-consistent potential V_e acting on the electron in a quantum dot containing one exciton, together with the corresponding approximate absorption spectra I, as a function of the parameter κ, i.e. the dot size L for (a) large $\kappa = \frac{1}{3}$, (b) medium $\kappa = \frac{1}{2}$ and (c) small dot $\kappa = \frac{2}{3}$.

ψ_1 and ψ_2. The explicit formulae for these states and their energies are given elsewhere [7]; as illustration, we shall only present the energies calculated for three different values of κ corresponding to the three frames in Fig. 4. All of the numbers given in the table below are in units of $\hbar\omega_e = 5.4$ meV.

	$\kappa = \frac{1}{3}$	$\kappa = \frac{1}{2}$	$\kappa = \frac{2}{3}$
E_1	-2.154	-1.292	-1.301
E_2	-1.024	-1.073	-0.812
$E_2 - E_1$	1.130	0.219	0.489

A crucial property of the pair of electron-hole states ψ_1 (the ground state) and ψ_2 (the first excited-state) is the vanishing of the dipole-matrix element between them, i.e.

$$d_{12} = \int d\boldsymbol{r}_e \int d\boldsymbol{r}_h\, \psi_2^*(\boldsymbol{r}_e, \boldsymbol{r}_h)(-\boldsymbol{r}_e + \boldsymbol{r}_h)\psi_1(\boldsymbol{r}_e, \boldsymbol{r}_h) = 0. \tag{87}$$

Since they are both axially symmetric (in these states, neither the electron nor the hole carry a non-zero orbital angular momentum). As a result, an optical transition (via absorption/emission of a far infra-red photon) between these states is forbidden and the excited state ψ_2 is meta-stable. The occurrence of a meta-stable state should be observed directly in the photo-luminescence experiments, where, at low temperatures, in the absence of any efficient relaxation mechanisms, a radiative recombination from the excited meta-stable state would lead to the appearance of an additional strong peak in the emission spectrum. As follows from the above table, the energy splitting $E_2 - E_1$ between the two major peaks depends strongly on κ, reaching a minimum around $\kappa \sim \frac{1}{2}$, corresponding to a situation in which the two minima in the potential $V_e^{(1)}$ have roughly the same depth. Increasing κ above $\frac{1}{2}$ or decreasing it below $\frac{1}{2}$ (i.e. introducing an asymmetry between the two minima in $V_e^{(1)}$) leads to an increase of the separation $E_2 - E_1$.

The ratio of the radiative recombination intensities I_1 and I_2, at temperature T, can be estimated using

$$\frac{I_2}{I_1} \sim \exp\left[-\frac{E_2 - E_1}{k_B T}\right]; \tag{88}$$

thus, we expect a strong enhancement of the second peak in the emission spectrum. Using the above formula for $T = 5$ K and the confinement $\hbar\omega_e = 5.4$ meV, we obtain

$$\frac{I_2}{I_1} \sim \begin{cases} 10^{-6} \text{ for } \kappa = \frac{1}{3}, \\ 10^{-1} \text{ for } \kappa = \frac{1}{2}, \\ 3 \times 10^{-3} \text{ for } \kappa = \frac{2}{3}. \end{cases} \tag{89}$$

Similarly, we estimate the widths of the two peaks τ_1^{-1} and τ_2^{-1} to be

$$\frac{\tau_2^{-1}}{\tau_1^{-1}} \sim \frac{|\phi_{e2}(r_0)|^2}{|\phi_{e1}(r_0)|^2} = \begin{cases} 321 \text{ for } \kappa = \frac{1}{3}, \\ 0.8 \text{ for } \kappa = \frac{1}{2}, \\ 84 \text{ for } \kappa = \frac{2}{3}. \end{cases} \tag{90}$$

where r_0 is the approximate radius of the hole orbit. Clearly, the enhancement of the second peak coincides with a decrease of its width. Around $\kappa = \frac{1}{2}$, the widths of the two lowest peaks are of the same order of magnitude.

Since the parameter κ is closely connected with the diameter of the dot, the characteristic evolution of the emission spectrum as a function of κ, i.e. the appearance of a doublet of peaks at $\kappa \sim \frac{1}{2}$ and the decay and smearing out of the higher energy peak at both lower and higher values of κ, seems to be a general property of quantum dots around a certain critical size (depending, e.g., on the curvature of the confining potential, effective mass etc.). A number of experiments have been reported in which we think this effect has been already seen [8].

The influence of the magnetic field on the above properties of the photo-luminescence spectrum of the dot is very interesting. A detailed analysis of this problem will be published elsewhere [9] and here we shall only note that the magnetic field enhances the confinement of both the electron and the hole. As a result, a double well can also appear in the effective potential of the hole. Combined with a pair of electron states this gives a set of four exciton states (a ground state and three meta-stable states), and, consequently, four distinct peaks in the PL spectrum. The double wells, and hence also the additional PL peaks, disappear at high magnetic fields and for very small dots. Similar behaviour has actually been observed in experiment [10].

3 Conclusions

A self-consistent theory of a many-electron quantum dot including the electron-electron (Coulomb) and spin-orbit interactions has been developed. The incorporation of quantum corrections into the ground-state energy of the system improves the classical result of Shikin *et al.* and compares well with the experiment of Demel *et al.* and Ashoori *et al.* The spin-orbit interaction has a strong effect on the electronic structure of the system, e.g. leading to a splitting of the resonance energy in the far infra-red (FIR) absorption. A predicted anti-crossing of the FIR absorption modes in a magnetic field describes very well similar behavior reported by Demel *et al.*

An electron-hole pair (exciton) confined in a quantum dot has also been studied within the effective mass approximation. For certain sizes of the dot, a special form of the effective potential acting on the hole (a combination of the electron-hole Coulomb attraction and the repulsive potential of the dot) leads to the occurrence of a weakly excited meta-stable exciton state. Correspondingly, an additional strong peak in the photo-luminescence spectrum appears, a peak that we think has already been observed in experiment.

Acknowledgments

This work was supported by the KBN grants PB 674/P03/96/10 and PB 1152/P03/94/06.

References

1. V. Shikin, S. Nazin, D. Heitmann and T. Demel, Phys. Rev. B, **43**, 11 903 (1991).
2. L. Jacak and J. Krasnyj, (to be published).
3. R. C. Ashoori, H. L. Stormer, J. S. Weiner, L. N. Pfeiffer, K. W. Baldwin and K. W. West, Phys. Rev. Lett. **71**, 613 (1993).
4. T. Demel, D. Heitmann, P. Grambow and K. Ploog, Phys. Rev. Lett. **64**, 788 (1990).
5. D. Heitmann, Physica B, **212**, 201 (1995); and references therein.
6. S. V. Vonsovskij, "Magnetism", Nauka–Moscow (1971), in Russian.
7. L. Jacak, J. Krasnyj, P. Hawrylak and A. Wójs, (to be published).
8. K. Brunner, U. Bockelmann, G. Abstreiter, M. Walther, G. Böhm, G. Tränkle and G. Weimann, Phys. Rev. Lett., **69**, 3216 (1992); U. Bockelmann, K. Brunner and G. Abstreiter, Solid State Electronics, **37**, 1109 (1994).
9. L. Jacak, J. Krasnyj, P. Hawrylak and A. Wójs, (to be published).
10. M. Bayer, A. Schmidt, A. Forchel, F. Faller, T. L. Reinecke,P. A. Knipp, A. A. Dremin, V. D. Kulakovskii, Phys. Rev. Lett., **74**, 3439 (1995).

Quantum Single Electron Transistor

Pawel Hawrylak

Institute for Microstructural Sciences, National Research Council of Canada
Ottawa, Canada K1A 0R6.

Abstract: We describe the basic physics of a Quantum Single Electron Transistor (QSET). The QSET is a quantum dot (QD) connected to two leads and a central electrode. The effect of the central electrode, electron-electron interaction, and a magnetic field on electron droplets in the QSET is studied using the Hartree-Fock approximation and exact diagonalization techniques. In a strong magnetic field, the central electrode induces electronic spin and charge transitions in the spin-polarized droplet. We show how these spin-related transitions can be observed in transport and coherent resonant tunnelling through the QSET.

PACS numbers:73.20Dx,71.50+t,71.45Gm

1 Introduction

A field effect transistor (FET) is a three-terminal device in which a source, drain, and a gate are contacted separately. The flow of charge from the source to the drain is classical and is a continuous function of the number of electrons in the base. A single electron transistor (SET) is a small, two-terminal device in which the flow of electrons is no longer a continuous function of the number of electrons in the electron gas. In a SET the current flow is sensitive to the number of electrons under the gate due to the finite charging energy U, or finite capacitance C. These are all classical concepts related to the electrostatic potential of the gate. A small metallic grain or a relatively large active area, of the order of microns ($10^4 \AA$), in a FET is a good example of a SET. In a much smaller (submicron) SET, the kinetic energy of the electrons is quantized, their mutual interaction and correlation energy is large, and quantum mechanics governs the physical properties of a Quantum SET. The active region of a QSET is a quantum dot (QD) with a relatively small number of electrons. This QD can be thought of as an artificial atom. While studies of QDs are numerous [1], only very recently [2] have attempts been made at realizing a true QSET, i.e. a very small three terminal device. We describe here some basic electronic properties of the quantum dot and of the effect of the central electrode on the operation of a QSET.

Our work [3] is motivated by two recent experimental developments, namely the work by Klein et al. [4] who undertook a very systematic study of the relationship between transport through a submicron QD with $N \approx 30$ electrons and its electronic properties, and by the work of Feng et al. [2]. Feng et al. incorporated an additional submicron electrode into electrostatically defined quantum dots and wires formed in a two dimensional electron gas. This additional electrode creates an attractive or repulsive impurity with tunable strength in the strongly interacting electron droplets [5, 6] with a well controlled number of electrons N ($N \approx 10 - 50$). The electron droplets in a QD present diverse many-electron effects, changing from chiral Luttinger liquid behaviour [7, 8, 9] in the Integer Quantum Hall regime to the incompressible liquids familiar in the context of the Fractional Quantum Hall Effect [10]. This behaviour is due to the competition of the electron-electron interactions, Zeeman, and kinetic energy [11, 12, 13]. The competing interactions in these artificial atoms are tunable in the applied magnetic field [8, 11, 12, 13] and lead to a series of incompressible ground states with "magic angular momentum" values. This in turn controls the thermodynamic [11], transport [14, 15] and optical properties [16, 17, 18] of the quantum dots.

We will discuss some essential concepts in transport and coherent resonant tunnelling, single-particle and collective excitation spectra, and results of Hartre-Fock calculations. We will show that the introduction of an attractive/repulsive electrode in a droplet induces electronic spin and charge transitions. These spin- and charge-related transitions modify the chemical potential μ of the dot and therefore can be observed in resonant tunnelling experiments. We limit our considerations to a spin-polarized, compact droplet because: a) this incompressible state is known exactly in the strong magnetic field [8]; b) the addition spectrum of this state is featureless [4], allowing for clear identification of electrode-induced effects; and c) recent theoretical [19, 20] and experimental [21] work predicts exotic spin-related excitations in a spin-polarized two-dimensional electron gas (2DEG) in a strong magnetic field.

2 The model

The QSET shown in Fig. 1 consists of a two-dimensional QD, weakly connected to a left and right electrode via tunnelling barriers. The central electrode is attached to the centre of the dot. We first neglect coupling of the dot to the leads and consider it as an isolated system.

In our model, the QD [12] contains N electrons confined by an effective parabolic potential with a characteristic energy ω_N. The frequency ω_N has a contribution from the externally imposed potential and from a positive charge $+Ne$ at a distance d away from the plane of the dot. The central electrode is characterized by a potential $U(r)$, which we assume to have cylindrical symmetry. The QSET is placed in the magnetic field B normal to the plane of the dot. The positive charge assures the charge neutrality of the dot and plays the role

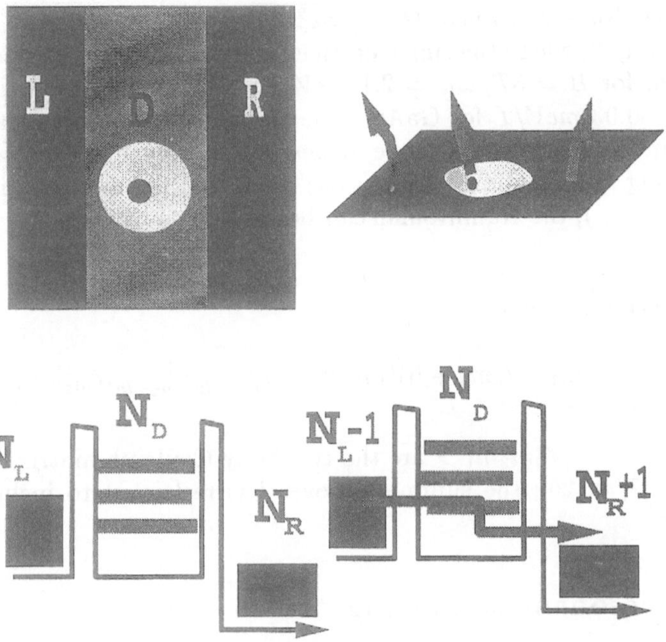

Fig. 1. A schematic picture of a model of a QSET and of the initial and final states in the tunnelling process.

of the gate. The single-particle Hamiltonian corresponds to a particle moving in a parabolic potential in the presence of the magnetic field. It is diagonalized [12, 18] by a transformation into a pair (a and b) of lowering and raising harmonic oscillator operators. The a and b oscillators evolve into the inter- and intra-Landau level oscillators with increasing magnetic field B. The single-particle energies are $E_{mn} = \Omega_+(n+\frac{1}{2}) + \Omega_-(m+\frac{1}{2})$; the eigenstates for spin σ, $|m,n;\sigma> = \sqrt{\frac{1}{m!n!}}(a^+)^m(b^+)^n|0,0;\sigma>$, and the two harmonic frequencies are $\Omega_\pm = [\Omega \pm \omega_c]/2$ ($\hbar = 1$ for the rest of this work). ω_c is the cyclotron energy, $l_0 = 1/(m\omega_c)^{1/2}$ the magnetic length, m the effective mass, and $\Omega = \sqrt{\omega_c^2 + 4\omega_0^2}$. The kinetic energy $\sim \Omega_-$ decreases with the magnetic field, while the Coulomb energy increases with the magnetic field. The Coulomb energy is measured in units of the exchange energy $E_0 = Ry\sqrt{2\pi}a_0/l_{\text{eff}}$, where Ry is the effective Rydberg, a_0 the effective Bohr radius, and $l_{\text{eff}} = l_0/(1+4\omega_0^2/\omega_c^2)^{1/4}$ is the effective magnetic length.

We restrict the Hilbert space to single particle states $|m,0;\sigma>$ originating from the lowest Landau level [7, 8]. The electrode potential U only shifts the single particle energies E_m by $<m|U|m>$ without changing the single particle states $|m>$. We model our electrode as a parabolic potential characterized by a characteristic frequency Ω_i, an effective radius m_i, and effective charge α, viz.

$< m|U|m >= \alpha\Omega_i(m_i - m)$, where $\Omega_i = [\sqrt{\omega_c^2 + 4\omega_i^2} \pm \omega_c]/2$, and $m \leq m_i$. This is illustrated in Fig. 2, where the single particle energies $e_m = E_m + < m|U|m > + \delta E_z$ are shown for $B = 3T$, $\omega_N = 2.1\ meV$, and $\omega_i = 3.0\ meV$ (δE_z is the Zeeman energy, $0.03\ meV/T$ for GaAs). The electrons are either attracted to, or repelled from, the centre of electrode, depending on the effective charge α.

After denoting the creation (anihilation) operators for electrons in states $|m;\sigma>$ by $c^+_{m,\sigma}(c_{m,\sigma})$, the Hamiltonian can be written as

$$H = \sum_{m\sigma} e_{m\sigma} c^+_{m\sigma} c_{m\sigma} +$$

$$1/2 \sum_{lm_1m_2\sigma\sigma'} < m_1 - l, m_2 + l|V|m_2 m_1 > c^+_{m_1-l\sigma} c^+_{m_2+l\sigma'} c_{m_2\sigma'} c_{m_1\sigma}, \quad (1)$$

where $< m_1 - l, m_2 + l|V|m_2 m_1 >$ are the two-body Coulomb matrix elements defined in Refs. [7, 8, 20]; the summation over l is restricted to include only positive integers.

3 Coherent resonant tunnelling

There have been a number of attempts at the description of transport through a quantum dot [23, 24, 25]. Most of these theories are fairly complicated due to the mixing of electron states in the dot with the states in the lead. This mixing results in Kondo resonances in the leads. These theories can be summarized by a simple formula [24] which relates the current through the QD to the density of states, i.e. the spectral function $A(\omega) \approx ImG(\omega)$, viz.

$$J = -2e/h \int d\omega [f_L(\omega) - f_R(\omega)] Im\{tr\{\Gamma(\omega)G(\omega)\}\}, \quad (2)$$

where $G(\omega)$ is the energy dependent Green's function of the dot, Γ is the tunnelling matrix, and $f_{L/R}(\omega)$ is the equillibrium distribution function of the left/right lead. Note that Eq. (2) depends not on the current-current correlation function but directly on the single-particle density of states. We do not completely understand all of the steps leading to Eq. (2), and so we will settle here on a simpler but transparent derivation of the coherent resonant tunnelling from the point of view of **spectroscopy**.

Let us consider a simple model of coherent resonant tunnelling (CRT). We start from three uncoupled systems as shown in Fig. 1: (a) the left electrode (L) with N_L electrons, (b) the quantum dot (D) with N_D electrons and (c) the right electrode (R) with N_R electrons. We neglect the electron-electron interaction in both electrodes but the QD states are treated exactly. The eigenstates of the noninteracting system can be characterized by the complete set of "intermediate" quantum numbers $\{i\}$ for a fixed number of particles, $|N_L>_i$, $|N_D>_i$ and $|N_R>_i$.

We now switch-on the coupling of the dot to the left (right) lead with single particle states k_L, i.e. $\Gamma^{LD} = \sum \Gamma^{LD}_{k_L,m}(d^+_{k_L} c_m + c^+_m d_{k_L})$. This allows the electrons

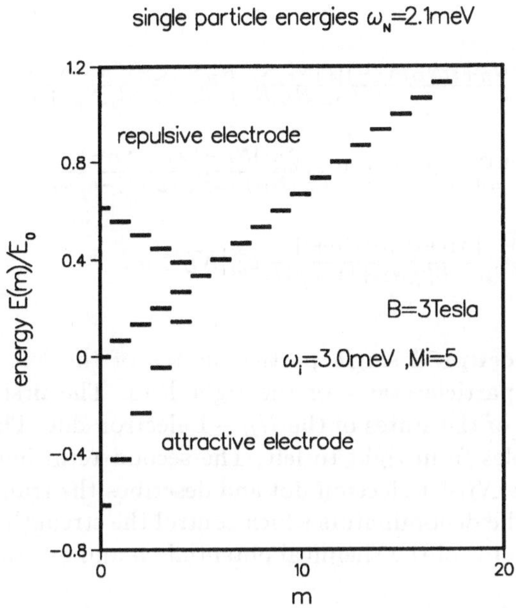

Fig. 2. A schematic picture of the single particle energies $E(m)$ of the states $|m>$ in the dot as a function of the angular momentum m in the the presence of an electrode. E_0 is the exchange energy.

to move coherently from left to right as shown in Fig. 1. We use Fermi's golden rule to calculate the transition probability, τ^{-1}, for the electron in the left lead to move to the right lead in response to the coupling $\Gamma = \Gamma^{\mathrm{LD}} + \Gamma^{\mathrm{RD}}$. Since each coupling moves an electron from one of the leads to the dot, the process of moving an electron from the left lead to the right lead is second order in Γ, i.e.

$$\tau^{-1} = 2\sum_f \times$$

$$|\sum_{N'_L,N'_D,N'_R,i} \frac{{}_f\langle N_R+1,N_D,N_L-1|\Gamma|N'_L,N'_D,N'_R\rangle_i\,{}_i\langle N'_R,N'_D,N_L|\Gamma|N_L,N_D,N_R\rangle_0}{E_0(N_L,N_D,N_R)-E_i(N'_L,N'_D,N'_R)+i\epsilon}|^2 \quad (3)$$

$$\delta(E_0(N_L,N_D,N_R) - E_f(N_L-1,N_D,N_R+1)).$$

The summation is over all of the final states $\{f\}$ with the proper number of electrons. The effective matrix element involves all intermediate states $\{i\}$. We can now express the transition probability in terms of the QD operators c, c^+, exact QD states and energies, as

$$\sum_{N_L',N_D',N_R',i} \frac{{}_f\!<N_R+1,N_D,N_L-1|\Gamma|N_L',N_D',N_R'>_i\,{}_i\!<N_R',N_D',N_L|\Gamma|N_L,N_D,N_R>_0}{E_0(N_L,N_D,N_R)-E_i(N_L',N_D',N_R')+i\epsilon}$$

$$= \sum_{m,n} \Gamma^{\mathrm{LD}}_{k_L,m}\Gamma^{\mathrm{RD}}_{k_R,n}\, {}_f\!< N_D|c_m^+ \frac{\sum_i |N_D-1>_i\,{}_i\!<N_D-1|}{E_0^D(N_D)-E_i^D(N_D-1)-e_f^R+i\epsilon} c_n|N_D>_0 \qquad (4)$$

$$+{}_f\!< N_D|c_n \frac{\sum_i |N_D+1>_i\,{}_i\!<N_D+1|}{E_0^D(N_D)-E_i^D(N_D+1)-e_f^L+i\epsilon} c_m^+|N_D>_0 \,.$$

Here e_f^L are the **occupied** single particle states of the left lead and e_f^R are the **empty** single particle states of the right lead. The first term involves a summation over all of the states of the N_D-1 electron dot. This term describes the transport of holes from right to left. The second term involves summation over all states of the N_D+1 electron dot and describes the transport of electrons from left to right. The denominators which control the strength of the resonances can be written in terms of the chemical potentials μ and excitations δE, namely

$$\begin{aligned} E_0^D(N_D) - E_i^D(N_D-1) - e_f^R &= \mu^D(N_D-1) - \mu^R - \delta E_i^D(N_D-1) - \delta e_f^R \\ E_0^D(N_D) - E_i^D(N_D+1) - e_f^L &= \mu^D(N_D) - \mu^L - \delta E_i^D(N_D+1) + \delta e_f^L. \end{aligned} \qquad (5)$$

We see that the "hole" term is resonant when $\mu^D(N_D-1) \approx \mu^R$ and the "electron" term is resonant when $\mu^D(N_D) \approx \mu^L$. In both cases the transition involves the creation and annihilation of electrons in single particle states n, m.

It is very illuminating to follow the standard approach from other resonant processes, such as resonant electronic Raman scattering, and approximate the summation over intermediate states as

$$\begin{aligned} &{}_f\!< N_D|c_n \frac{\sum_i |N_D+1>_i\,{}_i\!< N_D+1|}{E_0^D(N_D) - E_i^D(N_D+1) - e_f^L + i\epsilon} c_m^+|N_D>_0 \\ \approx\; &{}_f\!< N_D|c_n \frac{\sum_i |N_D+1>_i\,{}_i\!< N_D+1|}{\mu^D(N_D) - \mu^R + i\epsilon} c_m^+|N_D>_0 \\ \approx\; &\frac{{}_f\!< N_D|c_n c_m^+|N_D>_0}{\mu^D(N_D) - \mu^R + i\epsilon}. \end{aligned} \qquad (6)$$

We assumed a weak resonance between excited states of the dot and neglected them altogether. This permits the summation over all intermediate states to be performed and the tunnelling probability to be expressed as a product of the resonant term and a charge fluctuation term, viz.

$$\tau^{-1} \approx \frac{2}{|\mu^D(N_D)-\mu^R+i\epsilon|^2}\times$$

$$\sum_f |_f < N_D| \sum_{m,n} \Gamma^{\rm LD}_{k_L,m} \Gamma^{\rm RD}_{k_R,n} c_n c^+_m |N_D >_0 |^2$$

$$\delta(E_f(N_D) - E_0(N_D) + \delta e^L_f + \delta e^R_f - \delta\mu^{LR}) \tag{7}$$

$$\rightarrow$$

$$\tau^{-1}(\omega) \approx \sum_l \frac{\Gamma^4(l)}{|\mu^D(N_D)-\mu^R+i\epsilon|^2}$$

$$\sum_f |_f < N_D| \sum_m c_m c^+_{m+l} |N_D >_0 |^2 \delta(E_f(N_D) - E_0(N_D) + \omega).$$

The operator $\rho_l = \sum_m c_m c^+_{m+l}$ entering the transition probability is simply the charge-fluctuation operator. Hence, the tunnelling probability is determined by the spectrum of the collective charge density fluctuations, with resonant enhancement whenever the chemical potential of the electrode is close to the chemical potential of the dot. The spectral window ω is related to the bias through the difference in the chemical potentials, $\delta\mu^{LR}$, of the left and right electrodes. The tunnelling matrices give effective angular momentum selection rules. This is a simple and intuitive result. A tunnelling electron passes through the dot and polarizes the QD electrons via excitation of charge or spin fluctuations. The cross section is proportional to the spectrum of **collective excitations**. This is to be contrasted with Eq. (2) which expresses transport via the single particle spectral function.

To summarize, we must calculate the chemical potential of the QSET, its spectral function and the spectrum of collective excitations in order to understand the operation of the QSET.

4 Single-particle and collective excitations

We start by investigating the excitation spectrum of the quantum dot. The ground state $|G>$ is given by a single Slater determinant of all the lowest single particle states $|m>$ with spin σ down, i.e. $|G>= \prod_{m<m_{max}} c^+_{md}|0>$, ($m_{max} = N-1$). The excitation spectrum of the dot corresponds to the charge-density excitations $|m_e, m_h>_{cd}= c^+_{m_e,d} c_{m_h,d}|0>$ and spin-flip excitations $|m_e, m_h>_{SF}= c^+_{m_e,u} c_{m_h,d}|0>$. Note that charge-density excitations involve the removal of an electron from an occupied state $|m_h>$ to the outside of the droplet ($M>0$); while spin-flip excitations allow for the excited electron to move inside the electron droplet ($M<0$ and $M>0$ branches). The magneto-exciton picture used here can be related to the chiral bosonisation scheme [9, 27] where one defines chiral boson operators, B^+_M, in terms of the electron-hole pair operators $B^+_M = \sum_{m=0}^{\infty} c^+_{M+m,d} c_{m,d}$. The magneto-excitons corresponding to charge fluctuations can be related to the reconstruction of the charge density of the QD and we shall call them edge magneto-rotons (EMR).

Since the Hamiltonian conserves the total angular momentum, the angular momentum of the excitation $M = m_e - m_h$ is a good quantum number. For each M, the true excitations $|p, M>$ of the system are a linear combination of electron-hole pair excitations $|p, M> = \sum_{m_h} A^p_{m_h} |m_h + M, m_h>$. We diagonalize the Hamiltonian, Eq. (1), in the space of electron-hole pair states $|m_e, m_h>$ for a fixed value of $M = m_e - m_h$. The matrix elements $< m'_e, \sigma'_e, m'_h, \sigma'_h | H | m_e, \sigma_e, m_h, \sigma_h >$, describing the mixing of EMR, are given by

$$
\begin{aligned}
&< m'_e, \sigma'_e, m'_h, \sigma'_h | H | m_e, \sigma_e, m_h, \sigma_h >= \\
&\delta_{m_h m'_h} \delta_{m_e m'_e} \delta_{\sigma'_e \sigma_e} \delta_{\sigma'_h \sigma_h} \{< m_e|U|m_e > - < m_h|U|m_h > + \Sigma^{\mathrm{HF}}_{m_e \sigma_e} - \Sigma^{\mathrm{HF}}_{m_h \sigma_h}\} \\
&+ < m'_e m_h |V| m_e m'_h > \delta_{\sigma_e \sigma_h} - < m'_e m_h |V| m'_h m_e > . \qquad (8)
\end{aligned}
$$

The first two terms are the kinetic energy difference in promoting an electron from the state m_h to the state $|m_e>$ in the potential of the electrode. The remaining terms are contributions due to electron-electron interactions, the differences in the Hartree-Fock energies of the initial and final states and the mixing of magneto-excitons via Coulomb interactions (vertex corrections). Note that the vertex correction includes a diagonal term and contributes to the magneto-exciton energy.

To calculate the spectral function of the QSET, we calculate not only the one but also the two-pair excitation spectrum of a QD [18, 26]. This calculation is rather cumbersome and will not be described here in any detail. The spectral function $A(\omega) = Im(G(\omega))$ is divided into a "hole" term and an "electron" term. The "hole" term corresponds to the removal of an electron from the ground state of the N electron QSET, i.e.

$$
A_<(\omega) = \sum_f {}_f < N+1 | \sum_m c_m | N >_0 |^2 \delta(E_f(N-1) - E_0(N) + \omega); \qquad (9)
$$

the electron part corresponds to the addition of one electron to an N electron QSET, i.e.

$$
A_>(\omega) = \sum_f {}_f < N+1 | \sum_m c^+_m | N >_0 |^2 \delta(E_f(N+1) - E_0(N) - \omega). \qquad (10)
$$

In Fig. 3 we show the collective excitation spectrum of an $N = 10 - 1$ electron droplet and a hole spectral function of an $N = 10$ electron dot in a magnetic field of $B = 2T$. The ground state is a compact droplet. The excitation spectrum shows a low lying branch of edge magneto-rotons (EMR). The magneto-roton nature is visible in the softening of the dispersion for intermediate angular momenta. The higher energy excitations describe the interaction of pairs of EMRs. The removal of a hole couples strongly to collective excitations as is seen by the large peak in the density of states in the energy sector of EMR. This is not the case for electrons.

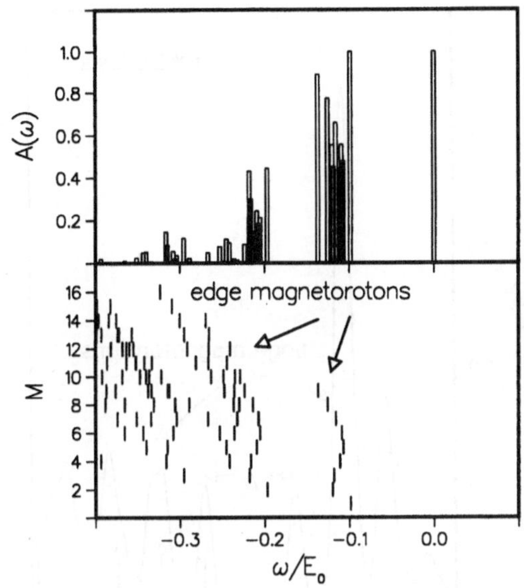

Fig. 3. The collective excitation spectrum and hole spectral function of an N=10 electron dot at B=2T.

In Fig. 4 we show the total spectral function of an N=10 electron dot. First, we see that there is a gap Δ across the Fermi level. The gap is related to the energy of adding/subtracting an electron to/from an N=10 electron droplet. The density of states is strongly modulated. The first excited states are the Centre of Mass (CM) excitations with energy Ω_- broadened by EMR. For the addition of electrons this interaction is weak but for holes a very strong enhancement of the density of states is present due to the large overlap of a single hole excitation with a many-body state. Therefore measurements of the density of states should concentrate on the hole part of the spectral function. A good spectroscopic tool is the recombination of acceptors [18]. We now turn to spin excitations and larger dots.

Let us examine the effect of a central electrode in larger dots with $N \approx 30$. We first study the stability of a compact droplet against the electrode potential. This stability is determined by spin-flip (SF) excitations. In Fig. 5a we show the spin-flip excitation (SF) spectrum of a compact dot at $B = 3T$ as a function of M, for $N = 27$ electrons on a disk with $N = 54$ states. The general shape of the excitation spectrum and the origin of the low lying excitations, can be understood by considering the removal of an electron either from the edge of the droplet or from the centre. The softening of the positive branch of the SF excitations around $M = 8$ is due to the loss of exchange as the electron is being

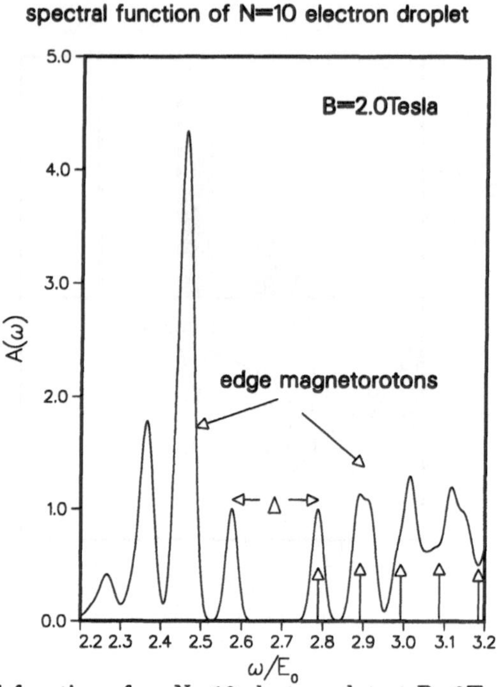

Fig. 4. The spectral function of an N=10 electron dot at B=2T.

removed from the dot. Note the dark areas in the figure; they correspond to peaks in the density of excited states. In Fig. 5b we show the effect of an attractive electrode on the SF excitation spectrum. An attractive electrode generates a soft mode in the negative M branch for $M = 5$. The soft mode indicates a transition from a spin-polarized droplet to a spin-density state which can be crudely characterized as an excess of spin and charge in the centre of the dot, accompanied by a hole in the spin down ground state. For larger electrode strength, the excess of spin and charge in the centre of the dot grows. This can only be treated approximately using HF.

5 Hartree-Fock calculation

The direct and exchange interaction of a large number of electrons can be treated in the Hartree-Fock approximation (HF) [7, 8, 18]. Because of the limited Hilbert space, the Hartree-Fock wavefunctions are exactly the same as the single particle wavefunctions $|m>$. The Hartree-Fock method is, therefore, equivalent to the minimization of the total HF energy $E_{\mathrm{HF}}^{\mathrm{tot}}$ in the space of HF configurations, given by a set of single particle occupations $\{f_{m,\sigma}\}$, with $f_{m,\sigma}$ restricted to 0 or 1, i.e.

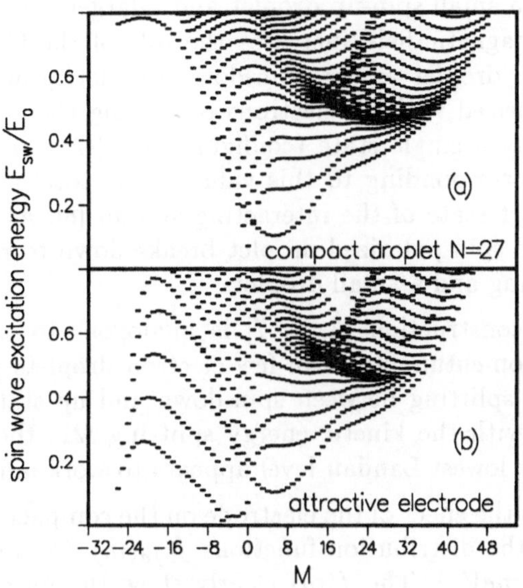

Fig. 5. Spin-flip excitation spectrum of a compact droplet of $N = 27$ electrons on a $0 < m < 53$ disk, (a) for $B = 3T$, no electrode (b) attractive electrode ($B = 3T$, $\omega_i = 3.1$)

$$E_{\rm HF}^{\rm tot}(\{f_{m,\sigma}\}) = \sum_{m\sigma} e_{m\sigma} f_{m\sigma} + 1/2 \sum_{m_1 m_2 \sigma\sigma'} < m_1 m_2 | V_{\sigma\sigma'}^{\rm HF} | m_2 m_1 > f_{m_1\sigma} f_{m_2\sigma'}, \tag{11}$$

where

$$< m_1 m_2 | V_{\sigma\sigma'}^{\rm HF} | m_2 m_1 > = < m_1, m_2 | V | m_2 m_1 > - < m_1, m_2 | V | m_2 m_1 > \delta_{\sigma\sigma'} \tag{12}$$

is the HF Coulomb matrix element and the factor of 1/2 corrects the overcounting of pair interactions.

Once the ground-state configuration is found, the HF quasiparticle energies $E_{\rm HF}(m\sigma)$ can be easily calculated using

$$E_{\rm HF}(m\sigma) = e_{m,\sigma} + \sum_{m_2\sigma'} < m m_2 | V_{\sigma\sigma'}^{\rm HF} | m_2 m > f_{m_2\sigma'}. \tag{13}$$

Fig. 6 shows the dispersion of HF energies $E_{\rm HF}(m\sigma)$ for a droplet of N=27 electrons and different values of the magnetic field. The energies are plotted in units of the exchange energy E_0. The thick bars correspond to energies of the

$n = 0$ states while the thin lines show the HF energies of the second Landau level ($n = 1$). The black squares denote the occupied states. For $B < 1T$, the droplet consists of a small spin-up droplet and a larger spin-down droplet. The increasing of the magnetic field increases the role of the Coulomb energy and shrinks the spin-up droplet as the spin-down droplet grows. At $B = 3T$ the droplet is spin-polarized with all electrons occupying the lowest single-particle states. There is only a single state (configuration) $|G>$ in the spin-polarized Hilbert subspace corresponding to this value of the total angular momentum; this state is an exact state of the interacting system [8]. As the magnetic field increases to $4T$, the spin-polarized droplet breaks down to reduce its Coulomb energy, forming a ring and a small droplet.

This droplet reconstruction [3, 8, 7] is analogous to transitions between "magic" angular momentum states in few-electron droplets [11, 12, 8, 13]. Because the exchange splitting between spin-down and up states, of the order of E_0, is comparable with the kinetic energy splitting Ω_+, the restriction of the Hilbert space to the lowest Landau level appears to work quite well here.

We now focus on the effect of the electrode on the compact spin-polarized dot. In Fig. 7, we show the distribution functions $f_{u/d}(m)$, for a strongly attractive electrode ($\omega_i = 10\ meV$). The $f(m)$ clearly show the formation of an excess of charge and spin in the centre of the dot, surrounded by a ring of holes, in a spin-polarized state. The situation is different for a repulsive electrode. The centre of the dot is empty and electrons form a ring of spin-polarized electrons. The electrode is, however, surrounded by a small ring of electrons with reversed spin.

Such electrode induced transitions should be observable in resonant tunnelling experiments. The tunnelling experiment in the Coulomb blockade regime measures the chemical potential, $\mu(N)$, of the dot [4, 13, 24]. In Fig. 8 we show the evolution of the chemical potential (the shift of the position of the conductivity peak) as a function of B in the absence of an electrode and as a function of electrode strength, ω_i, for both attractive and repulsive electrodes at the fixed magnetic field $B = 3T$ [3]. Fig. 8a shows the evolution of the chemical potential as a function of the magnetic field in the absence of an electrode. The oscillatory character for $B < 2T$ ($1 < \nu < 2$) corresponds to electrons flipping their spin while the spin-down droplet grows at the expense of the shrinking spin-up droplet. From $B = 2$ to $B = 3.4T$, the droplet is a spin-polarized compact droplet with an almost uniform charge density; for $B > 3.4T$, it enters the Fractional Quantum Hall regime, breaking down into inhomogeneous and strongly correlated parts. This breakdown corresponds to the introduction of holes into the compact droplet. The small number of holes immersed in the homogeneous droplet condense into highly correlated states in a way similar to a small number of electrons [22]. Since our main interest at the moment is in the effect of electrode, we concentrate on a well characterized compact droplet at $B = 3T$. Fig. 8b shows the chemical potential (the shift of the Coulomb blockade peak) as a function of electrode strength for a repulsive (attractive) electrode. For a repulsive (attractive) electrode, we expect the coulomb blockade peak to move

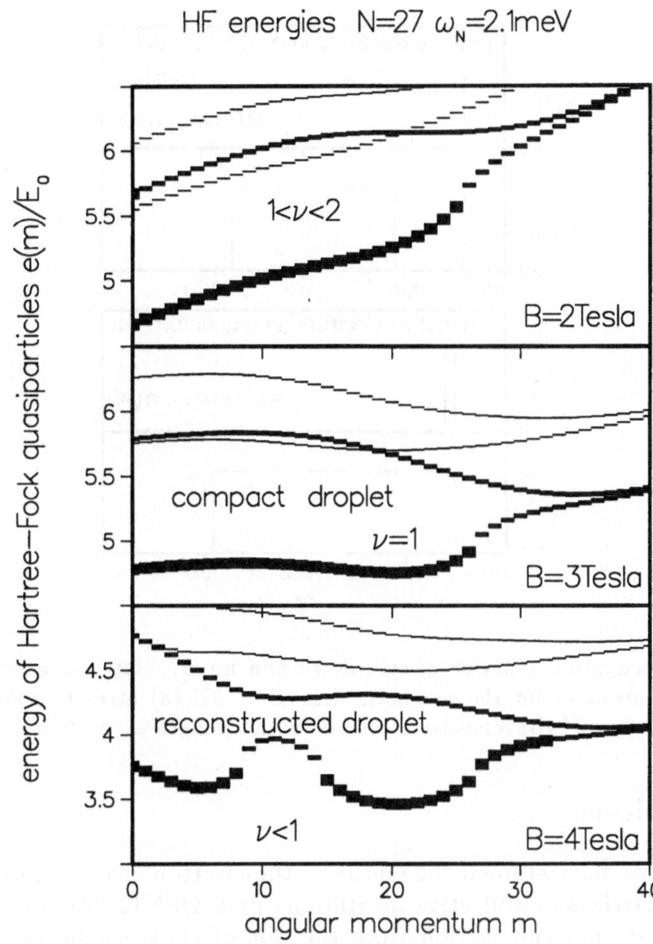

Fig. 6. Energies of the Hartree-Fock quasiparticles $E_{\mathrm{HF}}(m)$ as a function of the angular momentum m for values of the magnetic field B=2,3,4 T. The corresponding filling factors ν are indicated.

to higher (lower) energies (gate voltages). The shift of the chemical potential takes place in steps. These steps correspond to the discrete number of holes and spin-reversed electrons created in a compact droplet. The holes can come from the edge, or from the bulk, of the droplet. Both the spin-reversed electrons and holes can restructure as a function of the electrode strength, leading to additional structure in the charging curve (see, e.g., $\omega_i > 4$ region in Fig. 8b).

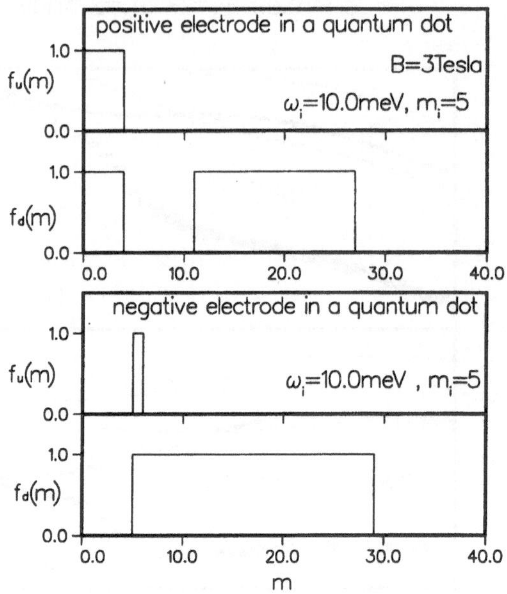

Fig. 7. The occupation function of spin-down and up HF states as a function of the angular momentum m for the magnetic field $B = 3T$ (a) attractive electrode with strength $\omega_i = 10\ meV$ (b) repulsive electrode with strength $\omega_i = 10\ meV$, insets show

6 Conclusions

In summary, we have studied the effects of the electron-electron interactions on the single-particle and collective excitations of a QSET. We concentrated on the addition to and subtraction from the dot of electrons by calculating the spectral function and on the coherent resonant tunnelling as a spectroscopy of the collective excitations. In particular, we have examined the effect of a tunable electrode on the ground and excited states of a strongly interacting electron droplet in a QSET. The effect of the electron-electron interactions in the droplet is tuned by the magnetic field and by the central electrode. We showed that both the magnetic field and the electrode induce spin and charge transitions in the ground state of the electron droplet. These transitions lead to structure in the chemical potential of the dot, and hence to shifts of the Coulomb blockade peak. The distinct structure of these shifts corresponds to a finite number of holes and spin reversed electrons created in the dot. This in turn should modulate the transport through the QSET.

Acknowledgments

We thank J.A. Brum, A. Wójs, A. Sachrajda, Y. Feng, and R. Williams for useful discussions.

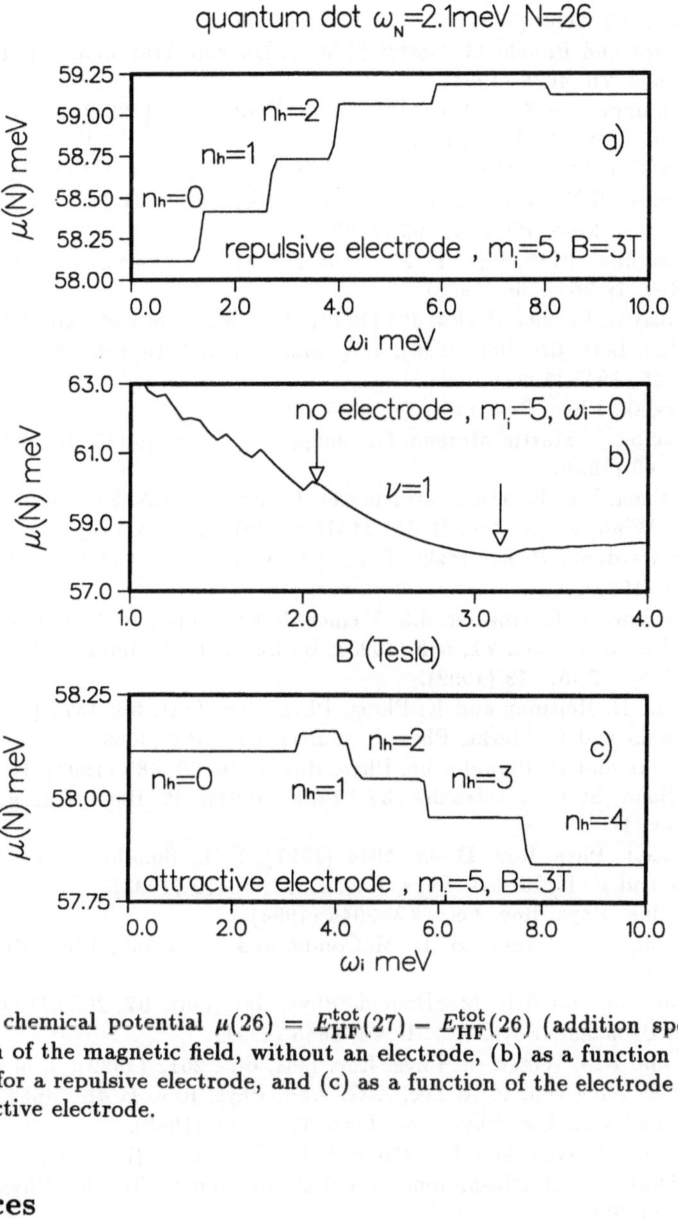

Fig. 8. The chemical potential $\mu(26) = E_{\mathrm{HF}}^{\mathrm{tot}}(27) - E_{\mathrm{HF}}^{\mathrm{tot}}(26)$ (addition spectrum) (a) as a function of the magnetic field, without an electrode, (b) as a function of electrode strength ω_i for a repulsive electrode, and (c) as a function of the electrode strength ω_i for an attractive electrode.

References

1. For recent reviews and references see M. Kastner, Physics Today, **24**, January 1993; T. Chakraborty, Comments in Cond. Matter Physics **16**, 35 (1992).
2. Y. Feng et al., Appl. Phys. Lett. 63, 1668 (1993); R.P. Taylor et al., J. Vac. Sci. Technol. **B11**, 628 (1993).
3. P. Hawrylak, Phys. Rev. **B 51**, 17 708 (1995).
4. O. Klein et al., Phys. Rev. Lett. **74**, 785 (1995).

5. A. Sachrajda et al., Phys. Rev. **B 50**, 10856 (1994); G. Kirczenow et al., Phys. Rev. Lett. **72**, 2069 (1994).
6. Yong S Joe and Ronald M. Cosby, M.W.C. Dharma-Wardana, Sergio E. Ulloa, J. Appl. Phys. **76**, 4676 (1994).
7. C. de Chamon and X.-G. Wen, Phys. Rev. **B 49**, 8227 (1994); J.J. Palacios et al., Europhys. Lett. **23**, 495 (1993).
8. A.H. MacDonald, S.R. Eric Yang and M.D. Johnson, Aust. J. Phys. **46**, 345 (1993).
9. M.D. Stone, H.W. Wyld and R.L. Schult, Phys. Rev. **B 45** , 14156 (1992); M. Stone, Int. J. Mod. Phys. **5**, 503 (1990).
10. R.B. Laughlin, Phys. Rev. **B 27**, 3383 (1983); SM. Girvin and Terrence Jach, Phys. Rev. **B 28**, 4506 (1983).
11. P.A. Maksym, Physica **B 184**, 385 (1993) ; P.A. Maksym and Tapash Chakraborty, Phys. Rev. Lett. **65**, 108 (1990); P.A. Maksym and Tapash Chakraborty, Phys. Rev. **B 45**, 1947 (1992).
12. P. Hawrylak, Phys. Rev. Lett. **71**, 3347 (1993).
13. J.J. Palacios, L. Martin-Moreno, G. Chiappe, E. Louis and C. Tejedor, Phys. Rev. **B 50**, 5760 (1994).
14. P.L. McEuen, E.B. Foxman, Jari Kinaret, U. Meirav, M.A. Kastner, N.S. Wingreen and S.J. Wind, Phys. Rev. **B 45**, 11419 (1992); A.S. Sachrajda, R.P. Taylor, C. Dharma-wardana, P. Zawadzki, J. A. Adams and P.T. Coleridge, Phys. Rev. **B 47**, 6811 (1993).
15. R. C. Ashoori, H.L. Stormer, J.S. Weiner, L.N. Pfeiffer, K.W. Baldwin and K.W. West, Phys. Rev. Lett. **71**, 613 (1993); Bo Su, V.J. Goldman and J.E. Cunningham, Science, **255**, 313 (1992).
16. B. Meurer, D. Heitman and K. Ploog, Phys. Rev. Lett. **68**, 1371 (1992).
17. Ch. Sikorski and U. Merkt, Phys. Rev. Lett. **62**, 2164 (1989).
18. P. Hawrylak and D. Pfannkuche, Phys. Rev. Lett. **70**, 485 (1993); D. Pfannkuche et al., Solid State Electronics **37**, 1221 (1994); P. Hawrylak, submitted to Phys. Rev. **B**.
19. E.H. Rezayi, Phys. Rev. **B 43**, 5944 (1991); S. L. Sondhi, A. Karlhede, S. A. Kivelson and E. H. Rezayi, Phys. Rev. **B 47**, 16419 (1993).
20. P. Hawrylak, Phys. Rev. Lett. **72**, 2943 (1994).
21. J.-P. Cheng, Y. J. Wang, B. D. McCombe and W. Schaff, Phys. Rev. Lett. **70**, 489 (1993).
22. M. D. Johnson and A.H. MacDonald, Phys. Rev. Lett. **67**, 2060 (1991).
23. C. W. J. Beenaker, Phys. Rev. **B 44**, 1646 (1991).
24. Y. Meir and N.S. Wingreen, Phys. Rev. Lett. **68** , 2512 (1992); J. M. Kinaret, Y. Meir, N. S. Wingreen, P. A. Lee, X-G. Wen, Phys. Rev. **B 46** , 4681 (1992).
25. T.K. Ng and P.A. Lee, Phys. Rev. Lett. **61** , 1768 (1988).
26. P. Hawrylak, A. Wojs and J.A. Brum, Solid St. Comm. (in press)
27. J. H. Oaknin, L. Martin-Moreno, J.J. Palacios and C. Tejedor Phys. Rev. Lett. **74**, 5120 (1995).

At the Limit of Device Miniaturization

Tomasz Dietl

Institute of Physics and College of Science, Polish Academy of Sciences
al. Lotników 32/46, PL-02668 Warszawa, Poland

Abstract: Some striking quantum phenomena discovered in small conductors, insulators and capacitors over the recent years are described. It is argued that while these phenomena may render further miniaturization ofthe present electronic devices difficult, they may constitute the principle of operation of the next generation of devices.

1 Introduction

It is widely appreciated that we witness an information revolution. The major driving force behind it is the enormous and steady progress in device miniaturization. Miniaturization increases the storage density of information, reducing simultaneously the price and time to access and process it. Indeed, today's microprocessors (μP) can execute one billion instructions per second (1 GIPS), and 100 million transistors packed on one 2 cm^2 chip of dynamic random access memory (DRAM) store 64 Mb of information - about 4 000 printed pages. Actually, according to Moore's law, which has been obeyed over the last 30 years, the storage density has increased by a factor of 2 every 18 months, as shown in Fig. 1. At the same time, scaling recipes which say how device parameters (supply voltage, gate thickness etc.) have to be changed with the overall transistor size, have successfully been employed down to the present critical dimension of 0.35 μm.

The natural question arises of when a breakdown of the Moore and scaling laws is to be expected or, in the other words, what will be the main constraints that may slowdown device miniaturization in the future. Of course, there are many kinds of such constraints identified by now: psychological (a barrier to be accustomed to ...), legal (copyrights, censorship against sex and violence, ...), financial (cost of a present-day DRAM factory is over 2 G$), technical (software, electrodiffusion, cross-talking, lithography resolution,....). While each of these issues is of potential importance, in the present paper we discuss still another constraint. We address the question of whether there exist physical phenomena

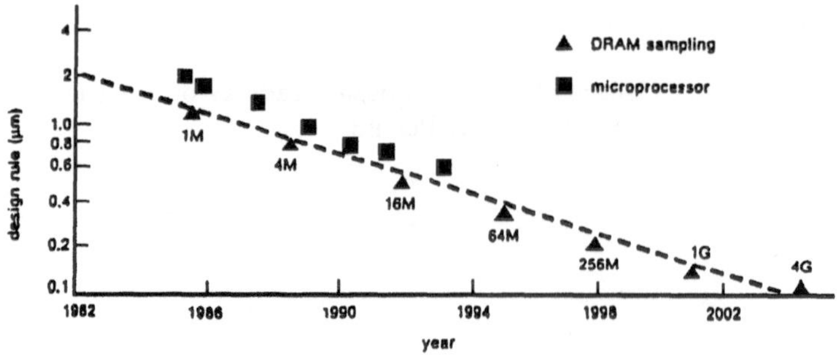

Fig. 1. Illustration of Moore's law: time evolution of the design rule for μPs and DRAMs together with the bit capacity of DRAM (dynamic random access memory).

that may limit further progress in the miniaturization of the devices. Since, without a doubt, the principal electronic device is the field effect transistor (FET), we describe the physics of its main building blocks, that is, the physics of small conductors, insulators, and capacitors. The results which we are going to present show that there exist indeed many novel phenomena in small structures. These phenomena may preclude the further miniaturization of today's devices. However, they may constitute the principle of operation of future devices.

2 Small conductors

The presence of many (10^8) transistors on one chip raises the question about statistical fluctuations of the conductance from one nominally identical small conductor to another one. According to the Boltzmann–Drude theory, in the absence of phonon scattering, the conductance G (inverse resistance) is inversely proportional to the number of defects in the structure, N. Of course, in the nominally identical structures the distribution and the actual number of defects undergo statistical fluctuations. For instance, in submicron wires of modulation-doped GaAs/AlGaAs heterostructures with the extremely high electron mobility of the order of 10^6 cm^{2}V/s there can be no defects at all (ballistic transport) or one or two defects (quasi-ballistic transport). In most cases, however, $N \gg 1$, so that the mean free path ℓ is much smaller than the conductor length L, and the transport becomes diffusive. For such a case the mean square root dispersion of N is $\Delta N = N^{1/2}$, and thus $\Delta G/G = N^{-1/2}$. Hence, for a conductor in the form of a cube of volume L^d, for which $G \propto L^{d-2}$, we arrive at $\Delta G \propto 1/L^{2-d/2}$. We see that in the framework of the Drude–Boltzmann theory, the statistical fluctuations of the conductance has the property of self averaging: $\Delta G \to 0$ for $L \to \infty$.

In view of the above discussion it was a surprise, when the independent diagrammatic quantum calculations of the conductance by Altshuler [1] and Lee and Stone [2] predicted that in any conductor

$$\Delta G \approx 0.5e^2/h, \tag{1}$$

independent of the material, space dimension d, and conductor size L. These

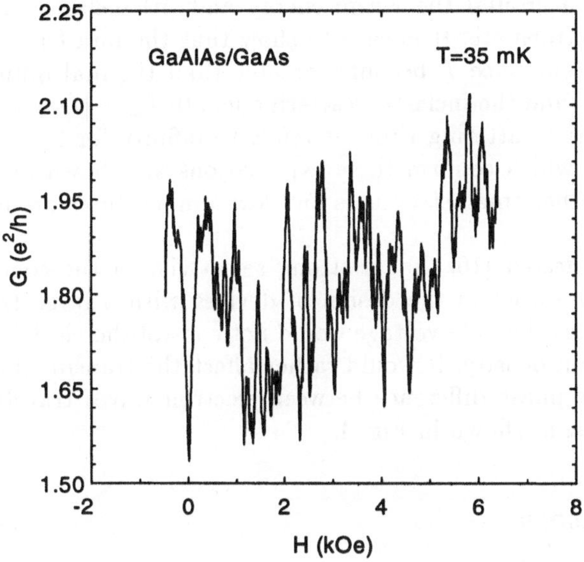

Fig. 2. Aperiodic conductance changes (in units of e^2/h) as a function of the magnetic field in wires of Si MOS-FET [5], Au film [6], GaAs/AlGaAs heterostructure [7], and HgCdMnTe bicrystal (two traces document the reproducibility of the fluctuations) [8].

remarkable properties of the variance ΔG explain why the phenomena are refereed to as universal conductance fluctuations (UCF). There are two quantum effects that account for the UCF. One is the interference of the de Broglie waves corresponding to different possible paths of the Fermi level electron through the disordered medium. Interference effects are not taken into account in the essentially classical Drude–Boltzmann description of the conductivity. The interference term contains detailed information on the impurity distribution in a given sample, and is responsible for large conductance fluctuations from one potential realization to another. Moreover, the interference contribution to the conductance can by varied in one sample by changing the wavelength of electrons at the Fermi level by, say, the gate voltage, or the magnetic field, as the vector potential affects the electron phase. The second effect, which must be considered to explain the universality of the UCF amplitude, is associated with the quantum repulsion between energy levels of the random systems. This repulsion leads to the so-called Wigner distribution of the energy levels, discussed in detail by Dyson [3] in the context of nucleus scattering spectra. The above theoretical suggestions [1, 2, 4] are confirmed by a number of experimental results, a few of them [5, 6, 7, 8] being displayed in Fig. 2.

The natural question arises of why these effects are not seen in large samples but only in nanostructures. There are two mechanisms accounting for the decrease of the UCF amplitude with L [4]: (i) thermal broadening of the electron distribution around the Fermi energy and (ii) inelastic scattering by phonons, other electrons etc. It is easy to show that the interference effect begins to be washed out once L becomes greater than thermal diffusion length $L_T = (hD/k_B T)^{1/2}$ and the inelastic scattering length $L_\varphi = [D\tau_\phi(T)]^{1/2}$, where $\tau_\phi(T)$ is the inelastic scattering time (it tends to infinity for $T \to 0$). Experimental findings [9] which confirm these expectations are shown in Fig. 3. It is seen that indeed rather small structures and low temperatures are necessary to observe the UCF.

It has been suggested [10] that a strong sensitivity of the conductance to interference could be used in the design of devices with a large transconductance. In such devices the gate voltage would not control the electric current by changing the electron density. It would rather affect the transmission probability by changing the phase difference between electron waves travelling by two conducting channels, as shown in Fig. 4.

3 Small insulators

The key requirement imposed on a logic device is a large contrast between its two states. In the case of FETs, this corresponds to a large resistivity difference for the two extreme values of the gate voltage. It has, however, been found [11] that on reducing the transistor length to the submicron range, the contrast in question tends to decrease. In particular, according to results presented in Fig. 5, even in the off-state, i.e. when the Fermi level is in the band-gap region, a small MOS-FET transistor can conduct current for some values of the gate voltage. It is now generally accepted that each sharp peak of the conductance in the off-region results from electron tunnelling between the source and the drain contacts via a defect that has a state in the band gap of the semiconductor. To contemplate the physics involved, note that the transmission coefficient T_r for electron tunnelling is exponentially small in both the heights of the energy barrier and the distance between the electrodes L. If, however, the electron energy becomes equal to that of a localized state, resonant tunnelling is possible. For such a process, the transmission probability is no longer determined by L but by $|L_{ds} - L_{dd}|$, where L_{ds} and L_{dd} are the distances of the defect to the source and drain, respectively [12]. We see that $T_r = 1$ for a defect that happens to be exactly in the middle of the structure. Thus, such a defect completely short-

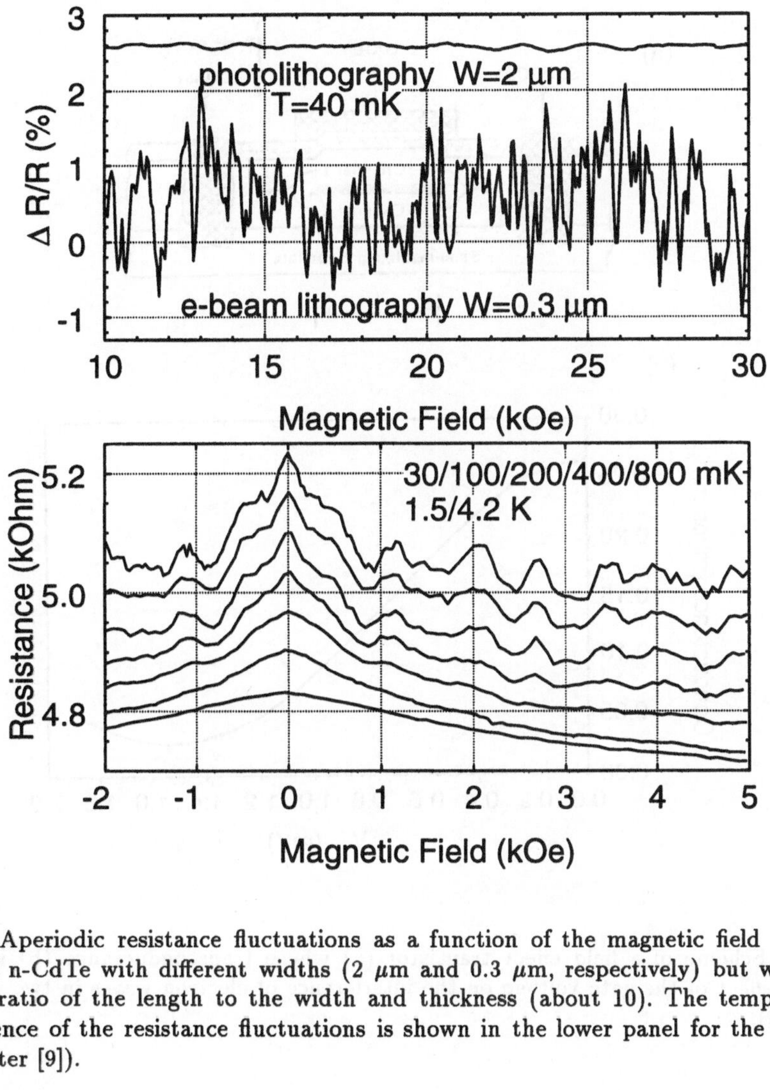

Fig. 3. Aperiodic resistance fluctuations as a function of the magnetic field for two wires of n-CdTe with different widths (2 μm and 0.3 μm, respectively) but with the similar ratio of the length to the width and thickness (about 10). The temperature dependence of the resistance fluctuations is shown in the lower panel for the smaller wire (after [9]).

circuits an insulator, provided that the energy of its states coincides with the Fermi level and, at the same time, its lifetime broadening Γ is greater than the thermal spread of electron energies, k_BT. For the state in question G decreases exponentially with $L/2$, hence the observation of the effect requires either small samples or rather low temperatures.

A manmade structure, the so-called resonant hot-electron transistor (RHET), is presented schematically in Fig. 6 [13]. It consists of a GaAs source, drain, and well regions separated by AlGaAs barriers. The two-dimensional state in the well serves as a resonant level for the electrons emitted by the source. It is known that such good quality layered structures with sharp and defect free interfaces can

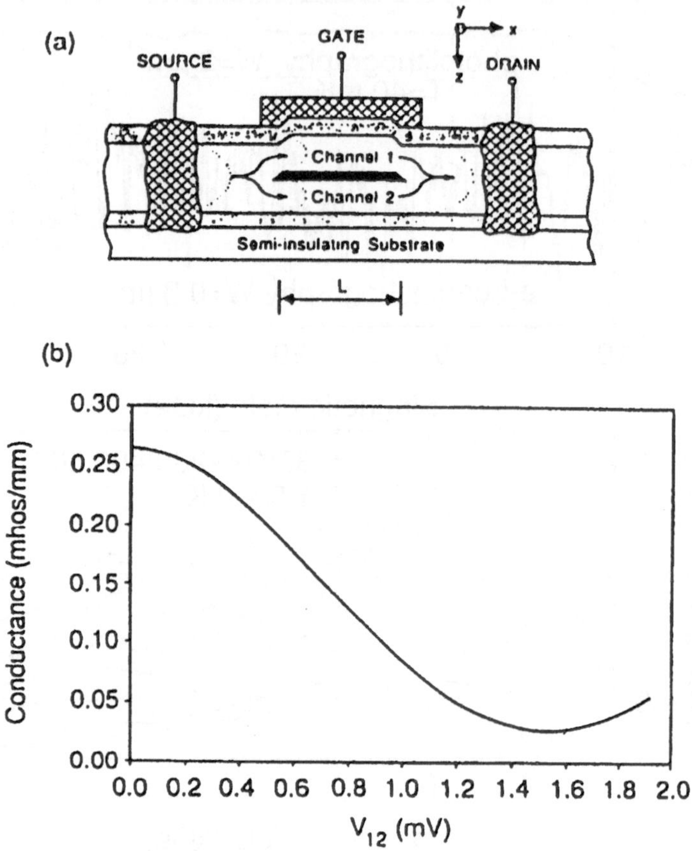

Fig. 4. Scheme of a field effect transistor (a) whose transconductance (b) is driven by the effect of the gate voltage on the interference of electron waves in two channels (after10).

be obtained by crystal growth at appropriately low temperatures, for instance, by molecular beam epitaxy (MBE) or metal organic chemical vapor deposition (MOCVD). This device has several interesting features, among them its characteristics exhibit three, not two, well-defined states, making simplifications of the logic-gate architecture possible. In addition, RHET seems to be a suitable device for chips, in which elements could be packed layer by layer, i.e. three dimensionally, not only in one plane. As shown in Fig. 7 [14], the current-voltage characteristics, specific to resonant tunneling, are visible even at room temperature in high quality structures.

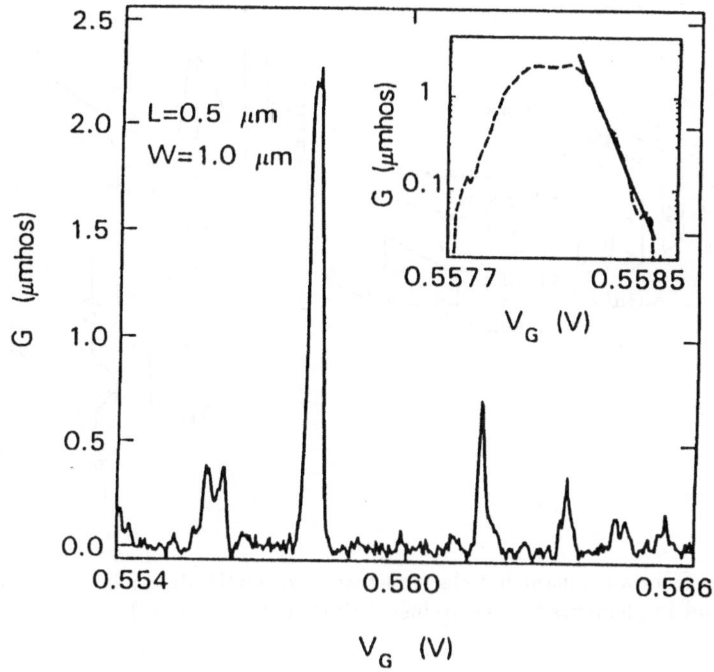

Fig. 5. Conductance of a small Si MOS-FET as a function of the gate voltage (after [11]).

4 Small capacitors

It is known that the equilibrium charge Q of a capacitor is determined by the competition between the positive energy of the Coulomb repulsion between accumulated charges, $Q^2/2C$, and a negative energy of the charge in the electrostatic potential, $-QV$. By minimizing the total energy,

$$E(Q) = Q^2/2C - QV, \tag{2}$$

with respect to Q we arrive to the text book conclusion that the accumulated charge increases linearly with the capacitor voltage, $Q = CV$.

Since, however, the charge is quantized we may suspect that the actual charging process does not proceed in a continuous way but in steps of the elementary charge e [15]. The charging of a capacitor plate which contains N electrons by a new electron will occur for such a voltage that $E[(N+1)e] = E(Ne)$. This condition gives $V = (N + 1/2)e/C$, and shows that if an increase of the voltage ΔV is too small, $\Delta V < e/C$, charging is not at all possible. This phenomenon is known under the name of Coulomb blockade. For a small plate-air-capacitor

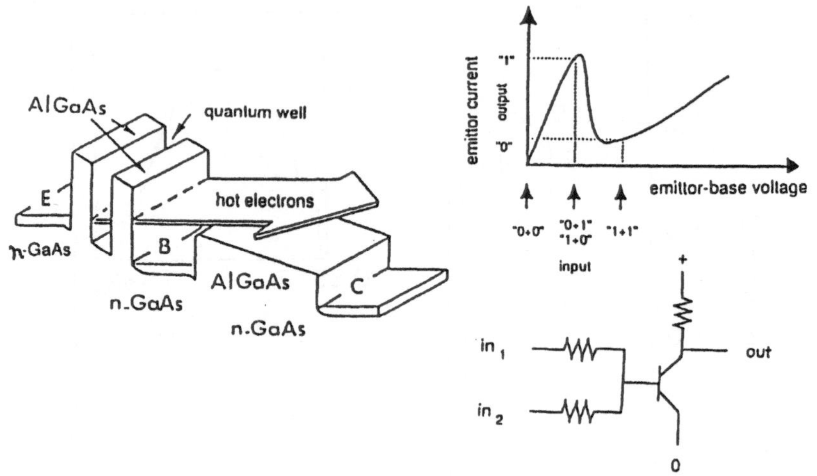

Fig. 6. Scheme for a resonant hot electron transistor (RHET), its current-voltage characteristics and implementation as exclusive-NOR gate (after [13]).

with the surface $S = 0.1$ mm^2 and the distance between the plates $d = 10$ nm, respectively, we evaluate $C = \epsilon_o S/d \approx 0.1$ fF, and thus e/C to be as large as 2 mV.

An example of the structure, in which the Coulomb blockade has been readily observed [16] is schematically presented in Fig. 8. The capacitor plates are formed by the back gate and the two-dimensional electron gas at the GaAs/AlGaAs interface. An additional gate, just at the top of the electron gas, is patterned in such a way that it permits the total depletion of the electron gas, except for a small dot-like region and the pads to it. The process of charging is probed by measuring the current through the dot, i.e. the dot conductance as a function of the capacitor voltage. A small voltage that drives the current as well as the negative potential of the upper gate are kept constant during this experiment. As shown in Fig. 9, the conductance shows a periodic multi-peak structure. According to the discussion above, each of the peaks marks the appearance of a new electron in the dot. The conductance modulation vanishes at higher temperatures $T > e^2/k_B C$, where thermal broadening of the distribution function means that the electrons to have energies sufficiently high to overcome the Coulomb barrier.

Several applications have been proposed for these *artificial atoms* - dots with a controlled number of electrons. A single electron transistor (SET), electrometer, and primary thermometer [17] constitute the most commonly discussed examples of possible devices. The Coulomb interaction *between* the charged dots has also been suggested as a basis for computer logic [18].

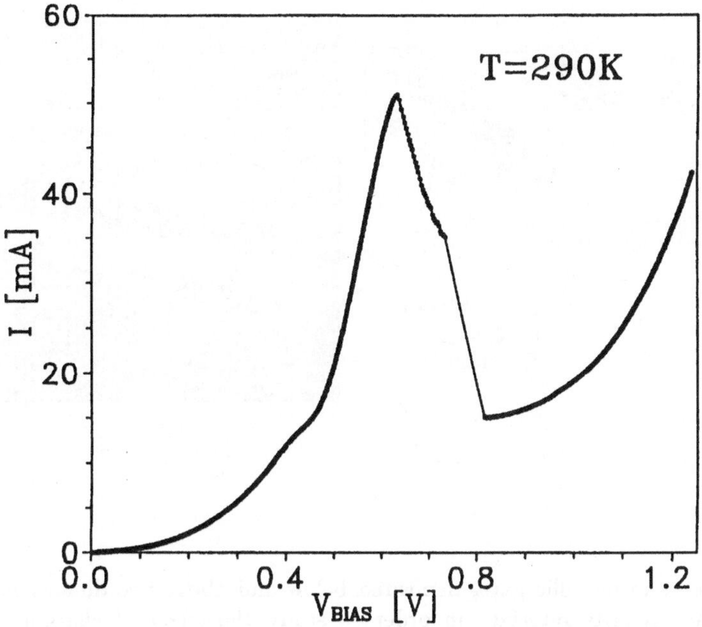

Fig. 7. Room temperature current voltage characteristics of resonant tunnelling structure of GaAs/AlGaAs (after [14]).

5 Concluding remarks

A traditional method that is employed in order to evaluate the lifetime of devices is to carry out degradation studies as a function of temperature well *above* room temperature and then to extrapolate the results downwards. In the physics of quantum devices we proceed other way around. Indeed, the examples discussed in the present paper demonstrate that a number of striking phenomena show up in the mesoscopic regime which is reached by *decreasing* the temperature and/or the sample size. Thus, we can find out at which level of integration these new phenomena might perturb the operation of the nowadays devices by extrapolating from the subkelvin region up to room temperature. In the same way we can evaluate the upper limit for the size at which the device whose principle of operation involves one of the newly discovered phenomena would work at room temperature. For instance, it is easy to show that the critical dimension of a room–temperature single–electron–transistor should be between 5 and 8 nm. The question of whether such small devices could be fabricated by advanced lithography, by direct deposition methods or by atom or molecule manipulation is under vigorous studies in the present time [19].

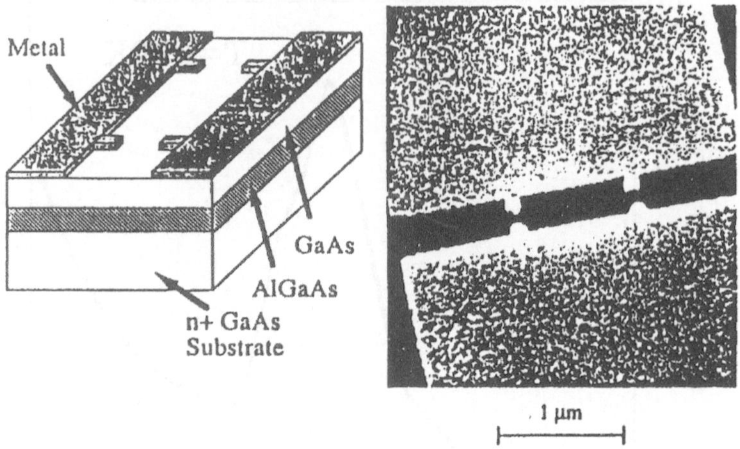

Fig. 8. Scheme of metallic gates deposited below and above two dimensional electron gas at GaAs/AlGaAs interface in order to study the effect of charging of a small capacitor (after [16]).

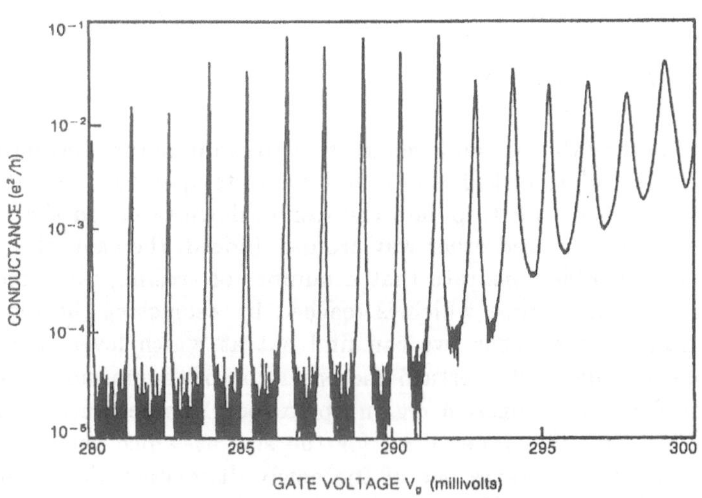

Fig. 9. Conductance of a function of the gate voltage for structure of Fig. 8 (after [16]).

References

1. B.L. Altshuler, Pisma Zh. Eksp. Teor. Fiz. **41**, 530 (1985) [JETP Lett. **41**, 648 (1985)].
2. P.A. Lee and A.D. Stone, Phys. Rev. Lett. **55**, 1622 (1985).
3. F.J. Dyson, J. Math. Phys. **3**, 157 (1962) and **3**, 166 (1962); F.J. Dyson and M.L. Mehta, J. Math. Phys. **4**, 701 (1963).
4. B.L. Altshuler and B.I. Shklovskii, Zh. Eksp. Teor. Fiz. **91**, 220 (1986) [Sov. Phys.: JETP **64**, 127 (1986)]; P.A. Lee, A.D. Stone and H. Fukuyama, Phys. Rev. B **35**, 1039 (1987).
5. W.J. Skocpol, P.M. Mankiewich, R.E. Howard, L.D. Jackel, D.N. Tennant and A.D. Stone, Phys. Rev. Lett. **56**, 2865 (1986); Phys. Rev. Lett. **58**, 2347 (1986).
6. S. Washburn and R.A. Webb, Adv. Phys. **35**, 375 (1986).
7. J. Wróbel, T. Brandes, F. Kuchar, B. Kramer, K. Ismail, K.Y. Lee, H. Hillmer, W. Schlapp and T. Dietl, Europhys. Lett. **29**, 481 (1995).
8. G. Grabecki, T. Dietl, W. Plesiewicz, A. Lenard, T. Skośkiewicz, E. Kamińska and A. Piotrowska, Physica B **194–196**, 1107 (1994).
9. J. Jaroszyński, J. Wróbel, M. Sawicki, E. Kamińska, T. Skośkiewicz, G. Karczewski, T. Wojtowicz, A. Piotrowska, J. Kossut and T. Dietl, Phys. Rev. Lett. **75**, 3170 (1995).
10. S. Datta, Superlattices and Microstructures **6**, 83 (1989).
11. A.B. Fowler, G.L. Timp, J.J. Wainer and R.A. Webb, Phys. Rev. Lett. **57**, 138 (1986).
12. B. Ricco and M.Ya. Azbel, Phys. Rev. B **29**, 1970 (1984).
13. N. Yokoyama, K. Imamura, S. Muto, S. Hiyammizu and H. Nishi, Japan. J. Appl. Phys. **24**, L853 (1985); also C. Weisbuch, in Low-Dimensional Electronic Systems, edited by G. Bauer, F. Kuchar and H. Heinrich (Springer, Berlin, 1992).
14. T. Figielski, T. Wosiński, A. Mąkosa, M. Kaniewska and K. Regiński, Lithuanian J. Physics **35**, 542 (1995).
15. J. Lambe and R. Jaklevic, Phys. Rev. Lett. **22**, 1371 (1969).
16. U. Meirav, M.A. Kastner and S.J. Wind, Phys. Rev. Lett. **65**, 771 (1990).
17. J.P. Pekola, K.P. Hirvi, J.P. Kauppinen and M.A. Paalanen, Phys. Rev. Lett. **73**, 2903 (1994).
18. C.S. Lent, P.D. Tougaw, W. Porod and G.M. Bernstein, Nanotechnology **4**, 49 (1993).
19. For a review of developments discussed in the present paper, see, e.g., Mesoscopic Phenomena in Solids, edited by B.L. Altshuler, P.A. Lee and R.A. Webb (Amsterdam, Elsevier, 1991); C.V.J. Beenakker and H. Van Houten, Solid State Phys. **44**, 1 (1991); C. Weisbuch and B. Vinter, Quantum Semiconductor Structures (Academic, Boston, 1991); Science and Technology of Mesoscopic Structures, edited by S. Namba et al. (Springer, Tokyo, 1992); Sheng S. Li, Semiconductor Physical Electronics (Plenum Press, New York, 1993); Physics of Low- Dimensional Semiconductor Structures, edited by. P. Butcher et al. (Plenum Press, New York, 1993).

Classical and Quantum Transport Calculations for Elastically Scattered Free Electron Gases in 2D Nanostructures when B = 0

P.N. Butcher

Department of Physics, University of Warwick,
Coventry CV4 7AL U.K.

Abstract: The Boltzmann equation in the relaxation time approximation is used to calculate formal expressions for the local electrical, thermal and thermoelectric transport coefficients of a strictly 2D, elastically scattered, free electron gas. The terminal transport coefficients for the same gas confined in a quantum wire are also calculated using the Landauer-Buttiker formalism. Both calculations are valid in a quantum wire structure when its width w is much greater than the Fermi wavelength and its length ℓ is much greater than the classical mean free path. Comparison shows that the sum of all the transmission coefficients through the system at the Fermi energy ε is therefore given by $T(\varepsilon) = (w/\ell)\varepsilon/\Delta\varepsilon$ where $\Delta\varepsilon = \hbar/\tau(\varepsilon)$ is the uncertainty in ε arising from the classical relaxation time $\tau(\varepsilon)$. A new way of calculating $T(\varepsilon)$ using wave functions is outlined. Numerical results for both $T(\varepsilon)$ and scattering wave functions are presented for two nanostructures: (i) a quantum wire with one hard wall finger pushed in and (ii) a quantum wire with two hard wall fingers pushed in so as to create a quantum dot.

1 Introduction

The approach to electron transport through the classical Boltzmann equation has had many successes in collision dominated systems [1,2]. In the last decade a great deal of attention has also been payed to non-interacting 2D electron gases (2DEGS) which are often almost collision free so that the motion of the electrons is quasi-ballistic. To understand their behaviour it is then necessary to use Schrodingers equation. The resulting quantum mechanical formulae for the electrical and thermal conductances and the thermoelectric coefficients are easily obtained [3,4].

In these lectures we concentrate on non-interacting free electrons in nanostructures with two terminals and suppose that there is no magnetic induction field applied. We show that the classical and quantum-mechanical formulae have almost identical structures. In the classical formulae the zero temperature conductance when the Fermi level is ε is denoted by $\sigma(\varepsilon)$. It contains all the relevant classical mechanics and determines all the classical transport coefficients. In the quantum-mechanical formulae $\sigma(\varepsilon)$ is replaced by $G(\varepsilon) = (2e^2/h)T(\varepsilon)$ where $T(\varepsilon)$ is the sum of all the transmission coefficients from any mode in one terminal to any mode in the other. The Landauer-Buttiker conductance $G(\varepsilon)$ contains all the relevant quantum mechanics and determines all the quantum- mechanical transport coefficients.

The similarity between the classical and quantum- mechanical formulae arises from the assumptions made in both calculations that the electrons move independently and are conserved. In some cases both methods of calculation are valid. A simple example is a rectangular 2DEG of width w, length ℓ, and having a classical relaxation time $\tau(\varepsilon)$ produced by elastic scattering. Quantum mechanics is always appropriate to this system. Classical mechanics is also valid provided that $\ell \gg \ell_c$, the classical mean free path, and $w \gg \lambda$, the de Broglie wavelength at the Fermi level ε. We may therefore relate $T(\varepsilon)$ to $\tau(\varepsilon)$. The result is $T(\varepsilon) = (w/L)\varepsilon/\Delta\varepsilon$ where $\Delta\varepsilon = \hbar/\tau(\varepsilon)$ is the width of the Fermi level produced by the elastic scattering.

We give a brief account of Boltzmann transport theory for a 2DEG in Section 2. Section 3 is devoted to the simple quantum-mechanical case of a 2D quantum wire in which there are no scattering centres. Then we may immediately derive the quantisation formula in the absence of a magnetic field: $G(\varepsilon) = N(2e^2/h)$ where N is the number of modes which can propagate in the wire at the Fermi level. It is also easy to understand that, if a fraction R of the particle current in a particular mode in one terminal is reflected by an obstacle into backward travelling modes, then its contribution to $G(\varepsilon)$ will be reduced by a factor $1-R = T$, where T is the total transmission coefficient (for the mode considered) into the other terminal. Summing over all the modes gives the Landauer-Buttiker formula quoted earlier: $G(\varepsilon) = (2e^2/h)T(\varepsilon)$, where $T(\varepsilon)$ is the sum of all the transmission coefficients from every mode on one side of the obstacle to every mode on the other side. We broaden out this intuitive discussion in Section 4 so as to write down the quantum- mechanical formulae for all the terminal transport coefficients.

Finally, in Section 5 we describe a new way to calculate $G(\varepsilon)$ [5,6] which also yields scattering wave functions. Results are given for a quantum wire with a hard wall finger pushed into it and for a quantum wire with two adjacent hard wall fingers push in. In the second case the space between the fingers forms a quantum dot which is accessed through the spaces above the fingers. We present numerical results for $G(\varepsilon)$ and scattering wave functions in both cases.

2 Boltzmann Transport Theory

We consider a strictly two-dimensional electron gas (2DEG) between parallel hard potential walls spaced a distance w apart. For a steady state situation Boltzmann's equation for the distribution function $f(\boldsymbol{k}, \boldsymbol{r})$ of the 2DEG is [1,2]

$$\mathbf{v} \cdot \boldsymbol{\nabla}_{\boldsymbol{r}} f - \frac{e}{\hbar} \boldsymbol{E} \cdot \boldsymbol{\nabla}_{\boldsymbol{k}} f = -(f - f_o)/\tau \tag{1}$$

In this equation all the vectors lie in the xy–plane and f_o is the Fermi-Dirac function:

$$f_o = [\exp\{(\varepsilon_{\boldsymbol{k}} - \mu)/k_B T\} + 1]^{-1} \tag{2}$$

where $\varepsilon_{\boldsymbol{k}}$ is the electron energy at wave vector $\boldsymbol{k}, \mu$ is the chemical potential (assumed constant) and T is the temperature which we suppose varies along the wire in the x–direction. On the right-hand side of Eq. (1) we have used the familiar relaxation time ansatz [1,2] for the rate of change of f due to collisions. The relaxation time is assumed to be a function of $\varepsilon_{\boldsymbol{k}}$ only and, in what follows, we consider free electrons in the lowest 2D subband with $\varepsilon_{\boldsymbol{k}} = \hbar^2 k^2/2m^*$.

To solve Eq. (1) we suppose that $\boldsymbol{E} = (E, 0)$ and $\boldsymbol{\nabla} T = (\partial T/\partial x, 0)$ in the (x, y)–plane and assume that both E and $\partial T/\partial x$ are small. Then we may write $f = f_o + f_1$ where f_1 is also small. Consequently, to first order in small quantities, the solution of Eq. (1) is given by

$$f_1 = \frac{\mathrm{d}f_o}{\mathrm{d}\varepsilon}\, \tau v_x \left[\mathrm{e}E + \frac{\partial T}{\partial x}\, \frac{\varepsilon - \mu}{T}\right] \tag{3}$$

The densities of electric current and heat flux along the wire are, respectively,

$$J = \frac{-\mathrm{e}}{2\pi^2} \int \mathrm{d}k_x \mathrm{d}k_y\; f_1\; v_x \tag{4a}$$

and

$$Q = \frac{1}{2\pi^2} \int \mathrm{d}k_x\; \mathrm{d}k_y\; f_1\; v_x(\varepsilon - \mu) \tag{4b}$$

where $v_x = \hbar k_x/m^*$ is the group velocity of an electron in the x direction Elementary manipulations allow us to carry out the angular integrations in Eqs. (4) immediately. We may also introduce the electron energy $\varepsilon = \hbar^2 k^2/2m^x$ as the remaining integration variable. The final results are:

$$J = \sigma\; E + L\; \frac{\partial T}{\partial x} \tag{5a}$$

$$Q = M\; E + N\; \frac{\partial T}{\partial x} \tag{5b}$$

where

$$\sigma = -\int \frac{\mathrm{d}f_o}{\mathrm{d}\varepsilon}\, \sigma(\varepsilon)\; \mathrm{d}\varepsilon \tag{6a}$$

is the electrical conductivity.

In Eq. (6a) $\sigma(\varepsilon) = n(\varepsilon)\mathrm{e}^2\tau(\varepsilon)/m^*$ denotes the conductivity when $T = 0$ and the chemical potential is set equal to ε [1]. The quantity $n(\varepsilon) = m^*\varepsilon/\pi\hbar^2$ is easily identified as the total 2D density of electron states in the lowest subband of the 2DEG with energy less than ε [5]. The remaining transport coefficients L, M and N are obtained from Eq. (5a) by multiplying $\sigma(\varepsilon)$ by $(\varepsilon - \mu)/\mathrm{e}T$, $(\varepsilon - \mu)/\mathrm{e}$ and $-(\varepsilon - \mu)^2/\mathrm{e}^2 T$ respectively. Thus we find that

$$L = -\int \frac{\mathrm{d}f_o}{\mathrm{d}\varepsilon}\, \sigma(\varepsilon)\; \frac{(\varepsilon - \mu)}{\mathrm{e}T}\; \mathrm{d}\varepsilon \tag{6b}$$

$$M = \int \frac{\mathrm{d}f_o}{\mathrm{d}\varepsilon}\, \sigma(\varepsilon)\, \frac{(\varepsilon-\mu)}{\mathrm{e}}\, \mathrm{d}\varepsilon \tag{6c}$$

and

$$N = \int \frac{\mathrm{d}f_o}{\mathrm{d}\varepsilon}\, \sigma(\varepsilon)\, \frac{(\varepsilon-\mu)^2}{\mathrm{e}^2 T}\, \mathrm{d}\varepsilon \tag{6d}$$

The coefficients in eqs.(5) and (6) are appropriate when we measure electric and heat current densities together with electric fields and temperature gradients. The quantum mechanical formulae involve the voltage drop $\Delta V = E\ell$, the temperature drop $\Delta T = \ell \mathrm{d}T/\mathrm{d}x$ and the total fluxes $J_T = wJ$ and $Q_T = wQ$ of electric current and heat along the quantum wire which we suppose has length ℓ and width w. In terms of these quantities Eqs. (5) become

$$J_T = \sigma_T \Delta V + L_T\, \Delta T \tag{7a}$$

and

$$Q_T = M_T\, \Delta V + N_T\, \Delta T \tag{7b}$$

where σ_T, L_T, M_T and N_T are obtained by multiplying the coefficients in Eq. (5) by w/L so that

$$(\sigma_T, L_T, M_T, N_T) = \frac{w}{\ell}(\sigma, L, M, N) \tag{8}$$

3 Electron Transport in a Quantum Wire

We consider a perfect 2D quantum wire with hard walls at $y = 0$ and $y = w$ and zero potential energy between the walls. The wave functions then take the form

$$\psi(x,y) = (2/w)^{1/2}\, \sin(r\pi y/w)\mathrm{e}^{ikx} \tag{9}$$

where r is a positive integer and we have normalised the mode functions $(2/w)^{1/2}\sin(r\pi y/w)$ to unity. It is convenient (for counting states) to introduce a length ℓ in the x–direction over which we apply periodic boundary conditions. Then the values of k are $s2\pi/\ell$ where s is an integer. Consequently the density of states per unit energy, per unit length, for one mode is:

$$N(\varepsilon) = \frac{1}{\ell}\, \frac{\Delta k}{2\pi/\ell} \cdot \frac{1}{\Delta\varepsilon} = \frac{1}{hv_x} \tag{10}$$

where Δk is a small increment of k, $\Delta\varepsilon$ is the corresponding increment of ε and

$$v_x = \frac{1}{\hbar}\, \frac{\mathrm{d}\varepsilon}{\mathrm{d}k} = \frac{\hbar k}{m^*} \tag{11}$$

is the group velocity. The corresponding density of current is

$$J_d = -2\mathrm{e}\, N(\varepsilon)\, v_x = -\frac{2\mathrm{e}}{h} \tag{12}$$

where -e is the electron charge. The factor of 2 in Eq. (12) allows for spin degeneracy (we ignore spin splitting throughout these lectures).

We see that J_d is equal to a constant involving only the fundamental quantities e and h. This is the origin of the quantisation of $G(\varepsilon)$ in quantum wires (and, more generally, in two terminal structures when $\mathbf{B} = 0$). To identify the conductance of the mode being considered we suppose that current is fed into the wire from ideal reservoirs at either end, both of which have $T = 0$, but also have chemical potentials which differ by $\Delta\mu$. Then the total current carried by the mode is $J_d \Delta\mu$. Moreover the voltage drop in the direction of current flow is $\Delta V = -\Delta\mu/\mathrm{e}$. Hence the conductance of the mode is

$$\frac{J_d \Delta\mu}{-\Delta\mu/\mathrm{e}} = \frac{2\mathrm{e}^2}{h} \tag{13}$$

and, if P modes can propagate at the Fermi level ε, the total conductance is

$$G = P\,\frac{2\mathrm{e}^2}{h} \tag{14}$$

Equation (14) obviously gives the maximum conductance that P propagating modes can have. As we argued in the introduction, the presence of any obstacle in the waveguide will mean that P is replaced in Eq. (14) by $T(\varepsilon)$ which is the sum of all the transmission coefficients through the obstacle from all propagating modes on the left-hand side to all propagating modes on the right-hand side (or vice versa, the two methods of calculation yield the same result when $\boldsymbol{B} = 0$). Thus we have

$$G(\varepsilon) = \frac{2\mathrm{e}^2}{h}\,T(\varepsilon) \tag{15a}$$

where

$$T(\varepsilon) = \sum_{\alpha}\sum_{\beta} T_{R\beta,L\alpha} \tag{15b}$$

in which $T_{R\beta,L\alpha}$ is the transmission coefficient from mode α on the left to mode β on the right. We note that the initial state $L\alpha$ and the final state $R\beta$ have the same energy in Eq. (15b) because the scattering is assumed to be elastic. More formal derivations of this result are given in references [3,4,5].

Equations (15) are the result of a calculation made at absolute zero. When the temperature is not zero we must weight the initial mode α in Eq. (15b) with the probability that it is occupied. This is just the Fermi function (2) with $T = T_L$ and $\mu = \mu_L$, the temperature and chemical potential of the reservoir feeding in electrons from the left. Similarly, the final mode β on the right in Eq. (15b) must be weighted with the probability that it is empty. This is just one minus the Fermi function (2) with $T = T_R$ and $\mu = \mu_R$. Finally, we suppose that the chemical potential and temperature differences between the terminals are both small. Then we easily arrive at the final linearised transport equations:

$$J = G(V_L - V_R) + L(T_L - T_R) \tag{16a}$$

$$Q = M(V_L - V_R) + N(T_L - T_R) \tag{16b}$$

where J and Q flow from left to right. The coefficients G, L, M and N are given by Eqs. (6) with $\sigma(\varepsilon)$ replaced by $G(\varepsilon)$ in Eq. (15) and the derivative of the Fermi function appears as a result of the linearisation of the theory.

4 The Relation Between $T(\varepsilon)$ and $\sigma(\varepsilon)$ when $T \rightarrow 0$

Consider a 2D free electron gas with $T = 0$ in a wire of length ℓ and width w and suppose that the electron scattering is elastic. Then the quantum theory worked out in the previous section is applicable and the conductance at Fermi energy ε is $G(\varepsilon) = 2e^2T(\varepsilon)/h$ with $T(\varepsilon)$ given by Eq. (15b). Moreover, if the length ℓ of the wire is many times the mean free path at the Fermi level and the width w of the wire is many times the Fermi wavelength then the classical treatment in Section 2 is also applicable. The conductance of the wire is therefore also given by σ_T in Eq. (8) which yields

$$\begin{aligned} \sigma_T &= w\,\sigma/\ell = wn(\varepsilon)\,e^2\,\tau(\varepsilon)/m^*\,\ell \\ &= w\,\varepsilon_F\,e^2\tau(\varepsilon)/\ell\,\pi\,\hbar^2 \end{aligned} \tag{17}$$

where we have used the result quoted just after Eq. (6a). Hence we have

$$\frac{2e^2}{h}\,T(\varepsilon) = \frac{w\,\varepsilon\,e^2\,\tau(\varepsilon)}{\ell\,\pi\,\hbar^2} \tag{18}$$

i.e.

$$T(\varepsilon) = \frac{w}{\ell}\,\frac{\varepsilon}{\Delta\varepsilon} \tag{19}$$

where

$$\Delta\varepsilon = \frac{\hbar}{\tau(\varepsilon)} \tag{20}$$

is the uncertainty in ε associated with the elastic scattering.

5 Calculations of $G(\varepsilon)$ and the Scattering Wave Functions

The calculation of $G(\varepsilon)$ involves a consideration of the scattering matrix of the nanostructure. In complicated 2D nanostructures the preferred numerical method has been a recursive Green's function approach [6-9]. It provides a good way to calculate $G(\varepsilon)$ but is not appropriate for calculating the scattering wave functions which provide additional insight into the behaviour of nanostructures. Recently a new method of calculation has been developed which is based entirely on wave functions [10,11]. In the terminals a general wave function can be written down immediately involving both incident and reflected propagating modes and evanescent modes which fall off away from the nanostructure. In each terminal an interface is specified which separates the terminal regions from the interior of the nanostructure which we refer to as the "cavity". The cavity may take many different forms in different nanostructures and consequently there is no general analytical way of dealing with it. Numerically, however, it does not present a serious problem. A set of "cavity functions" is calculated each one of which extends a particular terminal mode function continuously into the cavity and vanishes on the rest of the cavity boundaries including all the interfaces in other terminals. This makes it possible to set up a cavity wave function which is automatically continuous with any wave function in the terminals. The relationship between the amplitudes of reflected waves and damped evanescent waves to the amplitudes of the incident waves is determined by minimising the mean square discontinuity of the normal derivative of the wave function averaged over all the interfaces. The scattering matrix and the conductance matrix can then be determined by eliminating the evanescent modes.

The method has complete generality and can deal with nanostructures having any number of terminals. So far it has been used to study two two-terminal systems. The first is a quantum wire with one rectangular hard wall finger pushed in [10] and the second has two rectangular hard wall fingers pushed in with the space between them forming a quantum dot [11]. We describe below some of the results obtained.

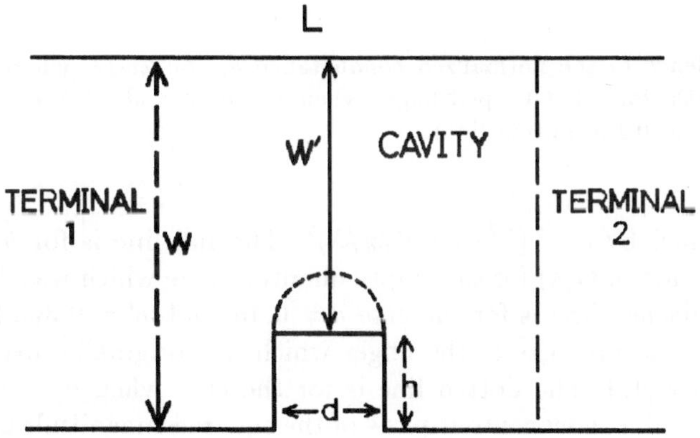

Fig. 1. The first nanostructure considered in the calculations. The dotted curve is a semicircular hard-wall cap.

The first structure and its terminal planes (dashed lines) are shown in Fig. 1 which illustrates the notation used for the dimensions of the structure. The dashed curve on the top of the finger shows a semicircular top for which calculations are also discussed in [10] but which are not included here. Fig. 2 shows a plot of $G(\varepsilon)$ against $2w/\lambda$ where λ is the de Broglie wavelength at the Fermi

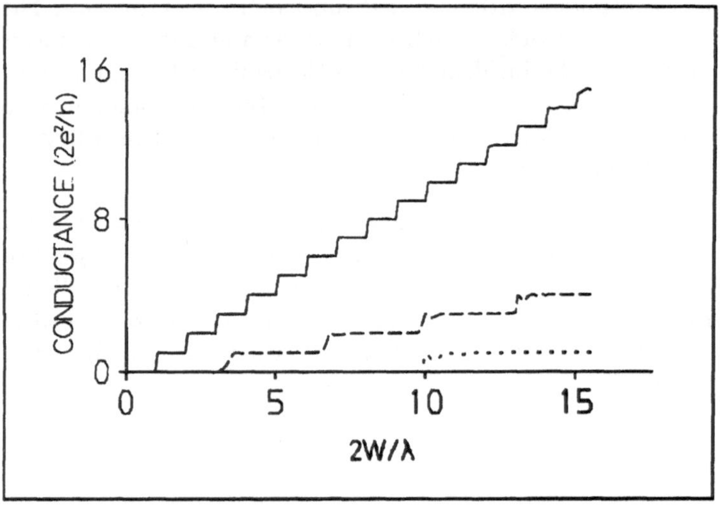

Fig. 2. The dependence of the normalized conductance g_{12} on $2w/\lambda$, where λ is the de Broglie wavelength, for a flat-topped finger when $d = 0.4w$ and $w' = w$ (full line), $0.3w$ (dashed line) and $0.1w$ (dotted line).

level which is defined by $\varepsilon = (\hbar^2/2m^*)(2\pi/\lambda)^2$. The full line is for $h = 0$. It shows the quantisation of $G(\varepsilon)$ for the empty quantum wire which was discussed in Section 3. The dashed line is for the case $d = 0.4w$ and $w' = 0.3w$. It shows the quantisation in the presence of the finger which was originally investigated by Stone and Szafer [12]. The dotted line is for the case when $d = 0.4w$ and $w' = 0.1w$. Figs. 3 and 4 show contour plots of the squared magnitude of a scattering wave function for the cases when $w' = 0.3w$ and $d = 0.4w$ when $2w/\lambda = 2$ (no transmission) and $2w/\lambda = 5$ (first plateau). In both cases the mode with the lowest cut-off energy is incident from the left.

The second structure and its dimensions are shown in Fig. 5. Two hard wall fingers are present and the space between them may be regarded as a quantum dot. In Fig. 6 the dashed line is a copy of the dashed line in Fig. 2 for $d = 0.4w$ and $w' = 0.3w$. The full line shows the result of the interference produced by

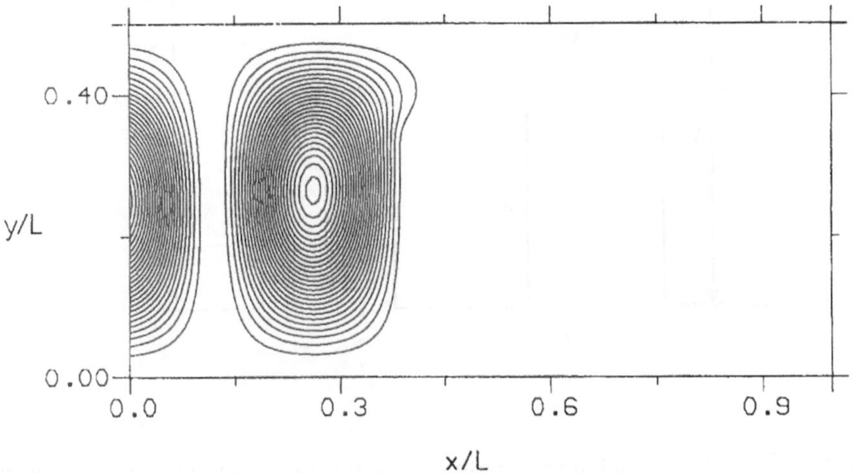

Fig. 3. Wave function contours when $d = 0.4w$, $w' = 0.3w$ and $2w/\lambda = 2$ (no transmission).

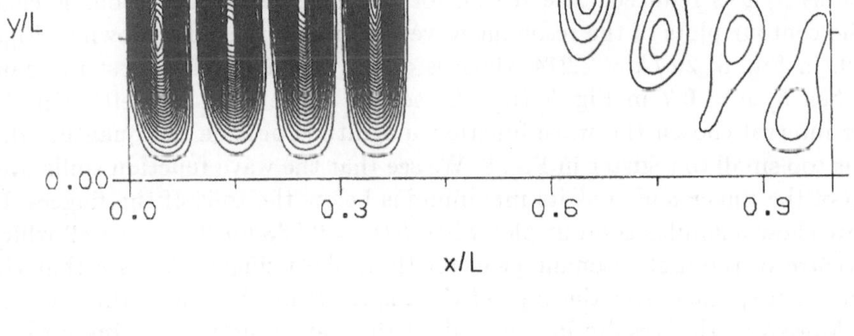

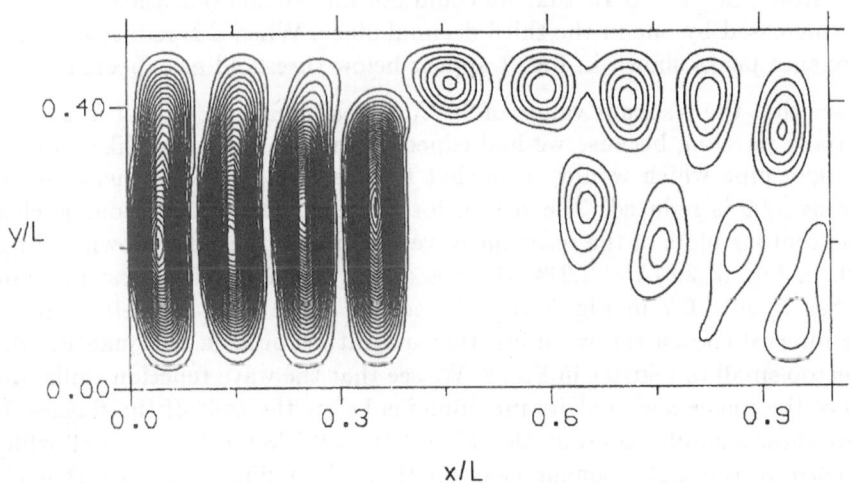

Fig. 4. Wave function contours when $d = 0.4w$, $w' = 0.3w$ and $2w/\lambda = 5$ (first plateau).

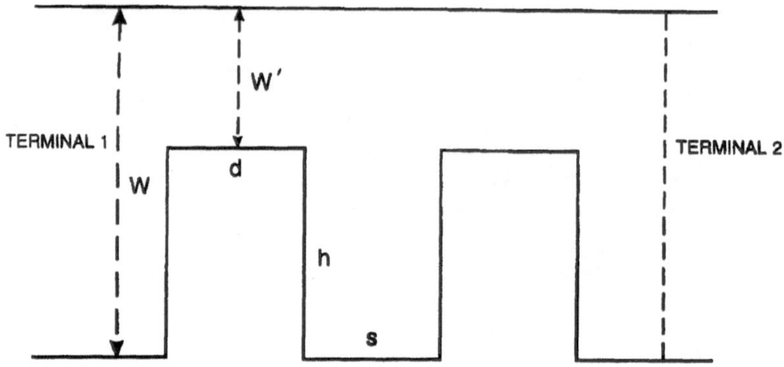

Fig. 5. A sketch of the second nanostructure giving the notation used for its dimensions. The full lines are hard potential walls.

the presence of two fingers spaced by $s = 0.467w$. We see that there are no resonant peaks produced by the quantum dot below the transmission threshold. One would expect to find some. Their absence can be traced to the very thick fingers ($d = 0.4w$) assumed in the calculations. The resonance peaks exist but are so narrow when $d = 0.4w$ that we could not find them even when $2w/\lambda$ was steadily increased by one in the third decimal place. When d is reduced to $0.1w$ the resonance peaks shown in Fig. 7 appear below threshold as expected.

We see that the resonant values of $2w/\lambda$ decrease with decreasing h/w. This was initially puzzling because we had expected the dot to behave like an open ended organ pipe which would mean that the resonant values of $2w/\lambda$ should increase as h/w is reduced. The reason for this unexpected behaviour is clear from the contour plots of the resonant wave function amplitudes shown in Figs. 8 and 9. In Fig. 8: $2w/\lambda = 2.214$ which is at the centre of the lowest resonant peak when $h/w = 0.7$ in Fig. 7 (i.e. the second curve from the left). On the contour interval chosen the wave function magnitude outside the quantum dot region is too small to register in Fig. 8. We see that the wave function spills over the top of the finger and that its maximum is below the tops of the fingers. In Fig. 9 we show a similar contour plot when $h/w = 0.533$ for $2w/\lambda = 1.89$ which is the centre of the first resonant peak for these short fingers. We see that the wave function spillage over the tops of the fingers is much greater than it is in Fig. 8. Moreover, this results in the peak of the wave function moving up to a point well above the fingers. The spillage of the wave function above the fingers is what controls the behaviour of the resonant value of $2w/\lambda$ as the finger height varies.

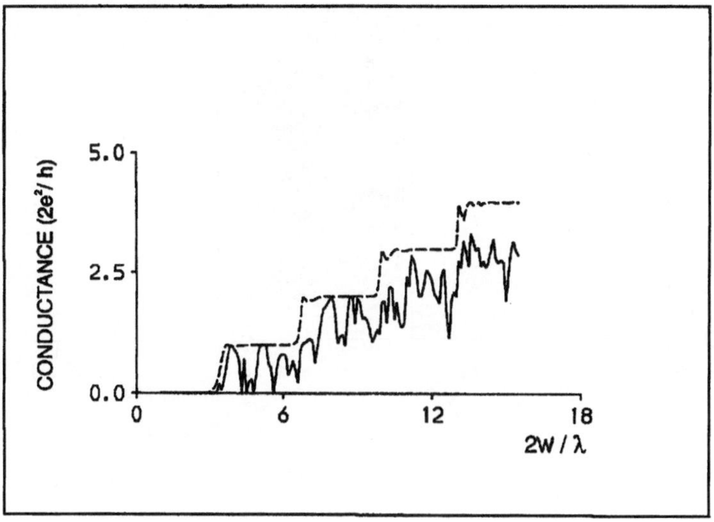

Fig. 6. A plot of conductance above the nominal threshold against $2w/\lambda$ when $d/w = 0.4$, $h/w = 0.7$, $s/w = 0.467$ and $w'/w = 0.3$. The full curve is calculated for the structure shown in figure 5. The dashed curve is for the case when one finger is removed.

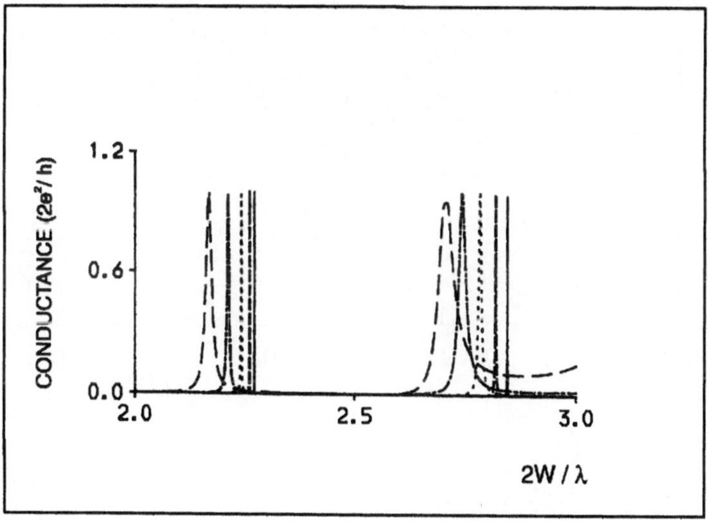

Fig. 7. The dependence of the resonance peaks on finger height for high fingers. The curves are drawn for $d/w = 0.1$, $s/w = 0.467$ and $h/w = 0.8$(——), 0.767(— — —), 0.733(- - -), 0.700(— - —) and 0.667(—— ——).

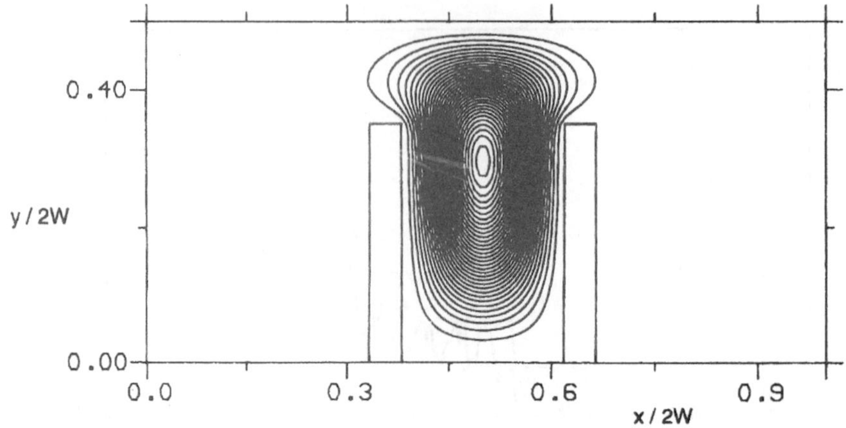

Fig. 8. Contours of $|\psi|^2$ for the structure described in figure 5. The contours are drawn for $2w/\lambda = 2.214$, i.e. the centre of the first resonance peak in Fig. 7 when $h/w = 0.700$ (second peak from the left).

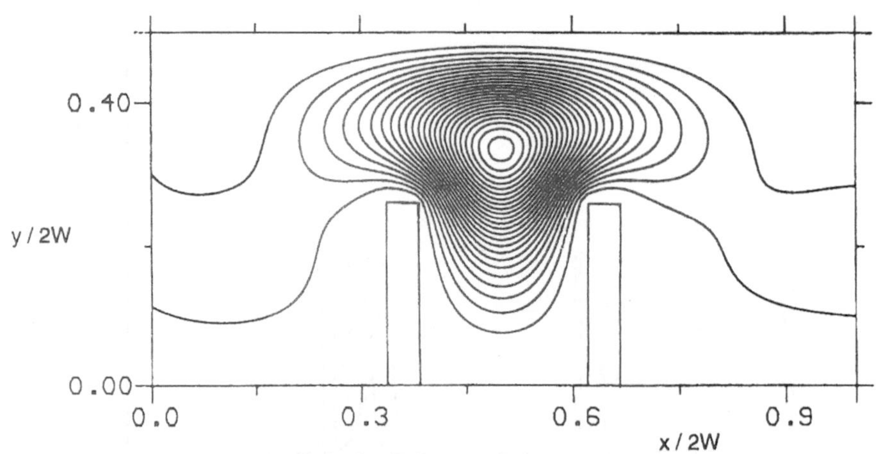

Fig. 9. Contours of $|\psi|^2$ for the structure described in figure 7 when $h/w = 0.533$, the contours are drawn for $2w/\lambda = 1.89$ which is at the centre of the first resonance peak in this case.

6 Conclusion

The close similarity between the structure of the formulae for the transport coefficients for systems which behave classically and those which behave quantum mechanically is surprising at first sight. It comes about because the structure of the equations is dominated by the independent particle model used in both cases together with particle conservation. It should also be noted that the classical equations are local: they involve electric fields and temperature gradients. Since these quantities cannot be measured inside nanostructures they are replaced by voltage and temperature differences between the electron reservoirs which are imagined to be connected to the terminals. Moreover $\sigma(\varepsilon)$ in the classical formulae is replaced by $G(\varepsilon) = (2e^2/h)\ T(\varepsilon)$ in the quantum mechanical formulae. Nevertheless, it remains easy to show that the Weidermann-Franz law is still valid in the quantal situation [3]. In an open quantum wire $T(\varepsilon)$ increases by unity every time a new mode begins to propagate. The Mott law for thermopower: $S \sim \sigma^{-1}(\varepsilon)\mathrm{d}\sigma(\varepsilon)/\mathrm{d}\varepsilon$ becomes $S \sim G^{-1}(\varepsilon)\mathrm{d}G(\varepsilon)/\mathrm{d}\varepsilon$ in a quantum wire. Consequently, when $T \rightarrow 0$, S is small in between the steps in $G(\varepsilon)$ and, has δ–function peaks whenever ε passes through the threshold of a new mode. For finite T the δ–functions are thermally broadened to a width in the order of $k_B T$. Similar behaviour is to be expected in quantum point contacts in which new modes switch on over small energy ranges. It has been observed in some beautiful experiments carried out by the Delft group [13, 14]. Further experimental work aimed at calibrating the temperature differences which arise in this work would be very valuable.

Acknowledgments

The author would like to thank J.A. McInnes for the numerical implementation of the method of calculation outlined in Section 5. Both of us are grateful to the SERC for their financial support.

References

1. Butcher P.N., 1973 Electrons in Crystalline Solids, 1973 (IAEA, Vienna).
2. Madelung O., 1978, Introduction to Solid State Theory, p.193 (Springer-Verlag, Berlin).
3. Butcher, P.N., J. Phys.: Condens. Matter **3**, 4896 (1990).
4. Buttiker M., 1992, Nanostructure Systems, Ed. M. Reed, p.191 (Academic Press, New York).
5. Butcher, P.N., 1993, Physics of Low-dimensional Semiconductor Structures, Eds. P.N. Butcher, N.H. March and M.P. Tosi, p. 95 (Plenum Press, New York).
6. D.S. Fisher, A. MacKinnon, E. Castonao and G. Kirczenow, 1992, Handbook on Semiconductors vol. 1, ed. P.T. Landsberg (Amsterdam: North-Holland) p. 863.
7. A. MacKinnon, Z. Phys. **B 59**, 385 (1985).
8. H.U. Baranger and A.D. Stone, Phys. Rev. **B 40**, 8169 (1989).
9. H.U. Baranger, D.P. DiVicenzo, R.A. Jalabert and A.D. Stone, Phys. Rev. **B 44**, 10637 (1991).

10. P.N. Butcher and J.A. McInnes, J. Phys.: Condens. Matter **7**, 745-758 (1995).
11. P.N Butcher and J.A. McInnes, J. Phys. Condens. Matter **7**, 6717-6726 (1995).
12. A.D. Stone and A. Szafar, Phys. Rev. Lett **62**, 300 (1990).
13. B.J. van Wees, H. van Houten, C.W.J. Beenakker, J.G. Williamson, L.P. Kouenhoven, D. van der Marel and C.J. Foxon, Phys. Rev. Lett. **60**, 848 (1988).
14. L.W. Moulenkamp, Th. Gravier, H. van Houten, O.J.A. Buijk, M.A.A. Mabusoone, Moulenkamp et al, Phys. Rev. Lett. **68**, 3765 (1992).

Kinetic Confinement of Electrons in Modulated Semiconductor Structures

M. Kubisa and W. Zawadzki

Institute of Physics,
Polish Academy of Sciences, 02-668 Warsaw, Poland

Abstract: A new type of electron confinement in modulated semiconductor systems is proposed. The confinement occurs when the effective mass of electrons in the central region of the structure is higher than that in the outside regions. This results in a 'kinetic well' produced by the transverse free motion. The calculated density of confined states is similar but not identical to that of the potentially bound 2D states. It is shown that the presence of an external magnetic field parallel to the growth direction stabilizes and controls the kinetically confined states. The resulting levels are strongly nonlinear functions of magnetic field intensity. We discuss specific structures in which the kinetic confinement of electrons would be possible.

1 Introduction

Potential confinement of charge carriers on the quantum scale, made possible by new fabrication technologies, has revolutionized semiconductor physics in recent years. The potential wells responsible for the confinement have been conventionally created by the differences in the energy gaps and charge transfer (in heterostructures), as well as by external voltages (in metal-insulator-semiconductor structures). Various parts of a heterostructure are often characterized by different effective masses of charge carriers and this fact has been accounted for in existing theories. The purpose of this lecture is to show that properly tailored difference of the effective masses in a double heterostructure causes in itself the kinetic confinement of charge carriers. In fact, the mass difference can confine carriers even in the region of a *higher* potential. The effect of an external magnetic field on such a heterostructure is described in some detail. We also consider examples of real systems, in which the kinetic confinement of electrons or holes can take place (Kubisa and Zawadzki, 1992).

2 Kinetic confinement: theory

Let us consider electrons in a double heterostructure shown in Fig. 1. The outside regions on both sides are characterized by the effective mass m_1 and the potential $V = 0$, the middle region of the width a by the mass m_2 and $V = V_0$. The Schrödinger equation for the system reads

$$
\begin{aligned}
-\frac{\hbar^2}{2m_1}\,\nabla^2\psi &= E\,\psi \qquad \text{for} \quad |z| > \frac{a}{2} \\
\left(-\frac{\hbar^2}{2m_2}\,\nabla^2 + V_0\right)\psi &= E\,\psi \qquad \text{for} \quad |z| < \frac{a}{2}\,.
\end{aligned}
\tag{1}
$$

Looking for a solution in the form $\psi = \exp(ik_x x + ik_y y)\ f(z)$, one obtains

$$
\begin{aligned}
\left(-\frac{\hbar^2}{2m_1}\frac{d^2}{dz^2} + \frac{\hbar^2 k_\perp^2}{2m_1}\right) f &= Ef \qquad \text{for} \quad |z| > \frac{a}{2} \\
\left(-\frac{\hbar^2}{2m_2}\frac{d^2}{dz^2} + \frac{\hbar^2 k_\perp^2}{2m_2} + V_0\right) f &= Ef \qquad \text{for} \quad |z| < \frac{a}{2}
\end{aligned}
\tag{2}
$$

where $k_\perp^2 = k_x^2 + k_y^2$. The solutions of Eqs. (2) must match at $z = \pm a/2$, according to Ando and Mori (1982) the boundary conditions are

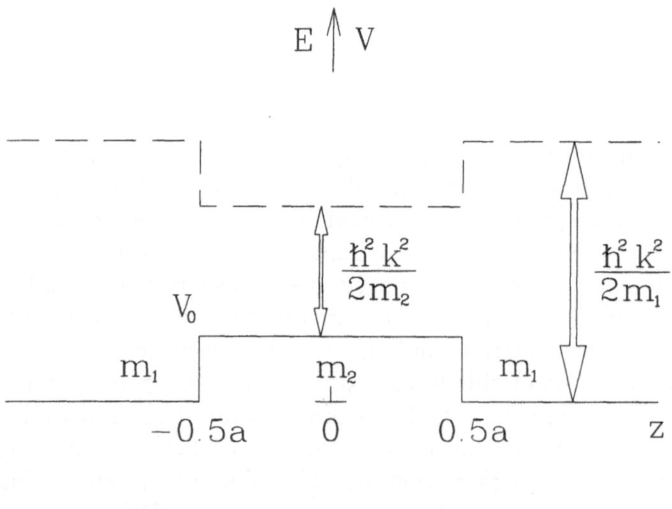

Fig. 1. The proposed double heterostructure for kinetic confinement. A larger effective mass in the central region ($m_2 > m_1$) produces a 'kinetic well', related to the transverse free motion (schematic).

$$f\left(\pm\frac{1}{2}a,+\right)=f\left(\pm\frac{1}{2}a,-\right)$$
$$\frac{1}{m_1}\frac{d}{dz}f\left(\pm\frac{1}{2}a,\pm\right)=\frac{1}{m_2}\frac{d}{dz}f\left(\pm\frac{1}{2}a,\mp\right). \tag{3}$$

We are looking for bound states along the z direction. They exist for the energies (see Fig. 1)

$$V_0+\frac{\hbar^2k_\perp^2}{2m_2}<E<\frac{\hbar^2k_\perp^2}{2m_1}. \tag{4}$$

For $V_0<0$ there always exists at least one bound state. For $V_0\geq 0$ the bound states exist for $\frac{\hbar^2k_\perp^2}{2m_2}>V_0(\eta-1)$, where $\eta=m_2/m_1$. The bound solutions have the general form

$$\begin{aligned} f(z)&=Ae^{\kappa z} && \text{for}\quad z<-\frac{a}{2}\\ &=Be^{i\lambda z}+Ce^{-i\lambda z} && \text{for}\quad |z|\leq\frac{a}{2}\\ &=De^{-\kappa z} && \text{for}\quad z>+\frac{a}{2}\end{aligned} \tag{5}$$

where $\hbar^2(\lambda^2+k_\perp^2)/2m_2+V_0=E$ and $\hbar^2(-\kappa^2+k_\perp^2)/2m_1=E$. This gives

$$\frac{\hbar^2}{2m_2}(\lambda^2+k_\perp^2)+V_0=\frac{\hbar^2}{2m_1}(-\kappa^2+k_\perp^2). \tag{6}$$

The boundary conditions (3) lead to a set of linear equations for A, B, C and D. The condition for a non-trivial solution gives

$$\cot(\lambda a)=\frac{\lambda^2-\eta^2\kappa^2}{2\eta\kappa\lambda} \tag{7}$$

subject to the relation (6) between κ and λ. The energies of the bound states are

$$E_l=V_0+\frac{\hbar^2}{2m_2}(\lambda_l^2+k_\perp^2) \tag{8}$$

for $l=0,1,2\ldots$, where λ_l is the lth root of Eq. (7). This equation is equivalent to

$$\eta\kappa=+\lambda\ \tan(\lambda a/2) \tag{9a}$$
$$\eta\kappa=-\lambda\ \cot(\lambda a/2). \tag{9a}$$

The roots of Eq. (9a) give the energies of the even states ($l=0,2,4,\ldots$), while Eq. (9b) gives the energies of the odd states ($l=1,3,5,\ldots$) with respect to the $z=0$ axis.

Equations (9) are generalizations of the well-known formulae for the energies of bound states in a finite rectangular well. The generalization accounts for the different masses at both sides of each interface. In our case the well is formed by the energy components of the transverse free motion. It follows from Eqs. (6) and (9) that λ_l itself is a function of $k_\perp$, so that the subband energies E_l given by Eq. (8) are non-parabolic functions of $k_\perp$. This is in contrast to the case of potential confinement, illustrating the general property of our system, in which the longitudinal motion is strongly coupled to the transverse motion.

Figure 2 shows subband energies as functions of $k_\perp a$, calculated for $\eta = m_2/m_1 = 2$ and $V_0 = 0$. The broken curve indicates the parabola $E = \hbar^2 k_\perp^2/2m_1$ (cf. Fig. 1). It can be seen that for $V_0 = 0$ there always exists at least one bound state. As $k_\perp$ increases and the kinetic well becomes deeper, a second bound state appears, then a third, and so on. One can see that the $E_l(k_\perp)$ dependences are indeed non-parabolic. The inset shows the calculated wavefunctions of two lowest bound states for $k_\perp a = 10$, demonstrating that kinetic confinement really takes place. Our system is like a moving bicycle: the kinetic forward motion gives it a lateral stability.

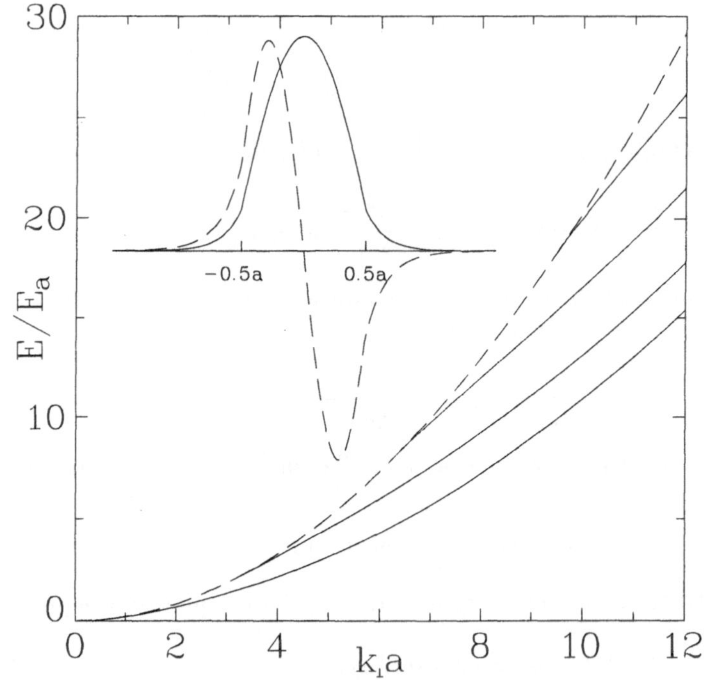

Fig. 2. Energies of the bound states (in units of $E_a = \hbar^2\pi^2/2m_2a^2$) resulting from the kinetic confinement (full curves) versus $k_\perp a$. The calculations were carried out for $m_2/m_1 = 2$ and $V_0 = 0$. The broken curve indicates the parabola $E = \hbar^2 k_\perp^2/2m_1$ (cf. Fig. 1). The inset shows the wavefunctions of the ground state and the first excited state, calculated for $k_\perp a = 10$.

Next we calculate the density of two-dimensional states (DOS) for a given subband, according to the formula $\rho_l(E) = (k_\perp/\pi)(dk_\perp/dE)$, and the total DOS $\rho(E) = \Sigma\rho_l(E)$. The results are shown in Fig. 3 (again for $\eta = m_2/m_1 = 2$ and $V_0 = 0$). The total DOS exhibits a step-like behaviour, demonstrating that we are dealing with a true 2D system. The striking feature is that the DOS 'jumps' to a finite value as soon as the energy E is non-zero. This is due to the fact that for $V_0 = 0$ the lowest bound state exists for any energy $E > 0$. When a given 2D

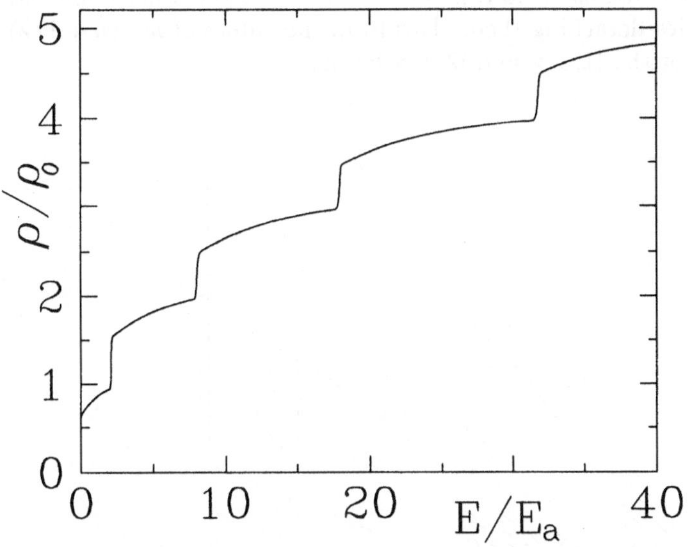

Fig. 3. The density of states for the kinetic subbands shown in Fig. 2 (in units $\rho_0 = m_1/\pi\hbar^2$ versus electron energy.

state appears, its initial step-like contribution to the DOS is $\rho_l(E) = m_1/\pi\hbar^2$, since the state is widely spread (predominantly over both m_1 regions). Thus we deal with the standard DOS for 2D system characterized by the mass m_1. With increasing energy, $\rho_l(E)$ asymptotically approaches the value of $m_2/\pi\hbar^2$, since the wavefunction is progressively confined in the region of m_2 (cf. also Fig. 5).

For $V_0 > 0$ it turns out that, in agreement with intuitive expectations based on Fig. 1, bound states appear as soon as the 'kinetic well' related to the mass difference overcomes the effect of the potential 'anti-well'. This takes place for sufficiently large $k_\perp$ values (and the corresponding energies E).

3 Effect of magnetic field

We consider the structure shown in Fig. 1 in the presence of an external magnetic field B parallel to the z direction. The effect of a magnetic field is of interest since it quantizes the transverse motion of the carriers. The above calculation can be directly used for the magnetic case, replacing $\hbar^2 k_\perp^2/2m_2$ by $\hbar\omega_2(n+1/2)$. Here $n = 0, 1, 2, \ldots$ is the Landau quantum number and $\omega_2 = eB/m_2$ is the cyclotron frequency (the spin is omitted). The kinetic well shown in Fig. 1 is now controlled for a given n by the value of B.

In Fig. 4 we show the resulting energies, calculated for $\eta = m_2/m_1 = 2$ and $V_0 = 0$. The broken curves indicate the energies $\hbar\omega_1(n+1/2)$, where $\omega_1 = eB/m_1$. These define the upper edges of the 'kinetic well' (cf. Fig. 1). It can be seen that there is always at least one bound state $|n, 0>$, attached to any given value of n. As B increases and the well becomes deeper other bound states appear, their energies detaching themselves from the values of $\hbar\omega_1(n+1/2)$. In Fig. 4 this is seen for the $|1, 1>$ and $|2, 1>$ levels.

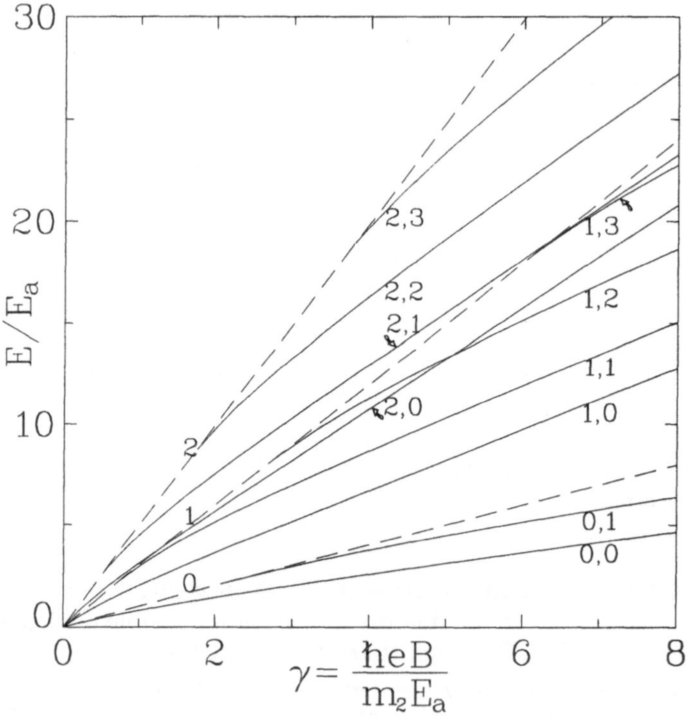

Fig. 4. Energies of the kinetically confined states (full curves) versus magnetic field (in units $(\hbar eB/m_2)/E_a$). The values of $(\hbar eB/m_1)(n+1/2)$, which represent the upper edges of the 'kinetic well' for $n = 0, 1, 2, \ldots$ (cf. Fig. 1) are indicated by the broken curves.

It can also be seen that the energies of the bound states are strongly nonlinear functions of B, in contrast to the 3D Landau levels and the 2D Landau levels of potentially confined electrons. This nonlinearity is related to the fact that at low B values the state is spread predominantly over the m_1 regions, so that the slope of the magnetic field dependence is governed by $\hbar eB/m_1$, while at higher fields, as the state is progressively confined in the m_2 region, the slope is governed by $\hbar eB/m_2$. The confinement is illustrated in Fig. 5. The nonlinear B dependence of the energies is probably the most striking signature of the kinetic confinement. This feature can be investigated in the cyclotron resonance experiments, for example between $|0,0>$ and $|1,0>$ states.

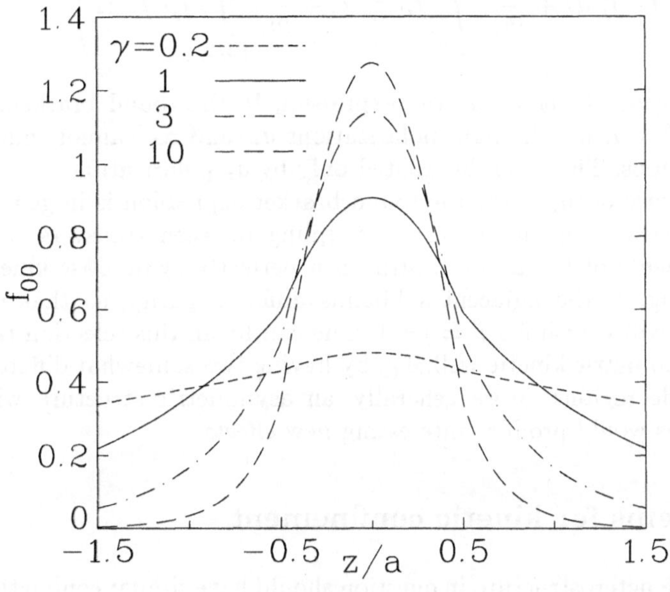

Fig. 5. The wave functions of the kinetically confined ground state $|0,0>$ for different magnetic fields. As the field increases, the state is progressively confined in the m_2 region.

In this connection we briefly consider intraband magneto-optical transitions between kinetically confined states. The vector potential of the magnetic field $B||z$ is taken as $\boldsymbol{A} = (-By, 0, 0)$. The wavefunctions have the general form: $\psi_{mj} = \exp(ik_x x)\Phi_n[(y-y_0)/L]f_j(z)$ where Φ_n is the harmonic oscillator function, $y_0 = k_x L^2$, the magnetic radius is $\boldsymbol{L} = (\hbar/eB^{1/2}$ and j numerates the kinetic subbands. The electron-photon interaction Hamiltonian reads

$$H_R = \left(\frac{eA_0'}{m_1}\right) \boldsymbol{a}\cdot\boldsymbol{P} \qquad \text{for } |z| > \frac{a}{2}$$

$$H_R = \left(\frac{eA_0'}{m_2}\right) \boldsymbol{a}\cdot\boldsymbol{P} \qquad \text{for } |z| < \frac{a}{2} \qquad (10)$$

where A'_0 and $\boldsymbol{a}$ are the amplitude and polarization of the radiation vector potential, and $\boldsymbol{P} = p + e\boldsymbol{A}$ is the generalized momentum. The scalar products can be factorized in the usual way: $a \cdot \boldsymbol{P} = a_+ P_- + a_- P_+ + a_z P_z$, where $a_\pm = (a_x \pm i a_y)/\sqrt{2}$. We consider the cyclotron resonance active light polarization a_-. Since $P_+ = (-\hbar/L)a^+$, where a^+ is the raising operator for the harmonic oscillator functions, the matrix element for the optical transition is

$$\langle \psi_{n'j'} | H_R^- | \psi_{nj} \rangle = -(eA'_0)(\hbar/L)\sqrt{n+1}\,\delta_{k'_x k_x}\,\delta_{n'_x n+1}$$

$$\times \left(\frac{1}{m_1} \int_{-\infty}^{-a/2} f_{j'}\, f_j\, dz + \frac{1}{m_2} \int_{-a/2}^{+a/2} f_{j'}\, f_j\, dz + \frac{1}{m_1} \int_{+a/2}^{\infty} f_{j'}\, f_j\, dz \right) . \tag{11}$$

In the standard case of $m_2 = m_1$ the expression in the round brackets is non- zero only for $j' = j$, i.e. the light polarizations a_- and a_+ cannot induce intersubband transitions. These can be excited only by a_z polarization.

However, in our case of $m_2 > m_1$ the round bracket expression is in general non-zero, which illustrates again the strong coupling between transverse and longitudinal components of the motion. Still, in a perfectly symmetric kinetic well, as shown in Fig. 1, the adjacent subbands differ in parity, so that the matrix element (11) will vanish for $j' = j \pm 1$. One can break this selection rule by producing an asymmetric kinetic well, e.g. by having two somewhat different masses in the outside regions. More generally, an asymmetric structure with three different masses would produce interesting new effects.

4 Possible systems for kinetic confinement

Materials forming the heterostructure in question should have similar conduction band symmetries, otherwise it is difficult to match the periodic Luttinger-Kohn amplitudes at the interface. On the other hand, according to the kp theory the electron effective mass is proportional to the gap value. Consequently, in order to have a sizable difference between the electron masses one needs a system with a corresponding gap difference, in which almost all of the band offset occurs in the valence band.

Zhang and Kobayashi (1992) studied ZnSe/GaAs quantum wells and showed that almost all bandgap discontinuity occurs in the valence band. Bertho et al. (1991) showed theoretically that in the ZnSe/ZnS system, although the gap difference for the two materials is around 1 eV, the conduction band offset is near zero.

Very promising systems are combinations of III-V compounds. As follows from diagrams of the conduction band energy versus lattice constants for III-V solid solutions, a system of $GaAs_{1-x}Sb_x$ for about $x \approx 0.2$ has the same lattice constant and the conduction band energy as $Ga_{1-x}In_xAs$ for $x \approx 0.2$. Also the $Al_{1-x}In_xAs$ system for $x \approx 0.7$ can be matched both in lattice constant and the conduction band energy with $GaAs_{1-x}Sb_x$ alloy for $x \approx 0.7$.

The offset at the interface of two semiconductors results from the gap difference and from the electric dipole related to dielectric properties. This dipole can be influenced by foreign atoms of both materials placed at the interface. It has been shown by Brabina et al. (1992) that one can match the conduction bands of GaAs and AlAs by placing a fraction of a monolayer of Ge or Si atoms at the interface between the two materials. Thus the offset can be tailored.

In a lower symmetry crystal having the conduction band in the form of an ellipsoid in k space, one could create the mass difference by growing the same material in two different crystal directions.

The situation for hole confinement is basically different since the valence band behaviour is in general determined by the heavy-hole dispersion (large density of states), which is much less sensitive to the gap variation. There exist systems, for example CdTe-$Cd_{0.9}Zn_{0.1}Te$ (Tuffigo et al. 1991), in which almost all band discontinuity occurs in the conduction band, but the heavy-hole masses in each material do not differ much.

However, interesting possibilities are provided here by the presence of biaxial strain, which occurs at the interfaces due to the lattice mismatch (cf. the review of Reilly (1989)). In tetrahedral semiconductors under biaxial tension the hydrostatic strain component reduces the average bandgap, while the axial component splits the degeneracy of the valence band maximum and introduces an anisotropic valence band structure, with the higher band being light along the strain axis ($k_\perp$), and relatively heavy perpendicular to that axis ($k_\parallel$). Under biaxial compression the average gap increases and the valence splitting is reversed, so the higher band is now heavy along the strain axis ($k_\perp$) and relatively light in the perpendicular direction ($k_\parallel$).

It is possible to create the confinement in the presence of a magnetic field not only by different effective masses but also by different spin g-values. The latter control the spin splitting, so that for spin-up or spin-down states one could create different energies in different parts of the system, which would lead to confinement. Semimagnetic materials of HgMnTe or CdMnTe type would be particularly suitable for such structures since by adding Mn paramagnetic ions one can strongly influence the spin g-values without much affecting the conduction (or valence) band edges at $B = 0$.

It should be mentioned that the kinetic confinement can be also realized in one material (rather than a heterostructure) possessing a strongly nonparabolic energy band. The electron can then change its effective mass by being further or closer to the band edge during its motion, which leads to the confinement, such system has been proposed and realized by Doezema and Drew (1986) using MOS structures on InAs, cf. also Radantzev et al. (1986).

This idea was further developed and applied to accumulation layers by Slinkman et al. (1989), Y. Zhang et al. (1991) and A. Zhang et al. (1991).

Finally, different effective masses were proposed for 'mass superlattices' (cf. Sasaki, 1984, and Milanovic et al., 1986). We do not discuss these developments since in the above lecture we have been mostly concerned with the confinement effects.

References

1. Ando T and Mori S Surf. Sci. **113**, 124 (1982).
2. Bertho D, Boiron D, Simon A, Jouanin C and Priester C 1991 Phys. Rev. **B44**, 6118
3. Bratina G, Sorba L, Antonini A, Biasol G and Franciosi A 1992 Phys. Rev. **B45**, 4528
4. Doezema R E and Drew H D 1986 Phys. Rev. Lett. **57**, 762
5. Kubisa M and Zawadzki W 1993 Semicond. Sci. Technol. **8**, S246; The present lecture is based on this paper.
6. Milanovi V, Tiapkin D and Ikoni Z 1986 Phys. Rev. **B34**, 7404
7. Radantzev V F, Derybina T I, Zverev L P, Kulayev G I and Khomutova S S 1986 Zh. Exp. Teor. Fiz. **91**, 1016 (Engl. transl. 1986 Sov. Phys.-JETP 64 598)
8. Reilly E P 1989 Semicond. Sci. Technol. **4**, 121
9. Sasaki A 1984 Phys. Rev. **B30**, 7016
10. Slinkman J, Zhang A and Doezema R E 1989 Phys. Rev. **B39**, 1251
11. Tuffigo H, Magnea H, Mariette H, Wasiela A and Merle d'Aubigné Y 1991 Phys. Rev. **B34**, 14629
12. Zhang A, Slinkman J and Doezema R E 1991 Phys. Rev. **B44**, 10752
13. Zhang S and Kobayashi N 1992 Appl. Phys. Lett. **60**, 883
14. Zhang Y, Zhang A, Slinkman J and Doezema R E 1991 Phys. Rev. **B44**, 10749

The Scaling Theory of the Integer Quantum Hall Effect

Bodo Huckestein

Institut für Theoretische Physik, Universität zu Köln, D–50937 Köln, Germany

Abstract:
A brief review is given of the present understanding of the transitions between integer quantized plateaus of the Hall conductivity in two-dimensional disordered systems in a strong magnetic field. The similarity to continuous thermodynamic phase transitions is emphasized. Results of numerical simulations for non-interacting electrons are presented and compared to experiment. The role of the Coulomb interactions at the integer quantum Hall transitions is studied.

1 Introduction

Two-dimensional electron systems at low temperatures have rather peculiar transport properties. These systems occur at the interfaces of semiconductor heterostructures or in field-effect transistors. When the amount of disorder in the samples and the temperature are low enough, the Hall resistance shows a series of plateaus as a function of magnetic field or carrier concentration. The Hall resistance on these plateaus is quantized to a very high accuracy to values of $R_i = (h/e^2)/i$, with integer i [1]. In fact, the accuracy of this quantization surpasses that of every other known resistor so that the national metrological laboratories use this integer quantum Hall effect as the standard of resistance. At the same time that the Hall resistance becomes quantized, the longitudinal resistance, measured in a four-probe geometry, vanishes. More precisely, at finite temperatures it becomes exponentially small and would vanish at zero temperature.

For systems with an even lower amount of disorder additional plateaus appear in the Hall resistance [2]. The Hall resistance is still quantized, but i is no longer an integer but a simple rational fraction with odd denominator, such as 1/3, 2/5, 2/3. This fractional quantum Hall effect is related to the occurrence of correlated groundstates due to the Coulomb interaction between the electrons.

One of the most remarkable aspects of the quantum Hall effect is the observation of the dissipationless state with quantized Hall resistance in a *disordered* system. While the origin of the quantization can be understood in the absence of disorder, the disorder is essential for the observation of the quantization [3, 4]. Without disorder there would be no plateaus in the Hall resistance. To show the quantization of the Hall resistance it is necessary to assume that the Fermi energy lies in a spectral gap of the system. For the integer quantum Hall effect this gap is the cyclotron gap between successive Landau levels. In the fractional quantum Hall effect this gap is the excitation gap of the correlated ground state. The quantization argument therefore holds only for very special values of the Fermi energy and a small change in parameters like the magnetic field or the carrier density moves the Fermi energy out of the spectral gap and the quantization argument no longer holds anymore. But if the Fermi energy moves into a region where all the states at the Fermi level are localized due to disorder then they do not contribute to the transport and the Hall resistance remains quantized. Furthermore the longitudinal conductivity will vanish for these states.

While this localization scenario explains the occurrence of plateaus in the Hall resistance, it does not yet provide a consistent picture. If all the states were localized by the disorder, then the system would be an insulator and would not show any Hall effect. In fact, the scaling theory of the disorder-induced metal-insulator transition states that in two-dimensional systems in the absence of spin-orbit interactions *all* states are localized [5]. To reconcile this result with the observation of the quantum Hall effect, we thus have to assume that in the presence of the strong magnetic field most states are localized but some states remain delocalized and give rise to the quantized Hall resistance. When the Fermi energy moves through the region of delocalized states the Hall conductivity can change and the longitudinal resistance is finite. We can thus identify these regions with the transition regions between the different quantized Hall plateaus. These simple ideas cannot answer the question of the width of this transition region. Experimentally, the widths of the transition regions become smaller with decreasing temperature and seem to vanish for zero temperature [6]. As we will see this is also compatible with numerical simulations and follows from theoretical arguments [7]. In contrast to the Hall conductivity, the longitudinal conductivity is solely determined by the properties of the states at the Fermi energy. When the Fermi energy lies in the plateau regions where the states are localized, the longitudinal conductivity σ_{xx} necessarily vanishes. Since the Hall conductivity σ_{yx} is finite, this leads to the somewhat surprising result that the longitudinal resistivity $\rho_{xx} = \sigma_{xx}/(\sigma_{xx}^2 + \sigma_{yx}^2)$ vanishes, too.

The transitions between successive quantum Hall plateaus share many features with thermodynamic phase transitions. At zero temperature the transitions take place at a singular energy. The regions in parameter space that are separated by the transitions are characterized by quantized values of the Hall resistance. But the analogy goes farther than this. Below we will see that the transitions are associated with a diverging length scale, the localization length, that plays the role of the correlation length in thermodynamic phase transitions. This localization length diverges with the distance from the transition as a power law. Close to the transitions physical quantities show scaling behaviour with the localization length as the single relevant length scale.

An understanding of the integer quantum Hall effect is thus incomplete without an understanding of the influence of disorder. In a sense the interplay of a strong magnetic field and disorder in a two-dimensional system gives rise to the series of plateaus that are the central feature of the quantum Hall effect. At present there exists no analytical theory of the quantum Hall effect that explains the quantization and allows the calculation of the scaling properties near the plateau transitions [8]. I will therefore present the results of numerical simulations that show that the picture sketched above indeed seems to be correct and that allow to calculate critical exponents and scaling functions. Most of these calculations are done for non-interacting electrons. While these calculations provide quantitative agreement with experiment for some quantities, they fail for others. This failure shows that the Coulomb interaction between the electrons can not completely be neglected even when it is weak compared to the disorder potential.

The rest of the paper is organized as follows. First a model for the integer quantum Hall effect is developed that is suitable for numerical simulations. Next results of these simulations are presented and analyzed in terms of single parameter scaling theory. The critical exponents obtained from this analysis are then compared to experimental results. Finally, the Coulomb interaction is considered and incorporated into the simulations within a self-consistent Hartree-Fock approximation. A fuller account of the topic of this paper has been published by the author [9]. For further reading a few books on the quantum Hall effect are suggested [10, 11, 12].

2 Random Landau Matrix

Our model for the integer quantum Hall effect will be non-interacting electrons confined to a two-dimensional plane and subject to a strong perpendicular magnetic field B and a disorder potential $V(\boldsymbol{r})$. Later we will discuss the influence of the Coulomb interaction. The system is thus described by the Hamiltonian

$$H = \frac{1}{2m}(\boldsymbol{p} - e\boldsymbol{A})^2 + V(\boldsymbol{r}), \tag{1}$$

where the first term is the kinetic energy in a magnetic field. In the absence of disorder the kinetic energy is quantized, $E_n = (n + 1/2)\hbar\omega_c$, $n = 0, 1, 2, \ldots$, and $\omega_c = eB/m$ is the cyclotron frequency. Each of these Landau levels is highly degenerate with the degeneracy per unit area given by the number of flux quanta $\Phi_0 = h/e$ per unit area,

$$n_B = B/\Phi_0 = eB/h = 1/2\pi l_c^2. \tag{2}$$

The length scale l_c defined by the magnetic field is called the magnetic length. A set of eigenfunctions of the kinetic energy for the Landau gauge $\boldsymbol{A} = Bx\hat{\boldsymbol{e}}_y$

and for a system of width L_y and periodic boundary conditions in y-direction are the Landau states

$$\psi_{nk}(\boldsymbol{r}) \equiv \langle \boldsymbol{r}|nk\rangle = \frac{1}{\sqrt{L_y l_c}} e^{-iky} \chi_n\left(\frac{x - kl_c^2}{l_c}\right), \tag{3}$$

where

$$\chi_n(x) = \left(2^n n! \sqrt{\pi}\right)^{-1/2} H_n(x) e^{-x^2/2} \tag{4}$$

are the harmonic-oscillator eigenfunctions, and $H_n(x)$ are Hermite polynomials. The degeneracy of the Landau levels reflects the fact that the eigenenergies do not depend on the quantum number k describing the eigenfunctions. Due to the periodic boundary conditions, k is an integer multiple of $2\pi/L_y$.

It is advantageous to exploit the quantization of the kinetic energy by the magnetic field. This is done by expanding the Hamiltonian in terms of the Landau states (3). As a major simplification the kinetic energy part of the Hamiltonian becomes diagonal in this basis and gives only additive contributions to the energies. The disorder potential $V(\boldsymbol{r})$ is transformed into a random matrix $\langle nk|V|n'k'\rangle$. In the limit of strong magnetic field, relevant to the quantum Hall effect, the matrix elements between different Landau levels will be much smaller than the matrix elements within each Landau level. It is then natural to adopt a single Landau level approximation where only matrix elements within the Landau level that contains the Fermi energy are considered.

The random Landau matrix elements $\langle nk|V|nk'\rangle$ can now be calculated for a given realization of the disorder potential $V(\boldsymbol{r})$, i.e.

$$\langle nk|V|nk'\rangle = \int d^2r \psi_{nk}^*(\boldsymbol{r}) V(\boldsymbol{r}) \psi_{nk'}(\boldsymbol{r}). \tag{5}$$

For numerical purposes this expression is not the most suitable. It is faster to directly generate matrix elements with correct statistical properties. We characterize these properties by correlation functions and assume that only the lowest order correlation functions are relevant for the critical behaviour. In particular, let us assume that the average $\overline{V(\boldsymbol{r})}$ of the potential vanishes, since a finite value would only shift the whole energy scale. Here, the overbar denotes the average with respect to the distribution of disorder. The strength and the range of the potential are determined by the two-point correlation function $\overline{V(\boldsymbol{r})V(\boldsymbol{r}')}$. We will assume gaussian correlations of the potential,

$$\overline{V(\boldsymbol{r})V(\boldsymbol{r}')} = \frac{V_0^2}{2\pi\sigma^2} \exp\left(-\frac{|\boldsymbol{r}|^2}{2\sigma^2}\right), \tag{6}$$

where V_0 is a measure of the strength of the potential and σ the correlation length or the range of the potential.

For a given correlation function of the disorder potential, it is possible to calculate the corresponding correlation function of the matrix elements

$\overline{\langle n_1k_1|V|n_2k_2\rangle\langle n_3k_3|V|n_4k_4\rangle}$ using the explicit form Eq. (3) of the eigenfunctions. For the lowest Landau level and our gaussian potential it is

$$\overline{\langle 0k_1|V|0k_2\rangle\langle 0k_3|V|0k_4\rangle} = \delta_{k_1-k_2,k_4-k_3}\frac{V_0^2}{\sqrt{2\pi}l_cL_y\beta}\exp\left(-\frac{1}{2}(k_1-k_2)^2l_c^2\beta^2\right)$$
$$\exp\left(-\frac{1}{2}(k_1-k_4)^2l_c^2\frac{1}{\beta^2}\right), \qquad (7)$$

where $\beta^2 = (\sigma^2 + l_c^2)/l_c^2$ is a dimensionless measure of the range of the potential. The matrices formed by the elements $\langle 0k|V|0k'\rangle$ constitute a random matrix ensemble but with statistical properties that set it quite apart from the more familiar Wigner-Dyson ensembles [13]. Unlike the latter the matrices under consideration, which we will call random Landau matrices, are bounded due to the exponential in eq. (7) involving $k_1 - k_2$, and their elements are correlated along the diagonals. These correlations survive even in the limit $\sigma \to 0$ where the real-space potential becomes δ-correlated. They have so far also spoiled attempts to obtain analytical results about the random Landau matrices.

A closer inspection of Eq. (7) tells us how to generate matrix elements with the proper statistical properties directly. Remembering that the Hamiltonian and hence the matrix $\langle nk|V|nk'\rangle$ is hermitian we see that the real and imaginary parts of the matrix elements on each diagonal form two mutually uncorrelated series of numbers with gaussian correlations. In terms of uncorrelated, complex random numbers $u_0(x,k)$ satisfying

$$\overline{u_0(x,k)u_0(x',k')} = \delta_{x,x'}\delta_{k,-k'}, \qquad (8)$$

the matrix elements can then be expressed as

$$\langle 0k_1|V|0k_2\rangle = \frac{V_0}{\left(\sqrt{2\pi}l_cL_y\Sigma\right)^{1/2}}\exp\left(-\frac{(k_1-k_2)^2l_c^2\beta^2}{4}\right)$$
$$\times\sum_i u_{2k_1+i,k_2-k_1}\exp\left(-\frac{\pi^2l_c^2}{L_y^2\beta^2}i^2\right), \qquad (9)$$

with

$$\Sigma = \sum_i \exp\left(-2\frac{\pi^2l_c^2}{L_y^2\beta^2}i^2\right). \qquad (10)$$

The limits on the summation over i can be chosen such that the influence of the neglected terms is less than the statistical fluctuation due to disorder.

In closing this section, it should be mentioned that similar expressions can be found for arbitrary Landau levels and correlations of the disorder potential. In principle the influence of higher order correlation functions should be considered. However, as will become apparent soon, even the form of the two-point correlation function is irrelevant for the critical behaviour of the model. The strength of the potential only sets the energy scale in the one Landau level approximation and the range of the potential enters only as an irrelevant parameter.

This approach to the random Landau level problem was developed by Huckestein and Kramer [14] and Mieck and Weidenmüller [15, 16, 17, 18]. A more detailed discussion is presented in Ref. [9].

3 Single Parameter Scaling

A central problem of any numerical study of critical behaviour is the extrapolation of results for finite systems to infinite system size. If there is a quantitative analytical theory for the phase transition than it is possible to compare numerical results to analytical predictions. For the plateau transitions in the quantum Hall effect no such theory exists. In its absence we will use an approach that turned out to be extremely useful in thermodynamic phase transitions: finite-size scaling theory [19]. In many cases, scaling theory can be justified by a renormalization group treatment of the problem. For the quantum Hall transitions no renormalization group analysis exists. We therefore take the pragmatic approach and assume that the finite-size scaling relations hold and check whether or not the numerical data are compatible with this assumption. The result of this analysis is that the data do indeed show universal single parameter scaling with the localization length entering as the only relevant length scale. The localization length diverges at the critical points in the centres of the Landau levels with a universal exponent $\nu = 2.35 \pm 0.03$. However, special care must be taken in order to consider the influence of irrelevant scaling fields.

At this point it is important to find the most suitable quantity for a finite-size scaling analysis. It should be as easy to calculate as possible but still be sensitive to the phase transition. A quantity that satisfies both conditions is the localization length of a finite system. The most suitable geometry for such a calculation is the cylinder geometry: a long strip with periodic boundary conditions in y-direction. We now exploit the fact that the Landau states $\psi_{nk}(\boldsymbol{r})$ are localized in the x-direction at the cyclotron centre coordinates $X(k) = kl_c^2$. Since our geometry is quasi-onedimensional all states in the system are exponentially localized [5]. The modulus of the single-particle Green's function $G(\boldsymbol{r}, \boldsymbol{r}'; E) = \langle \boldsymbol{r} | (E - H)^{-1} | \boldsymbol{r}' \rangle$ will therefore decrease, on average, exponentially with the distance $\boldsymbol{r} - \boldsymbol{r}'$. Since our system is periodic in one direction, the only direction where this distance can become very large is the x-direction along the cylinder. If we average the Green's function over the width L_y of the system for fixed x and x', we get the Green's function in k-space of a Landau level $G(k, k'; E) = \langle k | (E - H)^{-1} | k' \rangle$, where $x = kl_c^2$ and $x' = k'l_c^2$. We thus can directly calculate the Green's function as the inverse of the matrix $E - H$ without the need to calculate the real-space matrix elements.

A final problem that we face is related to the phase coherence of the system. Similarly to the universal conductance fluctuations in metallic systems [20, 21], the average Green's function fluctuates strongly, making it rather cumbersome for numerical purposes. On the other hand, we can define a localization length by considering the average of the logarithm of the modulus of the Green's function

$$\lambda(E)^{-1} = - \lim_{|k-k'| \to \infty} \frac{1}{|k - k'|} \overline{\ln |G(k, k'; E)|}. \tag{11}$$

This quantity does not fluctuate strongly and is self-averaging, i.e. it takes on the same value for almost all realizations of the disorder. It is therefore sufficient for numerical calculations to consider only a few realizations of the disorder and use these to estimate the error in the numerical value for ξ due to the finite distance $k - k'$ in the numerical simulation.

The random Landau matrix elements and hence the Green's function and the localization length $\lambda(E)$ depend on the width L_y of the system. $\lambda(E)$ is thus really a function of two variables $\lambda(E; L_y)$. We will now assume that, close to the critical energy E_c in the centre of each Landau level, the localization length $\xi(E) = \lim_{L_y \to \infty} \lambda(E; L_y)$ of the infinite two-dimensional system diverges according to a power law,

$$\xi(E) = \xi_0 \left| \frac{E - E_c}{\Gamma} \right|^{-\nu} , \tag{12}$$

where $\Gamma \propto V_0$ is a measure of the disorder broadening of the Landau level. The essence of finite-size scaling is that close to E_c, physical quantities that show single parameter scaling, depend on E and L_y only through the ratio $L_y/\xi(E)$. The scaling ansatz for the quasi-one dimensional localization length is thus

$$\lambda(E; L_y) = L_y \Lambda \left(L_y/\xi(E) \right) , \tag{13}$$

where $\Lambda(x)$ is a dimensionless scaling function. Since $\lambda(E_c; L_y)$ diverges as $L_y \to \infty$ and approaches $\xi(E)$ for $E \neq E_c$, one expects the asymptotic behaviour $\Lambda(x) \to$ const. for $x \to 0$ and $\Lambda(x) \to 1/x$ for $x \to \infty$.

Numerical results for different disorder potentials in the two lowest Landau levels are in fact compatible with the assumption of single parameter scaling as expressed in Eqns. (12) and (13). As an example, data for the lowest Landau level and zero correlation length of the potential are presented in Fig. 1. The corresponding scaling function is shown in Fig. 2 exhibiting the expected asymptotic behaviour. In order to obtain reliable results a statistical test of the assumed scaling behaviour has to be performed [22]. Such a test can discriminate which data are compatible with scaling behaviour and which are not due to too small a system size or an energy too far away from the critical point. The data that pass the test are depicted by the dotted area in Fig. 1. The fitted value for the localization length exponent ν can be read off from the slope of the lines in the inset of Fig. 2 and is given by $\nu = 2.35 \pm 0.03$.

Fig. 2 shows that not only the data from Fig. 1 fall onto the scaling curve but also data for $n = 0, \sigma = l_c$, and $n = 1, \sigma = l_c$, as well as data from the network model of Chalker and Coddington [23] that corresponds to the semi-classical limit $\sigma/l_c \to \infty$. All of these data are described by the same scaling function and the same exponent ν, strongly supporting the notion of universal critical behaviour in the integer quantum Hall effect. On the other hand, this notion has been questioned based on results for δ-correlated disorder in the first Landau level similar to those in Fig. 3. In this case no single parameter scaling behaviour is observed close to the critical energy [24]. For larger energies a strong energy dependence of the localization length is observed and has been interpreted as a failure of universal single-parameter scaling. However, the observed behaviour can be reconciled with universal single-parameter scaling by considering the influence of irrelevant scaling fields. Besides the relevant, i.e. diverging, length

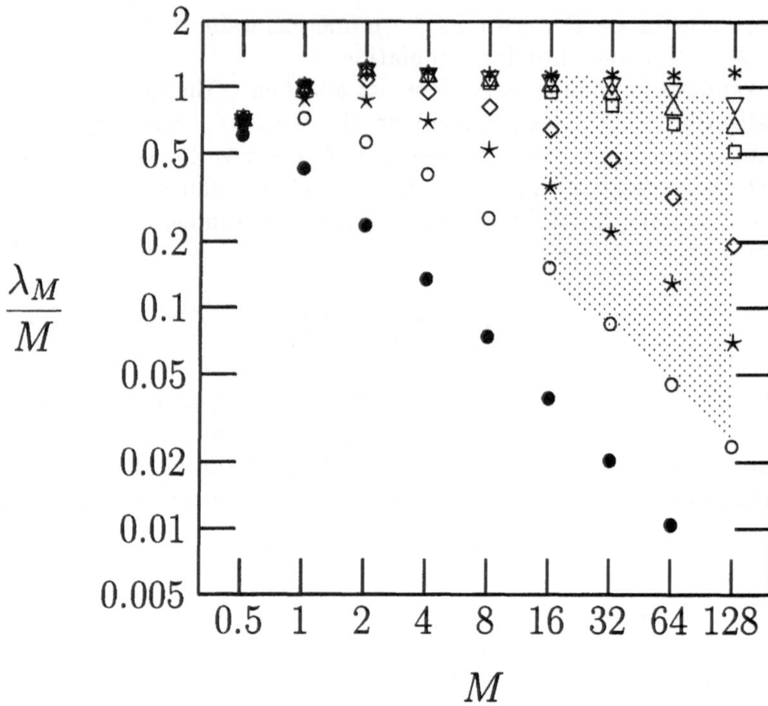

Fig. 1. Normalized localization length λ_M/M in the lowest Landau level for δ-correlated potential as a function of system width M (in units of $\sqrt{2\pi}l_c$) for energies 0.01 ($*$), 0.05 ($\triangledown$), 0.07 ($\vartriangle$), 0.1 ($\square$), 0.18 ($\diamond$), 0.30 ($\star$), 0.5 ($\circ$), and 1.0 ($\bullet$) (in units of Γ) (after [33]).

scale $\xi(E)$, there are other irrelevant length scales in the problem that depend on details like the range of the disorder potential or the Landau level index, but that are finite at the plateau transition. In the thermodynamic limit these length scales are negligible compared to $\xi(E)$, and hence are irrelevant, but for finite size systems they might be noticeable in numerical simulations. A scaling ansatz taking into account one irrelevant length $\xi_{\text{irr}}(\sigma)$ depending on the range σ of the potential is

$$\lambda(E;L_y;\sigma) = L_y \Lambda\left(L_y/\xi(E), L_y/\xi_{\text{irr}}(\sigma)\right). \tag{14}$$

Precisely at the critical energy E_c this leads to single parameter scaling as a function of $L_y/\xi_{\text{irr}}(\sigma)$ which is the behaviour observed in Fig. 4 where data for σ between 0 and l_c and different width L_y are fitted to a single scaling curve. Remarkably, the irrelevant length scale ξ_{irr} increases by four orders of magnitude when the correlation length of the disorder potential is reduced from $0.8l_c$ to 0. For $\sigma = 0$, it is thus much larger than the system width presently accessible to numerical simulations. The data in Fig. 3 do not reflect the asymptotic scaling behaviour described by the localization length exponent ν but the corrections to

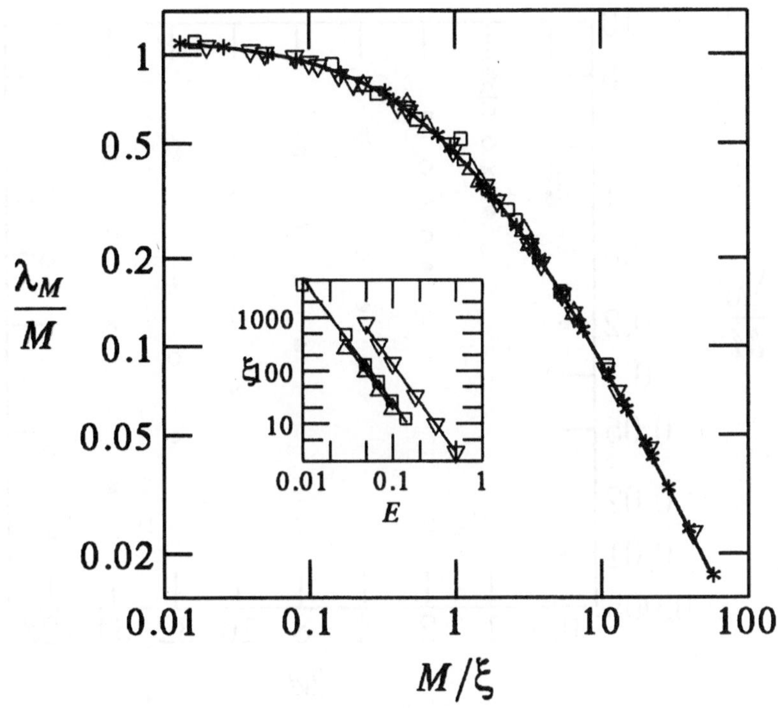

Fig. 2. The scaling function $\lambda_M(E)/M = \Lambda(L_y/\xi(E))$ (from [24]).

scaling due to irrelevant scaling fields. Close to the critical point, the corrections are characterized by an irrelevant scaling index y_{irr},

$$\Lambda(0, L_y/\xi_{\text{irr}}(\sigma)) = \Lambda^* + aL^{y_{\text{irr}}}\zeta_{\text{irr}}. \tag{15}$$

The numerical value of this scaling index is $y_{\text{irr}} = -0.38 \pm 0.04$ [25].

4 Comparison with Experiment

Not only in numerical simulations but also in experiments has single-parameter scaling been observed. It has been observed as a function of size of the systems, the temperature, and the frequency. In each case the broadening of the peaks in the longitudinal resistance due to the finite size, temperature or frequency is studied. For transport measurements the same information is obtained from the maximum slope of the Hall resistance curves between two plateaus.

The experiments that are most easily interpreted are the finite-size scaling experiments [26], where the underlying ideas are the same as in the numerical simulations described above. It is found that below a certain temperature, the width of the transition region between successive plateaus is temperature independent but depends on the size of the samples. According to our scaling ansatz, this width is given by the condition that $\xi(\Delta B) \approx L$, where ΔB is the

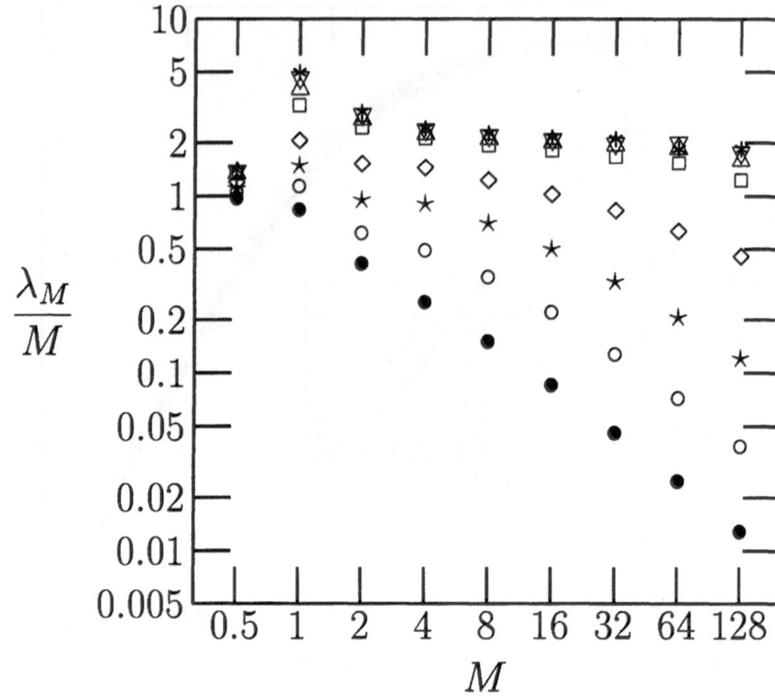

Fig. 3. Normalized localization length λ_M/M in the second ($n = 1$) Landau level for uncorrelated potential ($\sigma = 0$) as a function of system width M (in units of $\sqrt{2\pi}l_c$) for energies 0.01 ($*$), 0.1 ($\triangledown$), 0.18 ($\triangle$), 0.3 ($\square$), 0.5 ($\diamond$), 0.65 ($\star$), 0.8 ($\circ$), and 1.0 ($\bullet$) (in units of Γ) (from [24]).

distance to the critical point as function of magnetic field and L is the system size. Hence, the width of the transition region scales like $L^{-1/\nu}$. The experiments show $\nu = 2.4 \pm 0.1$ for different transitions as long as the spin splitting of the orbital Landau levels is resolved. This result is in very good agreement with the numerical result $\nu = 2.35 \pm 0.03$.

Frequency-dependent scaling was observed in the microwave absorption in a planar transmission line on the sample surface [27]. Here the width of the transition region scales with the microwave frequency as ω^γ with $\gamma = 0.41 \pm 0.04$ for spin-split Landau levels. To understand this result we have to generalize our scaling ansatz to dynamical responses. At finite frequencies a second diverging scale becomes important, the relaxation time $\tau(B)$. Its divergence at the critical magnetic field B_c in the centre of the Landau band defines a new exponent, the dynamical critical exponent z, $\tau(B) \propto \xi(B)^z \propto |B - B_c|^{-\nu z}$. The generalized scaling ansatz is then

$$\sigma_{xx}(L, f) = \frac{e^2}{h} S_{xx}(L/\xi(B), \omega\tau(B)). \tag{16}$$

For sufficiently high frequencies f, the width of the transition region is deter-

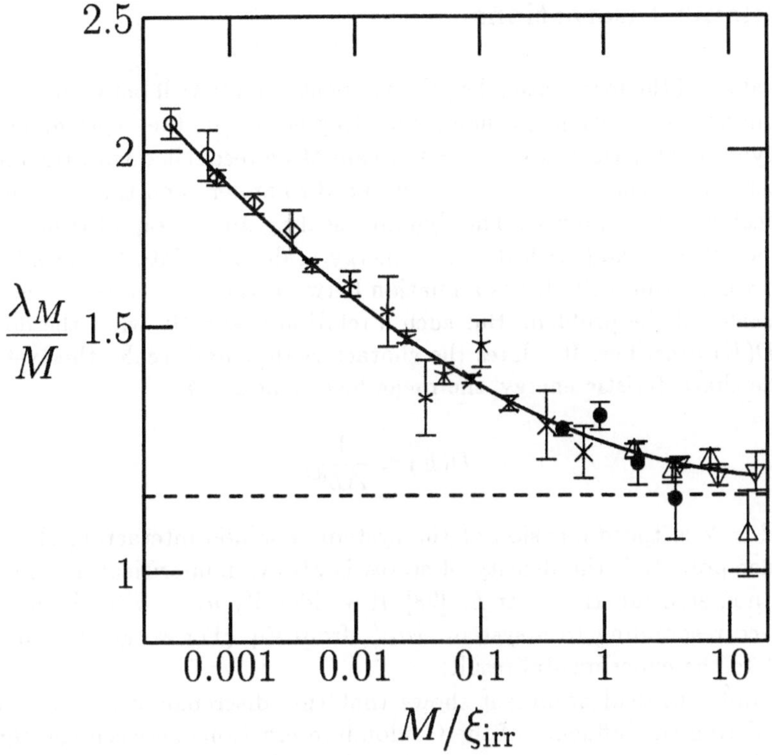

Fig. 4. Scaling towards the fixed point value $\Lambda_c = 1.14$ as a function of $M/\xi_{\rm irr}$ for different correlation length of the disorder potential (from [25]).

mined by the second argument of the scaling function S_{xx} and we can identify the exponent $\gamma = 1/\nu z$. The observed value of γ is compatible with the experimentally and numerically observed values of ν only if $z \approx 1$. We will come back to the significance of this result in the next section, when we discuss the influence of the electron-electron interactions.

Historically, scaling behaviour was first observed experimentally as a function of temperature [6]. The width of the transition region was found to scale as T^κ, with $\kappa = 0.42 \pm 0.04$. Within the framework of the scaling ansatz Eq. (16) there are two interpretations of this result. On the one hand, one can argue that at finite temperature inelastic processes lead to a phase coherence length $L_\Phi \propto T^{-p/2}$ that acts as an effective system size. The exponent κ is then identified as $\kappa = p/2\nu$. On the other hand, at a finite temperature T excitations with frequencies up to $\omega = k_B T/\hbar$ exist in the system. This leads to the identification $\kappa = 1/z\nu$. In this case, as for the frequency scaling, we are led to conclude that $z \approx 1$.

5 Coulomb Interactions

While values of the localization length exponent ν agree well between experiment and numerical simulations for non-interacting electrons, the experimental value of the dynamical critical exponent $z \approx 1$ cannot be reconciled with the picture of non-interacting electrons. To see this, we need to understand the meaning of the dynamical critical exponent. The dynamic scaling ansatz Eq. (16) introduces a new time scale, τ, and with it a new energy scale, τ^{-1}, into the problem. The relation $\tau \propto \xi^z$ thus establishes a relation between the characteristic energy and length scales of the problem. But such a relation is exactly what the density of states $D(E)$ describes. It relates the characteristic length scale, the system size L, to the characteristic energy, the mean level spacing Δ,

$$D(E) = \frac{1}{\Delta L^d}, \tag{17}$$

where $d = 2$ is the dimension of the system. For non-interacting electrons in a disorder potential, the density of states is always non-critical, i.e. $D(E)$ is a smooth non-zero function near E_c [28]. If we identify $\hbar\tau^{-1}$ with the mean level spacing corresponding to a system size ξ, from Eq. (17) we get $z = d = 2$, in contrast to the experimental result.

A simple classical argument shows that this discrepancy can be remedied by considering the influence of the Coulomb interactions between the electrons. To see this it is important to remember that all states in a Landau level are exponentially localized except for those at the critical energy. Efros and Sklovskii pointed out that in an insulator the classical electrostatic energy leads to the appearance of a gap in the density of states at the Fermi energy [29]. In two dimensions the density of states at the Fermi energy is proportional to the inverse linear size L^{-1} of the system. If these arguments hold for the quantum Hall system then from Eq. (17) it follows immediately that the dynamical critical exponent is changed to $z = 1$. Note that this effect is classical by nature and not related to the phase transition at the centre of the Landau level. In fact, the appearance of the Coulomb gap in the single particle density of states was indeed observed in numerical simulations [30]. In these calculations the Coulomb interactions was treated in a self-consistent Hartree-Fock approximation.

With the dynamical critical exponent changing due to Coulomb interaction, the question of the behaviour of other exponents like the localization length exponent arises. There are basically two possible answers to this question: yes, Coulomb interactions are relevant and change critical exponents; no, besides the change of the dynamical critical exponent due to the appearance of the Coulomb gap the interactions are irrelevant [31]. While no definite answer to the question exists, recent self-consistent Hartree-Fock calculations support the latter possibility [32]. In these numerical simulations no change was observed in the localization length exponent, the generalized dimension characterizing the inverse participation ratio, and the Thouless number estimate of the critical conductivity.

6 Conclusion

We have argued that the transitions between successive plateaus in the Hall resistance of disordered two-dimensional electron systems in a strong magnetic field can be described by a scaling theory similar to thermodynamic phase transitions. The theory shows one relevant length scale, the localization length ξ, with an associated critical exponent ν. Numerical simulations for non-interacting electrons were presented that are in quantitative agreement with experimental results. In addition to the localization length exponent, the dynamical critical exponent z characterizes the dynamical response of the system at finite frequencies and temperatures. We have shown that the theory of non-interacting electrons cannot explain the experimentally observed value of $z \approx 1$, making necessary the inclusion of Coulomb interactions between the electrons.

Acknowledgments

I like to acknowledge collaboration with B. Kramer, L. Schweitzer, S.-R. E. Yang, and A.H. MacDonald. I thank the organizers of the 32nd Karpacz Winter school for their invitation and hospitality, and the Deutsche Forschungsgemeinschaft for support through the Sonderforschungbereich 341.

References

1. K. von Klitzing, G. Dorda and M. Pepper, Phys. Rev. Lett. **45**, 494 (1980).
2. D.C. Tsui, H.L. Störmer and A.C. Gossard, Phys. Rev. Lett. **48**, 1559 (1982).
3. R.B. Laughlin, Phys. Rev. B **23**, 5632 (1981).
4. H. Aoki and T. Ando, Solid State Commun. **38**, 1079 (1981)
5. E. Abrahams, P.W. Anderson, D.C. Liciardello and V. Ramakrishnan, Phys. Rev. Lett. **42**, 673 (1979).
6. H.P. Wei, D.C. Tsui, M.A. Paalanen and A.M.M. Pruisken, Phys. Rev. Lett. **61**, 1294 (1988).
7. J.T. Chalker, J. Phys. C **20**, L493 (1987).
8. A.M.M. Pruisken, in *The Quantum Hall Effect* (Ref. [10]), Chap. 5, pp. 117–173.
9. B. Huckestein, Rev. Mod. Phys. **67**, 357 (1995).
10. *The Quantum Hall Effect, Graduate Texts in Contemporary Physics*, edited by R.E. Prange and S.M. Girvin (Springer, Berlin, 1987).
11. M. Janßen, O. Viehweger, U. Fastenrath and J. Hajdu, *Introduction to the Theory of the Integer Quantum Hall Effect* (VCH, Weinheim, 1994).
12. T. Chakraborty and P. Pietiläinen, *The Quantum Hall Effects*, Vol. 85 of *Springer Series in Solid-State Sciences*, 2nd ed. (Springer, Heidelberg, 1995).
13. M. Mehta, *Random Matrices and the Statistical Theory of Spectra* (Academic Press, New York, 1965).
14. B. Huckestein and B. Kramer, Solid State Comm. **71**, 445 (1989).
15. H.A. Weidenmüller, Nucl. Phys. B **290 [FS20]**, 87 (1987).
16. B. Mieck, Europhys. Lett. **13**, 453 (1990).
17. B. Mieck and H.A. Weidenmüller, Z. Phys. B **84**, 59 (1991).
18. B. Mieck, Z. Phys. B **90**, 427 (1993).
19. M.N. Barber, in *Phase Transitions and Critical Phenomena*, edited by C. Domb and J. Lebowitz (Academic Press, London, 1983), Vol. 8, pp. 146–266.

20. B.L. Altshuler, Pis'ma Zh. Eksp. Teor. Fiz. **41**, 530 (1985), [JETP Lett. **41**, 648 (1985)].
21. P.A. Lee and A.D. Stone, Phys. Rev. Lett. **54**, 1622 (1985).
22. B. Huckestein, Physica A **167**, 175 (1990).
23. J.T. Chalker and P.D. Coddington, J. Phys. C **21**, 2665 (1988).
24. B. Huckestein, Europhysics Lett. **20**, 451 (1992).
25. B. Huckestein, Phys. Rev. Lett. **72**, 1080 (1994).
26. S. Koch, R.J. Haug, K. von Klitzing and K. Ploog, Phys. Rev. Lett. **67**, 883 (1991).
27. L.W. Engel, D. Shahar, Ç. Kurdak and D.C. Tsui, Phys. Rev. Lett. **71**, 2638 (1993).
28. F. Wegner, Z. Phys. B **36**, 209 (1980).
29. A. Efros and B. Shklovskii, J. Phys. C **8**, L49 (1975). For a review see B.I. Shklovskii and A.L. Efros, *Electronic Properties of Doped Semiconductors*, Vol. 45 of *Springer Series in Solid-State Sciences* (Springer, Heidelberg, 1984)
30. S.-R.E. Yang and A.H. MacDonald, Phys. Rev. Lett. **70**, 4110 (1993).
31. D.-H. Lee and Z. Wang, Phys. Rev. Lett. **76**, 4014 (1996)
32. S.-R.E. Yang, A.H. MacDonald and B. Huckestein, Phys. Rev. Lett. **74**, 3229 (1995).
33. B. Huckestein and B. Kramer, Phys. Rev. Lett. **64**, 1437 (1990).

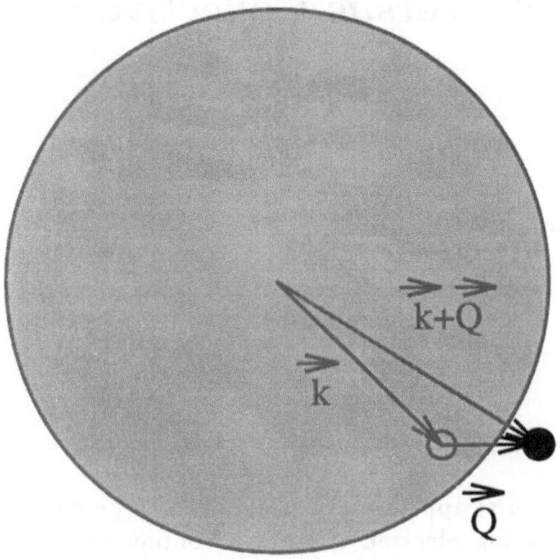

Fig. 1. An electron-hole pair excitation, denoted $|\boldsymbol{k}, \boldsymbol{k}+\boldsymbol{Q}>$, in a non-interacting electron gas.

collective excitation, the plasmon. This typically lies at quite a high energy. Fig. (2) shows the spectrum predicted in random phase approximation (RPA) for sodium metal. The plasmon has an energy $\hbar\omega_P(Q)$ which starts at ≈6 eV at $\boldsymbol{Q} = 0$ and disperses upwards in energy. Also shown in this picture is the continuous spectrum of e-h pairs. This spectrum is easily understood from Fig. (1). At any fixed value of $\boldsymbol{Q}$ with $|\boldsymbol{Q}| < 2k_F$, it is possible to find orbitals $\boldsymbol{k}$ just below the Fermi surface such that the corresponding orbital $\boldsymbol{k}+\boldsymbol{Q}$ lies just above the Fermi surface. This means that pair excitations exist with arbitrarily small excitation energies for $Q < 2k_F$, whereas for larger Q, a gap exists to the lowest single pair excitation. For any $\boldsymbol{Q}$ there is also a maximum energy e-h pair which can be created, namely when the hole state $\boldsymbol{k}$ lies just below the Fermi surface in the direction of $\boldsymbol{Q}$, and the corresponding electron state then lies as far as possible outside the Fermi sea. This excitation has energy $\hbar^2(k_F+Q)^2/2m - \hbar^2k_F^2/2m$. This formula gives the upper edge of the e-h pair continuum shown in Fig. (2). Surprisingly, in spite of the totally new plasmon found in RPA, nevertheless, the spectrum of e-h pairs is not altered from the free electron value in RPA.

This subject has been understood for more than 30 years. Nevertheless, the standard treatments in solid state and most many-body texts [1] do not discuss certain aspects which I find paradoxical. This article is intended as pedagogical, aiming to state and then to explain these paradoxes as clearly as possible. Of course, there is no actual paradox in the existing theory, which provides successful approximate methods for calculating many properties of metals, but the most popular approaches use language in which these interesting paradoxical issues are never apparent.

The paradoxes are forcefully apparent in two very interesting experimental

Single Particle *versus* Collective Electronic Excitations

Philip B. Allen

Department of Physics, SUNY,
Stony Brook, NY 11794, USA

Abstract: As a first approximation, a metal can be modelled as an electron gas. A non-interacting electron gas has a continuous spectrum of electron-hole pair excitations. At each wavevector $\boldsymbol{Q}$ with $|\boldsymbol{Q}|$ less than the maximum Fermi surface spanning vector ($2k_F$) there is a continuous set of electron-hole pair states, with a maximum energy but no gap (the minimum energy is zero.) Once the Coulomb interaction is taken into account, a new collective mode, the plasmon, is built from the electron-hole pair spectrum. The plasmon captures most of the spectral weight in the scattering cross-section, yet the particle-hole pairs remain practically unchanged, as can be seen from the success of the Landau Fermi-liquid picture. This article explores how even an isolated electron-hole pair in non-interacting approximation is a form of charge density wave excitation, and how the Coulomb interaction totally alters the charge properties, without affecting many other properties of the electron-hole pairs.

1 Introduction

The low-lying excitations of a non-interacting electron gas are simple rearrangements of the occupancy of the single electron plane-wave orbitals. Because real metals, according to Landau theory, have a lot in common with non-interacting quantum electron gases, this subject is well known to all who study solids. If we neglect band structure, the electron orbitals (labeled by quantum numbers $k = (\boldsymbol{k}, \sigma)$) have energy $\epsilon_k = \hbar^2 k^2/2m$. The ground state has all orbitals occupied which lie inside the Fermi surface, with wavevectors obeying $|\boldsymbol{k}| < k_F$ and energies obeying $\epsilon_k < \epsilon_F$, where k_F and ϵ_F are the Fermi wavevector and energy. The simplest excitation, known as an electron-hole (e-h) pair, consists of moving an electron out of a state $\boldsymbol{k}$ below the Fermi surface, putting it into a state $\boldsymbol{k} + \boldsymbol{Q}$ above. This is shown in Fig. (1).

The Coulomb interaction is numerically not small, but Landau theory argues that nevertheless, the consequences of the Coulomb interaction are less drastic than one might suppose. However, one drastic effect certainly happens, namely, out of the e-h pair excitation spectrum, the Coulomb interaction creates a new,

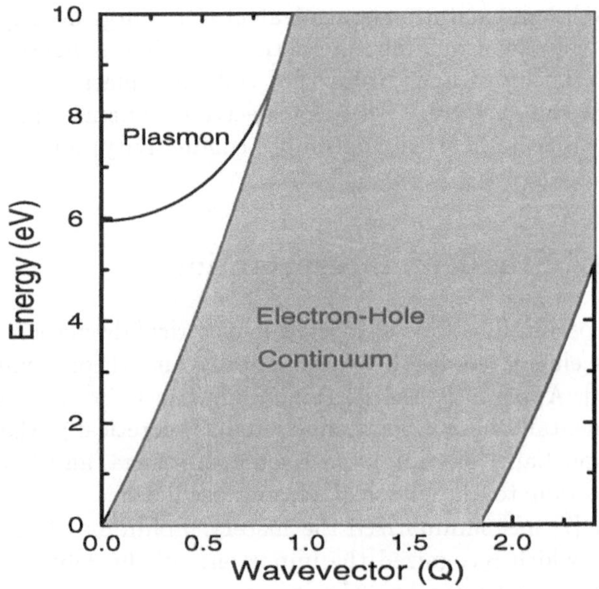

Fig. 2. The plasmon dispersion curve and the e-h pair spectrum as calculated in RPA using parameters appropriate to metallic sodium in free electron approximation. The wavevector is in units of $\mathring{A}^{-1}$.

studies of Raman scattering by electronic excitations, by Contreras, Sood, and Cardona [2]. The opening paragraph of the first of these papers reads:

"Metals and heavily doped (degenerate) semiconductors can scatter light either through single-particle or collective excitations of free carriers. The single-particle excitations correspond to charge-density fluctuations and, as such, they are screened at low frequencies in a self-consistent manner by the electrons themselves. Thus, in simple, free-electron-like carrier systems, no low-frequency scattering is observed. Instead, a peak at the plasma frequency is seen."

The "single-particle" excitations referred to above are just the e-h pair excitations. How close is this "correspondence" of e-h pairs to charge-density fluctuations? In my own case, the closeness was hard to appreciate at first. After all, the charge density difference between an excited e-h pair state and the ground state (elaborated later) must be just $\delta\rho = |\psi_{k+Q}|^2 - |\psi_k|^2$, which is zero for plane-wave states. I shall show later that the correspondence is actually perfect. To nuclear physicists this correspondence is quite familiar. Referring to excitations of the nucleus, Brown [3] says

"... we wish to talk about vibrations, which are density fluctuations or – in quantum-mechanical language – particle-hole excitations ..."

To summarize, the apparent paradoxes are these:

1. How can an e-h pair excitation be equivalent to a charge-density fluctuation?
2. If screening (by the Coulomb interaction) eliminates nearly all the low-frequency scattering of light in favor of collective plasma oscillations, how can it be that the low-lying (e-h pair) spectrum remains unaltered in other experiments (specific heat, susceptibility, conductivity) and in Landau theory?

2 History and Standard Interpretation

Electron density oscillations were suspected in electrical discharges in gases, and this subject was elucidated, both experimentally and theoretically, by Tonks and Langmuir [4]. Apparently the corresponding effect for electrons in metals was seen experimentally before being understood theoretically. Experiments by Ruthemann [5] and Lang [6] stimulated Kronig, Korringa, and Kramers to recognize the connection to the classical plasma oscillations seen by Tonks and Langmuir. Slater [7] has summarized the history. Bohm and Pines then wrote a series of papers which recognized the importance of the collective plasma degrees of freedom for understanding the interacting electron gas problem. In a review article, Pines [8] coined the term "plasmon", and ever since, the solid state texts have recognized the plasmon as one of the elementary excitations, or quasiparticles, of solid state physics.

The standard derivation of the frequency of plasma oscillations, appearing in all the texts, is identical to an argument from Tonks and Langmuir [4]. Consider a slab of thickness D and infinite transverse size, embedded in an infinite sample of metal with electron density n. Imagine that the electrons in this slab are all displaced by the same small amount u in the direction normal to the slab (call this the "upward" direction.) Then a thin layer of charge of thickness u and surface charge density $\sigma = -neu$ accumulates at the upper surface of the slab and $+neu$ at the lower surface. Therefore there is a capacitor-type E-field of magnitude $4\pi\sigma = 4\pi neu$ in the upward direction inside the slab. This field exerts a force $-eE$ in the downward direction on every electron in the slab. This is a restoring force $-Ku$ proportional to the displacement of each electron. This causes each electron in the slab to oscillate at frequency $\sqrt{(K/m)}$, namely,

$$\omega_P^2 = 4\pi ne^2/m \tag{1}$$

The plasmons in metals are quantized versions of these classical oscillators, with energy $\hbar\omega_P$.

The other standard textbook result is that the frequency of plasma oscillations is best understood or calculated by looking for the zero of the real part of the complex dielectric function $\epsilon_1(Q,\omega) + i\epsilon_2(Q,\omega)$. Even better, the Fourier transform $S(Q,\omega)$ of the density-density correlation function $< \rho(r,t)\rho(r',0) >$, gives the spectrum of density oscillations in a material. Van Hove showed that in Born approximation, the inelastic scattering cross section for particles (like x-rays and electrons) which couple to electron density is given by $S(Q,\omega)$, which is also called the inelastic structure factor. Finally, the same function, $S(Q,\omega)$, is directly proportional to $\mathrm{Im}(-1/\epsilon(Q,\omega)) = \epsilon_2/(\epsilon_1^2 + \epsilon_2^2)$. Thus the zeros of $\epsilon_1(Q,\omega)$ correspond to peaks in $S(Q,\omega)$ and to peaks in the inelastic scattering cross section, provided the damping, given by $\epsilon_2(Q,\omega)$ is small. When $\epsilon(Q,\omega)$ is calculated in RPA, the zeros of ϵ_1 give the plasmon dispersion shown in Fig. (2). The boundaries of the e-h continuum coincide with the region of (Q,ω) where ϵ_2 differs from zero. In that region, plasmons are very heavily damped, and merge smoothly into the e-h pair spectrum.

The best test of the ideas of collective plasmon excitations is experiment. In the past, inelastic electron scattering away from $Q = 0$ and also inelastic x-ray scattering with energy resolution better than 1eV have both been difficult. Recently, synchrotron x-ray sources have made the latter experiment much easier, and we can expect many new results on collective electron behavior [9]. A good example is the measured dispersion curve of plasmons in Na, as determined by inelastic electron scattering by vom Felde *et al.* [10] and shown in Fig. (3) These results demonstrate that at least in certain simple metals, sharp plasmons exist, and that the RPA is at least qualitatively very successful in explaining the spectrum.

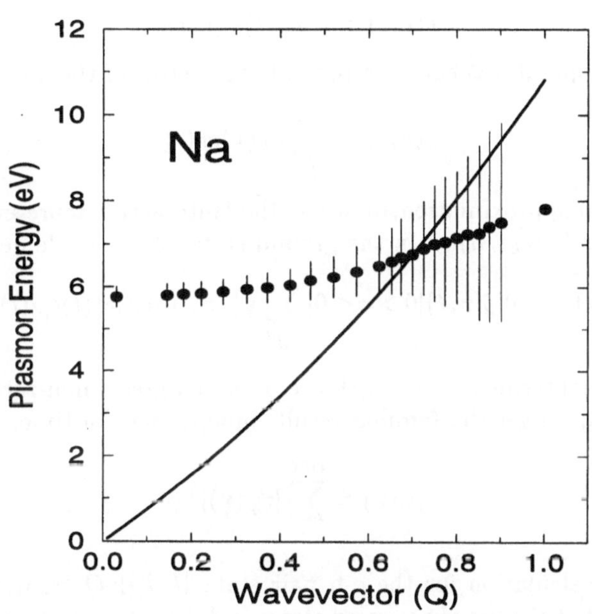

Fig. 3. Plasmon dispersion for sodium measured in Ref. [10]. The vertical bars are the measured widths of the plasmon resonances. Q is measured in units $\mathring{A}^{-1}$. The solid curve marks the edge of the e-h pair spectrum as given in Fig. (2). The *datum* at the largest Q had such a broad lineshape that no sensible width could be assigned.

3 Charge Density of an Electron-Hole Pair

Contreras *et al* [2] state that an e-h pair is a charge-density wave excitation. The simple truth of this statement does not emerge in any of the textbook treatments I know. Let us establish some notation. In a non-interacting gas, the ground state can be written as

$$|0>=\prod_{k}^{\mathrm{occ}} c_k^\dagger |\mathrm{vac}> . \tag{2}$$

Here I am using the notation k as a shorthand for all the quantum numbers of the orbital, that is, the wavevector $\boldsymbol{k}$ as well as the spin quantum number and (in a real material) the band index. An e-h pair excitation is denoted by

$$|k, k+Q>= c_{k+Q}^\dagger c_k |0> \tag{3}$$

This state vanishes unless the orbital k is occupied in the ground state, and the orbital $k+Q$ is empty.

The charge density is $-e$ times the electron density. I will leave out the factor $-e$ and refer to it as charge density anyway. To find the charge density of the e-h pair state, one would calculate the expectation value of the charge density operator,

$$\hat{\rho}(\boldsymbol{r},t) = \hat{\psi}^\dagger(\boldsymbol{r},t)\hat{\psi}(\boldsymbol{r},t) \tag{4}$$

where the field operator $\hat{\psi}$ can be represented in terms of the one electron states as

$$\hat{\psi}(\boldsymbol{r},t) = \sum_k \psi_k(\boldsymbol{r}) c_k(t) \tag{5}$$

In a one-electron approximation, or else in the "interaction representation", $c_k(t)$ has the form $\exp(-i\epsilon_k t/\hbar)c_k$. In the ground state, the particle density is

$$\rho_0(\boldsymbol{r},t) =< 0|\hat{\rho}(\boldsymbol{r},t)|0>=< 0|\sum_{p,p'} \psi_p^*(\boldsymbol{r})\psi_{p'}(\boldsymbol{r}) c_p^\dagger(t) c_{p'}(t)|0> \tag{6}$$

Only the diagonal terms $p = p'$ which are occupied give a non-zero contribution to this sum, and we get the familiar result, independent of time,

$$\rho_0(\boldsymbol{r}) = \sum_{p}^{\mathrm{occ}} |\psi_p(\boldsymbol{r})|^2 \tag{7}$$

Repeating the calculation for the e-h pair state $|k, k+Q>$, we get the same answer except that the list of occupied states is different. If we look at the change in the charge density between the ground state and the e-h pair state, we get

$$\delta\rho(\boldsymbol{r},t) = |\psi_{k+Q}(\boldsymbol{r})|^2 - |\psi_k(\boldsymbol{r})|^2 . \tag{8}$$

For free electrons, the orbitals $\psi_k(\boldsymbol{r})$ are plane waves $\exp(i\boldsymbol{k}\cdot\boldsymbol{r})$, and $\delta\rho$ vanishes. For the Bloch states of a real metal, the orbitals are plane waves modulated by periodic functions $u_k(\boldsymbol{r})$. For small Q, the periodic function u_{k+Q} is almost the same as u_k, so the change in charge density vanishes as Q goes to zero. There is no evidence of a charge-density wave.

Nevertheless, Contreras *et al.* [2] are right. There is a charge-density wave hidden in the pair state. The way to see it is to consider what would happen if a small amount of e-h pair excitation were mixed with the ground state. In other words, consider the state

$$|\Psi_{k,k+Q}> \equiv |0> + \eta|k, k+Q> = (1 + \eta c^\dagger_{k+q} c_k)|0> \tag{9}$$

where η is a small complex number, and it is assumed that the pair excitation does not vanish, *i. e.* that $\boldsymbol{k}$ is inside the Fermi distribution and $\boldsymbol{k} + Q$ outside. The calculation now yields (to first order in the small parameter η)

$$\delta\rho = \eta\psi^*_{k+Q}(\boldsymbol{r})\psi_k(\boldsymbol{r})e^{-i(\epsilon_{k+Q}-\epsilon_k)t/\hbar} + \text{c.c.} \tag{10}$$

where c.c. stands for "complex conjugate." For small $\boldsymbol{Q}$, the electron-hole pair energy $(\epsilon_{k+Q} - \epsilon_k)/\hbar$ is $\boldsymbol{Q}\cdot\boldsymbol{v}_k$ where $\boldsymbol{v}_k$ is the group velocity, $\partial\epsilon_k/\partial\boldsymbol{k}$. At the same level in smallness of $\boldsymbol{Q}$, we can write the product of wavefunctions as

$$\psi^*_{k+Q}(\boldsymbol{r})\psi_k(\boldsymbol{r}) = |\psi_k(\boldsymbol{r})|^2 e^{i\boldsymbol{Q}\cdot\boldsymbol{r}} \tag{11}$$

This is just the zeroth order result of a "$\boldsymbol{k}\cdot\boldsymbol{p}$" perturbation theory. Substituting these changes into Eq. (10), we get the result

$$\delta\rho(\boldsymbol{r}, t) = 2|\eta||\psi_k(\boldsymbol{r})|^2 \cos[\boldsymbol{Q}\cdot(\boldsymbol{r} - \boldsymbol{v}_k t) + \phi] \tag{12}$$

where ϕ is the phase of η. Now we see that the e-h pair excitation, relative to the ground state, is a charge-density wave! The propagation direction is of course given by $\boldsymbol{Q}$, and the phase velocity is the component $\hat{Q}\cdot\boldsymbol{v}_k$ of the electron group velocity in the direction of $\boldsymbol{Q}$. The group velocity, however, is $\boldsymbol{v}_k$. The simple cosine oscillation of electron density is very reminiscent of a phonon. However, unlike phonons which have 3 branches for each atom in the cell, the number of e-h pairs at a fixed $\boldsymbol{Q}$ is a macroscopic number. Specifically, for free electrons with $Q > 2k_F$, the number of pairs with wavevector $\boldsymbol{Q}$ is the same as the number of occupied electron states, whereas for $Q < 2k_F$, the number is reduced by a factor $(3/2)(Q/2k_F)[1 - (Q/2k_F)^2/3]$.

It is evident from the electrostatic argument of Tonks and Langmuir that once the Coulomb interaction is taken into account, there will be a large additional restoring force on the charge oscillation, and the result will be that the charge density will oscillate at the plasma frequency rather than the non-interacting frequency $\boldsymbol{Q}\cdot\boldsymbol{v}_k$. This will be shown explicitly in the next section. One should pause to ask what property of the medium permits wave propagation (Eqn. (12)) of electron charge before the Coulomb interaction is turned on. If there is no interaction, a classical gas does not support propagating waves. The answer is the Fermi degeneracy, which forces an energy cost if the local density is altered.

4 Influence of Coulomb Interaction on Charge of the e-h Pair

It is convenient to formulate this as a linear response problem. This way we expect the Coulomb interaction to enter nicely as a dielectric screening. Therefore, we want a perturbing field which will create the e-h pair and which can be represented as a term in the Hamiltonian. The following operator does the trick:

$$H' = \hbar\eta c^{\dagger}_{k+Q} c_k \delta(t). \tag{13}$$

This inserts an electron-hole pair at time zero. The dimensionless strength of the insertion field, η, will be taken to be a small number. Linear response theory provides formal answers for the alterations of system properties which can be measured at a later time, to first order in η. In this section, we explore the resulting charge-density response, that is, the expectation value of the operator $\hat{\rho}$ of Eqn. (4). Later we look at the current response.

Standard techniques of linear response theory [11] tell us that the response is

$$\delta\rho(\boldsymbol{r},t) = \frac{1}{i\hbar}\int_{-\infty}^{t} dt' \langle 0|\hat{\rho}(t)H'(t') - H'^{\dagger}(t')\hat{\rho}(t)|0\rangle. \tag{14}$$

This is a standard Kubo-type formula, except that since the Hamiltonian H' is not Hermitean, so the commutator is replaced by a slighly more complicated form. Inserting Eqn. (13) for H', the charge density response is

$$\delta\rho(\boldsymbol{r},t) = 2\mathrm{Im}\left\{\eta D(t)\right\}\theta(t) \tag{15}$$

$$D(t) = < 0|\hat{T}\hat{\rho}(\boldsymbol{r},t)c^{\dagger}_{k+q}c_k|0 > \tag{16}$$

where $\theta(t)$ is 1 if $t > 0$ and zero otherwise. D is a type of two-particle Green's function, and can be evaluated perturbatively in powers of the Coulomb interaction by standard Feynman methods using time-ordered Green's functions. The time-ordering operator $\hat{T}$ has been inserted into Eqn. (16). Since ultimately we only need to know D for positive times, this doesn't change anything but facilitates the perturbation theory.

First, evaluate the charge response function without any Coulomb interactions, D_0. By standard methods one gets

$$D_0(t) = e^{i\boldsymbol{Q}\cdot\boldsymbol{r}}\int_{-\infty}^{\infty}\frac{d\omega}{2\pi}e^{-i\omega t}D_0(\omega) \tag{17}$$

$$D_0(\omega) = i\left\{\frac{f_k(1-f_{k+Q})}{\omega + i\eta - \boldsymbol{Q}\cdot\boldsymbol{v}_k} - \frac{f_{k+Q}(1-f_k)}{\omega - i\eta - \boldsymbol{Q}\cdot\boldsymbol{v}_k}\right\} \tag{18}$$

For positive t, the contour is closed in the lower complex ω plane, and only the first term in $D_0(\omega)$ contributes. The answer is

$$\delta\rho(\boldsymbol{r},t) = 2|\eta|\sin(\boldsymbol{Q}\cdot\boldsymbol{r} - \boldsymbol{Q}\cdot\boldsymbol{v}_k t + \phi)|\psi_k(\boldsymbol{r})|^2 f_k(1-f_{k+Q}) \tag{19}$$

This is just the same as the previous result for the charge density change which occurs when an e-h pair is present with amplitude η at all times, Eqn. (12), except for a phase change of $\pi/2$. This is because our earlier pair oscillator started at time zero with given amplitude η whereas the present oscillator at time zero was given "velocity" by the impulsive insertion. The present version also contains explicitly the Fermi factors f_k and $(1-f_{k+Q})$ which are zero or one depending on whether the orbital is within or without the Fermi distribution.

The Coulomb interaction has the form

$$\hat{V} = \frac{1}{2}\sum_{Q'} \hat{\rho}_{-Q'} v(Q')\hat{\rho}_{Q'} \tag{20}$$

where the operator $\hat{\rho}(Q)$ is the Fourier transform of the charge density operator Eqn. (4). For free electrons this is just

$$\hat{\rho}_Q = \sum_k c^{\dagger}_{k+Q} c_k \tag{21}$$

To evaluate the response function D exactly in the presence of Coulomb interactions is of course impossible. Diagrammatic perturbation theory allows a classification of the correction terms. The RPA is a standard approximation which keeps an infinite set of leading diagrams (which contain the leading small $\boldsymbol{Q}$ divergence, or the long-range part of the Coulomb interaction). The result at RPA level is

$$D_{RPA}(t) = e^{i\boldsymbol{Q}\cdot\boldsymbol{r}} \int_{-\infty}^{\infty} \frac{d\omega}{2\pi} \frac{D_0(\omega)e^{-i\omega t}}{1 - v(\boldsymbol{Q})\chi_{0T}(\boldsymbol{Q},\omega)} \tag{22}$$

where χ_0 is the usual non-interacting susceptibility "bubble" diagram, and the subscript T indicates that it is the "time-ordered" rather than the usual "retarded" or causal response function χ_{0R} that is needed. The factor $v(Q)$ is $4\pi e^2/Q^2$, and the denominator $1 - v(Q)\chi_0$ is the dielectric function $\epsilon(Q,\omega)$. These formulas are written for the uniform electron gas with no background of actual atoms. Generalizing to real electrons interacting with a crystalline array of atoms is not hard, but complicates the notation because of the matrix nature of ϵ. At this stage it is also easy to include the dominant impurity effects, simply by including them in ϵ. If the mean free path $\ell = v_F\tau$ is short enough that $Q\ell \ll 1$, the standard Drude result applies,

$$\epsilon_R(Q,\omega) = 1 - \frac{\omega_p^2}{\omega(\omega + i/\tau)} \tag{23}$$

It is important to convert this to the retarded form, which can be done using the general relations

$$\mathrm{Re}\chi_R(\omega) = \mathrm{Re}\chi_T(\omega) \tag{24}$$

$$\mathrm{Im}\chi_R(\omega) = \tanh\left(\frac{\hbar\omega}{2k_BT}\right)\mathrm{Im}\chi_T(\omega) \tag{25}$$

At low T, the hyperbolic tangent is $\approx \pm 1$ and to convert a retarded function to a time-ordered function, one just changes the sign of the imaginary part when $\omega < 0$. To a good approximation the Drude response function can be written

$$\frac{1}{\epsilon_R} \cong \frac{1}{2\omega_P}\left[\frac{\omega + i/\tau}{\omega - \omega_P + i/2\tau} - \frac{\omega + i/\tau}{\omega + \omega_P + i/2\tau}\right] \tag{26}$$

Because $\omega_P \gg 1/\tau$, the first term of Eqn. (26) is only important when $\omega > 0$ and the second term is only important when $\omega < 0$. Therefore this is converted into a time-ordered response function by the simple approximate expedient of taking the complex conjugate of the second term,

$$\frac{1}{\epsilon_T} = \frac{\omega}{2\omega_p}\left[\frac{\omega + i/\tau}{\omega - \omega_p + i/2\tau} - \frac{\omega - i/\tau}{\omega + \omega_p - i/2\tau}\right] \tag{27}$$

Putting this into Eqn. (22) and closing the contour in the lower half of the complex ω plane, the result is

$$\begin{aligned} D_{RPA}(t) &= e^{i(\boldsymbol{Q}\cdot\boldsymbol{r} - \omega_p t)} e^{-t/2\tau}(f_k - f_{k+Q})/2 \\ &+ e^{i(\boldsymbol{Q}\cdot\boldsymbol{r} - \boldsymbol{Q}\cdot\boldsymbol{v}_k t)} f_k(1 - f_{k+Q}) \times \mathcal{O}\left(\frac{Qv_F \text{ or } 1/\tau}{\omega_p}\right)^2 \end{aligned} \tag{28}$$

For simplicity, the Bloch wavefunction $|\psi_k|^2$ has been dropped. Only free electron results are given for the remainder of this section. The second term of Eqn. (28) is negligible compared to the first at small Q. Therefore, using Eqn. (15), the result for the charge disturbance after accounting for the long range Coulomb field is

$$\delta\rho(\boldsymbol{r}, t) = |\eta| \sin(\boldsymbol{Q}\cdot\boldsymbol{r} - \omega_p t) e^{-t/2\tau}(f_k - f_{k+Q}). \tag{29}$$

This is almost what one would have naively guessed. The amplitude is reduced by 2, and the frequency is now the plasma frequency, ω_p, as expected, with only a negligible component left oscillating at the original frequency $\boldsymbol{Q}\cdot\boldsymbol{v}_k$. Compensating for what looks like an arbitrary reduction by half of the amplitude is a new effect, coming from the difference between the factors $f_k(1-f_{k+Q})$ and $(f_k - f_{k+Q})$. In the interacting system, it is possible to insert an electron-hole pair even when the state k is above the Fermi surface ($f_k = 0$) and the state $k+Q$ is below ($f_{k+Q} = 1$). The factor $f_k(1 - f_{k+Q})$ of the non-interacting response is zero, but the factor $(f_k - f_{k+Q})$ of the interacting system is -1 (the density disturbance has a phase shift of π.) Where does this new term come from? The Coulomb interaction Eqn. (20) continuously creates "vaccuum fluctuations" which are pairs of electron-hole pairs of equal and opposite momentum Q' and $-Q'$, as shown in Fig. (4). Thus there is the possibility that an electron with $k > k_F$ and a hole with $k + Q < k_F$ are already present at time zero because of a vaccuum fluctuation (which also created a pair $k' < k_F$ and $k' + Q > k_F$). Then the "insertion" operator $c^\dagger_{k+Q}c_k$ removes the already present electron-hole pair, leaving its mate k',$k'+Q$. This process has equal amplitude as the simpler insertion process, but each is reduced by 2 compared to the non-interacting case.

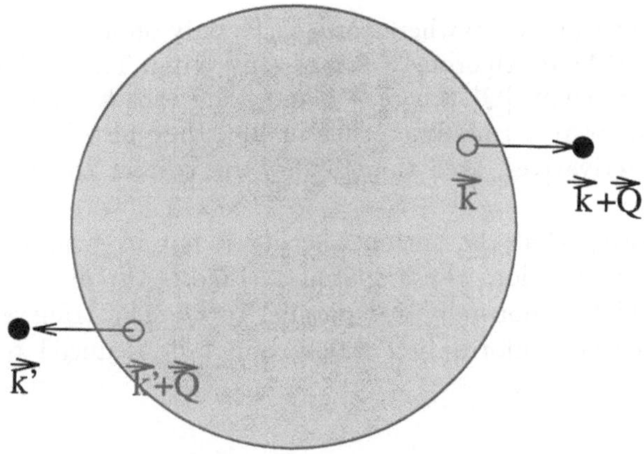

Fig. 4. A typical vacuum fluctuation, consisting of two e-h pairs with momentum $\boldsymbol{Q}$ and $-\boldsymbol{Q}$.

5 Current Density of an Electron-Hole Pair in a Metal

It is also interesting to ask what is the current density associated with an electron-hole pair excitation. The current density operator, analogous to Eqn. (4), is

$$\hat{\boldsymbol{j}}(\boldsymbol{r}) = \frac{\hbar}{2mi}\left[\hat{\psi}^\dagger(\boldsymbol{r})\left(\boldsymbol{\nabla}\hat{\psi}(\boldsymbol{r})\right) - \left(\boldsymbol{\nabla}\hat{\psi}(\boldsymbol{r})\right)^\dagger \hat{\psi}(\boldsymbol{r})\right] \tag{30}$$

The ground state carries no current. To first order in the amplitude η, the current density in the state $|\Psi_{k,k+Q}>$ (Eqn. 9) is

$$\delta\boldsymbol{j}(\boldsymbol{r}) = \frac{\hbar}{2mi}\eta\left[\psi_k^*(\boldsymbol{r})\left(\boldsymbol{\nabla}\psi_{k+Q}(\boldsymbol{r})\right) - \left(\boldsymbol{\nabla}\psi_k^*(\boldsymbol{r})\right)\psi_{k+Q}(\boldsymbol{r})\right] e^{-i(\epsilon_{k+Q}-\epsilon_k)t/\hbar} + \text{c.c.} \tag{31}$$

Because the wavevector $\boldsymbol{Q}$ is small, the factor inside the square brackets of Eqn. (31) can be written as

$$\frac{\hbar}{2mi}[\quad] = e^{i\boldsymbol{Q}\cdot\boldsymbol{r}}\boldsymbol{j}_k(\boldsymbol{r}) \tag{32}$$

$$\boldsymbol{j}_k(\boldsymbol{r}) = \frac{\hbar}{2mi}\left[\psi_k^*(\boldsymbol{\nabla}\psi_k) - (\boldsymbol{\nabla}\psi_k^*)\psi_k\right] \tag{33}$$

where $\boldsymbol{j}_k(\boldsymbol{r})$ is the current density associated with the single particle Bloch state $\psi_k(\boldsymbol{r})$. Since the corresponding density $|\psi_k(\boldsymbol{r})|^2$ is time-independent, the vector field $\boldsymbol{j}_k(\boldsymbol{r})$ must be divergenceless. The spatial average of this current density is $\boldsymbol{v}_k/\Omega$ where $\boldsymbol{v}_k$ is the group velocity of the state and Ω is the volume of the crystal. Thus we get

$$\delta\boldsymbol{j}(\boldsymbol{r}) = 2|\eta|\boldsymbol{j}_k(\boldsymbol{r})\cos[\boldsymbol{Q}\cdot(\boldsymbol{r}-\boldsymbol{v}_k t)+\phi]f_k(1-f_{k+Q}) \tag{34}$$

The electron-hole pair conserves charge while it propagates, that is $\boldsymbol{\nabla}\cdot\boldsymbol{j}+\partial\rho/\partial t$ is zero. However, Eqs. (12) and (34) do not rigorously obey this law, since the current $\boldsymbol{j}_k$ is not equal everywhere to $\boldsymbol{v}_k|\psi_k|^2$, only on average. The more exact forms (10) and (31) are rigorously charge-conserving. The derivatives $\nabla \cdot \boldsymbol{j}$ and $\partial\rho/\partial t$ are first order in $|\boldsymbol{Q}|$, whereas $\boldsymbol{j}$ and ρ are zeroth order. Since Eqs. (12) and (34) throw away terms first order in $|\boldsymbol{Q}|$, they also lose some first order parts of their derivatives, even though they are correct to the order that they are written.

It is interesting that the current density is not in general longitudinal. It is polarized in a direction whose spatial average is the direction of the group velocity $\boldsymbol{v}_k$, which is normally not parallel to $\boldsymbol{Q}$. The transverse part of the current contains new information which cannot be deduced from the density alone.

6 Influence of Coulomb Interaction on Current of the e-h Pair

The same linear response proceedure used earlier for the charge can be generalized for the current. Analogous to Eqns. (15, 16) we get

$$\delta \boldsymbol{j}(\boldsymbol{r},t) = 2\mathrm{Im}\left\{\eta \boldsymbol{J}(t)\right\}\theta(t) \tag{35}$$

$$J_\alpha(t) = <0|\hat{T}\hat{j}_\alpha(\boldsymbol{r},t)c^\dagger_{k+q}c_k|0> \tag{36}$$

When this is evaluated without Coulomb interactions, the result is

$$\delta \boldsymbol{j}(\boldsymbol{r}) = 2|\eta|\boldsymbol{j}_k(\boldsymbol{r})\sin[\boldsymbol{Q}\cdot(\boldsymbol{r}-\boldsymbol{v}_k t)+\phi]f_k(1-f_{k+Q}), \tag{37}$$

analogous to Eqn. (19).

When Coulomb interactions are accounted for by perturbation theory, we discover that transverse and longitudinal parts of the current are treated very differently. Sums over $\boldsymbol{k}'$ occur in which there are factors of the type

$$(\boldsymbol{v}_{k'\perp} + \boldsymbol{v}_{k'\parallel}) \times F(\epsilon_{k'}, \epsilon_{k'+Q}). \tag{38}$$

The velocity $\boldsymbol{v}_{k'}$ has been split into a transverse part (perpendicular to $\boldsymbol{Q}$) and a longitudinal part. The transverse contribution vanishes, because it is always possible, for free electrons or for band electrons as long as $\boldsymbol{Q}$ has high enough symmetry, to find two states $\boldsymbol{k}'$ which have identical values of $\epsilon_{k'}$, $\epsilon_{k'+Q}$ and $v_{k'\parallel}$ and opposite values of $v_{k'\perp}$. Therefore, all Coulomb corrections to the transverse current cancel, and the transverse current continues to oscillate at its unrenormalized frequency.

The longitudinal part of the current gets screened in RPA the same way the charge does. A diagrammatic analysis yields the result

$$\boldsymbol{Q}\cdot\boldsymbol{J}(t) = e^{i\boldsymbol{Q}\cdot\boldsymbol{r}}\int \frac{d\omega}{2\pi}e^{-i\omega t}\frac{\omega D_0(\omega)}{1-v(Q)\chi_{0T}(\boldsymbol{Q},\omega)} \tag{39}$$

The final result for the current is

$$\delta \boldsymbol{j}(\boldsymbol{r},t) = \frac{2|\eta|}{V}\boldsymbol{v}_{k\perp}\sin(\boldsymbol{Q}\cdot\boldsymbol{r} - \boldsymbol{Q}\cdot\boldsymbol{v}_k t + \phi) f_k(1 - f_{k+Q})$$
$$+ \frac{|\eta|\omega_P}{VQ}\hat{\boldsymbol{Q}}\sin(\boldsymbol{Q}\cdot\boldsymbol{r} - \omega_P t + \phi)e^{-t/2\tau}(f_k - f_{k+Q}) \quad (40)$$

This answer was calculated using the Drude response in the same way that Eqn. (29) was calculated.

We have now learned something interesting about e-h pair excitations. Even though the Coulomb interaction totally alters the charge oscillations of every e-h pair, other properties can remain completely immune to alteration, such as the transverse part of the current oscillation. This is part of the explanation of the second "apparent" paradox. A more complete explanation is given in the next section.

7 What Is the Wavefunction of a Plasmon?

Another way of understanding the broad immunity of e-h pairs to Coulombic alteration is to recognize that it is easy to combine e-h pairs so that the charge is hidden. A trivial way is to make a random combination of pairs. Rather than a single coherent cosine of charge oscillation as in Eqn. (12), one now expects an incoherent sum of many cosines with different phases, adding up to a disturbance whose charge is smaller than a coherent sum by a factor $1/\sqrt{N}$ where N is the number of e-h pairs combined. Then the Coulomb interaction will still alter the charge oscillation, but this is only a minor aspect of the complete excitation.

A way of picturing this more elegantly is to consider the space of all single e-h pair states. The Hamiltonian can couple through the Coulomb interaction only states of the same $\boldsymbol{Q}$. Therefore we restrict attention to states of the form

$$|\Psi> = \sum_k A_k |k, k+Q> \quad (41)$$

where the sum goes over all k's such that k is below the Fermi surface and $k+Q$ is above. In this subspace the Hamiltonian matrix is

$$\hat{H}_{\text{1e-h}} = \begin{pmatrix} \epsilon_{k_1+Q} - \epsilon_{k_1} + v(Q) & v(Q) & \cdots \\ v(Q) & \epsilon_{k_2+Q} - \epsilon_{k_2} + v(Q) & \cdots \\ \vdots & \vdots & \end{pmatrix} \quad (42)$$

The kinetic energy of the pair $(k, k+Q)$ appears on the diagonal, and the Coulomb interaction $v(Q) = 4\pi e^2/Q^2$ couples each pair to every other pair. The exchange term $-v(k-k')$ also couples pairs to each other, provided their spins are the same, but this is a smaller and a complicating factor which is omitted in first approximation.

We look for eigenstates of this truncated Hamiltonian. This proceedure is called the Tamm-Dancoff approximation by nuclear physicists [1], and is an example of what chemists call a "configuration interaction" calculation. The eigenfunctions obey

$$\hat{H}_{1\text{e-h}}|A>=\lambda|A> \tag{43}$$

which is equivalent to

$$(\epsilon_{k+Q}-\epsilon_k-\lambda)A_k=-v(Q)\sum_{k'}A_{k'} \tag{44}$$

Solving for A_k =const $\times v(Q)/(\epsilon_{k+Q}-\epsilon_k-\lambda)$, we compute $\sum_k A_k$ and find the self-consistency condition

$$1=-v(Q)\sum_k \frac{f_k(1-f_{k+Q})}{\epsilon_{k+Q}-\epsilon_k-\lambda} \tag{45}$$

This is equivalent to the secular equation $\det(\hat{H}_{1\text{e-h}}-\lambda\hat{1})=0$.

The nature of this equation can be appreciated by considering the form of the right hand side for a small system with a discrete spectrum of e-h pairs of energy $\epsilon_{k+Q}-\epsilon_k$. Between every adjacent two pair energies, a root λ of Eqn. (45) is found, but the largest root $\lambda=\hbar\omega_{\text{TD}}$ is split off to higher energy as illustrated in Fig. 5. This root is the collective electronic excitation in Tamm-Dancoff approximation. Its wavefunction is

$$|\Psi>=\sum_k \frac{Cv(Q)}{\hbar\omega_{\text{TD}}-(\epsilon_{k+Q}-\epsilon_k)}|k,k+Q> \tag{46}$$

Because $\hbar\omega_{\text{TD}}$ lies above the energy of the pair spectrum, all the coefficients in Eqn. (46) are positive. This coherent sum has a large oscillating charge at $t\approx 0$, and looks quite a lot like a plasmon. The other eigenstates of Eqn. (42) are orthogonal to (46), which means that at $t\approx 0$ the charge must largely cancel out. These other states account for the persistence of non-interacting properties in the e-h pair spectrum of the electron gas with Coulomb interactions.

Unfortunately, the Tamm-Dancoff approximation, as is well-known in nuclear physics, does not yield an accurate answer for the collective mode spectrum. The RPA answer, which is apparently surprisingly accurate, is equivalent to a modified secular equation

$$1=-v(Q)\sum_k \frac{f_k(1-f_{k+Q})-f_{k+Q}(1-f_k)}{\epsilon_{k+Q}-\epsilon_k-\lambda} \tag{47}$$

which is found by setting the real part of the RPA dielectric function to zero. Eqn. (45) is tantalizingly close to Eqn. (47). However, the additional term which occurs in Eqn. (47), having the factor $f_{k+Q}(1-f_k)$ in the numerator, is quite foreign to the Tamm-Dancoff approximation, because it seems to refer to states $|k,k+Q>$ where the hole state k is above the Fermi surface and the electron state $k+Q$ is below.

The pair creation operator $c^{\dagger}_{k+Q}c_k$ operating on the non-interacting ground state cannot create anything if k is above the Fermi surface and $k+Q$ below, but if it acts on the interacting ground state with vaccuum fluctuations included, then the pair creation operator can destroy a pair of net momentum $-Q$ by destroying a virtual electron in state k and a virtual hole in state $k+Q$. This effectively releases a pair state in some other state $(k', k'+Q)$ as shown in Fig. 4, which had previously been part of the same vaccuum fluctuation as the destroyed pair.

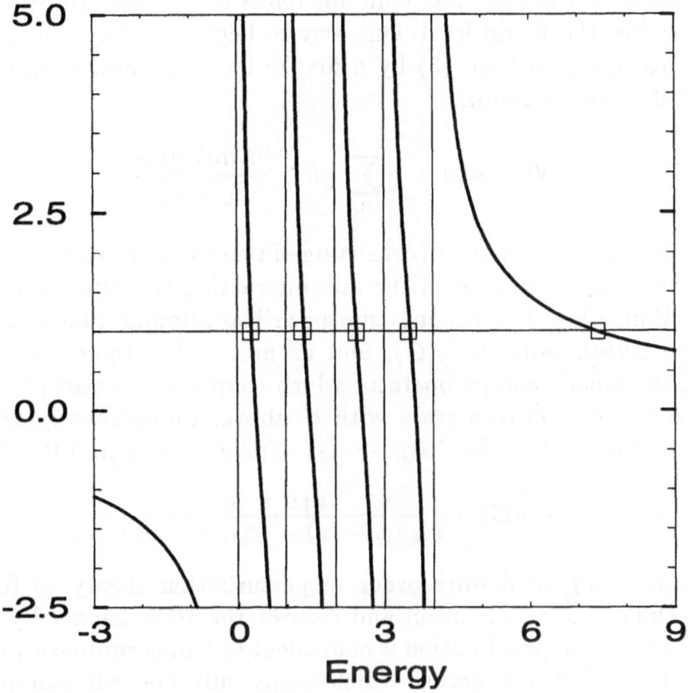

Fig. 5. **Finding the roots of the secular equation. The curve is the right hand side of Eqn. (45). The vertical lines mark the energies of e-h pair states, with a root pinned between each two neighboring e-h states. One root (which becomes the plasmon) splits off to higher energy.**

In order to produce the modified secular Eqn. (47) in place of Eqn. (45) from our wavefunction argument, we need an enlarged subspace that includes e-h pairs where the hole is above and the electron below the Fermi level. The kinetic energy part of the Hamiltonian will have the same form $\epsilon_{k+Q} - \epsilon_k$ on the diagonal in both the previous and the new parts of the subspace, whereas the potential energy term must have the form

$$\hat{V}_{e-h} = \left(\begin{array}{ccc|ccc} v(Q) & v(Q) & \cdots & -v(Q) & -v(Q) & \cdots \\ v(Q) & v(Q) & \cdots & -v(Q) & -v(Q) & \cdots \\ \vdots & \vdots & & \vdots & \vdots & \\ \hline -v(Q) & -v(Q) & \cdots & v(Q) & v(Q) & \cdots \\ -v(Q) & -v(Q) & \cdots & v(Q) & v(Q) & \cdots \\ \vdots & \vdots & & \vdots & \vdots & \end{array} \right) \tag{48}$$

where the second set of rows and columns refers to the new states with k above and $k + Q$ below the Fermi level. One way to begin building this is to redefine the pair states given in Eqn. (3) by a first-order perturbative improvement of the ground state or vaccuum,

$$|0> \rightarrow |0> + \sum_m |m> \frac{< m|\hat{V}|0 >}{E_0 - E_m} \tag{49}$$

where $|m>$ is short for any state (consisting of two e-h pairs of opposite momenta $\boldsymbol{Q}'$, -$\boldsymbol{Q}'$) which can be created in the non-interacting vaccuum by the Coulomb interaction Eqn. (20). The original subspace Hamiltonian matrix Eqn. (42) is preserved to zeroth order in $v(Q)$, and to first order, there are new matrix elements of the kinetic energy operator which couple an e-h pair $(k, k+Q)$ with k below the Fermi level to a state with k' above. Unfortunately, the coupling matrix element is not just the term $-v(Q)$ appearing in Eqn. (48), but instead

$$- v(Q) \times \frac{\epsilon_{k+Q} - \epsilon_k}{(\epsilon_{k+Q} - \epsilon_k) - (\epsilon_{k'+Q} - \epsilon_{k'})} \tag{50}$$

It will require going to infinite order in perturbation theory to fully correct the Tamm-Dancoff approximation and recover the RPA answer by this route. The Tamm-Dancoff approximation is equivalent to a diagrammatic perturbation theory for the dielectric function which keeps only one e-h pair propagating forward in time, and neglects the backward-in-time propagation which enters the usual Dyson series when the vaccuum fluctuations are included properly in the same order of perturbation theory. The full RPA treatment can have arbitrarily many additional vaccuum fluctuations with wavevectors $(\boldsymbol{Q}, -\boldsymbol{Q})$, but our approximate improvement allows at most one vaccuum fluctuation term. It might be nice to find explicitly the formula for the plasmon wavefunction which corresponds properly to the RPA formula, but I have not done this. The subject is treated in texts [1], [3] on the nuclear many-body problem.

8 Conclusion

The apparent paradoxes now seem largely answered. The answers did not involve any new physics, but instead required thinking about the problem in a way familiar to nuclear physicists but less familiar to solid state physicists. An e-h pair, even before adding the Coulomb interaction, is seen to carry a charge density wave, once the interference between the pair and the ground state is considered in the wavefunction. The Coulomb interaction totally alters the charge-carrying part of this state, but does not affect the transverse part of the current. In the effort to find approximate eigenstates of the interacting problem, the Tamm-Dancoff proceedure, even though ultimately inaccurate, offers a nice way of seeing how a single plasmon-like mode can split off from the e-h continuum, leaving the rest of the continuum largely unaltered.

When this paper was presented in Karpacz, C. P. Enz commented that his instinct was that there was still more to this subject than had been uncovered so far. This seems to me true of all enquiries. Unfortunately I am not able to suggest the next questions which should be asked.

Acknowledgments

I thank G. E. Brown, M. Cardona, V. J. Emery, A. D. Jackson, and especially J. K. Jain for helpful discussions. This work was supported in part by NSF grant no. DMR-9417755. I thank Prof. R. Car for hospitality at IRRMA in Lausanne, where the first half of this work was done.

References

1. A particularly complete treatment of the RPA and plasmons is given by A.L. Fetter and J.D. Walecka, *Quantum Theory of Many-Particle Systems* (McGraw-Hill, New York, 1971).
2. G. Contreras, A.K. Sood, and M. Cardona, Phys. Rev. **B32**, 924-9 (1985); Phys. Rev. **B32**, 930-3 (1985).
3. G.E. Brown, *Unified Theory of Nuclear Models and Forces*, 3rd Ed. (North-Holland, Amsterdam, 1971) p. 42.
4. L. Tonks and I. Langmuir, Phys. Rev. **33**, 195-210 (1929).
5. G. Ruthemann, Ann. Physik **2**, 113-34 (1948).
6. W. Lang, Optik **3**, 233-246 (1948).
7. J.C. Slater, *Insulators, Semiconductors, and Metals* (McGraw Hill, New York, 1967) pp.119-125.
8. D. Pines, Rev. Mod. Phys. **28**, 184 (1956).
9. E.D. Isaacs and P. Platzman, Physics Today, February 1996, pp.40-45.
10. A. vom Felde, J. Sprösser-Prou, and J. Fink, Phys. Rev. **B40**, 10181 (1989).
11. P. Nozières, *Theory of Interacting Fermi Systems* (W. A. Benjamin, New York, 1964).

Review of the Physics of High-Temperature Superconductors

Charles P. Enz

Départment de Physique Théorique, Université de Genève
1211 Geneva 4, Switzerland

Abstract: For a better appreciation of the revolutionary aspect of the discovery in 1986 of the first high-temperature superconductor, the main stations in the 80 years old history of superconductivity are revisited. It is emphasized that the first breakthrough in this history came only in its 5th decade when Bardeen, Cooper and Schrieffer succeded in formulating a microscopic theory of this phenomenon. The progress made in the following 3 decades then is described mentioning the names of Little, Ginzburg, Bednorz and Müller who were among the few physicists who had the optimism to consider another breakthrough possible. After Bednorz and Müller's discovery, progress leaped forward and produced an avalanche of publications which can be faced only by being selective. Thus after presenting the main characteristics of the different cuprate families of the new superconductors the main theoretical models are presented, which all make explicit use of the fact that the strongly correlated mobile holes in the copper-oxygen layers carry most of the interesting physics. Some of the new ideas like the separation of spin and charge of the holes and the possible existence of a "spin liquid" instead of a Fermi liquid are considered. Also discussed are the main pairing mechanisms proposed to describe the superconducting state and, in particular, the alternative of s- versus d-wave symmetry. Finally, possible explanations of the strange linear resistivity as function of the temperature observed parallel to the layers in the normal state and the technical applications in the form of wires, squids and filters for medicine and the communications industry are reviewed.

1 The early history of superconductivity

In 1908 Kamerlingh Onnes succeeded in liquefying helium at a temperature of 4.2 degrees Kelvin (K), that is at $-268.9C$. Three years later he discovered that at this temperature mercury completely lost its electrical resistance [1]. In the course of the years many other metals were discovered to be superconductors namely, by increasing atomic number, Al; Ti, V, Zn, Ga; Zr, Nb, Mo, Tc, Ru, Cd, In, Sn; La; Hf, Ta, W, Re, Os, Ir, Hg, Tl, Pb; Th, Pa, U.

It turned out, however, that zero resistance is not a sufficient condition for superconductivity. In 1933 Meissner and Ochsenfeld discovered that a magnetic field H is expelled from a metal when it is in the superconducting state, i.e. when its temperature T is below the *critical temperature* T_c, provided that H is also smaller than a critical field H_c. The two situations of expulsion are shown in Fig.1 [1].

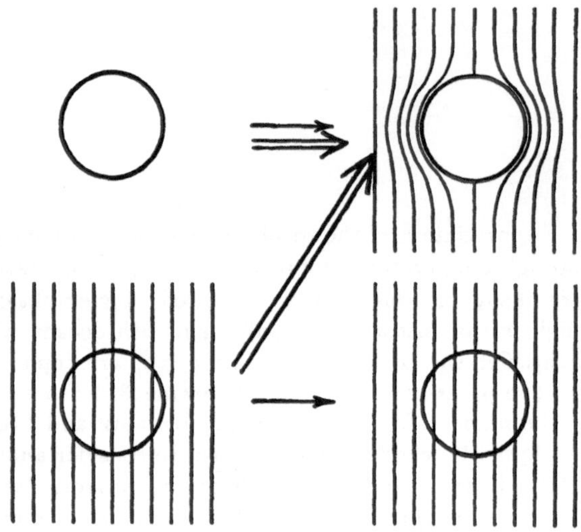

Fig. 1. Field expulsion from a superconductor. Above: $T < T_c$ and $H = 0 \rightarrow H_c > H > 0$. Below: $H_c > H > 0$ and $T > T_c \rightarrow T < T_c$. $\longrightarrow$: perfect conductor; $\Longrightarrow$: superconductor.

The explanation of this "Meissner effect" is that an external magnetic field $H < H_c$ induces a surface current distribution j perpendicular to the field. j, in turn produces a field that exactly compensates H in the interior, i.e. $\boldsymbol{B} = 0$, leading to perfect diamagnetism, $\chi = -(4\pi)^{-1}$. The existence of a *critical field* H_c is then understood to be due to a critical value of the surface current density, above which superconductivity breaks down.

While phenomenological understanding increased steadily [1], the breakthrough in the microscopic theory came only in 1957 with the famous work of Bardeen, Cooper and Schrieffer (BCS) [2]. The idea of the BCS theory is that electrons form pairs with zero momentum (and also zero angular momentum), and these "Cooper pairs" condense at T_c into the superconducting state. This condensation of Cooper pairs is analogous to the Bose-Einstein condensation of "bosons" (see below), e.g. helium atoms [1]. The difference, however, is that in BCS theory condensation occurs simultanously with pairing and the diameter of the pair, i.e. the *coherence length* ξ, is large compared to the average distance between electrons.

Pairing requires an attractive force between the electrons which in BCS theory is mediated by the interaction of the electrons with the vibrations of the crystal lattice, the phonons [see, e.g. Section 1 of Ref. [3]]. The resulting effective electron-electron interaction V determines the critical temperature according to the equation [2]

$$T_c = 1.14\omega e^{-1/N(0)V}. \tag{1}$$

Here ω is a limiting phonon frequency (Debye frequency) and $N(0)$ the density of states at zero energy, counted from the filling level (Fermi energy) of the electrons. T_c and ω are measured in energy units.

Since phonon frequencies vary with the ionic mass M as $M^{-1/2}$, Eq. (1) gives the same M-dependence for T_c. As noted by BCS, this explains the isotope effect which had been discovered 7 years earlier [4]. It was soon realized, however, that the M-dependence of T_c is considerably more complicated because not only ω in Eq. (1) but also V depend on the ionic mass. This latter dependence essentially comes from the Coulomb repulsion which was not mentioned above while the part of V mediated by the phonons is practically independent of M.

Pair breaking requires an energy 2Δ where $\Delta(T)$ is the *order parameter* which depends on the temperature, and $\Delta_o = \Delta(0)$ is the *gap* in the excitation spectrum. BCS theory gives the following relations for $\Delta(T)$ near T_c and at $T = 0$:

$$\Delta(T) = 3.06T_c(1 - T/T_c)^{1/2}, \Delta_o = 1.76T_c. \tag{2}$$

2 The search for new superconductors

The obvious interest of superconductivity, namely to allow electrical currents without loss, as well as the intrinsic interest led to a competitive search for materials with ever higher T_c's and/or higher H_c's. After the elements many alloys were tested, among which NbTi became of particular importance for cryogenic applications. But besides alloys, interest turned to compounds with precise stoichiometry, first binaries then ternaries and quaternaries and beyond.

The most important binary superconductors were the so-called A15 compounds discovered in 1954, namely V_3Si with a T_c of $17K$ and Nb_3Sn which has $T_c = 18K$ [5]. Apart from the mentioned alloy $NbTi$, Nb_3Sn has been the main material from which wires were produced. Nb_3Ge which had the highest T_c of $23.2K$ until the recent discovery of the "high-T_c superconductors" (see below), had proven too difficult to be produced on an industrial scale.

Two classes of ternary compounds were subsequently found to contain some interesting superconductors among their members, first the "Chevrel phases" named after the French discoverer of this class. The first superconductors in this class were discovered by Matthias in 1972 [6], namely $M_xMo_6S_8$ with $M = Cu, Ag, Zn, Sn, Pb$. While the T_c's are all lower than $15K$, some of the Chevrel superconductors such as $PbMo_6S_8$ have the highest H_c's, as is seen in Fig. 2 [7] which reviews the materials discussed so far.

The *upper critical field* H_{c2} refered to in Fig. 2 is the field at which superconductivity breaks down. However, already at a *lower critical field* $H_{c1} < H_{c2}$

non-superconducting cylindrical regions in the field direction start to penetrate into the superconductor forming a *mixed state*. This behaviour is characteristic of *type II* superconductors for which the coherence length ξ is smaller than a critical value [see, e.g. Section 27 of Ref. [3]].

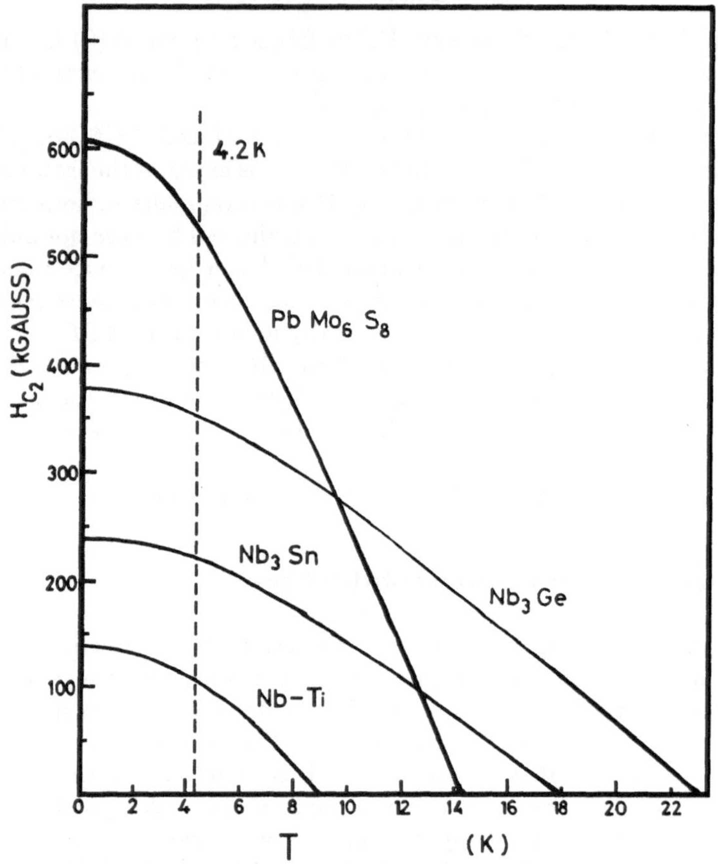

Fig. 2. Upper critical field H_{c2} versus T for the most interesting "old" superconductors.

The second class of ternaries are the "heavy-fermion compounds", so called because they have a linear electronic specific heat at low temperatures [3]

$$c_v = \gamma T \qquad (3)$$

with a γ-value 100 to 1000 times larger than for other metals, indicating an anomalously large effective mass of the conduction electrons. The first superconducting heavy fermion compound $CeCu_2Si_2$ was discovered in 1979 [8]. However, it has a T_c of only about $0.5K$.

A quite different type of superconductor was discovered in 1980, namely an organic compound of quasi-linear structure with the abbreviated name of

$(TMTSF)_2PF_6$ [9]. Although this substance becomes superconducting only under high pressure, some $12kbar$, and with a T_c of about $1K$, this discovery showed for the first time that superconductivity is not restricted to ordinary metals. This is all the more important in view of the "high-T_c superconductors" discovered since and also because many other quasi-linear organic substances have been found to be superconducting at atmospheric pressure.

The suggestion that organic molecules could be superconducting, however, is much older. Indeed, in 1964 Little [10] discussed a model molecule consisting of a spine to which many side chains are attached. Based on BCS theory he concluded that such a system had a phonon spectrum allowing T_c's much higher then room temperature! Unfortunately, his work did not obtain the attention it is retrospectively thought to have deserved.

During three decades the barrier of $T_c \sim 23K$ had not been broken, in spite of considerable efforts worldwide. This barrier thus became psychological, and various theoretical attempts were made to make plausible that there was an intrinsic limit. The history of this question is discussed in the books by V.L. Ginzburg [11] who emphasized, however, that theory allowed for much higher values of T_c. He therefore should be considered the pioneer of high-temperature superconductivity.

But it still came as a shock when in 1986 J. G. Bednorz and K. A. Müller from the IBM Zurich laboratory announced the discovery of a $Ba-La-Cu-O$ compound with a T_c of some $30K$ [12], close enough to the barrier to be credible! The next decisive step came less than a year later when the group of C. W. Chu from the University of Houston (Texas) and of M. K. Wu from the University of Alabama discovered the $Y-Ba-Cu-O$ compound with $T_c \sim 93K$ [13]. This meant superconductivity above the boiling point of nitrogen ($-195.8C$) which now could be substituted for the expensive helium as a coolant.

3 The new "high-temperature" superconductors

The new materials discovered by Bednorz and Müller [12] and by Chu and his group [13] are quaternaries whose crystal structure is built up from unit cells which are elongated along the c-axis and have an essentially quadratic base in the $a-b$ plane perpendicular to the c-axis. The first class consists of the lanthanum compounds $La_{2-x}M_xCuO_4$ or $LSCO$ with variable impurity content $x \leq 0.15$ and $M = Ba, Sr$, while the second is $YBa_2Cu_3O_{6+x}$ or $YBCO$ with variable oxygen content $x \leq 1$. Many more have been discovered since [14, 15, 16]. All have as a basic structure pairs of pyramids pointing in the positive and negative c-direction with an O-atom in each corner and a Cu-atom at the center of their base. In some cases (the La-compound and the Ca-compounds with $n = 1$ in the chemical formulae (4) below) the two pyramids touch to form an octahedron with one Cu in the center (see Figs. 3 and 4).

The following is a non-exhaustive list of the *cuprate families* whose outstanding common feature are the CuO_2-layers in the $a-b$ planes [14, 15, 16]:

$$\begin{aligned}
&La_{2-x}M_xCuO_4\ ; M = Sr, Ba, Ca\ ; T_c \leq 35K \\
&YBa_2Cu_3O_{6+x}\ ; T_c \leq 95K \\
&Bi_2Sr_2Ca_{n-1}Cu_nO_{4+2n+x}\ ; n = 1,2,3,4\ ; T_c \leq 110K \\
&TlBa_2Ca_{n-1}Cu_nO_{3+2n+x}\ ; n = 1,2,3,4\ ; T_c \leq 125K \\
&HgBa_2Ca_{n-1}Cu_nO_{2+2n+x}\ ; n = 1,2,3\ ; T_c \leq 135K \\
&R_{2-x-y}Ce_xSr_yCuO_4\ ; R = Nd, Pr, Gd, Sm; T_c \leq 24K\ .
\end{aligned} \tag{4}$$

All these families are hole conductors, with the exception of the last one which for $y = 0$ is an electron conductor.

As shown, e.g. in Fig. 1 of A.W. Hewat et al. on p. 221 of Ref. [17], T_c increases with the number n of CuO_2-layers per unit cell. This dependence has been explained in the framework of Ginzburg-Landau theory [see, e.g. Section 26 of Ref. [3]] as a relation $T_c \propto (n/c)^{2/3}$ where c is the lattice constant along the c-axis [see T. Schneider on p. 351 of Ref. [17]].

In many materials there is a structural phase transittemperature far above T_c and which has no obvious connection with superconductivity. In the lanthanum compounds the transition, with decreasing temperature, is from tetragonal to orthorhombic, as shown in Fig. 3 [18], in $YBCO$ it is the reverse, as seen in Fig. 4 [19].

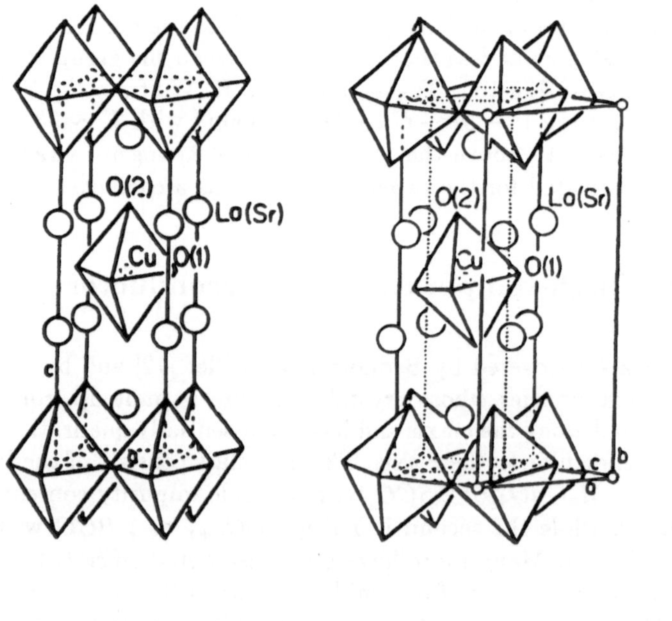

Fig. 3. Unit cells of tetragonal (left) and orthorhombic (right) $La_{1.85}Sr_{0.15}CuO_4$.

Since $Cu^+ \sim 3d^{10}$ and $O^{2-} \sim 2p^6$ are closed atomic shells one would expect the undoped compounds La_2CuO_4 and $YBa_2Cu_3O_6$ to be insulators. However,

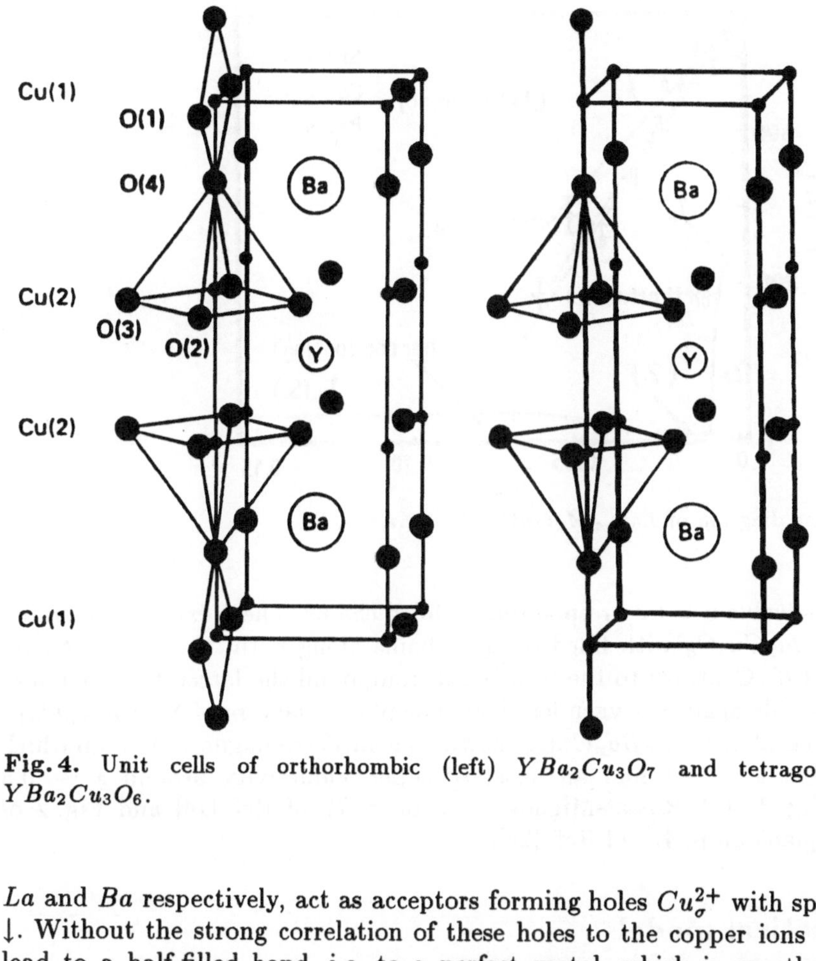

Fig. 4. Unit cells of orthorhombic (left) $YBa_2Cu_3O_7$ and tetragonal (right) $YBa_2Cu_3O_6$.

La and *Ba* respectively, act as acceptors forming holes Cu^{2+}_{σ} with spin $\sigma =\uparrow$ or $\downarrow$. Without the strong correlation of these holes to the copper ions this would lead to a half-filled band, i.e. to a perfect metal, which is exactly what the first band calculations predicted [15, 20]. Experimentally however, the undoped compounds are indeed found to be insulators. Since they cannot be ordinary band-gap insulators they must be insulators by correlation, i.e. *Mott insulators* [3]. In addition, it is observed that at low temperature the spins of the copper holes show a 3-dimensional *antiferromagnetic (AF) order* [21].

Free carriers are created either by doping, indicated by the index x in the chemical formulae of (4) or by oxygen disorder which implies changes in the valence of the metal ions [see the conclusion of A.W. Hewat et al. on p. 224 of Ref. [17]]. Doping gives rise to a *phase diagram* as shown for the *La*-compound in Fig. 5 [22] [see also Fig. 2 of K. Kitazawa on p. 202 and in Fig. 1 of M. Kataoka on p. 357 of Ref. [17]].

For $La_{2-x}Ba_xCuO_4$ this means, e.g. that the antiferromagnetism of the undoped system rapidly disappears at $x \sim 0.01$ and, after a spin-glass-type intermediary phase, is followed at $x \sim 0.02$ by superconductivity. $YBa_2Cu_3O_{6+x}$, on the other hand, contains chains of oxygens along the b-axis which act as

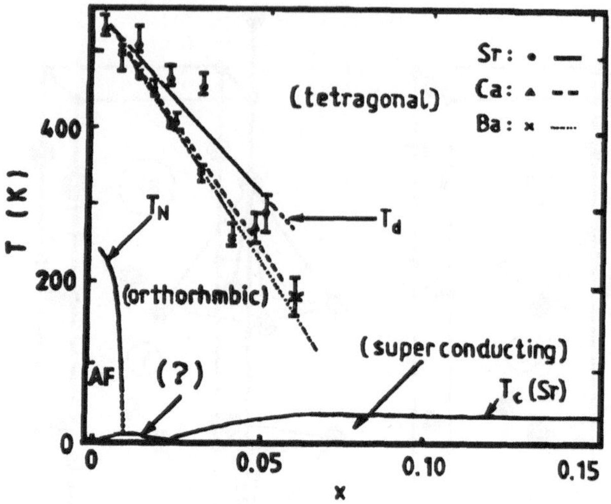

Fig. 5. Phase diagram of $La_{2-x}M_xCuO_4$, $M = Sr, Ca, Ba$.

dopants, the value $x = 1$ corresponding to filled chains. There exists also a modification $YBa_2Cu_4O_8$ with filled oxygen chains along both, the a- and b-axes and $T_c \sim 76K$. Contrary to the former 123-compound the latter 124-version of $YBCO$ is stable against oxygen loss [23]. The phase diagram of $YBa_2Cu_3O_{6+x}$ is similar to that of $La_{2-x}Ba_xCuO_4$ showing an antiferromagnetic domain which disappears at $x \sim 0.4$ and an onset of superconductivity around $x \sim 0.5$ [see, e.g. Fig. 3 of J. Rossat-Mignot et al. on p. 77 of Ref. [24] and Fig. 2 of J.M. Tranquada on p. 424 of Ref. [25]].

4 Theoretical models

The antiferromagnetic state mentioned earlier means that the spins of Cu_σ^{2+} in the square lattice of the $a-b$ planes alternate, as shown in Fig. 6, and there is also an alternation in the consecutive $a-b$ planes [21].

The CuO_2-planes may then be described by a one-band *Hubbard model* [see, e.g. Section 33 of Ref. [3]]

$$\mathcal{H}_H = -t \sum_{<ij>,\sigma} c_{i\sigma}^+ c_{j\sigma} + U \sum_i n_{i\uparrow} n_{i\downarrow} \tag{5}$$

where $n_{i\sigma} \equiv c_{i\sigma}^+ c_{i\sigma}$ and $< ij >$ means nearest neighbour sites i, j. Strong on-site repulsion, $U \gg t$, then splits the half-filled metallic band into the filled lower and the empty upper Hubbard bands [see Fig. 5 of K. Kitazawa on p. 204 of Ref. [17]] and the metal becomes an insulator with no double occupancy.

At half-filling, charge neutrality then implies that the charged holes Cu_σ^{2+} on all the sites i are replaced by neutral spins. Thus the antiferromagnetism

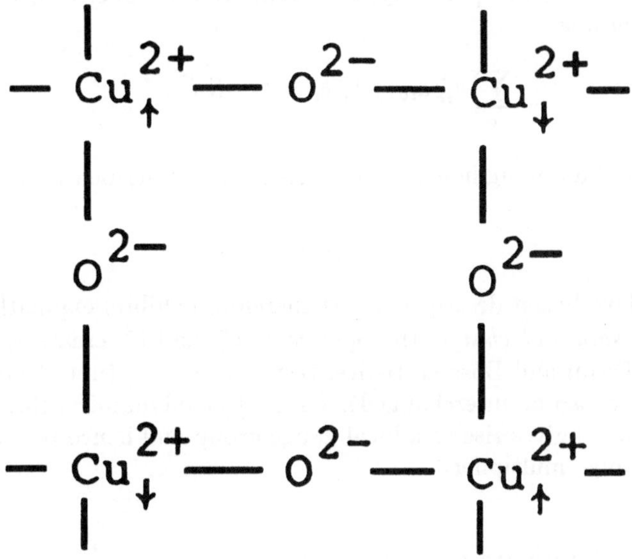

Fig. 6. Antiferromagnetic spin arrangement of the $Cu-O$ lattice in the $a-b$ planes.

in the CuO_2-planes (see Fig. 6) may be described by a 2-dimensional spin-1/2 Heisenberg model

$$\mathcal{H}_S = J \sum_{<ij>} \boldsymbol{\sigma}_i \cdot \boldsymbol{\sigma}_j, \tag{6}$$

where $\boldsymbol{\sigma}$ is the vector of Pauli matrices and $J > 0$. According to the Mermin-Wagner theorem this model cannot order antiferromagnetically at finite temperature [26] so the groundstate must have spin-flip fluctuations. This led Anderson to propose the *resonating valence bond* (RVB) groundstate in which all spins are randomly paired into singlets but such that on average the antiferromagnetic order is maintained [Ref. [27], see also Ref. [28]]. However, the existence of a true Néel groundstate is still under debate.

Doping has the effect of adding a hole on some sites i which nominally transforms a Cu_σ^{2+} into a singlet Cu^{3+}, i.e. the spin is replaced by a charge. Physically the doping holes, instead of being localized on the copper sites, will rather be distributed on the four neighbouring oxygens [29]. In the limit $U \gg t$ the Hubbard repulsion in Eq. (5) then is equivalent to the Heisenberg term Eq. (6) with $J = 4t^2/U$, thus giving rise to the $t-J$ model [29, 30]

$$\mathcal{H}_{tJ} = -t \sum_{<ij>,\sigma} c_{i\sigma}^{+} c_{j\sigma} + J \sum_{<ij>} \boldsymbol{\sigma}_i \cdot \boldsymbol{\sigma}_j \ . \tag{7}$$

In this model the limit $U \gg t$ now implies that *all* the sites i are occupied either by a spin or by a charge. Describing the creation of a spin or a charge at site

i by operators $a_{i\sigma}^+$ or b_i^+, respectively, the occupation of all sites gives rise to a *local conservation law*

$$\sum_\sigma a_{i\sigma}^+ a_{i\sigma} + b_i^+ b_i = 1 \ ; all\ i \tag{8}$$

and the creation of a doping hole on a charge-site i is described by the operator

$$c_{i\sigma}^+ = a_{i\sigma}^+ b_i \ . \tag{9}$$

This so-called slave-boson description [31] therefore exhibits explicitly the *separation between spin and charge*: the operators $a_{i\sigma}^+$ and b_i^+ create *spinons* and *holons* obeying Fermi and Bose statistics, respectively [27] (note that formally, the statistics may also be interchanged). As in electrodynamics, this local conservation law Eq. (8) gives rise to a local gauge group and hence to a *gauge field* acting as a Lagrange multiplier.

5 Fermi liquid versus spin liquid

In the above description using spinons and holons, there are no longer any holes and hence no Fermi liquid of holes but instead there is a *spin liquid* [Ref. [27], see also P.W. Anderson in Ref. [32]]. This is a fundamental change in the description which must be tested experimentally.

The most stringent test is provided by high-resolution angle-resolved photoemission in the vicinity of the Fermi energy [15, 33, 34]. Such measurements were done on cleaved surfaces of single crystals of $Bi_2Sr_2CaCu_2O_8$ (this compound is known for its stability against oxygen loss). The results are consistent with band structure calculations and strongly suggest the existence of a Fermi liquid. In addition, comparison of the data taken above and below the transition temperature T_c clearly show the opening of a gap of the order $\Delta \simeq 4k_BT_c$ as compared to the weak-coupling (BCS) value $1.76k_BT_c$ [33, 34].

Nuclear magnetic resonance, on the other hand, indicates that the Fermi liquid still possesses antiferromagnetic correlations [35]. Furthermore, optical and transport properties in the normal state show universal anomalies which may be fitted by a particular form of the polarizability. This semi-phenomenological description goes under the name of "marginal" Fermi liquid [36].

On the other hand, the rich symmetry content of the Hubbard model has been argued to provide a sensitive test of the validity of this model for the description of the cuprates [37]. No experimental realisation of this test is known, however.

Alternate theoretical models either assume mobile Cu-holes in an AF background hopping via the O-neighbours [38]-[41] or mobile O-holes hopping via neighbouring Cu-sites [42, 43] or band holes interacting with the commensurate *spin density wave (SDW)* [3] describing the AF spin arrangement shown in Fig. 6 [44]. In all of these models superconductivity requires a pairing mechanism in order to obtain boson-like objects from the holes which are fermions.

6 The superconducting state

The most striking feature of the new superconductors, apart from the high T_c-values are the *short coherence lengths* $\xi_a/a \sim \xi_b/b \sim 6$, $\xi_c/c \sim 0.2$ [Ref. [45] and G. Deutscher in Ref. [17]]. This means that the hole pairs are rather localized, particularly along the c-axis, i.e. they resemble more Schafroth pairs (local bosons) than extended Cooper pairs [see Section 20 of Ref. [3]]. Hence, these superconductors are strongly type II (see above) and resemble a Bose fluid. The latter similarity is manifest in their scaling properties and in the fact that at $T = 0$, T_c is a function of the carrier density n [46]. Such a $T_c - n$ relation had been derived in an effective extended Hubbard model with on-site ("negative U") and intersite attraction in mean-field approximation [47] but also in a boson exchange model in strong-coupling approximation [41].

In the models based on the assumption of hopping Cu-holes, two different pairing mechanisms have been proposed. In the first, the AF background acts like a "spin bag" which preferentially attracts two holes with opposite spin on nearest-neighbour sites and thus leads to pairing. In this case superconductivity is obtained by a conventional BCS formalism [38, 39].

The second type of model makes use of the fluctuating valence reaction [19, 40, 41]

$$Cu^{3+}_{\uparrow\downarrow} - O^{2-} - Cu^{3+}_{\uparrow\downarrow} \Rightarrow Cu^{2+}_{\uparrow} - O^{o}_{\uparrow\downarrow} - Cu^{2+}_{\downarrow} \tag{10}$$

which also explains the oxygen loss mechanism because of the neutral oxygen. This reaction leads to a coupling between the two holes on the Cu^{3+} and the two holes on O^o. Approximating the latter by a doubly charged spin-zero boson, the zero-momentum projection of this interaction has the form [40, 41]

$$\mathcal{H}_{int} = W \sum_{\boldsymbol{k}} a^{+}_{\boldsymbol{k}\uparrow} a^{+}_{-\boldsymbol{k}\downarrow} b + herm.conj\,. \tag{11}$$

Here $a^{+}_{\boldsymbol{k}\sigma}$ is the creation operator for a hole of momentum $\boldsymbol{k}$ and b the annihilation operator for the doubly charged boson on the O-site. The coupling constant W may be expressed in terms of an extended Hubbard model in which the O-ions are explicitly taken into account. Treated in the strong coupling approximation this model yields a T_c which, in contradistinction to most of the other proposals, does not have the exponential upper bound inherent in all of the expressions of the type of Eq. (1) [41].

In Emery's model [42, 43], pairing of the hopping O-holes is obtained from the attraction resulting from the exchange of a hole between two neighbouring O-ions via the intermediate Cu site which here is assumed to be delocalized [42]. This process is in a certain sense the reverse of the fluctuating valence reaction Eq. (10). It may indeed be written as the 2-step process

$$O^{2-} - Cu^{2+}_{\downarrow} - O^{-}_{\uparrow} \Rightarrow O^{2-} - Cu^{3+}_{\uparrow\downarrow} - O^{2-} \Rightarrow O^{-}_{\sigma} - Cu^{2+}_{-\sigma} - O^{2-} \tag{12}$$

where, however, Cu^{3+} is supposed to be delocalized [29]. Again the coupling constant for this reaction follows from an extended Hubbard model [42]. But superconductivity is derived in the framework of conventional BCS theory.

Pairing of the band-holes in Schrieffer's model is again based on the mechanism of spin bag attraction [44]: An added hole interacts with the SDW mentioned earlier by creating its own spin bag which then attracts a second hole. Again superconductivity is derived along the lines of BCS.

In view of the still open question of the mechanism of superconductivity in the cuprates, the information that the superfluid phase consists of carrier *pairs* is most important. The detection of trapped flux in a ring of $YBCO$ has clearly demonstrated quantization in units of $h/2e$ [see Fig. 1 of C.E. Gough on p. 145 of Ref. [25]] thus demonstrating the existence of pairs.

7 s- versus d-wave pairing

More recently, indications have been accumulating that the energy gap may depend on the wavevector in the $a-b$ plane, $\Delta(k_x, k_y)$. Such a dependence was predicted by Bickers, Scalapino and collaborators already in 1987 [48]. Their result was that the holes pair in a *d-state*, and not as in BCS theory in an *s-state*, the explicit dependence being

$$\Delta(k_x, k_y) = \Delta_o[\cos(k_x a) - \cos(k_y a)] \ . \tag{13}$$

This expression, which changes sign at the nodal lines $k_y \pm k_x = 0$ is represented in Fig. 7. An interesting pairing mechanism leading to an order parameter with d-wave symmetry is the *exchange of antiferromagnetic spin fluctuations* [49].

Experimental hints of an anisotropic Δ are numerous but contradictory since some experiments confirm the d-state dependence Eq. (13), while others favour of an anisotropic s-state pairing without nodal lines [50, 51]. Because of this situation the idea of the simultaneous presence of both, s- and d-wave order parameters but only one transition temperature T_c seems quite natural [52]. Models with mixed s- and d-wave order parameters have also been discussed elsewhere, e.g. in Refs. [53] and [54].

An alternative to the paramagnon exchange mechanism of Ref. [49] may be obtained as follows: The short coherence lengths and the proximity of a Mott transition suggest the picture of hole pairs in real space (CuO_2-planes) described by a pair wavefunction of the size of the coherence length, that is, essentially, by considering only nearest neighbour pairings which, however, must respect the square symmetry of the CuO_2-lattice. This situation corresponds to the local *charge transfer mechanism* of Eq. (10) and is described by coupling the hole pair to a doubly charged local boson as in Eq. (11) but now generalized to a square-symmetric form [55].

The Fourier-transformed pair wavefunction then is found to be

$$\varphi_{\boldsymbol{k}} = \varphi_{s\boldsymbol{k}} + \varphi_{d\boldsymbol{k}} \tag{14}$$

where

$$\left.\begin{matrix} \varphi_{s\boldsymbol{k}} \\ \varphi_{d\boldsymbol{k}} \end{matrix}\right\} = \left\{\begin{matrix} \alpha_s \\ \alpha_d \end{matrix}\right\} [\cos(k_x a) \pm \cos(k_y a)] \tag{15}$$

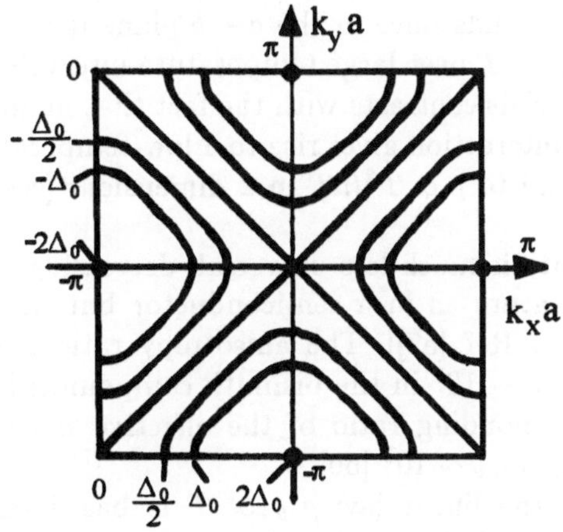

Fig. 7. Constant Δ-contours of the d-state gap function Eq. (13).

and the coefficients α_s, α_d may be chosen to be real. Writing the singlet pair operator associated with the wavefunction Eq. (14) as [see Section 20 of Ref. [3]]

$$B_q^+ = \sum_k \varphi_{k-q/2} a_{k\uparrow}^+ a_{q-k\downarrow}^+ , \tag{16}$$

the effective attraction between holes giving rise to such pairs is obtained by elimination of the local boson field in the usual way [see Section 20 of Ref. [3]],

$$\mathcal{H}_{attr} = -g \sum_q B_q^+ B_q \tag{17}$$

where $g = |W|^2/\Omega_o$ and Ω_o, the energy of the local boson, is assumed to be large compared to the band energy of the holes. As in BCS theory $\mathcal{H}_{attr}$ then determines gap equations whose solutions are the order parameters $\Delta_{\mu k} = \varphi_{\mu k}\overline{\Delta}$ ($\mu = s, d$) with s- and d-wave symmetry, and the d-wave order parameter is seen to have the form of Eq. (13). This result is self-consistent in the sense that the wavefunction defining the pair operator Eq. (16) also determines the symmetry of the order parameter.

8 The resistivity problem of the normal state

In view of the difficulty of understanding the mechanism of superconductivity in the cuprates much effort has been directed to the normal properties above T_c. While much of this activity concerns effects due to a magnetic field we here concentrate on the ordinary resistivity. For a review of normal properties of the cuprates see Ref. [56].

The temperature dependence of the resistivity is one of the puzzling normal state properties of these new materials since in the $a-b$ plane it is found universally to follow a linear law $\rho_{ab} \propto T$ over large temperature intervals above T_c [see Ref. [57] for a short review]. This contrasts with the fact that in an ordinary Fermi liquid, the carrier-carrier interaction gives rise to a low-temperature resistivity $\rho \propto T^2$ in 3 dimensions and to $\rho \propto T^2 lnT$ in 2 dimensions [see Ref. [58] and Section 17 of Ref. [3]].

Along the c-axis the resistivity is much less universal, decreasing in general strongly with increasing temperature as in a semiconductor but may also be parallel to ρ_{ab} [see, e.g. Fig. 2 of Ref. [57]]. The anisotropy ratio ρ_c/ρ_{ab} may vary between $\sim 10^2$ for $YBCO$ to $\sim 10^5$ in the bismuth compounds [see Fig. 2 of Ref. [59]] for which the corresponding ratio of the effective masses in the Ginzburg-Landau equation is $m_c/m_{ab} \sim 10^3$ [60].

The obvious explanation of the linear law $\rho_{ab} \propto T$ is based on a high-temperature expansion of the Bloch-Grüneisen law due to electron-phonon scattering [see, e.g. Section 16 of Ref. [3]],

$$\rho^{ph} = const\left(\frac{T}{\Theta}\right)^5 \int_0^{\Theta/T} \frac{x^5 dx}{e^x - 1} \, . \tag{18}$$

Here Θ is defined in terms of the Debye temperature Θ_D by the relation that $(\Theta/2\Theta_D)^3$ is the number of carriers per unit cell per spin. Indeed, for $\Theta/T \ll 1$ the integrand in Eq. (18) may be developed in powers of x and yields a linear T-dependence. Since carrier concentrations in the cuprates are small, Θ may be substantially lower than Θ_D and quite comparable with the high T_c's. The problem, however, arises in the case of the low-T_c compound of Ref. [59] where this expansion is no longer justified.

As can be seen from Eq. (18), the development of the integrand actually is a development of the Bose distribution function of the phonons,

$$n_o(\omega_q) \simeq 1 + n_o(\omega_q) \simeq \frac{k_B T}{\omega_q} - \frac{1}{2} + O\left(\frac{\omega_q}{k_B T}\right) \gg 1 \, . \tag{19}$$

It is characteristic of this development that the constant (second) term is negative and hence may simulate an unphysical residual resistivity. This is what one finds in the case of $YBa_2Cu_4O_8$ [see Fig. 4 of Ref. [23]], where a Bloch-Grüneisen curve fits perfectly [see Fig. 2 of Ref. [61]]. But the high-temperature expansion Eq. (19) is valid not only for phonons but for any zero-mass boson emitted and absorbed by the holes. Examples are spin fluctuations, photons and gauge bosons. While photon exchange is always negligible, gauge bosons, i.e. the quanta of the field associated with the local conservation law Eq. (8) are thought to be the explanation of the linear law in Refs. [62] and [63]. More realistic is the exchange of antiferromagnetic spin fluctuations which indeed gives a crossover $\rho_{ab} \propto T^\alpha$ from $\alpha > 1$ at low T to $\alpha \simeq 1$ in a large temperature interval [49, 64]. However, the most natural explanation of the linear law would seem to be in terms of the distortive electron-lattice interaction discussed in Ref. [54].

Another qualitatively correct crossover mechanism is obtained in a model of two bands separated by a small gap [65]. The contribution of the two bands to the conductivity may be written as

$$\sigma = -2e^2 \sum_{i=1,2} \tau^i \sum_{\boldsymbol{k}} (v^i_{\boldsymbol{k}})^2 f'_o(\varepsilon^i_{\boldsymbol{k}}) \ . \tag{20}$$

Taking the Fermi energy in the lower band $i = 1$, one has $|v^1_{\boldsymbol{k}}| = v_F$ but $(v^2_{\boldsymbol{k}})^2 \propto T$, corresponding to diffusion. Assuming a Fermi-liquid dependence $\tau^i \propto T^{-2}$ for both bands one then finds at low temperatures so that band 2 is empty, $\sigma \propto T^{-2}$. At higher temperatures, however, where band 2 fills up, the latter contributes a term $\propto T^{-1}$ which eventually dominates.

9 Technical applications of high-T_C superconductors

The discovery of $YBCO$ in 1987 [13] with a critical temperature of up to $95K$ brought a tremendous impulse to the technical applications of superconductivity. The reason was that it now became possible to replace expensive liquid helium (boiling point $\sim 4K$) as a coolant by liquid nitrogen whose boiling point is at $\sim 77K$. For a while the enthousiasm was so great that magnetically levitating trains and superconducting long-distance power lines were thought to be just around the corner. However, the cuprate families (4) turned out to be not only difficult to understand theoretically but also very hard to be produced in sufficiently high quality. While in the beginning the preparation of $LSCO$ and $YBCO$ was so easy that most third world laboratories were able to take part in this great adventure, the increasing demands for quality both in basic and in applied research unfortunately left laboratories with modest equipment and restricted funds out in the cold.

It was particularly hard to produce wires of these materials because the latter resemble more granular and brittle insulators than smooth metals like copper. Therefore, wires are produced today by pressing the cuprate powder on ribbons or into tubes of a metal like silver [66]. Due to the granularity of these materials, another serious problem has been the inability to obtain sufficiently high current densities, because of the resistivity between the grains.

Another problem stems from the fact already mentioned that all cuprates are strongly of type II. The consequence is an extra resistivity caused by the motion of the magnetic flux lines in the mixed state of the superconductor where the magnetic field penetrates by forming a network of these flux lines. In spite of these obstacles the production of wires for magnets and power lines is progressing steadily [66].

More spectacular is the progress in the thin film technology and in the electronics applications for communications systems, medical instrumentation and radar. An important example is provided by the so-called squids which are instruments for ultrasensitive measurements of magnetic fields and electric potentials. Here an important application is the magneto-encephalography of the brain. Today squids made of $YBCO$ have attained the quality of those fabricated with conventional superconductors. Other applications are rf or microwave filters for cellular telephones and radar. For more details see Ref. [67].

References

1. F. London, *Superfluids. Macroscopic Theory of Superconductivity* (Dover, New York 1961), vol 1.
2. J. Bardeen, L. Cooper and J. Schrieffer, Phys. Rev. **108**, 1175 (1957).
3. C.P. Enz, *A Course on Many-Body Theory Applied to Solid-State Physics* (World Scientific, Singapore, 1992).
4. E. Maxwell, Phys. Rev. **78**, 477 (1950); C. A. Reynolds et al., Phys. Rev. **78**, 487 (1950); H. Fröhlich, Phys. Rev. **79**, 845 (1950).
5. G. F. Hardy and J. K. Hulm, Phys. Rev. **93**, 1004 (1954); B. T. Matthias et al., Phys. Rev. **95**, 1435 (1954).
6. B. T. Matthias et al., Science **175**, 1465 (1972).
7. Ø. Fischer, Appl. Phys. **16**, 1 (1978).
8. F. Steglich et al., Phys. Rev. Lett. **43**, 1892 (1979).
9. D. Jérôme et al., J. Phys. Lett. **41**, 397 (1980).
10. W. A. Little, Phys. Rev. A**134**, 1416 (1964).
11. V. L. Ginzburg and D. A. Kirzhnits, eds. *High-Temperature Superconductivity*, p.1 (Consultants Bureau, New York 1982); see also *Superconductivity, Superdiamagnetism, Superfluidity*, ed. V. L. Ginzburg (MIR Publishers, Moscow 1987), p. 11.
12. J.G. Bednorz and K.A. Müller, Z. Phys. B **64**, 189 (1986). See also Nobel Lectures 1987, Rev. Mod. Phys. **60**, 585 (1988).
13. C. W. Chu et al., Phys. Rev. Lett. **58**, 908 (1987).
14. K.A. Müller, Z. Physik B**80**, 193 (1990).
15. W.E. Pickett, Rev. Mod. Phys. **61**, 433 (1989).
16. Physics Today, **46**(7), 20 (1993).
17. Proc. Oberlech Workshop on High-T_c Superconductors, ed. K.A. Müller and J.G. Bednorz, IBM J. Res. Develop. **33**, no. 3, May 1989.
18. M. François et al., Solid State Communic. **63**, 35 (1987).
19. P. P. Edwards et al., Chemistry in Britain **23**, 962.
20. L.F. Mattheiss, Phys. Rev. Lett. **58**, 1028 (1987); A.J. Freeman et al., Phys. Rev. Lett. **58**, 1035 (1987).
21. D. Vaknin et al., Phys. Rev. Lett. **58**, 2802 (1987); J.M. Tranquada et al., Phys. Rev. Lett. **60**, 156 (1988); T.R. Thurston et. al., Phys. Rev. Lett. **65**, 263 (1990).
22. M. Kato et al., Physica C**152**, 116 (1988).
23. J. Schoenes, J. Karpinski, E. Kaldis, J. Keller and P. de la Mora, Physica C **166**, 145 (1990).
24. *Int. Seminar on High Temp. Superconductivity* (Dubna, 1989), eds. V.L. Aksenov, N.N. Bogolubov and N.M. Plakida (World Scientific, Singapore, 1989).
25. *Earlier and Recent Aspects of Superconductivity* (Erice, 1989), eds. J.G. Bednorz and K.A. Müller Springer Series in Solid-State Sciences 90 (Springer, Berlin, 1990, 1991).
26. N.D. Mermin and H. Wagner, Phys. Rev. Lett. **17**, 1133, 1307 (1966).
27. P. W. Anderson, Science **235**, 1196 (1987).
28. S. Tang and J.E. Hirsch, Phys. Rev. B **39**, 4548 (1989).

29. F.C. Zhang and T.M. Rice, Phys. Rev. B **37**, 3759 (1988).
30. F.C. Zhang and T.M. Rice, Phys. Rev. B **41**, 7243 (1990); W. Stephan and P. Horsch, Phys. Rev. Lett. **66**, 2258 (1991).
31. S.E. Barnes, J. Phys. F **6**, 1375 (1976); P. Coleman, Phys. Rev. B **29**, 3035 (1984).
32. Special Issue on High-Temperature Superconductivity, Physics Today **44**(6)(1991).
33. C.G. Olson et al., Science **245**, 731 (1989); Phys. Rev. B **42**, 381 (1990); Phys. Rev. Lett. **65**, 3056 (1990).
34. Z.-X. Shen et al., Phys. Rev. Lett. **70**, 1553 (1993).
35. A.J. Millis, H. Monien and D. Pines, Phys. Rev. B **42**, 167 (1990).
36. C.M. Varma, P.B. Littlewood, S. Schmitt-Rink, E. Abrahams and A. E. Ruckenstein, Phys. Rev. Lett. **63**, 1996 (1989).
37. S.C. Zhang, Phys. Rev. Lett. **65**, 120 (1990); C.N. Yang and S.C. Zhang, Int. J. Mod. Phys. B **5**, 977 (1991).
38. T. Izuyama, J. Phys. Soc. Japan **56**, 4247 (1987).
39. A. Aharony et al., Phys. Rev. Lett. **60**, 1330 (1988).
40. C. P. Enz and Z. M. Galasiewicz, Solid State Communic. **66**, 49 (1988).
41. C.P. Enz, Helv. Phys. Acta **62**, 122 (1989).
42. V. J. Emery, Phys. Rev. Lett. **58**, 2794 (1987).
43. J. E. Hirsch, Phys. Rev. Lett. **59**, 228 (1987).
44. J. R. Schrieffer et al., Phys. Rev. Lett. **60**, 944 (1988).
45. Y. Iye, Int. J. Mod. Phys. B **3**, 367 (1989).
46. T. Schneider and H. Keller, Phys. Rev. Lett. **69**, 3374 (1992); Physica C **207**, 366 (1993); Int. J. Mod. Phys. B **8**, 487 (1993).
47. R. Micnas, J. Ranninger and S. Robaszkiewicz, Rev. Mod. Phys. **62**, 113 (1990).
48. N.E. Bickers, D.J. Scalapino and R.T. Scalettar, Int. J. Mod. Phys. B **1**, 687 (1987); N.E. Bickers, D.J. Scalapino and S.R. White, Phys. Rev. Lett. **62**, 961 (1989).
49. P. Monthoux and D. Pines, Phys. Rev. **47**, 6069 (1993).
50. B. Goss Levi, Physics Today **46**(5), 17 (1993) and **49**(1), 19 (1996).
51. R.C. Dynes, Solid State Commun. **92**, 53 (1994); J.R. Schrieffer, Solid State Commun. **92**, 129 (1994).
52. K.A. Müller, Nature **377**, 133 (1995).
53. Yong Ren, Ji-Hai Xu and C.S. Ting, Phys. Rev. Lett. **74**, 3680 (1995).
54. A. Bill, *Distortive Elektron-Gitter-Wechselwirkung in Hochtemperatursupraleitern: Paarungsmechanismus, Gap-Anisotropie und Phononenrenormierung* (ph.d. thesis, University of Stuttgart), Verlag Shaker (Aachen, 1995).
55. C.P. Enz, to appear in Phys. Rev. **B**, 1 August 1996.
56. K. Levin et al., Physica C **175**, 449 (1991).
57. Y. Iye, Int. J. Mod. Phys. B **3**, 367 (1989).
58. O. Entin-Wohlman and Y. Imry, Phys. Rev. B **45**, 1590 (1992).
59. S. Martin et al., Phys. Rev. B **41**, 846 (1990).
60. D.E. Farrell et al., Phys. Rev. Lett. **63**, 782 (1989).
61. G. Triscone et al., Physica C **168**, 40 (1990).
62. N. Nagaosa and P.A. Lee, Phys. Rev. Lett. **64**, 2450 (1990).
63. L.B. Ioffe and P.B. Wiegmann, Phys. Rev. Lett. **65**, 653 (1990).
64. D. Ihle, M. Kasner and N.M. Plakida, Z. Phys. B **52**, 193 (1991); D. Ihle and N.M. Plakida, Physica C **185-189**, 1637 (1991); J. Magn. Mat. **104-107**, 511 (1992). T. Moriya, Y. Takahashi and K. Ueda, J. Phys. Soc. Japan **59**, 2905 (1990).
65. G.A. Levin and K.F. Quader, Phys. Rev. B **46**, 5872 (1992).

66. B. Hensel, G. Grasso and R. Flükiger, Phys. Rev. B **51**, 15 456 (1995).
67. G.B. Lubkin, Physics Today **48**(3), 20 (1995).

Charge Dynamics in Cuprate Superconductors

E. Tutiš [1], *H. Nikšić* [2] *and S. Barišić* [2]

[1] Institute of Physics of the University of Zagreb,
P.O.B. 304, Zagreb, Croatia
[2] Department of Physics, Faculty of Sciences,
P.O.B. 162, Zagreb, Croatia

Abstract: In this lecture we present some interesting issues that arise when the dynamics of the charge carriers in the CuO_2 planes of high temperature superconductors is considered. Based on the qualitative picture of doping, set by experiments and some previous calculations, we consider the strength of the various inter- and intra-cell charge transfer susceptibilities, the question of Coulomb screening and charge collective-modes. The starting point is the usual *p-d* model extended by the long-range Coulomb (LRC) interaction. Within this model it is possible to examine the case in which the LRC forces frustrate the electronic phase separation, the instability which is present in the model without a LRC interaction. While the static dielectric function in such systems is negative down to arbitrarily small wavevectors, the system is not unstable. We consider the dominant electronic charge susceptibilities and possible consequences for the lattice properties.

1 Introduction

A decade of experimental and theoretical research has indeed showed that the physics of the high temperature copper-oxide superconductors is both complex and intriguing. A major source of complexity lies in the fact that the electronic correlations in these materials are strong. Antiferromagnetism in undoped $LaCuO_4$ and $YBa_2Cu_3O_6$ is clear evidence of this. A second source may be found in experiments which suggest that the electron-phonon interaction may be strong as well. As far as the effects of the strong electron-electron interactions on the electronic properties of doped superconducting materials are concerned, two different viewpoints exist. From the one point of view, the spin correlations, although much weakened upon doping, still predominantly determine the carrier dynamics. From the other point of view, the superconducting cuprates are doped charge transfer insulators, in which the tendency towards spin ordering may be an important, but still secondary, effect. This second approach focusses on the charge degrees of freedom and charge fluctuations, and on the effects that they may have on superconductivity [1, 2], the electronic properties in the metallic phase, and the crystal lattice dynamics [3]. Some of these issues are the subject of this lecture. We consider the dominant static charge susceptibilities, the collective modes and the dielectric properties of a three dimensional system of copper-oxide planes. The features that emerge are related to the properties of these materials observed in experiments, in particular, the anomalies found in the modes of oscillations of in-plane oxygen atoms, the apex oxygen position and movement and the incommensurate and commensurate deformations in the direction parallel to the planes. In this lecture, we consider a system that is close to the doped charge transfer insulator. However, some of the conclusions remain qualitatively valid away from this regime. In particular, this is true for the discussion related to the oxygen-oxygen charge transfer modes and the discussion regarding the frustrated phase separated instability.

2 The p-d model with long-range Coulomb interaction

The presence of CuO_2 planes is the major feature of all cuprate high temperature superconductors. The electronic hopping between the planes is relatively small. The 'insulating' layer between the planes may be composed of atoms of various elements. Another generic feature, as shown by many experiments (e.g. EELS, X-ray absorption, photoemission), is that upon doping the excess holes predominately populate the $p_{x,y} - \sigma$ orbitals of the in-plane oxygen atoms. These orbitals further hybridize with the, mainly $d_{x^2-y^2}$, orbitals on the copper atoms. Together with the fact that correlation effects are important for the in-plane dynamics of the electrons, this constitutes the basis of the *p-d* model.

The original *p-d* model [5, 1] consists of a tight-binding part and a part accounting for the short range Coulomb interaction,

$$H_{pd} = H_{0tba} + H_{SRC} . \tag{1}$$

In the tight-binding part, the orbital energy levels (ε_d, ε_p, $\Delta_{pd} \equiv \varepsilon_p - \varepsilon_d$) are specified and various hybridization terms (copper-oxygen, t_0, oxygen-oxygen, t', etc.) included. H_{SRC} contains the terms describing the Coulomb repulsion on the copper, U_d, and oxygen site, U_p, as well as the terms describing the interaction between electrons on neighbouring sites (copper-oxygen V_{pd}, the oxygen-oxygen V_{pp}, etc.)

In this lecture we will consider an extension of this model which includes long-range Coulomb forces. This extension is necessary in order to examine the ability of the strongly correlated electrons to participate in the screening and to determine the effects that strong local forces have on the electronic plasmon. A second reason for the inclusion of the long-range Coulomb forces is to stabilise the system against phase separation, an instability that occurs in a part of the parameter space of the original *p-d* model.

The introduction of long-range Coulomb forces into the *p-d* model is relatively straightforward. However, a few points should be emphasized. Firstly, as soon as long-range Coulomb forces are considered, it is natural to extend the model from one plane to a three-dimensional collection of parallel planes. The electron-electron interaction is described[3] by

$$H_{coul} = \frac{1}{2} \sum_{\boldsymbol{R},\boldsymbol{R}',\alpha,\alpha'} n_\alpha(\boldsymbol{R}) V_{\alpha\alpha'}(\boldsymbol{R} + \boldsymbol{a}_\alpha - \boldsymbol{R}' - \boldsymbol{a}_{\alpha'}) n_\beta(\boldsymbol{R}') \qquad (2)$$

where $\boldsymbol{R} = (R_\parallel, \boldsymbol{R}_\perp)$ represents the cell index and α denotes the orbital of the atom positioned at $\boldsymbol{a}_\alpha$ with respect to the origin of the unit cell. The potential $V_{\alpha\beta}(\boldsymbol{R} + \boldsymbol{a}_\alpha - \boldsymbol{a}_\beta)$, describes the interaction of electrons on different atoms in unit cells separated by $\boldsymbol{R}$. It is proportional to $1/|\boldsymbol{R} + \boldsymbol{a}_\alpha - \boldsymbol{a}_\beta|$ at long distances and reflects the relative position and charge distribution in the orbitals at short distances. The Fourier transform, $V_{\alpha,\beta}(\boldsymbol{k}) = \sum_{\boldsymbol{R}} V_{\alpha\beta}(\boldsymbol{R} + \boldsymbol{a}_\alpha - \boldsymbol{a}_\beta)\exp(i\boldsymbol{k}\boldsymbol{R}) \equiv \tilde{V}_{\alpha\beta}(\boldsymbol{k})\exp(-i\boldsymbol{k}\boldsymbol{a}_\beta - i\boldsymbol{k}\boldsymbol{a}_\alpha)$ behaves as $1/\boldsymbol{k}^2$ at long wavelengths, while the short wavelength dependence reflects the structure and strength of the local forces. This may be emphasized by writing $\tilde{V}_{\alpha\beta}(\boldsymbol{k})$ in the form[4]

$$\tilde{V}_{\alpha,\beta}(\boldsymbol{k}) = \frac{4\pi e^2}{(a^2 d_\perp \varepsilon_\infty)\boldsymbol{k}^2} + C_{\alpha\beta}(\boldsymbol{k}) \qquad (3)$$

where $a^2 d_\perp$ is the volume of the unit cell (a is the lattice constant of the CuO_2 plane and $d_\perp$ is the distance between neighbouring planes). The high frequency atomic screening is included through ε_∞. The terms $C_{\alpha\beta}(\boldsymbol{k})$ have finite values[5] at $\boldsymbol{k} = 0$.

As usual, the $\boldsymbol{k} = 0$ term of the $1/\boldsymbol{k}^2$ part of the Hamiltonian cancels out because of the neutrality of the crystal. Further, the 'passive' orbitals on the ions with fixed valence may be excluded from the electronic Hamiltonian, leaving only the active orbitals in the copper oxide planes of the model.[6]

The Coulomb part of the p-d model, with the long-range forces included, is

$$H_{coul} = \sum_{\boldsymbol{k}\in B.Z.,\alpha,\beta} \frac{1}{2}\left[(1 - \delta_{\boldsymbol{k},0})\frac{4\pi w t_0}{a^3 \boldsymbol{k}^2} + C_{\alpha\beta}(\boldsymbol{k})\right] \tilde{n}_\alpha(\boldsymbol{k})\tilde{n}_\beta(-\boldsymbol{k}) \qquad (4)$$

---

[3] As in the original p-d model, we neglect the Fock terms. Physically, this is correct for the long-range, but not necessary correct for the short range Coulomb forces.

[4] A somewhat more complicated, but less suggestive form would be preferable for large $\boldsymbol{k}$, close to the zone boundary. For example, an additional $\exp(-\boldsymbol{k}^2 a^2)$ factor may be introduced in order to make the cutoff at the zone boundary soft. Also, it is possible to replace $\boldsymbol{k}^2$ by $\sum_{l=x,y,z}[2 - \cos(k_l d_l)]/d_l^2$ in order to have zero derivatives at the zone boundaries.

[5] It is an instructive exercise to reformulate the calculation of the Madelung energy of a pure ionic crystal following the outlined procedure. The contribution of the $1/\boldsymbol{k}^2$ term explicitly cancels out due to the neutrality of the crystal, $\sum_\alpha n_\alpha = 0$, and the Madelung energy per unit cell takes the form $E_M/N = (1/2)\sum_{\alpha,\beta} C_{\alpha\beta}(0) n_\alpha n_\beta$.

[6] The terms neglected should be kept in mind as the source of a shift, presumably small, of the energy levels of the active orbitals with doping. More importantly, these terms reappear as the electron-phonon coupling terms when lattice movement is considered.

where $\tilde{n}_\alpha(\boldsymbol{k}) \propto \sum_{\boldsymbol{R}} n_\alpha(\boldsymbol{R}) \exp(-i\boldsymbol{k}\boldsymbol{R} - i\boldsymbol{k}\boldsymbol{a}_\alpha)$ and the dimensionless parameter $w = e^2/d_\perp t_0 \varepsilon_\infty$, measuring the strength of the long-range Coulomb interaction is introduced. The approximate value of w for the cuprate superconductors may be estimated by taking $d_\perp \sim 12$Å, $\varepsilon_\infty \sim 3$, $t_0 \approx 1.3$eV. This gives $w \sim 0.25$.

In the following, we will consider the cases with and without the $1/\boldsymbol{k}^2$ term in the Hamiltonian. In the latter case, $C_{\alpha\beta}(\boldsymbol{k})$ corresponds to the U_d, V_{pd}, V_{pp} usually used in the literature. Physically, they represent the effective, screened electron-electron interaction. In the case where $1/\boldsymbol{k}^2$ is explicitly included we continue to use the notation U_d, V_{pd}, V_{pp}, etc., although the physical meaning of the $C_{\alpha\beta}$'s in this case is somewhat different. They represent the local field corrections to the $1/\boldsymbol{k}^2$ interaction, presumably large for short distances.

3 Qualitative features of the electronic spectrum of the doped charge transfer insulator

Various estimates of the local Coulomb terms in the p-d model emphasize the large repulsion on the copper site, $U_d \sim 10$ eV. The limit $U_d \to \infty$ is frequently considered in the literature, and we will work in this limit as well. The $n_d \leq 1$ constraint forced by the infinite U_d is often dealt with by introducing the auxiliary (slave) boson field b and the auxiliary fermion field f in terms of which the real electron operator on the copper site becomes the composite object, $d_{\boldsymbol{R}} = b^\dagger_{\boldsymbol{R}} f_{\boldsymbol{R}}$. The next step is usually to consider the mean field approximation for the boson field (the *saddle point approximation* in the partition function/path integral language). Usually, the corrections beyond the mean field (the Gaussian fluctuations around the saddle point) are also considered. While rigorous justification for this procedure is normally sought in the $1/N$ expansion of the generalised model (i.e. N spin components instead of two), its physical appeal comes from the fact that it captures several major physical features. In particular, this applies to related models [4], that may be more accurately investigated by other means (for instance, the appearance of the Abrikosov-Suhl resonance in the Anderson model). For the p-d model, this procedure qualitatively reproduces some basic experimental facts in the cuprates, for instance the appearance of the in-gap resonance [6] and the large Fermi surface [7] upon doping. A pedagogical survey of the method and some of its results may be found in Ref. [8].

What basically happens at the mean field level [9] for the slave boson in the p-d model with large U_d is: a) the boson field b acquires a static component B_0 (with $\bar{n_d} = 1 - |B_0|^2 \leq 1$) and an oscillatory component at the frequency λ; b) λ also represents the shift of the copper orbital energy on changing from the original electron to auxiliary fermion fields, $\varepsilon_f = \varepsilon_d + \lambda$; the copper-oxygen hybridization in the auxiliary fermion Hamiltonian is reduced with respect to the original hybridization, $t = t_0 B_0$.

The band that occurs at $\varepsilon \approx \varepsilon_f$ (f-band) in the auxiliary fermion density of states becomes the in-gap resonance when the real electronic spectrum is considered. Quite simply, the majority of the spectral weight for the copper site that lies in the auxiliary fermion f-band is transfered back [10, 4] to original energy $\varepsilon_d = \varepsilon_f - \lambda$ upon calculating the auxiliary fermion - auxiliary boson convolution in the real electron Green function $(-i)\langle T(b^\dagger f_\sigma)(t)(f^\dagger b)(0)\rangle$. The spectra are illustrated in Fig. 1.

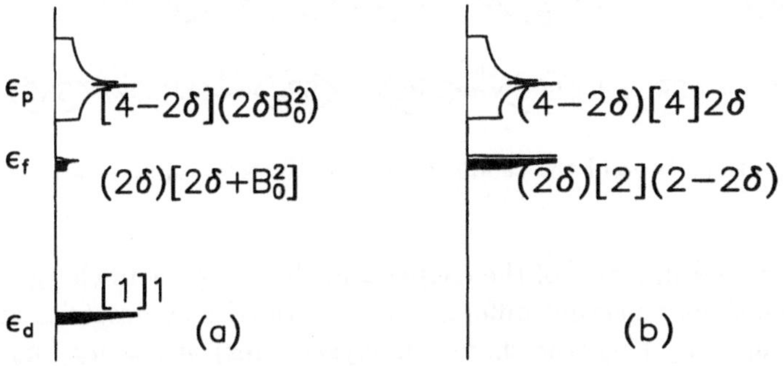

Fig. 1. a) The mean-field density of states in the doped *p-d* system. b) The density of states for the auxiliary fermions. The labels $(p)[t](d)$ indicate the total weight of the particular part of the spectrum $[t]$ and the shares (p) and (d) of the oxygen and the copper orbitals, respectively. The resonance at $\varepsilon \approx \varepsilon_f$ in (a) and the corresponding band in (b) are approximately half filed for small doping level δ. of the oxygen (p) and the copper (d).

The Fermi energy stays in the resonance. The Fermi surface states are predominantly comprised of oxygen orbitals, especially in the doped charge transfer insulator (CTI) regime ($\Delta_{pd} \gg t_0$). In that regime the mean field value B_0 vanishes when the doping δ (measuring the hole concentration with respect to the level of one hole per cell) goes to zero, $B_0^2 \propto \delta$. The renormalized auxiliary fermion bandwidth $\tilde{W} \propto B_0^2$, being also the effective width of the resonance, and the number of states in the resonance are also proportional to doping level δ.

It will be useful to keep this qualitative picture[7] in mind when the results for the charge susceptibilities are discussed.

4 Charge transfer susceptibilities and Coulomb screening

The results for the charge susceptibilities that we present are calculated upon taking into account the first fluctuation correction beyond the mean field approximation. While for large U_d, the slave boson approach is used, for other, smaller Coulomb terms we use the Hartree[8] approximation [8, 12] as the mean field approximation. The fluctuation correction for the long-range Coulomb interaction

[7] This picture may be refined by a calculation of the auxiliary field propagators beyond the saddle point level [4, 8, 11] and a calculation of the electronic spectrum beyond the decoupling approximation, leading to a simple convolution.

[8] The Hartree approximation and the RPA fluctuation correction does not seem to provide an adequate treatment of the short range Coulomb interaction terms as is

corresponds to the usual random phase approximation (RPA). Fig. 2 reflects the simple structure of our calculations. The charge transfers that we will consider

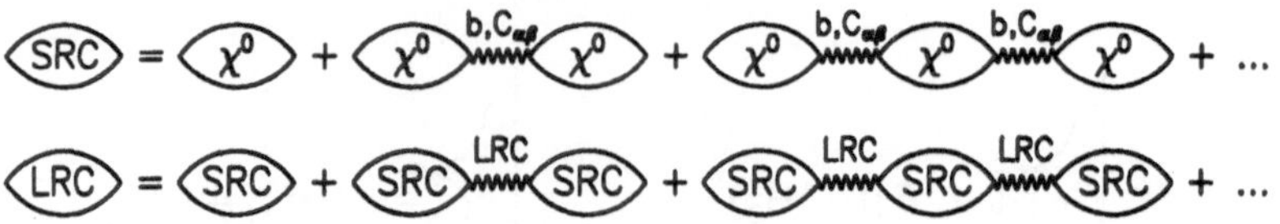

Fig. 2. Corrections beyond the saddle point and Hartree approximations.

may be expressed in terms of the copper and the oxygen site charges n_d, n_x, n_y. The combinations reflecting different charge transfer symmetries and different physics are $n_p = n_x + n_y$ (the charge on oxygen sites), $n_{pp} = n_x - n_y$ (the charge transfer between inequivalent oxygen sites), $n_{pd} = n_p - n_d$ (the copper-oxygen charge transfer) and $n_n = n_p + n_d$ (the total charge in the cell). The susceptibilities $\chi_\alpha = \langle\langle \tilde{n}_\alpha \tilde{n}_\alpha \rangle\rangle$, $\alpha = d, p, pp, pd, n$ will be considered. We will distinguish the susceptibilities χ^0 of the tight-binding fermions with the mean field renormalized band parameters; the susceptibilities χ^S calculated for the model with short range forces; the susceptibilities χ calculated for the model in which the long-range Coulomb forces are included as well.

In Figs. 3 various static charge transfer susceptibilities are shown as a function of the wave vector $\boldsymbol{k} = (k, 0, 0)$ (hereafter we will use the bare copper-oxygen hybridization energy t_0 as the unit of energy and the CuO_2 plane lattice constant a as the unit of distance in order to simplify the notation). From these figures the following may be noted:

a) The inclusion of the fluctuation of the slave boson field, corresponding to $U_d = \infty$, suppresses the charge fluctuation on the copper sites for all wavevectors, i.e. $\chi_d^0 \gg \chi_d^S$. The copper-oxygen charge transfer susceptibility is also diminished. It can be also seen that the Fermi surface enhancement[9] seen in χ_d^0 at $k_x \approx 0.1\pi$ is lost in χ_d^S.
b) The fluctuation of the oxygen charge n_p is not too greatly affected by the auxiliary boson fluctuations. However, the Fermi surface effect disappears in χ_p^S as well.
c) The inclusion of the long-range Coulomb forces further suppresses the total charge fluctuations at small wavevector, $\chi_n \ll \chi_n^S$. While this is expected to happen, it is important to realize that the charge fluctuations on the

the V_{pd} term. However, we will really consider only some qualitative features that this interaction brings in. These features are expected to remain present even if some more adequate treatment like the full Hartree-Fock approximation is used.

[9] Several types of the Fermi surface enhancements of the static susceptibilities in the two dimensional tight-binding model occur at $\boldsymbol{k} \approx 0$ and $\boldsymbol{k} \approx (\pm\pi, \pm\pi)$: van Hove effects, as well as the nesting or the sliding Fermi surface effects, if some flat parts of the Fermi surface exist. The degeneracy of these effects for the square Fermi surface is lifted for some more complicated Fermi surface shapes.

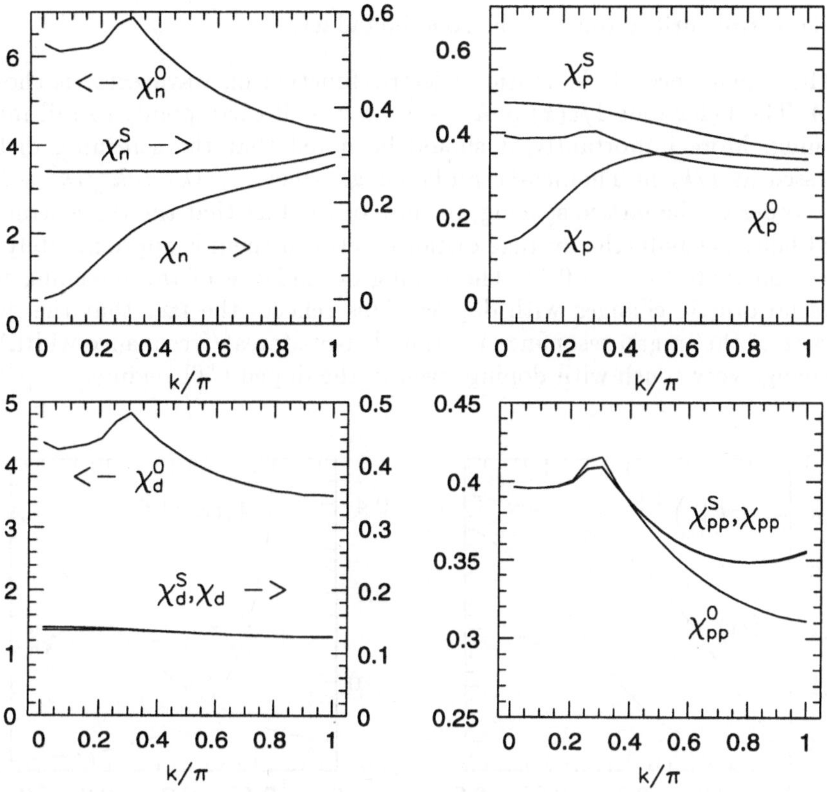

Fig. 3. Static charge susceptibilities for $\Delta_{pd} = 4$, $\delta = 0.1$ as functions of the wavevector, $\boldsymbol{k} = (k, 0, 0)$. The displayed susceptibilities reflect the transfers of a) the total cell charge, b) total oxygen charge in the cell, c) the copper charge and d) the intracell oxygen-oxygen charge transfer.

oxygen sites are mostly affected. This implies that the holes on the oxygen orbitals predominantly participate in the Coulomb screening and that their dynamics determines the dielectric properties of the system. This issue will become more clear when we consider the charge fluctuation spectra.

d) The *pp* susceptibility related to the charge transfer between inequivalent oxygen atoms (x and y) in the unit cell is unaffected both by the long-range Coulomb interaction and by the slave boson fluctuations. The reason lies in the particular symmetry of the *pp* charge transfer. In particular, the Fermi surface effects are pronounced in χ_{pp}^S and χ_{pp}, in contrast to the their absence in the susceptibilities χ_p^S and χ_p related to the total oxygen charge. In fact, all van Hove and nesting enhancements present in the tight binding's χ^0 remain in χ_{pp} whereas they are suppressed in all other channels by the large U_d. This, in particular, means that the $\boldsymbol{k} = 0$ charge transfer transitions [3, 13, 14] and $\boldsymbol{k} = (\pi, \pi)$ charge density wave should be searched for mainly in the *pp* channel. A particularly favourable doping level is the one for which the Fermi surface touches the zone boundary.

The dependence of the static dielectric function on wavevector is shown in Fig. 4. The behaviour $1/\varepsilon(\boldsymbol{k}) \propto \boldsymbol{k}^2 \to 0$ as $\boldsymbol{k} \to 0$ corresponds to full metallic screening. More importantly, it should be noted that the screening distance, expressed as $1/k_s$ in Thomas-Fermi language, where $1/\varepsilon(\boldsymbol{k}) = \boldsymbol{k}^2/(\boldsymbol{k}^2 + k_s^2)$, is of the order of the lattice spacing, in spite of the fact that the concentration of doped holes is small (the average distance between them is approximately three lattice constants for $\delta = 0.1$). The doping dependence of the screening length shows no drastic changes with doping. This reflects the fact that the density of states of the in-gap resonance i.e. (number-of-states)/(resonance-width) does not change very much with doping, even in the doped CTI regime.

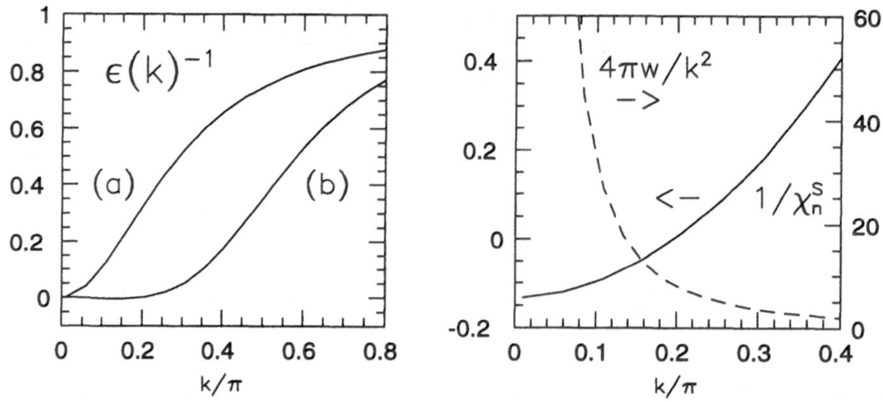

Fig. 4. The figure on the left shows the dielectric function for (a) $V_{pd} = 0$ and (b) $V_{pd} = 1.25$. The second case (as discussed in section 6.) is rather close to the border of the 'normal' ($V_{pd} < 1.15$) and FEPSI ($V_{pd} > 1.15$) situation. The figure on the right shows the total charge susceptibility in the unstable $V_{pd} = 1.25$ system without long-range forces, $\chi_n^S(0) < 0$. Other parameters are $\Delta_{pd} = 4$, $\delta = 0.1$ and $V_{pp} = 0$.

5 Dynamic correlation functions and collective modes

More detailed information on the charge dynamics is obtained upon considering the dynamic charge correlation-functions. They further clarify which orbitals participate in the various intra and inter-cell charge transfers and how they are affected by the short- and long-range Coulomb forces. Some aspects of the spectrum of charge fluctuations for the system with large U_d were considered by Kotliar, Castellani and coworkers [8, 15]. They discussed the appearance of a high frequency exciton collective mode ($\omega = \omega_{pd}$) as well as the zero sound mode in the large U_d p-d model. The spectrum of the oxygen charge-fluctuations, Fig. 5, shows the appearance of these modes in the spectrum together with

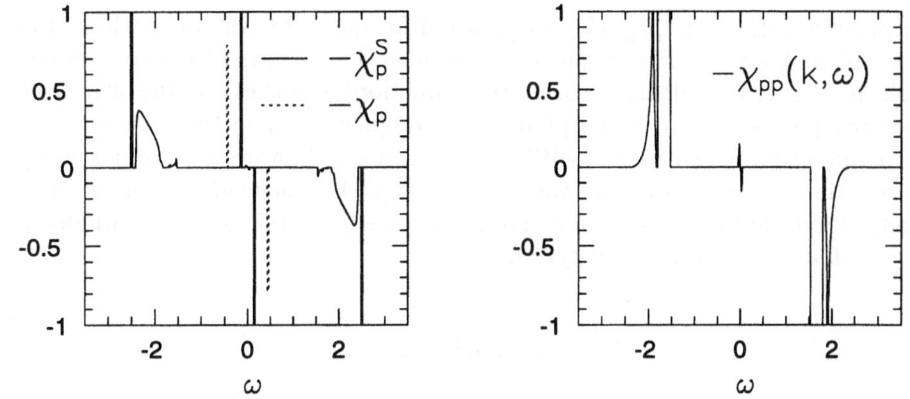

Fig. 5. Left: The spectrum of fluctuations of the charge on the oxygen orbitals, $n_p = n_x + n_y$. The full line corresponds to the case without a long-range interaction. The inter-cell and the intra-cell (Cu-O) charge fluctuations, corresponding to the zero sound mode and the Cu-O exciton mode, may be distinguished. The long-range Coulomb interaction changes the zero sound to the intraband plasmon mode, pushing its frequency to finite frequency at low wavelengths. Right: In the intra-cell oxygen-oxygen charge fluctuation spectrum, $n_{pp} = n_x - n_y$, there are no collective modes present. The strength of the interband contribution to χ_{pp} is large relative to those in other spectra. Parameters as before, namely $V_{pd} = 0$, $k/\pi = (0.11, 0, 0)$.

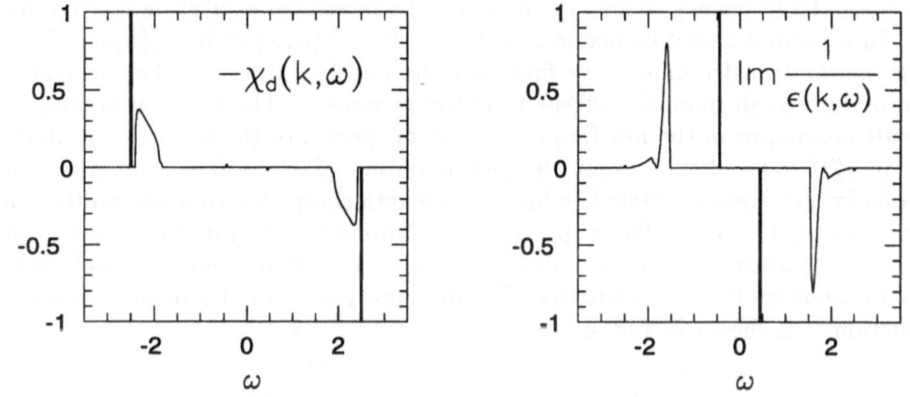

Fig. 6. Left: The spectrum of charge fluctuations on the copper site. The Cu-O exciton mode and the interband excitations are pronounced. Right: The total charge fluctuation spectrum as seen through $1/\varepsilon(k,\omega)$. The intraband plasmon and the intra-cell dipolar excitation, which lie within the interband continuum, dominate the spectrum.

the intraband and interband continua. Fig. 5 also shows the effect of the long-range Coulomb interaction on the spectra. It is interesting to note that the frequency of the exciton mode is not affected at long wavelengths. This shows the quadrupolar character of the corresponding charge excitations. On the other hand, the zero sound mode changes to the intraband plasmon mode with the dispersion $\omega_1^2(\boldsymbol{k}) = \omega_{1p}^2 k_\parallel^2/(k_\parallel^2 + k_\perp^2)$ usual for quasi-2d systems with negligible inter-plane hopping. The frequency (as well as the strength of the corresponding pole in the density-density correlation function) vanishes for the wave vector directed perpendicular to the planes. The frequency ω_{1p} of the long-wavelength in-plane oscillations increases with increasing strength of the long-range Coulomb interaction w, but always remains in the gap. The dependence on w may be understood through a simple equation for the Coulomb collective modes in a two (infinitely narrow) band system,

$$1 - \frac{\Omega_1^2}{\omega^2} - \frac{\Omega_2^2}{\omega^2 - \Delta_{pf}^2} = 0. \tag{5}$$

Here $\Delta_{pf} = \varepsilon_f - \varepsilon_p$ measures the renormalised gap while Ω_1^2 and Ω_2^2 are products of the Coulomb factor $w \propto e^2$ and the intraband and the interband oscillator strengths, respectively. Ω_1, reflecting the intraband charge transfer oscillator strength, assumes a particularly simple doping dependence $\Omega_1^2 \propto \delta$ in the doped CTI regime. This is indeed expected from the picture that we have developed up until now.

The second solution of the equation (corresponding to the third collective mode in the calculation) represents the intracell dipolar mode. It starts from finite frequency $\omega \approx \Delta_{pf}$ for small w and becomes the 'big plasmon' as w is further increased. However, in our calculations a substantial $w > 1$ is required to push the frequency of this mode above the interband excitation continuum.

In a more detailed inspection of the spectra $\chi''_\alpha(\boldsymbol{k},\omega) \equiv \mathrm{Im}\chi_\alpha(\boldsymbol{k},\omega)$ of various charge transfer modes, we find that the intraband part of the spectrum is suppressed in all channels, except from the pp mode.[10] The zero sound/plasmon mode dominates in the low frequency part of spectra of the total charge fluctuation $\chi''_n(\boldsymbol{k},\omega)$ and the oxygen charge fluctuation $\chi''_p(\boldsymbol{k},\omega)$. The strength of this mode in the channels related to long-wavelength charge fluctuations on the copper site is rather small. These spectra are dominated by the interband excitations and the quadrupolar exciton mode. The latter is absent from the total charge fluctuations at long wavelengths. The imaginary part of the inverse dielectric function,[11] is shown in Fig. 6.

6 Frustrated electronic phase separation instability (FEPSI)

The dynamics and dielectric properties of the system with frustrated electronic phase separation instability is even more interesting to consider. Experimentally, the proximity of the phase separation instability (PSI) and superconductivity in cuprates and related materials is rather well documented [16]. The possibility

[10] None of the collective modes appears in $\chi''_{pp}(\boldsymbol{k},\omega)$ for $\boldsymbol{k} = 0$. For finite $\boldsymbol{k}$ their strength is negligible.

[11] We extract the 'macroscopic' dielectric function from the matrix of charge susceptibilities by considering the coupling of the system to an external potential.

for a PSI in a model with a strong electron-electron interaction was addressed by several authors. In particular, V.J. Emery and S.A. Kivelson considered the case of a lightly doped antiferromagnet and discussed the possible consequences of FEPSI on the normal state properties and superconductivity in the cuprates [17].

The phase separation instability in the *p-d* model with short-range forces occurs as the copper-oxygen Coulomb term V_{pd} is increased in the hole doped system, as pointed out in Ref. [12]. However, the enhancement of the copper-oxygen charge transfer by V_{pd}, often emphasized by some authors [1, 2], is not an important issue here.[12] [13]

The main reason that a system with strong correlations chooses the phase separated phase is to diminish its kinetic energy. The kinetic energy diminishes on doping since the frustration of the electron hopping caused by the strong coulomb interaction becomes less effective.[14] On approaching the thermodynamic instability, the derivative $(\partial\mu/\partial n) = (\partial\mu/\partial\delta)$ (the rate of change of the chemical potential with doping) approaches zero and becomes negative in the unstable phase. Also, the susceptibility $\chi_n^S(\boldsymbol{k} = 0, \omega = 0)$ (as well as all other χ_α^S's) diverges at this point, $1/\chi_n^S(\boldsymbol{k} = 0, \omega = 0) = (\partial\mu/\partial n)$. An example of $1/\chi_n^S(k)$ in the unstable system is shown in Fig. 4.

Of course, once the long-range Coulomb interaction is introduced into the model, the true phase-separated state is not likely to appear because of the huge costs in the Coulomb energy. Speculating on this issue, various authors stop at this point and suggest that the charge density wave (CDW) state will replace the phase separated state (i.e. that the instability will be observed as a divergence of χ_n at finite wavevector instead of at $\boldsymbol{k} = 0$). However, the pure electronic CDW (assuming that the lattice is too rigid to participate in the formation of the CDW) corresponds to an unrealistically weak value of the Coulomb force for the cuprate superconductors, namely $w \ll 1$. We find that the stable FEPSI state with $1/\chi_\alpha(\boldsymbol{k}) > 0$ and homogeneous electronic density is more probable. However, the static dielectric function,

$$\frac{1}{\varepsilon(\boldsymbol{k})} = \frac{1}{1 + (4\pi w/\boldsymbol{k}^2)\chi_n^S(\boldsymbol{k}_\parallel)}, \tag{6}$$

is negative in FEPSI systems at long wavelengths since $\chi_n^S(\boldsymbol{k})$, which accounts for the local forces, turns negative, $1/\chi_n^S(0) \propto (\partial\mu/\partial n) < 0$, as shown in Fig. 4.

---

[12] The value of V_{pd} for which the significant enhancement of the copper-oxygen charge fluctuation takes place (possibly related to the softening of the corresponding exciton mode) is much larger than the value required for PSI. For example, in our calculations for $\Delta_{pd} = 4t_0$, PSI occurs already at $V_{pd} = 1.15t_0$, while a value approximately two times larger is required for the cooper-oxygen charge transfer instability to occur.

[13] The *p-d* model in the parameter range $(U_d - \Delta_{pd}) \ll \Delta_{pd} \ll U_d$ exhibits the phase separation instability[10] even for $V_{pd} = 0$.

[14] V.J. Emery originally discussed and repeatedly emphasized this issue in the framework of the lightly doped *t-J* model.

The figure suggests the following formula to describe the dependence of the dielectric function on the wavevector,

$$\frac{1}{\varepsilon(k)} \approx \frac{1}{1-b}\left(\frac{k^2}{k^2+k_{s1}^2} - \frac{k^2 b}{k^2+k_{s2}^2}\right), \tag{7}$$

with $1/k_{s2} \ll k_{s2}$ ($1/k_{s2}$ is at least of the order of a few lattice constants in the CuO_2 plane) and $b < 1$. The formula resembles the Thomas-Fermi formula, but has two characteristic wavevectors instead of one. In order to get a qualitative feeling for the system, one may try to play with a three-dimensional toy model with a dielectric function of the form Eq. (7). The feature that readily emerges is the overscreening of the test charge at distances $1/k_{s1}$ followed by the complete screening at distances beyond $1/k_{s2}$. This implies the attraction of two equal test charges down to distances $1/k_{s2}$. Also, the limit $1/\varepsilon \to 0$ as $k \to 0$ accounts for the total metallic screening. It is easy to see that a piece of the material exposed to the static electric field behaves just like an ordinary metal, developing a finite surface charge in order to ensure $\boldsymbol{E} = 0$ in the bulk.

Returning to our quasi-two dimensional FEPSI metal, we examine its static charge susceptibility. It is given in a form that may be easily analysed (see again Fig. 4),

$$\frac{1}{\chi_n(\boldsymbol{k})} = \frac{4\pi w}{k_\perp^2 + \boldsymbol{k}_\parallel^2} + \frac{1}{\chi_n^S(\boldsymbol{k}_\parallel)}. \tag{8}$$

The most pronounced features (see Fig. 7) are the maximum at finite $k_\parallel = k_c$ (not related to any particular Fermi surface wave vector – k_c decreases with decreasing w) for $k_\perp = 0$, and the increase of χ_n towards the $k_\perp = \pi/d_\perp$ zone boundary. It may be noted here that the divergence of $\chi_n(k_\parallel = k_c)$ for sufficiently small w is the usually mentioned CDW instability which comes as the alternative to the homogeneous FEPSI in the CuO_2 plane. This instability corresponds to the softening of the intraband plasmon branch at $k_\parallel = k_c$. However, the Cu-O exciton mode is not affected at this point.

7 Lattice properties and the stability of FEPSI systems

Finally, we may turn to the question of the stability of the FEPSI system, considering in particular the influence of the electron gas on the ionic lattice (which was considered infinitely rigid up to now). First, the question of the total compressibility of a FEPSI system may be addressed. It may be shown in a number of ways that the short-range forces between the ions stabilize the system. One way is to consider the longitudinal sound velocity and the corresponding frequency $\omega^2(\boldsymbol{k}) = c^2\boldsymbol{k}^2 = c_O^2\boldsymbol{k}^2 + \Omega_{p,ion}^2/\varepsilon_{el}(\boldsymbol{k})$ consisting of the contribution of the short-range forces between the ions and the long-range forces screened by the electrons. For FEPSI systems, the second term is negative. In terms of the total bulk modulus B which determines the sound velocity $c^2 = B/\rho$, the contribution of the second term equals $n^2(\partial\mu/\partial n) < 0$. This, however, does not jeopardize the stability of the system, since size of the electronic contribution

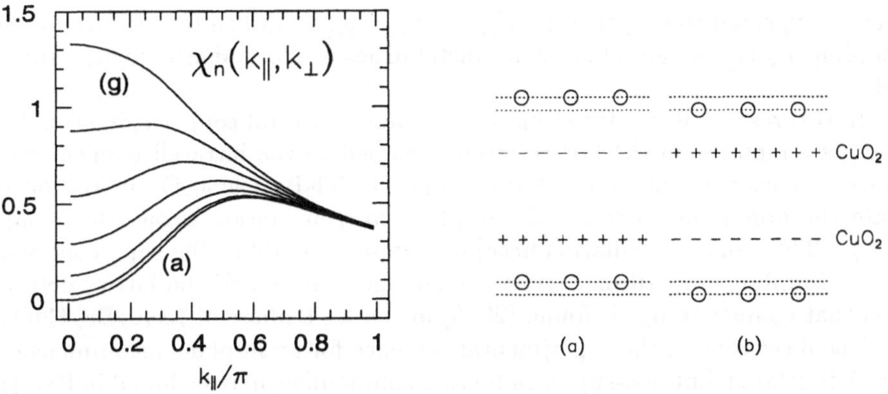

Fig. 7. Left: The dependence of the total charge susceptibility on $k_\perp$, $k_\perp/\pi = 0(a), 0.1, .., 0.6(g)$. Right: The possible effect of the large inter-plane charge transfer susceptibility on the apex oxygen atoms in systems with CuO_2 bilayers, like $YBa_2Cu_3O_{7-x}$.

is two orders of magnitude smaller than the total bulk modulus measured for cuprates,[15] $B \sim 10^{-12}\text{dyn/cm}^2$. The reason for the small absolute value of the electronic contribution may ultimately be traced back to the small density of the electron gas.[16]

More interesting than the compression of the whole system are the phonon modes that couple directly to the charge fluctuations that our model calculations distinguishes as relatively large. In summary,[17] these are: a) the in-cell oxygen-oxygen charge fluctuations, b) the $k_\perp = \pi$ inter-plane charge transfer (this becomes $k_\perp = 0$ when systems with CuO_2 bilayers are considered) and c) the incommensurate charge density fluctuations inside the CuO_2 planes. These modes

[15] A correct order of magnitude for the crystal compressibility may by be obtained from simple Madelung calculations.

[16] In jellium language, $r_s \sim 8$, if we count one hole per unit cell; counting only doped holes we get $r_s \sim 16$ for $\delta = 0.1$. At this point, it seems interesting to note that the theoretical search for high temperature superconductivity before the discovery of cuprate superconductors indeed pointed towards the low density metals [18]. One of the basic properties of these systems is the appearance of the negative static dielectric function. That the negative static dielectric function is a feature desirable for the electronic mechanism of superconductivity was rather pedagogically explained by Littlewood in Ref. [2]. However, as emphasized by Kirzhnits at al. [18], this is not a sufficient condition for superconductivity to occur - the interaction between test charges is not the one that enters the gap equation - the local field corrections and correlation effects that themselves lead to negative static dielectric function should be taken care of, as well.

[17] The possibility of a soft Cu-O charge-transfer exciton is beyond the focus of this lecture. However, the formalism outlined here is perfectly suitable for its consideration since, as already explained, one almost inevitably works 'deeply' inside the FEPSI phase.

may be expected to become soft, $\tilde{\omega}_{ph}^2 = \omega_{ph,0}^2 - g_{ep}^2 \chi_{el}$, when the electron-phonon coupling g_{ep} is substantial or lattice instabilities and/or anharmonicities may result.

In this respect it is interesting to note that in several copper oxide superconductors anomalies in the lattice modes coupled to the intracell oxygen-oxygen charge transfer are observed. For example, in $Tl_2Ba_2CaCu_2O_8$ in the normal state the dominant mode for the in-plane oxygen motion is one that couples to n_{pp}. This kind of oscillation becomes suppressed [19] in the superconducting phase. Similar competition between superconductivity and the lattice deformation that couples to n_{pp} is found [20, 3] in the LTT phase of $La_{2-x}Ba_xCuO_4$.

The discussion of the experimental evidence for an in-plane incommensurate CDW in related, but non-superconducting compounds, may be found in Ref. [17].

Here we would like to point out one possible sign of the FEPSI situation in $YBa_2Cu_3O_{7-x}$, the compound which contains two CuO_2 layers per unit cell. From our discussion we expect to find anomalies for ion movements that couple to the charge transfer between these layers. The IR-active mode of the apex oxygen atoms is a natural candidate for a such coupling. The coupling, being strong enough, results in a situation with two equivalent, minimum energy positions for the apex oxygen atoms (see Fig. 7 for an illustration). Experimentally, this picture, supported by other measurements, first emerged from the analysis of the EXAFS [21] in $YBa_2Cu_3O_{7-x}$.

8 Conclusion

The strong short-range Coulomb forces substantially complicate the charge dynamics in doped copper-oxide planes. It is difficult to envisage an effective band picture that would simultaneously account for the static (k_s) and dynamic (ω_p) screening as well as the Fermi surface effects which we pointed out. Also, the strong Coulomb forces may tend to cause an electronic phase separation, the vicinity of which was pointed out by several authors as the possible source of superconductive pairing and the anomalous normal state properties. At this point, the introduction of the long-range Coulomb forces into the model seems crucial, which we did for the *p-d* model. We showed that, while the long-range Coulomb forces suppress the instability, the negative static dielectric function and, more importantly, the rather large total charge susceptibilities in some parts of the *k*-space remain as the characteristic signs of FEPSI systems. Some of these signs were found in cuprate superconductors and related materials. While the lattice instabilities which may occur in order to exploit the large electronic susceptibilities probably do not contribute to the appearance of superconductivity, the effective electronic interaction which drives FEPSI may be favourable in that sense.

Acknowledgments

We gratefully acknowledge the discussions and remarks by I. Batistić and D. K. Sunko.

References

1. C.M. Varma, S. Schmitt-Rink, E. Abrahams, Solid State Commun. **62** (1987) 681; C.M. Varma et al., Phys. Rev. Lett **63** (1989) 1996; C.M. Varma, Phys. Rev. Lett **75**, (1995) 898.
2. P.B. Littlewood in: V.J. Emery(ed.), *Correlated electron systems*, World Scientific (1993), *p. 1*
3. S. Barišić in: B.K. Chakraverty (ed.), *Critical Trends in High T_c Superconductivity*, World Scientific, 1991.
4. P. Coleman, Phys. Rev. B **35** (1987) 5072.
5. V.J. Emery, Phys. Rev. Lett. **58** (1987) 2794.
6. N. Nucker et al., Phys. Rev. B **37** (1988) 5158; T. Takahashi et al., Nature **334** (1988) 691; T. Watanabe et al., Phys. Rev. B **44** (1991) 5316.
7. J.C. Compuzano et al., Phys. Rev. Lett. **64** (1990) 2308; D.S. Dessau et al., Phys. Rev. Lett. bf 71 (1993) 2781; Aebi P. et al., Phys. Rev. Lett. **72** (1994) 2757.
8. G. Kotliar in: V.J. Emery (ed.), *Correlated electron system*, World Scientific (1993), *p. 118.*
9. G. Kotliar, P.A. Lee, N. Read, Physica C **153-155** (1988) 538.
10. E. Tutiš, Ph.D. thesis, University of Zagreb, 1994.
11. H. Nikšić, E. Tutiš, S. Barišić, Physica C. **241** (1995) 247-256.
12. M. Grilli, R. Raimondi, C. Castellani, C. Di Castro, G. Kotliar, Phys. Rev. Lett. **67** (1991) 259.
13. S. Barišić, J. Zelenko, Solid State Commun. **74** (1990) 367.
14. S. Barišić, E. Tutiš, Solid State Commun. **87** (1993) 557.
15. C. Castellani, G. Kotliar, R. Riamondi, M. Grilli, Z.Wang, M. Rozenberg, Phys. Rev. Lett. **69** (1992) 2009.
16. K.A. Müller and G. Bedenek (eds.), *Phase Separation in Cuprate Superconductors*, World Scientific, (1993).
17. V.J. Emery and S.A. Kivelson, Physica **C 209** (1993) 567; **C 235-240** (1994) 189.
18. O.V. Dolgov, D.A. Kirzhnits and E.G. Maksimov; V.L. Ginzburg and D.A. Kirzhnits in: V.L. Gizburg (ed.), *Superconductivity, Superdiamagnetism, Superfluidity*, Mir, Moscow 1987.
19. B.H. Toby, T. Egami, J.D. Jorgensen, M.A. Subramanian, Phys. Rev. Lett. **64** (1990) 2414.
20. M. Sera, Y. Ando, K. Fukuda, M. Sato, I. Watanabe, S. Nakshima, K. Kumgai, Solid State Commun. **69** (1990) 851.
21. J. Mustre de Leon et al., Phys. Rev. Lett. **65**, (1990) 1675; Phys. Rev. B **45**, (1992) 2447.

Coherent Precession and Spin Superfluidity in ^{3}He

Yu.G. Makhlin

Landau Institute for Theoretical Physics
2 Kosygin st., 117940, Moscow, Russia

1 Introduction

Spin (magnetic) superfluidity is an interesting phenomenon having much in common with ordinary mass superfluidity. On the other hand, it has a different nature which allows, for example, direct measurements of the usually unmeasurable phase of the condensate. In addition, spin superfluidity has some specific features which deserve careful study.

Phase coherence over macroscopic samples is an essential feature of mass superfluidity. The corresponding feature of spin superfluids is the coherence of the phase of the precession of the magnetization in an external magnetic field, the so-called coherent precession.

The phenomenon of spin (magnetic) superfluidity is related to the possibility of spin transport without dissipation, i.e. a spin supercurrent. This phenomenon was discovered during a nonlinear NMR-study of spin dynamics in superfluid ^{3}He-B in 1984 by an experimental group at the Kapitza Institute [1] in collaboration with I. Fomin.

We begin by briefly recalling the foundations of the physics of superfluid ^{3}He (see [3] for a review). Superfluid ^{3}He was the first material for which unconventional superfluidity was established. The Cooper pairs of ^{3}He have angular momentum $L = 1$ and spin $S = 1$ (triplet state). The superfluid order parameter is given by the anomalous spin average

$$d(k) \sim \langle \hat{a}_{-k\alpha}(i\sigma_y \boldsymbol{\sigma})_{\alpha\beta}\hat{a}_{k\beta}\rangle, \tag{1}$$

where $d_\alpha = A_{\alpha i}\hat{k}_i$ is a linear function of the momentum k. The exact form of the order parameter $A_{\alpha i}$ can be found, e.g., from the BCS equations or Ginzburg-Landau theory. In bulk ^{3}He, two different phases in different regions of T–P phase diagram are known to be stable, namely the A- and B-phase.

This spin structure, Eq. (1), results in a new type of magnetic ordering and in unusual magnetic properties of the fluid. Consequently the investigation of spin dynamics of superfluid ^{3}He is very interesting. Further, these investigations are very important for the study of various properties, including the non-magnetic properties, of the liquid, where nuclear magnetic resonance (NMR) is used extensively.

The spin supercurrent leads to the formation of unusual non-uniform precessing spin structures in NMR experiments and ensures their stability. A number of phenomena of spin superfluidity analogous to those of the usual mass superfluidity/superconductivity have been investigated: the Josephson effect, phase slippage, the formation of spin vortices etc.

At very low temperatures in the collisionless regime $\omega_L \tau \gg 1$,[1] a similar phenomenon was recently discovered in ^{3}He-^{4}He mixtures and in normal ^{3}He [4]. In this case non-dissipative transport of magnetization is also possible due to Fermi-liquid effects [5], typical for systems with a significant exchange-interaction. In normal Fermi-liquids and superfluids close to T_c, precessing spin structures in the collisionless regime are now under experimental investigation.

In this lecture we present the idea of spin superfluidity. In Section 2, we describe the typical NMR experiment and discuss the case of the hydrodynamic regime in the B-phase. In Section 3, the A-phase is discussed. In Section 4, spin dynamics and coherent precession in normal Fermi-liquids is considered. Section 5 deals with the collisionless regime in superfluids near transition temperature. We also compare the behaviour of the system in different regimes (hydrodynamic and collisionless) and in different phases (A-, B- and normal phases).

2 Coherent precession in the B-phase: hydrodynamic regime

In order to demonstrate coherent precession, let us consider a typical experimental situation, namely a magnetic field H applied to a sample in a container induces the magnetization parallel to the field. The spin density S is deflected by a pulse of a perpendicular field from its "vertical" direction and begins to precess about the field (Larmor precession) according to

$$\frac{\partial S}{\partial t} = S \times \omega_L, \qquad (2)$$

where $\omega_L = \gamma H$ is the Larmor frequency of the external field and γ is the gyromagnetic ratio of a ^{3}He atom.

Suppose now that external field is slightly non-uniform (here and later we suppose that the magnetic field and its gradient are parallel to the z-axis and that all of the variables are independent of x and y. In this case the Larmor frequency depends on the z-coordinate and a difference in phases of precession in different points appears. Experimentally the integral of the horizontal magnetization $\int S_\perp dV$ is measured and the corresponding signal vanishes after a time interval $t_\nabla \sim 2\pi/\delta\omega_L$, where $\delta\omega_L$ is the characteristic difference in the Larmor frequencies in the container.

[1] ω_L is the Larmor frequency of an external magnetic field, and τ is the quasiparticle relaxation time.

Provided the temperature is not too low, this simple picture agrees with experiments in normal ^{3}He [6]; however, in experiments with $t_\nabla \sim 1$ms in the B-phase of superfluid ^{3}He, the detected signal decreases gradually during several hundred milliseconds! Moreover, the frequency of precession is uniform in spite of the spatial dependence of the Larmor frequency. This phenomenon is usually referred to as a long-lived induction signal (LLIS). Coherent precession, when induced by a small horizontal rf magnetic field at some frequency ω_P, was observed also in the continuous wave (CW) NMR regime

To understand why the magnetization precesses coherently, let us consider the energy of the system. The phenomenological hamiltonian is given by the sum [7]

$$\mathcal{E} = \frac{1}{2}\frac{\gamma^2}{\chi}S^2 - \gamma SH + U_D\{d\} + E_\nabla \tag{3}$$

of the internal magnetic energy (χ is the magnetic susceptibility), the Zeeman energy E_Z, the energy U_D of the dipole interaction of the magnetic moments of the atoms of ^{3}He, which are ordered due to the spin structure of Cooper pairs, and the energy E_∇ of the inhomogeneity of the distribution of the order parameter. The longitudinal component of the magnetization S_z is a conserved quantity.[2] In addition, it can be shown that the structures with the equilibrium *absolute value* of the magnetization realize a local minimum of the energy with a depth of the order of the dipole energy. Thus we shall consider $|S|$ as being fixed. Therefore, to get the stable precessing state, we should minimize the energy at a fixed value of the longitudinal magnetization, or equivalently, to minimize

$$\tilde{\mathcal{E}} = \mathcal{E} + \omega_P S_z \cong (\omega_P - \omega_L)S_z + U_D + E_\nabla, \tag{4}$$

where ω_P is a Lagrange multiplier.[3] The first term on the right-hand-side of Eq. (4) is known as the spectroscopic energy E_{sp}.

For the structure with minimal energy, we have $\delta\tilde{\mathcal{E}}/\delta S = 0$ if we only allow the variations $\delta S \perp S$, conserving $|S|$. Therefore, $(\delta\mathcal{E}/\delta S) + \boldsymbol{\omega}_P \parallel S$, where $\boldsymbol{\omega}_P \equiv \omega_P \hat{z}$. On the other hand, using the well-known commutation relations for the components of the spin

$$[S_i, S_j] = ie_{ijk}S_k, \tag{5}$$

we get the equation of motion for magnetization, viz.

$$\dot{S} = i[\mathcal{E}, S] = \frac{\delta\mathcal{E}}{\delta S} \times S. \tag{6}$$

So, $\dot{S} = S \times \boldsymbol{\omega}_P$, i.e., in a stationary configuration the magnetization *should precess with a uniform frequency* $\boldsymbol{\omega}_P$. Let us find this frequency and the corresponding spin distribution.

---

[2] can lead to a non-conservation of S_z, but the amplitude of variations is small and, if averaged over fast precession of the structure, vanishes. The time of longitudinal spin relaxation due to other spin-non-conserving interactions is very long at millikelvin temperatures.

[3] We have neglected a constant term on the right-hand-side of Eq. (4).

In the B-phase the effective dipole energy is [8]

$$U_D^B = \frac{8}{15}\frac{\chi}{\gamma^2}\Omega_B^2\left(\frac{1}{4}+\cos\beta\right)^2\cdot\Theta(\beta-\beta_L), \tag{7}$$

where Ω_B is the frequency of the longitudinal NMR in the B-phase and β is the tipping angle of magnetization, i.e. the angle between the direction of the magnetization and the magnetic field. $\Theta(\beta-\beta_L)$ is a step function and the Leggett angle is $\beta_L = \arccos(-1/4) \approx 104°$. In a typical experimental situation,

$$E_{sp} \ll U_D \ll E_Z$$

or

$$(\omega_L - \omega_P)\omega_L \ll \Omega_B^2 \ll \omega_L^2.$$

In particular, $|\omega_L - \omega_P| \ll \omega_L$.

Simple considerations show that the minimum of the sum of the dipole energy and the spectroscopic energy is given by the following distribution of the magnetization:

$$\beta = 0 \quad \text{for} \quad \omega_L > \omega_P, \tag{8}$$

$$\beta = \beta_L \quad \text{for} \quad \omega_L < \omega_P. \tag{9}$$

So, a two-domain structure is formed in an inhomogeneous magnetic field (Fig. a), the domain wall being situated at the point where $\omega_L = \omega_P$. These two domains are referred to as a homogeneously precessing domain (HPD, $\beta \approx 104°$, $\omega_L < \omega_P$) and a non-precessing domain (NPD, $\beta = 0$, $\omega_L > \omega_P$), respectively. The position of the domain wall is determined by ω_P in CW NMR and by the longitudinal magnetization in pulsed NMR. The gradient energy [3] smoothes the abrupt domain wall leading to a thickness of the intermediate layer

$$\lambda \sim \left(\frac{c_{sp}^2}{\omega_P \nabla \omega_L}\right)^{1/3}, \tag{10}$$

where c_{sp} is the spin wave velocity. In this layer the tipping angle of the magnetization changes continuously from 0 to β_L. In addition the gradient energy ensures that the *phase of precession* in the HPD *is uniform* over the volume of the domain. This two-domain spin distribution is stable because it corresponds to a local minimum of the energy. As the gradient of the field decreases, $\nabla H \to 0$ and E_{sp} vanishes, the thickness of the wall $\lambda \to \infty$, i.e. in an almost homogeneous field (when λ exceeds the size of the container) uniform precession is stable. Dissipation leads to the slow growth of a NPD and determines the life-time of the structure.

It is interesting to interpret this effect as the formation of unusual off-diagonal long-range order (ODLRO) [9]. In the usual mass superfluids, the order parameter describing ODLRO is an expectation value of the annihilation operator $\langle\hat{\psi}\rangle \cong |\Delta|\exp(i\phi)$ where the annihilation operator changes the number of particles. In the case of uniform precession, the rôle of the number of particles is played by the longitudinal magnetization S_z, which is a conserved quantity if one neglects the very slow longitudinal relaxation, and the creation-annihilation operators are $S_\pm = S_x \pm iS_y$. The absolute value $|S_+| = |S_\perp|$ corresponds to the absolute value of superfluid order parameter $|\Delta|$, and the phase of precession corresponds to the phase ϕ. In this sense the HPD is a "spin superfluid" state and the NPD is a "spin non-superfluid (normal)" state.

ODLRO is possible only if it is supported by the rigidity of the order parameter; in ordinary superfluids/superconductors, the gradient energy $K(\nabla\phi)^2$ takes the minimum value for uniform distributions of phase. The analogous term in spin superfluidity in ^{3}He-B comes from the stiffness of the superfluid order parameter.

This interesting analogy is full of consequences; a number of effects typical for mass superfluidity can be observed in spin superfluid systems.

Consider two containers with a HPD connected by a narrow channel (weak link). In CW NMR experiments, the phases of precession in the two containers can be adjusted independently and a phase difference between them can be established. This difference leads to a supercurrent which is now spin supercurrent. This simple setup allows the observation of the critical spin supercurrent with phase slippage events, stationary (the phase difference is fixed) and non-stationary (different precession frequencies in two HPDs) Josephson effects and even the formation of a spin vortex [10]. Unlike the usual superfluidity and superconductivity where the phase of the condensate cannot be observed, the phase of precession can be directly controlled and adjusted. In experiments on spin superfluidity, one can measure the phase difference and distribution along the channel and even the position of phase slippage centres, which are unmeasurable in mass superfluidity. The experiments are in semiquantitative agreement with theory.

3 Coherent precession in the A-phase

The difference between the A- and B-phases lies in the different form of the dipole energy. In ^{3}He-A, the effective dipole energy is [11]

$$U_D^A = -\frac{3}{16}\frac{\chi}{\gamma^2}\Omega_A^2\left(\cos\beta + \frac{1}{3}\right)^2 . \qquad (11)$$

Here Ω_A is the frequency of the longitudinal resonance in the A-phase.

Unlike the case of the B-phase, the dipole energy is not a concave but a convex function of the longitudinal magnetization $S_z = S\cos\beta$. This leads to the very different behaviour of the A-phase in NMR experiments. In pulsed NMR, where the longitudinal magnetization ("number of particles") is a conserved quantity, the uniform precession is unstable to a separation into domains even in a homogeneous external field. A two-domain structure may appear, the magnetization in one domain being parallel and in the other being antiparallel to the external field (Fig. b). The position of the domain wall is determined by the integral of the longitudinal magnetization. The relative position of these domains (which one is "upper") is arbitrary in a homogeneous field, but even a small gradient in the magnetic field distinguishes between two structures; the equilibrium domain is situated in the region with higher external field. Standard energy considerations give $\lambda_D \sim c_{sp}/\Omega_A$ for the length of the intermediate layer. This two-domain solution of the equations of spin dynamics has not yet been seen; although the initial stage of the decay of a uniform precession has been investigated (see [12] and references therein).

4 Coherent precession in normal Fermi-liquids

Recently an analogous LLIS was discovered during NMR measurements in normal Fermi-liquids, namely in ^{3}He-^{4}He mixtures and in pure ^{3}He. It was suggested by Nunes et al. [13] that a two-domain precessing state is responsible for this LLIS. This suggestion has been confirmed in recent experiments by Dmitriev and coworkers [4, 14].

We know from previous sections that the spin supercurrent plays an important rôle in the formation and stabilization of two-domain structures. It turns out that due to Fermi-liquid effects there is an analogue to the spin supercurrent in a normal liquid .

The expression for the spin current in a normal Fermi-liquid, first obtained by Leggett [5] from the kinetic equation, is

$$J_i = -\frac{D_0}{1+\mu^2 S^2}\left\{\frac{\partial S}{\partial x_i} - \mu S \times \frac{\partial S}{\partial x_i} + \mu^2\left(S\frac{\partial S}{\partial x_i}\right) S\right\}, \tag{12}$$

where

$$\begin{aligned} D_0 &= \frac{1}{3}w^2\tau, \quad w^2 = v_F^2(1+F_0^a), \\ \mu &= \frac{\gamma^2}{\chi}\tau\frac{\kappa}{1+F_1^a/3}, \quad \kappa = \frac{F_1^a/3 - F_0^a}{1+F_0^a}, \end{aligned} \tag{13}$$

v_F is Fermi velocity, F_0^a, F_1^a, the Fermi-liquid parameters and τ is the quasiparticle collision time. The first term in this equation corresponds to an ordinary spin diffusion current.

In the collisionless regime, $\omega_L\tau \gg 1$ corresponding to the just mentioned experiments, $\mu S \gg 1$ and the main contribution to the spin current is from the last two terms in Eq. (12). The leading last term is responsible for a quick smoothing of inhomogeneities in the distribution of the absolute value of the magnetization [15, 16]. Later we consider this absolute value as uniform in space (not necessarily equal to its equilibrium value because of the absence of the dipole energy in the normal state). The dynamics of the direction of the spin density is governed by the second term in Eq. (12), i.e.

$$J_i \cong \frac{\chi}{\gamma^2}\frac{w^2}{3\kappa S^2} S \times \nabla_i S. \tag{14}$$

This term represents a non-dissipative contribution to the spin current. It is this term which is responsible for spin superfluidity in normal Fermi-liquids.

If one neglects the effects of dissipation, the equation of motion for the spin density

$$\dot{S} = S \times \omega_L - \nabla_i J_i \tag{15}$$

can be obtained from the *formal* hamiltonian[4]

$$\mathcal{H} = \frac{\gamma^2}{2\chi} S^2 - \omega_L S - \frac{\chi}{\gamma^2} \frac{w^2}{6\kappa S^2} (\nabla_i S)^2 \tag{16}$$

and the standard commutation relations Eq. (5) for the spin components. It is the well-known Landau-Lifshitz equation in external magnetic field. An interesting feature of the hamiltonian Eq. (16) is that the gradient term is *negative* (!) for $\kappa > 0$ (as in ^{3}He). Stable spin distributions are the extrema of the formal energy Eq. (16) with the additional Lagrange term $\omega_P S_z$. In the hydrodynamic regime (see the previous sections), the stability of the minima of the energy was maintained by energy conservation; the system cannot go away from a minimum because it would change its energy. Similarly, the maxima are stable due to energy conservation. A peculiarity of the collisionless regime is that the effective energy (16) has no local minima; one can decrease the energy of an arbitrary structure a considerable amount by imposing very low-amplitude and short-wavelength spatial variations of the magnetization.

The maximization of the spectroscopic energy shows that a two-domain structure is formed in an inhomogeneous magnetic field (Fig. c). Contrary to the situation in the B-phase, a NPD with the equilibrium direction of the magnetization is situated in the region of lower external field. In the non-equilibrium domain, the magnetization is antiparallel to its equilibrium value (antiparallel domain, APD). The domain-wall thickness is $\lambda_n \sim (w^2/\omega_P \nabla\omega_L)^{1/3}$.

Both of these domains are non-superfluid in the sense of spin superfluidity ($S_\perp = 0$; see the discussion in Section 2) but the intermediate region *is* spin-superfluid, i.e. the direction of spin density in the domain wall changes continuously from its direction in the NPD to its direction in the APD with a constant absolute value. In this region the phase of precession is *uniform*. So, this structure also produces a long-lived induction signal and can be used in investigations of spin superfluidity.

The existence of the two-domain structure was predicted theoretically by Fomin and was investigated experimentally by Dmitriev et al. [4, 6, 14].

5 Coherent precession in superfluid phases: collisionless regime

In the previous sections we have studied two different regimes: a) the hydrodynamic regime in a superfluid where the main contribution to spin transport is

[4] This hamiltonian is valid in the regime where dissipation is small. It is not related to the energy of the system and its value can grow due to dissipation.

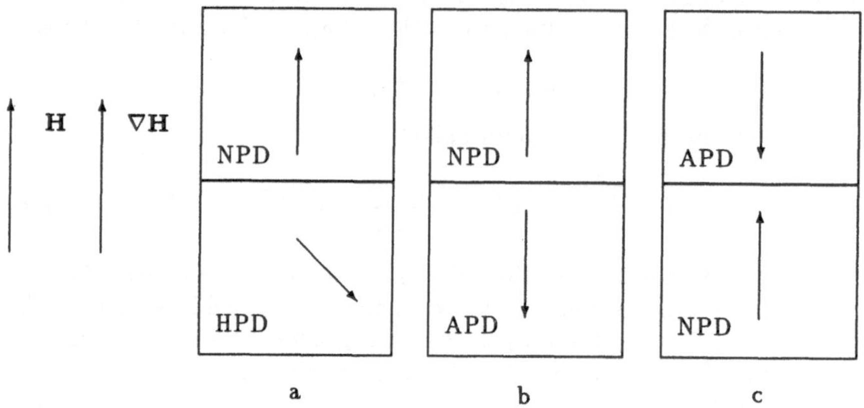

Fig. 1. Two-domain coherently precessing structures in the hydrodynamic regime in a) the B-phase and b) the A-phase and in c) the collisionless regime (normal state, A- or B-phases). In the latter case the domain-wall thickness is different in different phases (see text).

due from the spin supercurrent and the dipole energy is essential; and b) the collisionless regime in a normal fluid with a Fermi-liquid non-dissipative spin current. It was found recently [16, 17] that in the collisionless limit, $\omega_L\tau \gg 1$, just below T_c ($T > 0.85T_c$) a new regime exists where the Fermi-liquid spin current prevails (at lower temperatures the spin supercurrent exceeds it) and U_D plays a rôle.[5]. Under these conditions a new treatment of the possible precessing structures is necessary [16, 17]

It was shown above that the stable distributions of magnetization in this situation are given by the maxima of the effective energy. Contrary to the analysis in the hydrodynamic regime, uniform precession in a homogeneous field is stable in the A-phase where the dipole energy is a convex function of S_z and unstable in the B-phase where this function is concave.

As a function of β, the dipole energy in the B-phase has a plateau and the position of the *minimum* (Sec. 2) of the sum $U_D + E_{sp}$ changes abruptly from $\beta = 0$ at $\omega_L > \omega_P$ to $\beta = \beta_L$ at $\omega_L < \omega_P$. In the A-phase there is no plateau and the position of the *maximum* changes continuously from 0 to π over a characteristic length of the order of $\lambda_n^3/\lambda_D^2 > \lambda_n$, where $\lambda_D \sim v_F/\Omega_A$ is the dipole length. As $\nabla H \to 0$, the thickness of the domain wall goes to infinity, i.e. the precession becomes uniform.

In the B-phase the thickness of the intermediate layer is of the order of λ_D, i.e. the position of the maximum changes abruptly from 0 to π (cf. Sec. 3). This structure should exist even in a homogeneous external field.

[5] If we are not too close to T_c, $\lambda_D < \lambda_n$ (see below) or $T_c - T > 10^{-3}T_c$.

The experimental situation in this range of temperatures is very complicated. A number of different states of coherent precession have been discovered in ^{3}He-B near T_c as well as a number of states were predicted theoretically (taking into account the complicated structure of the order parameter, i.e. other degrees of freedom besides the one described by β; see, e.g. [8]). Nevertheless, these states are not identified yet.

One should note also that a new coherently precessing state with an extremely long life-time (~30s) was discovered in ^{3}He-B at ultra low temperatures $T \sim 0.1T_c$ [18].

References

1. A.S. Borovik-Romanov, Yu.M. Bunkov, V.V. Dmitriev, Yu.M. Mukharskii, K. Flakhbart *ZhETF*, **88**, 2025 (1985) [*Sov. Phys. JETP*, **61**, 1199 (1985)].
2. I.A. Fomin *ZhETF*, **88**, 2039 (1985) [*Sov. Phys. JETP*, **61**, 1207 (1985)].
3. D. Vollhardt, P. Wölfle, *The Superfluid Phases of Helium 3*, Taylor & Francis, London-New York-Philadelphia, 1990.
4. V.V. Dmitriev, I.A. Fomin, *Pis'ma v ZhETF* **59**, 352 (1994) [*JETP Letters*, **59**, 378 (1994)].
5. A.J. Leggett, *J.Phys.*, **C3**, 448 (1970).
6. V.V. Dmitriev, S.R. Zakazov, V.V. Moroz *Pis'ma v ZhETF*, **61**, (1995) [*JETP Letters*, **61**, 324 (1995)].
7. A.J. Leggett, *Ann. Phys. (NY)*, **85**, 11 (1974).
8. Yu.M. Bunkov, G.E. Volovik, *ZhETF*, **103**, 1619 (1993) [*Sov. Phys. JETP*, **76**, 794 (1993)].
9. Yu.M. Bunkov, O.D. Timofeevskaya, G.E. Volovik, *Phys. Rev. Lett.*, **73**, 1817 (1994)
10. A.S. Borovik-Romanov, Yu.M. Bunkov, V.V. Dmitriev, Yu.M. Mukharsky, D.A. Sergatskov, *Proc. of Symp. on Quantum Fluids and Solids*, AIP CONF. PROC. 194 / Eds. G.Ihas, Y.Takaho, N.Y., 1989. P.27
11. Yu.M. Bunkov, G.E. Volovik, *Europhys. Lett.*, **21**, 837 (1993).
12. I.A. Fomin, *Pulsed NMR and the Spatially Nonuniform Precession of Spin in the Superfluid Phases of ^{3}He*. In: *Helium 3*, ed. by W.P. Halperin and L.P. Pitaevskii. Elsevier Science Publishers B.V., 1990. Chapter 9. P.609.
13. G. Nunes, Jr., C. Jin, D.L. Hawthorne A.M. Putnam, D.M. Lee, *Phys. Rev.*, **B46**, 9082 (1992).
14. V.V. Dmitriev, V.V. Moroz, S.R. Zakazov, *J. Low Temp. Phys.*, **101**, 141 (1995); V.V. Dmitriev, V.V. Moroz, A.S. Visotsky, S.R. Zakazov, *Physica B*, **210**, 366 (1995).
15. I.A. Fomin, *J. Low Temp. Phys.*, **101**, 749 (1995).
16. Yu.G. Makhlin, V.P. Mineev, *ZhETF* [*Sov. Phys. JETP*] to appear (1996).
17. V.P. Mineev, Yu.G. Makhlin, *Pis'ma v ZhETF*, **62**, 576 (1995) [*JETP Letters*, **62** (1995)].
18. Yu. Bunkov, S. Fisher, *J. Low Temp. Phys.*, **101**, 123 (1995).

Diamagnetic Domains in Beryllium as Seen by Muon Spin Rotation Spectroscopy

G. Solt[1], C. Baines[1], V.S. Egorov[2], D. Herlach[1], E. Krasnoperov[2], U. Zimmermann[1]

[1] Paul Scherrer Institut, CH-5232 Villigen PSI, Switzerland

[2] Russian Research Center "Kurchatov Institute", Moscow 123182, Russia

Abstract: Condon's theory, predicting the periodic formation and disappearence of dia- and paramagnetic domains in non-magnetic metals under the conditions of a strong de Haas-van Alphen effect, is reviewed. Earlier experiments, evincing domains in silver and supporting the conjecture of domains in Be are discussed, and the recently obtained first spectroscopical evidence for the domain phase in beryllium is presented. The principle of muon spin rotation spectroscopy, used in these investigations, is described in some detail.

1 Introduction

The periodic formation and disappearence of dia- and paramagnetic domains in a normal, non-magnetic metal in strong magnetic fields at low temperatures, that is, under the conditions of the de Haas-van Alphen (dHvA) effect, was predicted by Condon [1]. The 'Condon' domains are spectacular manifestations of the *collective behaviour* of the electrons in quantized cyclotron orbits (Landau states): the effective magnetic field determining the individual Landau orbitals contains, besides the applied external field **H**, the oscillating dHvA-component from the responding electron system itself. This 'magnetic interaction' between electrons in Landau levels can be observed, in general, by the presence of strong higher harmonics in the dHvA signal for the magnetization $M(H)$ and susceptibility κ ('Shoenberg effect' [11, 3]). But when, for a particular orientation of the single crystal with respect to **H**, the *amplitude* of the dHvA oscillation for κ becomes sufficiently large, the magnetic induction **B** in the sample shows *qualitatively new* features. In these conditions B as a function of H behaves just like the specific volume v of a real (van der Waals) gas as a function of p in the range of the gas-liquid equilibrium: some intervals $I_j \equiv (B_{j,1}, B_{j,2})$ for B are forbidden and the local value of the induction makes a 'jump' between $B_1 and B_2$, as H varies continuously. Moreover, at a particular geometry and for certain intervals of H, uniformly magnetized regions, domains with one of the *two fixed values* $B_{j,1}, B_{j,2}$ coexist in the sample, and their *relative amount* varies with H. In the *diamagnetic* domains with $B = B_1 < H$, the magnetization $\mathbf{M}$ is opposite to $\mathbf{H}$; whereas the domains with $B = B_2 > H$ are *paramagnetic* ($\mathbf{M}||\mathbf{H}$).

Besides the 'thermodynamical' similarity to the gas-liquid equilibrium, the domain phase has an even deeper analogy with the *mixed state* of type 1 superconductors [4]. In both cases the not sufficiently 'flexible' quantum mechanical ground state of the electron system comes into conflict with electrodynamical boundary conditions, and the way out of this situation is the *macroscopic* break-up of the homogeneous state. Here lies the interest in the somewhat exotic Condon domains: an example of a 'tunable' many-electron ground state which undergoes, at some points, macroscopic 'breaks' while the tuning parameter H continuously varies. That the break-up occurs periodically with H, implying a *recurrence* of phase transitions within an extended range of H, where the dia- and paramagnetic domains return and subsequently disappear, is a *unique property* of Condon domains.

In view of these highly interesting features, it may surprise that, until very recently, direct evidence for Condon domains could be obtained only in the single case of silver (NMR, 1968 [5]). And not for lack of trying: beryllium which, for reasons explained in Ch. 3, has been a preferred 'domain'-candidate, was the subject of several studies, and a large body of *macroscopic* data has been collected in the dHvA regime (susceptibility, magnetoresistance and thermopower), each of them indeed consistent with the hypothesis of domain formation. However, spectroscopic studies by the method of NMR failed for Be, since "...the nuclear thermalization time of $\sim 1/2$ h and the inherent quadrupole splitting made the data collection and interpretation difficult" [5].

In Ch. 2 some basic facts about electron orbital magnetism in metals are reviewed, with emphasis on the *amplitude* of the dHvA effect, the key parameter in Condon's theory. After discussing the case of Be as candidate for a domain-forming metal, in Ch. 3 the principles of Muon Spin Rotation Spectroscopy (μSR) and our findings for Be are presented, Ch. 4 contains the conclusions.

2 Oscillating orbital magnetism: dHvA effect and Condon domains

The oscillating magnetic response of a non-magnetic metal, known as the de Haas - van Alphen (dHvA) effect, was discovered in bismuth in 1930 and first thought to be a mere curiosity, for "...bismuth has always been a black sheep because of its anomalous behaviour as regards electrical and other properties..." [6]. Yet, it turned out that the effect is present in *all metals* at sufficiently low temperatures, and that oscillations appear not only in the magnetization but in all physical properties related to conduction electrons, such as resistance, thermopower, sound velocities, specific heat. The dHvA effect has become a standard method of Fermi surface studies and is a textbook subject; here only a brief review of the features relevant for magnetic interaction and domain formation is given ([3]).

2.1 The dHvA oscillations of the magnetic susceptibility

The physical properties of the electron gas in a magnetic field show oscillations (Landau, 1930) [7] due to: 1) the quantization of the helicoidal orbits in planes normal to a homogeneous magnetic field **H** and 2) the existence of a sharply defined Fermi surface separating occupied from unoccupied states in wavenumber **k**-space.

For free, non-interacting electrons the quantum numbers **k**, labelling plane-wave states, are uniformly distributed. In the presence of a magnetic field **H** the projections of the orbits in a plane normal to **H** become closed and the components of **k** in this plane cease to be good quantum numbers. The orbital energies are

$$\epsilon_{nk_z} = \hbar\omega_0(n+1/2) + \hbar^2 k_z^2/2m; \qquad n = 0, 1, 2, ... \tag{1}$$

with the *cyclotron frequency* $\omega_0 = eH/mc$. The degeneracy or multiplicity of levels for each n, for an interval of length dk_z of the longitudinal wavenumber, is

$$D = (eV/\pi\hbar c) \cdot H \cdot (dk_z/2\pi) \tag{2}$$

where V is the volume of the system. On increasing H the levels move upwards, but their population changes: the electrons 'flow' into lower levels, owing to the increased multiplicity. The net result is an *oscillation* of $E_{total}(H)$ and $M(H) = -\partial E/\partial H$ [3], with a period corresponding to subsequent crossings of the Fermi surface by the Landau levels. Between two crossings at $k_z = 0$, according to Eq. (1), the quantity $\epsilon_F/(2\hbar\mu_B H) \equiv F/H$ changes by unity; the magnetic field F, which is the reciprocal of the period P for the variable $(1/H)$, is the 'dHvA frequency'. This picture, with some modification, remains true also for a non-spherical Fermi surface. The allowed orbitals in **k**-space are, quite generally, such that the areas $\mathcal{A}$, enclosed by their projection in a plane normal to **H**, satisfy the 'Onsager condition' [8]

$$\mathcal{A}(\epsilon, k_z) = 2\pi eH(n+\gamma)/\hbar c; \qquad n = 0, 1, 2... \tag{3}$$

where the constant $\gamma < 1$, in general, slightly departs from its value $\gamma = 1/2$ for a parabolic band. Equation (4) leads to quantized energies for the 'transverse' motion like Eq. (1), but with the cyclotron frequency $\omega_c = eH/m^*c$, where the 'cyclotron mass' belonging to a certain orbit is

$$m^* = \frac{\hbar^2}{2\pi}\left(\frac{\partial\mathcal{A}}{\partial\epsilon}\right)_{kz}. \tag{4}$$

The 'dHvA frequency' F in this general case is

$$F = \hbar c\mathcal{A}_{ext}/2\pi e \tag{5}$$

where $\mathcal{A}_{ext}$ is a maximum or minimum cross section area of the Fermi surface cut by planes normal to **H**. The *extremal* areas determine F since, by varying

H, the number of allowed orbits crossing the Fermi surface at $\mathcal{A}_{ext}$ is singularly large.

In all these expressions the homogeneous *magnetic field, keeping the electrons in cyclotron orbits*, has been traditionally denoted by **H** ([8, 3]). The microscopic magnetic field **h** is, however, position dependent. It can be seen that **H** in Eqs. (1-3) is actually the *spatial average* $\overline{\mathbf{h}(\mathbf{r})}$ within the sample, which is by definition the *induction* **B**. In fact, a cyclotron orbit area $\mathcal{A}$ in **k**-space corresponds to an enclosed area $A_r = (eH/\hbar c)^2 \mathcal{A}$ in *real space*, and the criterion Eq. (3) can be equivalently formulated as $A_r H \equiv \Phi = hc/e(n+\gamma)$: *it is the quantization of the magnetic flux* Φ, transversing A_r, which selects the allowed orbitals; $hc/e = 4.136 \cdot 10^{-7}\mathrm{Gcm}^2$ is the universal *magnetic flux quantum*. But the radius of an orbital $r_c = v_F/\omega_c = \hbar k_F c/eH = 6.6 \cdot 10^{-8}(k_F/H)$ is, for normal metals and even for magnetic fields of 10^5 G, larger than some thousands of Å (k_F is in cm^{-1}, H in G). Thus, the variation of $\mathbf{h}(\mathbf{r})$ on the scale of the lattice constant is 'averaged out', leaving the mean value **B**. From here on the field acting on the electrons will consequently be denoted by **B**, and **H** will stand for the applied field.

The exact expression for the 'longitudinal' (parallel to **H**) component M of the magnetization **M** is given by the 'Lifshitz-Kosevich (LK)' formula [9],

$$M = -M_0 \sin(2\pi F/B + \phi) + \sum_{p=2}^{\infty} -M_p \sin(2\pi p F/B + \phi_p), \tag{6}$$

by which the differential susceptibility $\kappa = \partial M/\partial B$ is

$$\kappa = \kappa_0 \cos(2\pi F/B + \phi) + \sum_{p=2}^{\infty} \kappa_p \cos(2\pi p F/B + \phi_p). \tag{7}$$

The oscillations are periodic in the variable F/B, as expected, but the amplitudes also vary, though slowly, with B; for κ_0 one has

$$\kappa_0 = 2\pi F M_0/B^2 = \left\{ \left(\frac{e}{\hbar c}\right)^{1/2} \frac{\hbar^2}{4\pi^4 m} \right\} \mathcal{A}_{ext}^2 \left(\frac{2\pi}{|\mathcal{A}''|}\right)^{1/2} \frac{m}{m^*} B^{-3/2} G_s R(T, x_D) \tag{8}$$

(only the terms rapidly varying with B were differentiated). The amplitudes of the higher harmonics have a similar form: κ_p contains an additional factor of $p^{-1/2}$ and has a *multiplier* p in the arguments of the 'reduction factors' G_s and R,

$$G_s = \cos(\pi g m^*/(2m)) \tag{9}$$

$$R(T, x_D) = \left\{2\pi^2 u/\sinh(2\pi^2 u)\right\} \exp(-2\pi^2 v); \qquad u = kT/\hbar\omega_c;$$
$$v = k x_D/\hbar\omega_c. \tag{10}$$

Here G_s accounts for the two spin directions and R for the phase smearings due to a smoothed Fermi edge at temperature T and to the electron relaxation time τ caused by scattering events; g is the spin-splitting factor for the conduction electrons and the 'Dingle temperature' is $x_D = \hbar/(2\pi k\tau)$. When the Fermi surface has more than one extremal cross section $\mathcal{A}^{(i)}_{ext}$ normal to the given direction of **H**, the contributions are superposed (this is the case for Be, Sect. 2.4).

While in usual Fermi surface studies the absolute value of the amplitudes, in particular κ_0, is less important, *for domain formation* precisely κ_0 is *the key factor.* By Eq. (8), κ_0 is determined by B, T and x_D and by the parameters of the Fermi surface at the extremal cross section normal to **H**. Besides $\mathcal{A}_{ext}$ and m^*, κ_0 depends also on the 'longitudinal' variation of the cross section $\mathcal{A}'' = \partial^2\mathcal{A}_{ext}/\partial dk_z^2$ (the derivative is taken along **H**). As Eq. (8) shows, κ_0 may be very large for a Fermi surface which is 'cylinder-shaped' near $\mathcal{A}_{ext}$, i.e. varying principally in *only two dimensions* normal to **H**. The numerical values of κ_0 will be considered in Sect. 2.2.

Normally, in Eq. (7) for κ only the first term has to be retained, since even for good crystal qualities ($v << 1$), at not too low temperatures and not too high fields (that is, unless $2\pi^2 kT/\hbar\omega_c < 1$), the factors R damp efficiently all $p > 1$ harmonics.

It is important that the argument B of the LK formulae is *not a control variable* (like the applied field H is), it contains also the response of the electron system. This leads to curious consequences, discussed in the Sect. 2.2.

2.2 Magnetic instability and forbidden B intervals

Considering first a long, thin sample, oriented along the homogeneous applied field, **H** does not change on penetrating the sample and the induction **B** is not restricted by boundary conditions. Therefore, within the sample

$$\partial H/\partial B = 1 - 4\pi\partial M/\partial B = 1 - 4\pi\kappa_0 \cos(2\pi F/B + \phi). \qquad (11)$$

Clearly, if the 'amplitude' $a = 4\pi\kappa_0$ is sufficiently large,

$$a \equiv 4\pi|\partial M/\partial B|_{max} = 4\pi\kappa_0 > 1, \qquad (12)$$

the derivative $\partial H/\partial B$ is *negative* for the entire *range* $\kappa > 1/(4\pi)$ within *each* dHvA cycle. A plot of $H(B)$, Fig. 1, shows that B becomes a *multivalued* function of H. However, *the states with* $\partial H/\partial B < 0$ *are thermodynamically unstable* [10] and never realized. The inequality $\partial H/\partial B > 0$ is the magnetic analogue of the stability condition $-dp/dv > 0$, true for any real gas, liquid or solid. It is known what happens when the homogeneous phase of the real (van der Waals) gas, for $T < T_c$, does not satisfy this requirement: the $v(p)$ function is then multivalued and, on, e.g., decreasing p, the specific volume v of the *real* system has a jump $v_{liq} \rightarrow v_{gas}$ (boiling), 'leaving out' the range $v_{liq} < v < v_{gas}$ which contains the instability interval. Exactly this happens for the long thin sample discussed here. Free energy arguments [11] show that, on varying H, the induction B jumps along the straight lines in Fig. 1, leaping over the 'forbidden intervals' with the instability range.

But in contrast to the van der Waals gas, our magnetic system can be described in fairly simple analytical terms. First of all, M and κ are periodic

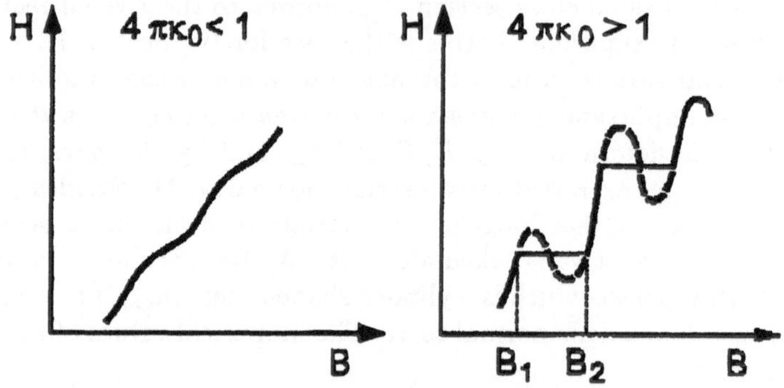

Fig. 1. Schematic H-B diagram for a long cylinder oriented along **H** for (a) $a < 1$ and (b) $a > 1$ [7]. The induction B 'jumps' along the straight lines, connecting B_1 and B_2 with the same free energy, similarly to the p-v diagram of a real gas.

functions of $1/B$ (Eqs. (6) and (7)), but for not too large B intervals they are approximately periodic also in *the variable* B. Indeed, by choosing $B = B_0 << F$, for convenience such that $M(B_0) = 0$ and $\kappa(B_0) = -\kappa_0$, the Taylor expansion of the phases in the sine and cosine of Eqs. (6) and (7) gives

$$M \approx M_0 \sin\{2\pi(B - B_0)/\Delta H - \pi\}; \qquad \kappa \approx \kappa_0 \cos\{2\pi(B - B_0)/\Delta H - \pi\} \quad (13)$$

where the period for the variable B,

$$\Delta H = B_0^2/F \quad (14)$$

is conventionally denoted by ΔH and called the 'dHvA period'. This approximation is good for B near B_0, i.e. for $|B - B_0| = \lambda B_0$ with $\lambda << 1$; this range may contain several dHvA cycles. (One sees also that $\kappa_0 = 2\pi M_0/\Delta H >> M_0/B$ in this case.)

The *magnetic interaction* of the electrons is immediately apparent in the structure of Eqs. (6) and (13). Since $B = H + 4\pi M$, the equations contain M on *both sides*, expressing the fact that the size of *each* cyclotron orbit depends on the resulting *collective* dHvA magnetization. In terms of the *control variable* H, Eq. (13) has the form

$$M = M_0 \sin\{2\pi(H - H_0 + 4\pi M)/\Delta H - \pi\} \quad (15)$$

since $M(H_0) = 0$. This, in the reduced variables [3] $2\pi(H - H_0)/\Delta H - \pi = x$ and $8\pi^2 M/\Delta H = y$ becomes

$$y = a \cdot \sin(x + y) \quad (16)$$

to be solved for $y = y(x)$ (in the original variables $M = M(H)$); the *amplitude of the magnetic interaction* a is $a = 8\pi^2 M_0/\Delta H = 4\pi\kappa_0$, met already in Eq. (12).

For $a < 1$, the exact solution $y = y(x)$ was obtained by Lagrange (1770) [12]; in modern notation it is [3]

$$y = \sum_{\nu=1}^{\infty}(2/\nu)J_\nu(\nu a)\sin\nu x \tag{17}$$

where J_ν is the ν-th Bessel function. For $a \geq 1$ the series is divergent, which is not surprising, since together with $B(H)$ also $y \sim 4\pi M(H) = B(H) - H$ has to be multivalued. The result by Eq. (17) is plotted for some $a < 1$ in Fig. 2a, while in Fig. 2b a graphical solution for $a > 1$ is shown, constructed by shifting the abscissa of the point $y = a\sin x$ (i.e. of $M = M(B)$, Eq. (15)) by an amount of y to the left. One sees that, for $a > 1$, $M(H)$ is three-valued in periodically situated 'critical' intervals.

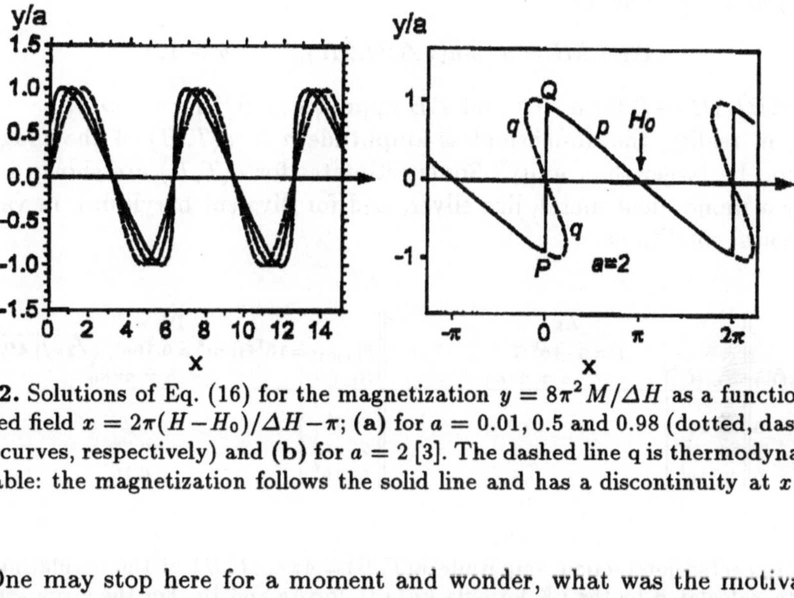

Fig. 2. Solutions of Eq. (16) for the magnetization $y = 8\pi^2 M/\Delta H$ as a function of the applied field $x = 2\pi(H - H_0)/\Delta H - \pi$; **(a)** for $a = 0.01, 0.5$ and 0.98 (dotted, dashed and solid curves, respectively) and **(b)** for $a = 2$ [3]. The dashed line q is thermodynamically unstable: the magnetization follows the solid line and has a discontinuity at $x = 0$ [3].

One may stop here for a moment and wonder, what was the motivation of Lagrange for dealing with Eq. (16), two centuries before the discovery of the dHvA effect? No surprise, the *physical problem* leading to the mathematical formulation of Eq. (16) at that time was different. But it was of capital interest and related, quite amusingly, also to 'orbits': Eq. (16) is just 'Kepler's equation' [12, 13] of astronomy, an implicit relation for the *variation of the angular position y with time x of a planet along its elliptical orbit* of eccentricity $e = a$. More precisely, $y = u - 2\pi t/T$, where the 'eccentric anomaly' u is the angular coordinate of the *projection* M' of the revolving planet M onto the parent circle of the ellipse; $x = 2\pi t/T$ is scaled time and T is the period of revolution. Thus, y is the instantaneous *advance* of the planetary phase u over the phase of a *hypothetical* planet m, revolving *uniformly* on the parent circle of the ellipse with the *mean* angular velocity $2\pi/T$. Kepler's equation is the integral of the equations for planetary motion [14], but implicit, and its solution, Eq. (17), was required for the *explicit* time dependence $u = u(t)$, to *predict* the position of the planet. Since in the astronomical problem $a = e < 1$, only this case was considered at the time. In the solution of Lagrange J_ν was given in power series form, constituting the first piece of what will be called the literature on Bessel functions.

Returning to Fig. 2a, y is nearly sinusoidal for $a = 0.01$: $M(H)$ and $M(B)$ are almost the same. With increasing a, $M(H)$ becomes more and more asymmetric; for $a = 0.98$ it has almost vertical 'steps'. This *triangular* shape and the related strong higher harmonics in the Fourier expansion of $M(H)$, due exclusively to the self-consistent contribution of M to B, is the 'Shoenberg effect' (1962, [11]).

For $a > 1$, with increasing H the magnetization *opposite* to $\mathbf{H}$ at $M = M_P < 0$ suddenly *turns*: $M \to M_Q$, becomes *parallel* to $\mathbf{H}$ and begins to *decrease* again. Thus, by the jump $M_P \to M_Q$ and the corresponding discontinuous change $B \to B + 8\pi M_Q$, the part with *paramagnetic differential susceptibility* is completely 'left out'of the dHvA cycle. From $\Delta B = 8\pi M_Q$ and $M_Q = M_0 \sin 8\pi^2 M_Q$ (Eq. (15) at $x = 0$) one has

$$\pi\Delta B/\Delta H = a \cdot \sin(\pi\Delta B/\Delta H); \qquad a > 1, \tag{18}$$

so that $\Delta B/\Delta H \to 0$ for $a \to 1$ and ΔB approaches ΔH for $a >> 1$.

Can, in reality, the dimensionless amplitude $a = a(T, B)$ of the magnetic interaction be larger than unity? Some estimates for $a(T, B)$ are shown in Table 1 for a monovalent metal, like silver, and for divalent beryllium, in various experimental conditions.

		Ag		Be	
		$F = 5 \cdot 10^8$ G		$F_{[0001]} = 10^7$ G; $m^* = 0.16m$; $\sqrt{2\pi/\lvert\mathcal{A}''\rvert} = 5$	
T(K)	x_D(K)	H_m(G)	$a = 4\pi\kappa_0$	H_m(G)	$a = 4\pi\kappa_0$
1	0	$6 \cdot 10^4$	10.0	10^4	1.8
	1	$13 \cdot 10^4$	2.0	$2 \cdot 10^4$	0.35
5	0	$3 \cdot 10^5$	0.9	$5 \cdot 10^4$	0.16

Table 1. Expected interaction amplitude $a(T, B) = 4\pi\kappa_0(T, H)$ of the oscillating susceptibility, calculated by the LK formula Eq. (8), for Ag and Be. For the given temperature T, sample purity x_D, and the particular field direction determining the 'dHvA frequency' $F = \hbar c \mathcal{A}_{ext}/2\pi e$ (Eq. (5)), H_m is the field at which κ_0 is at maximum, as determined by Eq. (8). (For Ag, $m = m^*$ and $|\mathcal{A}''| = 2\pi$ were taken and the anisotropy of F was neglected.)

One realizes that $a > 1$ is *by no means exceptional* for a sufficiently perfect single crystal at low temperatures. However, to actually reach these values, all *other* sources of phase smearing, in the first place inhomogeneity of the applied field, must be unimportant.

In conclusion, with growing interaction amplitude $a(T, B)$, the relation $M(H)$ becomes strongly anharmonic ($M(B)$, by Eq. (13), is *always* sinusoidal). For $a > 1$ a new phenomenon occurs: M and B make periodical *jumps* as H varies, some intervals for B become forbidden. It was seen that the geometry of Fermi surface for typical common metals *a priori* allows the condition $a > 1$ to be fulfilled.

2.3 Dia- and paramagnetic domains

The presence of forbidden intervals for $a > 1$ lead to spectacular consequences when additional constraints exist for B. Instead of the long, thin sample, consider now a single crystal *plate*, set *normal* to $\mathbf{H}$. In this case $\mathbf{H}$ *does* change when penetrating the sample: inside one has $H_{in} = H - 4\pi nM$, where $n \equiv n^z$ is the demagnetizing coefficient for the field in normal, z-direction. For the 'ideally' large, thin plate $n = 1$, and a *uniform* induction inside the sample satisfies the boundary condition $B = H_{ex}$ (recall that in our notation $H_{ex} \equiv H$).

But we can vary $H \equiv |\mathbf{H}|$ *continuously*, whereas B *cannot* take values in the interval $B_1 < B < B_2$! The conflict is clear: by the boundary condition, the uniform induction B should always 'follow' the applied H however this varies, but the cooperative Landau states resist, forbidding some ranges for B. For the electron system, uniformity of B in the plate and the boundary conditions *together* are, therefore, unacceptable. In this situation, similarly to the formation of the 'intermediate state' of a type 1 superconductor plate in a perpendicular field $H < H_c$, the conflict between the boundary conditions for a uniform state and the 'inflexible' quantum mechanical ground state is avoided by giving up *uniformity*. In respecting the 'forbidden interval' rules, imposed by the electron system, the only possibility is the breaking up of the homogeneous magnetic state and formation of a mixture of *macroscopic regions* with the two different, allowed B-values. This mixture should be such that, instead of the uniform condition $B = H$, the boundary condition be satisfied for the *sample average* $\overline{B}$, in the form $\overline{B} = H$, as the continuity of flux requires.

By virtue of the simplicity of the LK formulae, it is easy to follow the variation of M. Assume that H is tuned upwards from B_0; together with it H_i also increases. The state of the sample is *uniform and diamagnetic* ($M < 0$) until M has reached $M = M_P$ (Fig. 2b). At this point H_{in} has just increased up to the 'critical' value $H_{in} = H_{cr} = (B_1 + B_2)/2$, for which $x(H_{cr}) = 0$, and $H = B_1$. By tuning H further upwards, $H_{in} = H - 4\pi M$ remains *pinned* at H_{cr}, and M makes the jump $M_P \to M_Q$ only in *some domains* in the sample, within strips with walls running parallel to the field lines. The state is no more uniform, dia- and paramagnetic strips alternate. The pinning of H_{in} means that $H - 4\pi\overline{M} = H_{cr}$, i.e. the average magnetization must grow *linearly* from $\overline{M} = -M_Q$ to $\overline{M} = M_Q$, as $\overline{M} = (H - H_{cr})/4\pi$. The increase of $\overline{M}$ occurs by the growth of domains with paramagnetic ($M = M_Q$) polarization,

$$\overline{M} = (1 - \alpha_{||})M_P + \alpha_{||}M_Q; \quad \text{with} \quad \alpha_{||}(H) = (H - B_1)/(B_2 - B_1). \qquad (19)$$

Here $\alpha_{||}$ is the volume fraction of paramagnetic domains, increasing linearly with H. At $H = B_2$ all diamagnetic domains have disappeared, the magnetic state becomes uniform as it was for $H < B_1$, and $B = H$ can again be *locally* (and not only on the average) satisfied, until H reaches the *next* forbidden interval in the following dHvA cycle. The phase with the dia- and paramagnetic 'Condon domains' is drawn schematically in Fig. 3a, which also shows how the average boundary conditions are locally realized: domains and field lines are distorted in regions near the surface. In Fig. 3b the dependence of M on B is plotted for the plate geometry, showing that κ is constant all through the 'originally' oscillating $\kappa > 0$ region. The average domain thickness b can be estimated in the

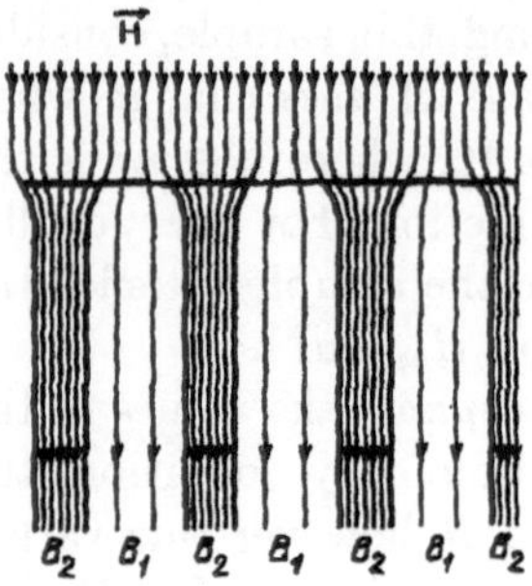

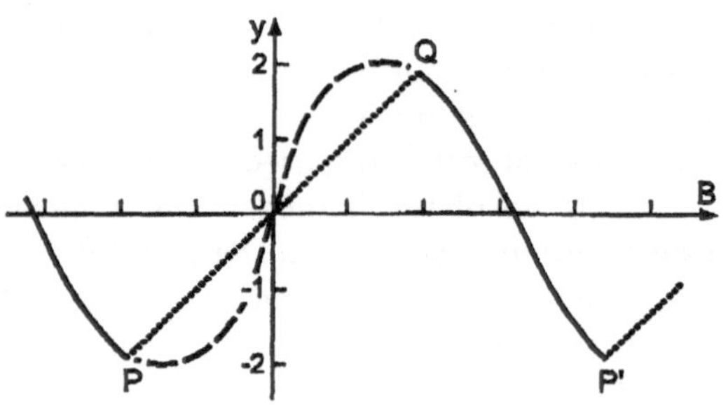

Fig. 3. **(a)** Dia- and paramagnetic domains in a single crystal plate for $B_1 < H < B_2$ [15]. Locally, B takes only one of the two values B_1, B_2, but $\overline{B} = H$. In the diamagnetic domains $M = M_P < 0$, in the paramagnetic ones $M = -M_P > 0$ (Fig. 2b). Note the inhomogeneous field range and the distortion of domains near the surface. **(b)** Reduced magnetization $y = 8\pi^2 M/\Delta H$ as a function of the induction B for $a = 2$ [3]. The dashed line is never realized, the dotted line shows the average magnetization $\overline{y}$ in the domain phase.

usual way, by minimizing the sum of domain wall and magnetostatic energies of the magnetized regions [3]. For a plate of thickness D, the minimum is at $b = C(a)(wD)^{1/2}$, the factor C is of the order of unity, except for $a - 1 << 1$, where it approaches zero. The wall thickness w should be on the scale of the cyclotron radius r_c (Sect. 2.1). For $r_c = 10^{-4}$ cm, corresponding to A_{ext} in Be

(Table 1), for $H = 10^4$ G and $D = 1$ mm, one has $b \approx 0.003$ cm. Theoretically, therefore, the walls are 'thin' compared to domain size, $w << b$. What does the experiment say?

2.4 Domains in silver. Domains in beryllium?

By Table 1, a good noble metal single crystal satisfies the domain-forming condition $a > 1$ (Eq. (12)) in fields of the order of $H \sim 10^5$ G. Out of the noble metals, a large, dominating single oscillation has been observed only for silver [11]. Also, unlike copper and gold, Ag nuclei have no electric quadrupole moment ($I = 1/2$), which makes silver more convenient for a NMR study. In their experiment, Condon and Walstedt [5] tuned H about $H_0 = 9 \cdot 10^4$ G at $T = 4.2$ K, and the NMR signal from the skin-depth layer of a single crystal Ag plate was detected. The dHvA period at this field is (Eq. (14)) $\Delta H = 16.8$ G. Within about 2/3 of the dHvA cycle *two distinct frequencies*, that is two values B_1 and B_2 were resolved; the two frequencies did not vary with H, and the line intensities followed the predicted linear change, Eq. (19), of the domain volume fractions. This beautiful experimental *tour de force confirmed* the existence of Condon domains in silver, showing at the same time the limitations of NMR in this field: for beryllium, ($I = 3/2$) the experiment was not conclusive, the quadrupole broadening turned out to be prohibitively large [5].

But Condon domains have been expected *in the first place* precisely in beryllium, on the basis of indications from various *macroscopic* data (magnetization and magnetothermal oscillations [1], magnetoresistance [16], susceptibility [17], thermopower [18]). The aboundance of data for the magnetic instability in Be is due to its particular Fermi surface, making the magnetic interaction observable already at the relatively low fields of $\sim 2 \cdot 10^4$ G (Table 1.) Consider first the experimental information available on $a(T, B) = 4\pi\kappa_0$. The relevant cross sections $\mathcal{A}_{ext}$ of the Fermi surface, normal to the hexagonal axis [0001], are the 'waist' and the two 'hips' seen in Fig. 4. The difference betweeen F_w and F_h, given by Eq. (5), is only about 3%. Therefore $\kappa(B)$, as a superposition of *two* sine functions (Eq. (13)), has a beating, a sinusoidally modulated amplitude κ_0 with nodes at every $F/\Delta F \approx 33$-th period. This is an example of where, although the fundamental amplitudes in Eq. (8) are indeed slowly varying with B, the amplitude κ_0 resulting from *two* dHvA oscillations changes radically within some tens of dHvA periods. The data for κ [17] at $T = 4.2$ K, Fig. 5a, show that $a(T, B)$ at this temperature and field ($H_0 \approx 20700$ G) is *below* the critical value, $a \approx 4\pi \cdot 0.023 = 0.29$. The picture changes dramatically upon lowering T, Fig. 5b. Judging from the *minima* of κ, the largest κ_0 at $T = 1.4$ K has become ~ 10 times larger, by a factor of ≈ 2.7 *above* the critical $1/4\pi$. However, apart from the range near the nodes of κ_0, the corresponding peaks for $\kappa > 0$ are *missing*, κ_{max} has the constant value $0.075 \approx 1/(4\pi)$ along the beat cycle, as if the parts of the dHvA cycles with $4\pi\kappa > 0.94$ were cut down ! This is a strong argument for the predicted domain phase: according to Fig. 3b and Eq. (19), for $a > 1$ one has a sinusoidal shape only for $\kappa < 0$, and a constant $\kappa = 1/4\pi$

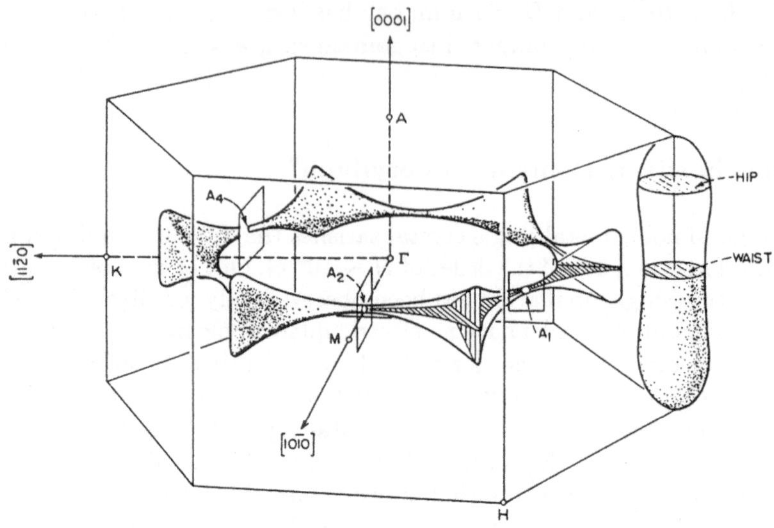

Fig. 4. The Brillouin zone and Fermi surface of Be. The third zone contains the electron 'cigar', of which the two extremal sections normal to [0001] determine, by Eq. (5), the dHvA frequency for the 'hips' $F_h = 9.8 \cdot 10^6$ G and the ≈3% lower one for the 'waist', $F_w = 9.5 \cdot 10^6$ G [16].

instead of the sinusoidal $\kappa > 0$ half-cycle. The measured temperature dependence $a/(4\pi) = \kappa_0(T)$ in Fig. 5c indicates that, for the given Be crystal, Condon domains can exist only for $T < 3$ K.

Similar lineshape analyses for other macroscopic properties of beryllium in the dHvA regime added [16, 3, 18] further 'circumstantial evidence' [1] for the domain phase. The long expected *direct observation* of dia- and paramagnetic Condon domains in Be has become possible only recently, with the development of Muon Spin Rotation spectroscopy.

3 Direct evidence for domains by Muon Spin Rotation (μSR) spectroscopy

Spectroscopy by μ particles, implanted into a solid and 'reporting' about the local magnetic fields, is a relatively new technique, exploiting parity violation in the π and μ decay processes. Here only some principal features of the method can be presented; for a detailed description, see [19, 20, 21].

3.1 Principles of the μSR method

The possibility of using muons as spectroscopical probes in condensed matter is based on the non-conservation of parity in weak interactions. In the decay of a

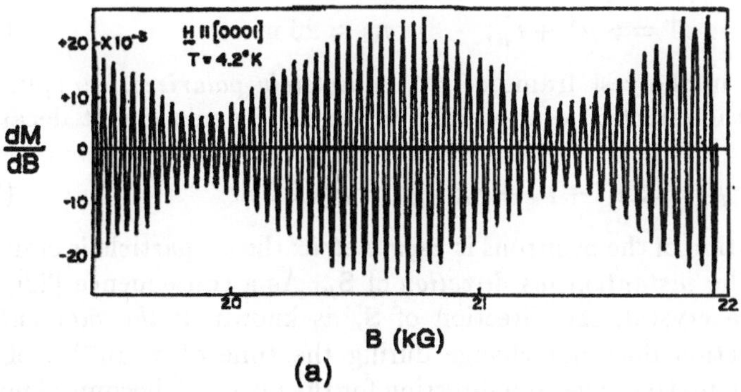

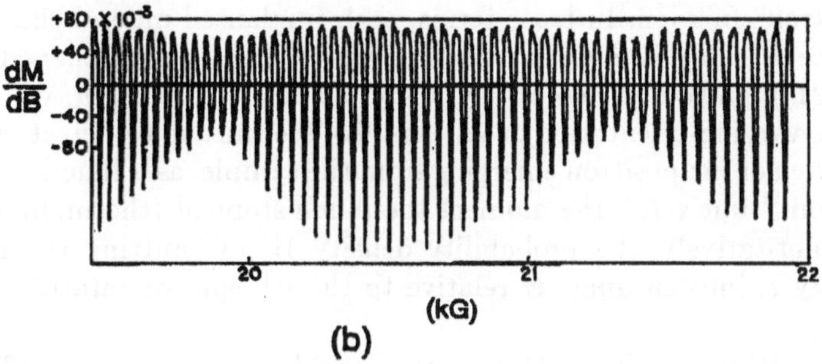

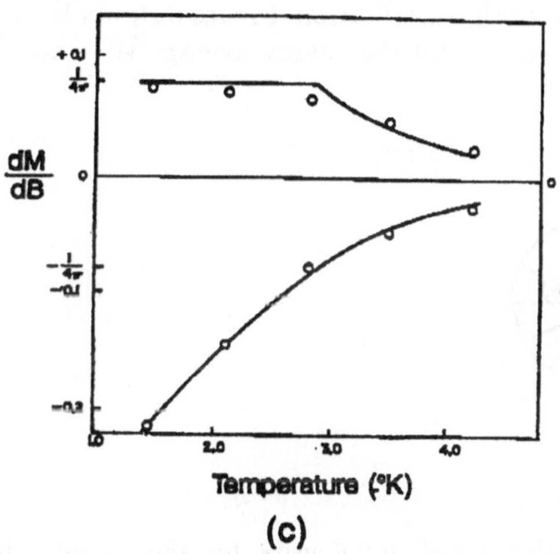

Fig. 5. Differential susceptibility κ single crystal Be, **H**||[0001] [17]. The beating is due to the *two* dHvA frequencies F_w and F_h, see Fig. (4). **(a)** For $T = 4.2$ K the dHvA oscillations are symmetrical. **(b)** At lower temperature, $T = 1.4$ K, the amplitude κ_0 becomes larger, owing to a factor R (Eqs. (8) and (10)) nearer to unity. But the $\kappa > 1/4\pi$ part of each cycle is 'cut short', as expected by Fig. 3b. **(c)** Observed amplitudes $\kappa_0(T)$ for $\kappa < 0$ and $\kappa > 0$. No domain formation is indicated for $T > 3$ K.

π^+ particle,

$$\pi^+ \Longrightarrow \mu^+ + \nu_\mu; \qquad \tau_{pi} \approx 26 \text{ ns} \tag{20}$$

the emitted μ^+ is, in the rest frame of π^+, *completely polarized:* its spin $\mathbf{S}_\mu$ points opposite to the direction of motion. Further, in the subsequent decay of the *muon*,

$$\mu^+ \Longrightarrow e^+ + \nu_e + \overline{\nu_\mu}; \qquad \tau_\mu \approx 2.2 \ \mu\text{s} \tag{21}$$

the angular distribution of the positrons is *asymmetric*: the e^+ particle is emitted preferentially into the *instantaneous direction* of $\mathbf{S}_\mu$. As a consequence [22], for a μ^+ absorbed in a crystal, the direction of $\mathbf{S}_\mu$ is known *at the moment of incidence*; this direction does not change during the time of $\approx 10^{-9}$ s of its *thermalization* in the matter. The spin direction for the given μ^+ becomes known again for a *second time*, at the moment of the μ^+ *decay*: it is revealed by the asymmetry of positron emission. In loose terms, each implanted muon behaves as a magnetized 'torch', only it is not a pencil of visible light but of energetic e^+ particles (Eq. (21)) emitted preferentially *along the axis* of the torch, which allows us to *see* the varying orientation of its axis. To 'see' means to detect the varying *flight direction* of the positron emerging from the sample, as a function of the time elapsed from 'time zero', the moment the muon stopped (thermalized) in the sample. Quantitatively, the probability density W of emitting the e^+ particle, with energy ϵ, into an angle Θ relative to the μ^+ spin orientation, is [19]

$$W_\epsilon(\Theta)d\Theta d\epsilon = c_\epsilon(1 + A_\epsilon \cos\Theta)d\Theta d\epsilon \tag{22}$$

where A is called the 'asymmetry' of the distribution. Figure 6 shows W for an e^+ emitted with the highest energy ϵ_m and for the energy average $\overline{W} \sim 1 + \overline{A}\cos\Theta$

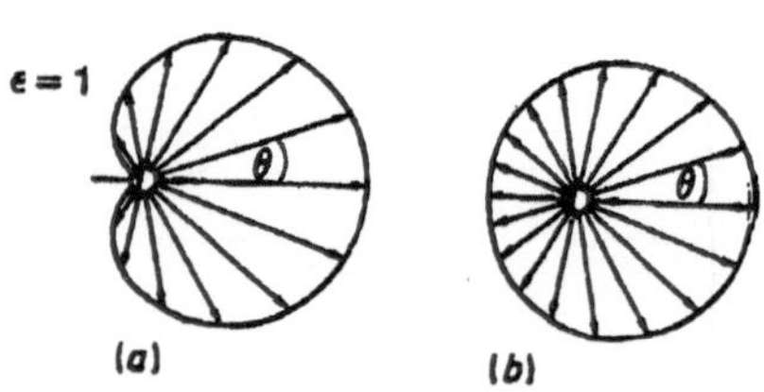

Fig. 6. Polar diagram $W(\Theta)$ of the probability density for the angular distribution of emitted positrons in μ^+ decay [19]; **(a)** for the most energetic positrons $W(\epsilon_{max}) \sim 1 + \cos\Theta$, **(b)** the average over all positron energies, $\overline{W} \sim 1 + (1/3)\cos\Theta$.

observed when positrons of all energies are counted. For ϵ_m (52.8 MeV) the 'pencil' of emitted e^+'s is narrower ($A_m = 1$), but part of the asymmetry persists also in the average, $\overline{A} = 1/3$. For the asymmetry to be detected, the positrons need to *escape* from the sample and reach the detectors, but this is what happens, since for a mean e^+ energy of ≈ 36 MeV the sample is, as a rule, transparent.

Clearly, μSR has many similarities to other methods, in particular to the 'free induction decay' in NMR, PAC (perturbed angular correlation) and Mössbauer spectroscopies, which are also used to investigate local magnetic fields by observing precession of *nuclei*. Since the implanted μ^+ stops at *interstices* and, as a light particle ($m_\mu \approx m_p/9$), may diffuse in the lattice, μSR is also adequate to study classical and quantum diffusion. As compared to NMR, μSR differs in that *no field is needed* to produce the *initial* polarization of the probe (this is set by the 'elementary' property of Eq. (20)), and also that, ideally, *each* individual μ^+ decay is detected, making the method very sensitive. At the same time, the implanted μ^+ *modifies* its environment, distorts the lattice and perturbs the local electronic structure; this has to be kept in mind when analyzing the results. As for the *time scale*, limited from above by the life-time τ_μ to $\leq 10^{-5}$ s, it extends down to $\sim 10^{-12}$ s. In Table 2, some 'elementary' characteristics of the μ^+ particle are shown. The μ^+ beam is produced at high energy (~ 600 MeV)

property	value
mass m_μ	$206.768 \cdot m_e$
charge	$+e$
spin I	1/2
gyromagnetic ratio $\gamma_\mu/2\pi$	13.5538 kH/G
lifetime τ_μ	2.197 μs

Table 2. Some properties of the μ^+ particle, relevant for μSR spectroscopy

proton accelerators. By hitting a production target, the protons generate π-mesons, which begin to decay according to Eq. (20). The 'surface muons', best adapted for most solid state research, come from those pions stopped in the target near its surface; the polarization of surface μ^+'s is nearly 100%. The μ^+ beam ($E = 4$ MeV, $p = 29.8$ MeV/c) is transported to the sample, surrounded by e^+ detectors covering different directions. The penetration length of surface muons is $\approx (150 \pm 10)$ mg cm$^{-2}/\rho$ for a sample density ρ; for Be this is ≈ 0.8 mm.

From the different variants of the μSR technique [19, 20, 21], only the 'primary', *time differential – transverse field* μSR is of interest here. In this method **H** is (ideally) *perpendicular* to the incident μ^+ polarization. Each μ^+ is (in principle) detected *on entering* the sample ($t = t_0$), a clock is started and the positron detectors become ready to count. Within the next $\approx 10^{-5}$ s, at $t = t_0 + \Delta t$, the muon decays, and the emitted e^+ is detected by, say, the i-th detector. The signal from the detector stops the clock, and the event is counted in a 'bin', labelled by the the time delay Δt, of a rate-vs-time *histogram* corresponding to the direction i. For a periodic motion of the μ^+-spin $\mathbf{S}_\mu(t)$ with period T, the rate N of events in the bins Δt and $\Delta t + T$ will be statistically identical (when corrected for the factor e^{-t/τ_μ}): the precession is manifested by an *oscillating*

$N(\Delta t)$ function. The arrival of a *second* μ^+ within $10-20$ μs makes Δt ambiguous, therefore the incident muon rate cannot be too high, and standard nuclear electronics has to be used to reduce background and spurious counts of diverse origin. An experimental set-up is schematically drawn in Fig. 7. To understand

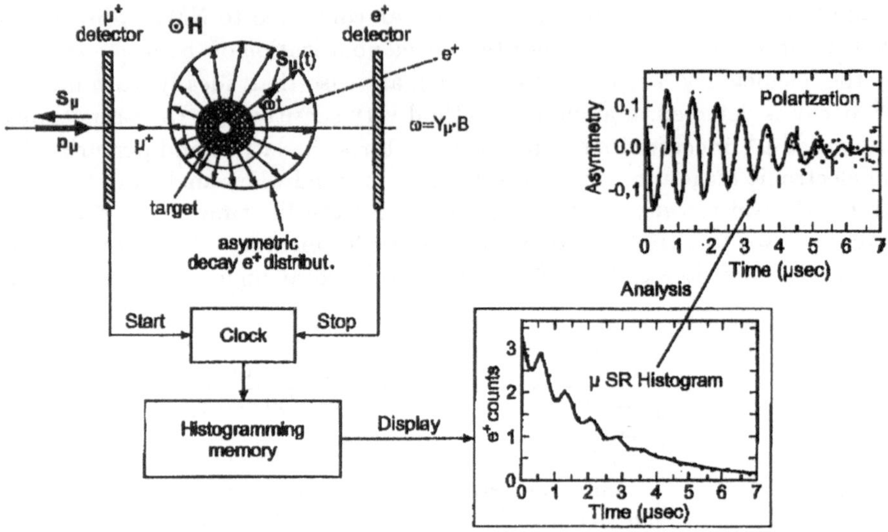

Fig. 7. Principle of the time-differential transverse field μSR experiment. At incidence, the μ^+ spin $\mathbf{S}_\mu$ is opposite to the μ^+ momentum, $\mathbf{H}$ is perpendicular to the plane of the drawing. The precession of $\mathbf{S}_\mu$ about $\mathbf{B}$ is seen as a damped oscillatory signal (Eqs. (24-27)) in the 'forward' detector (as well as, with different phases δ_i, in the detectors 'backward', 'up' and 'down', omitted in the figure).

the form of the μ^+SR signal, consider first a single μ^+ subjected to the field B. The probability that the decay e^+ is emitted in the forward, $\Theta = 0$, direction at the moment $t_0 + \Delta t$ is, from Eq. (22),

$$\overline{W}(\Theta = 0) \sim e^{-\Delta t/\tau_\mu}(1 + \overline{A}\cos(\gamma_\mu B \Delta t + \pi)) \tag{23}$$

since, for this geometry, the angular distance between the precessing $\mathbf{S}_\mu$ and the 'forward' direction is $\Theta = \pi + \gamma_\mu Bt$. Apart from the lifetime factor, a pure cosine-like variation is predicted for the number of positrons counted by the 'forward' detector. What we actually observe is the signal from a *statistical ensemble* of muons. The rate of positrons counted by detector i at time $t \equiv \Delta t$ is

$$N_i(t) = N_{0i}e^{-t/\tau_\mu}(1 + \overline{A}f_i G(t)\cos(\omega_B t + \delta_i)) + b_i; \quad \text{with} \quad \omega_B = \gamma_\mu \overline{B}. \tag{24}$$

where N_{0i} depends on the beam intensity and on the position, size and efficiency of the i-th detector, the ideal asymmetry $\overline{A} = 1/3$ is somewhat reduced by the factor $f_i < 1$ due to incomplete initial polarization and the finite solid angle of the detector, δ_i is a phase and b_i the background.

In the hypothetical case of a unique **B** for all muons, Eq. (23) would contain the 'whole truth', and for the *relaxation function* G one would have $G(t) \equiv 1$. In reality, even for a constant *applied field* **H**, each μ^+ experiences a different local field **B**, corresponding to its 'random' stopping in a particular magnetic environment. Random local field contributions arise from disordered nuclear magnetic moments, and in the presence of non-uniform electron magnetism, like flux-lines in superconductors and spin density waves. Besides varying *spatially*, **B** varies, in general, also with *time*: the muon may diffuse through interstices and 'see' a variable environment, but also the dynamics of the neighbouring magnetic moments *at a given site* may become manifest during the μ^+ lifetime. For the μ^+ ensemble, this random variation in environments means that the *amplitude* of the observed polarization $\mathbf{P}(t) = 2 < \mathbf{S}_\mu >$, precessing at a frequency $\omega_1 = \gamma_\mu \overline{B}$, decays with the time, since however small the deviations $B - \overline{B}$ are for the individual muons, the result is an increasingly large *random* phase difference $\gamma_\mu \int (B - \overline{B})dt$ as time elapses. The function $G(t)$ describes this relaxation due to dephasing; $G(t{=}0) = 1$, and $G(t) < 1$ is determined by the dephasing mechanisms in the given physical situation. For example, for static, *random* local fields with a normal (Gaussian) distribution the theory gives [23]

$$G(t) = \exp(-\sigma^2 t^2/2); \qquad \text{with} \quad \sigma^2 = \gamma_\mu^2 \overline{(B - \overline{B})^2}, \tag{25}$$

while for a mobile μ^+ with a mean residence time τ_c at a site one has 'motional narrowing' with an exponential G-function,

$$G(t) = \exp(-\lambda t); \qquad \text{with} \quad \lambda = \sigma^2 \tau_c \tag{26}$$

if only $\sigma\tau_c << 1$. Though each local static field is the result of a given *discrete* configuration of the neighbouring spins, and therefore the static field distribution is certainly *not* exactly Gaussian, the above expressions for G are in many cases sufficiently near to reality. The Fourier transform of $N(t)e^{t/\tau_\mu}$, $\mathcal{F}(\omega)$, has a peak at $\omega = \omega_B$, with a width $\delta\omega$ and a shape corresponding to $G(t)$. A Gaussian G implies $\mathcal{F}$ to be also Gaussian with full width of $\delta\omega = 2\sigma$, while for of exponential G, Eq. (26), the peak in $\mathcal{F}$ is Lorentzian, $\delta\omega = 2\lambda$.

Apart from the quasi-continuous distribution of fields about a single $\overline{B}$, several *components* with different $\overline{B}$'s occur when, for example, more than one magnetically inequivalent site for the μ^+ are available, or if domains with different magnetization coexist in the sample. Then instead of Eq. (24) one has a sum for all $j = 1, 2, ..., n$ components,

$$N_i(t) = N_{i0} e^{-t/\tau_\mu} \sum_j \left\{ 1 + a_i^{(j)} G^{(j)}(t) \cos(\omega^{(j)} t + \delta_i^{(j)}) \right\} + b_i; \quad \text{with} \quad \omega^{(j)} = \gamma_\mu B^{(j)} \tag{27}$$

where $G_i^{(j)}(0) = 1$ and $\sum_j a_i^{(j)} = \overline{A} f_i$.

The spectroscopical information, to be unravelled from the observed $N_i(t)$ spectrum, is contained in the quantities $\omega^{(j)}$ and $G^{(j)}$, giving the mean values

$\overline{B}^{(j)}$ and the spectral lineshapes for the j-th 'mode' of field distribution, and in the 'partial' asymmetries a_i, revealing the proportion of muons precessing in this mode. (A study of $\delta_i^{(j)}$ may also be useful in some problems.)

3.2 New μSR results for beryllium

The idea is simple: when Condon domains are formed with $\overline{B}^{(1)} = B_1$ and $\overline{B}^{(2)} = B_2$, two μ^+ precession frequencies have to appear, according to Eq. (27), just as in the NMR study for Ag. We have seen that the NMR experiment for Be failed mainly because of large line broadening – does μSR any better?

There are reasons for the answer to be yes, as recognized already some time ago [24]. First, even the 'slow' surface muons have a penetration depth of some tenths of a millimetre, thus the 'sampling' of the local fields is not restricted by the skin depth. Next, because $S_\mu = 1/2$, the μ^+ 'test' particle has no quadrupolar moment, unlike most of the nuclei including ^{9}Be, and the line broadening comes only from the random *magnetic* fields at the μ^+ site. Apart from electron orbital magnetization, which is the subject under study, the unique sources for such fields are the ^{9}Be nuclei ($I = 3/2$), with a dipole moment of $\mu_{Be} = 0.38 \cdot 10^4 \hbar I$ (CGS units). The disordered moments exert a field of $\sim \delta B \sim \mu_{Be}/r_{i\mu}^3$ at the μ^+ site, where $r_{i\mu}$ is the nucleus-muon distance. For $r_{i\mu} \sim 2.5\text{Å}$ one obtains $\delta B \sim 1$ G. This corresponds to a line broadening of $\delta\omega = \gamma_\mu \delta B$, which is *much smaller* than $\omega^{(2)} - \omega^{(1)} = \gamma_\mu \Delta B$ (we have seen that ΔB is comparable to the dHvA period which, for $H = 27400$ G is, by Eq. (14), ≈ 78 G). (The electric quadrupole interactions of the Be nuclei *modify* the precise value of δB, but they do not change this order of magnitude.) Therefore the dia- and paramagnetic frequencies should appear *distinctly* in the μSR signal if, first, the muons do not diffuse out of the domain where they have been stopped, and second, if the fraction of muons precessing in a field B *in between* B_1 and B_2 is very small. The first proviso is not severe, since during the time of some τ_μ the muon is displaced at most by some lattice constants. For the second, the muon can observe a field B between B_1, B_2 only if it is stopped in a domain wall, where B changes *smoothly* from B_1 to B_2. In a spectroscopical method like μSR or NMR, observing two distinct frequencies means, therefore, that the ratio of domain wall to domain thickness is sufficiently small.

The estimated value of $\delta\omega$ and the adequacy of the *exponential* relaxation function G at He-temperatures was confirmed by preliminary experiments. However, while the above value of H ensures a conveniently large dHvA period, it is necessary to modify the set-up of Fig. 7: in high fields perpendicular to $\mathbf{v}_\mu$, the μ^+ beam becomes strongly deflected. To avoid this, $\mathbf{S}_\mu$, primarily collinear with the beam, is turned 'upwards' by an angle of (ideally) 90^0 before implantation, on passing through a 'spin-rotator' (a region with crossed electric and magnetic fields, leaving the velocities unchanged), and $\mathbf{H}$ is set *parallel* to $\mathbf{v}_\mu$. The beam is no more deflected, but $\mathbf{H}$ is still 'transverse' for $\mathbf{S}_\mu$, and the precession of $\mathbf{S}_\mu$ is observed now in the 'l' (left) and 'r' (right) detectors, Fig. 8.

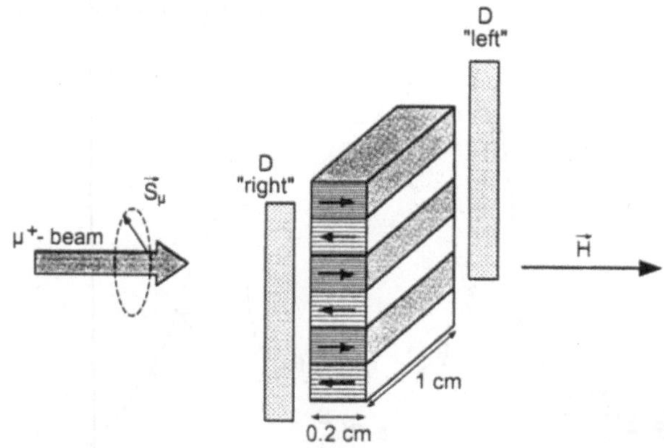

Fig. 8. Observation of Condon domains by μSR. The scheme in Fig. 7 is modified by rotating, before incidence, $\mathbf{S}_\mu$ 'upwards' with respect to the beam direction, and setting $\mathbf{H}$ parallel to the beam. The precession is observed in detectors 'left' and 'right'.

For the experiment [25] the Be plate, cut normal to the [0001] direction, was set perpendicular to $\mathbf{H}$. The field H was first scanned at $T = 0.8$ K from 27480 G downwards, by steps of 7 G, and the *oscillating part* of the rates N in detectors 'l' and 'r' were described by the formula

$$\tilde{N}_i = N_0 \sum_{j=1,2 \text{ or } j=1} a_i^{(j)} e^{-\lambda_j t} \cos(2\pi\nu^{(j)}t + \delta_i^{(j)}); \qquad (i = l, r) \qquad (28)$$

(this is the oscillating part of Eq. (27) without lifetime factor and constants, $2\pi\nu = \omega$ and for G the exponential Eq. (26) was used). One single frequency should result for the uniformly magnetized state, and *two* are expected for the domain phase. The tentative use of Eq. (27) with a *single frequency* ($j = 1$) all over the dHvA cycle gives for the damping rate λ, the result plotted in Fig. 9. The figure shows what one can call the *dHvA oscillations of the* μ^+ *depolarization rate*: the period in $\lambda(H)$ *coincides* with the dHvA period ΔH! The large, periodic increase of λ is what one expects in this 'one-frequency' representation, since without introducing a second frequency, the *splitting* due to domains should appear as a large line *broadening* of the single 'allowed' line.

Although this 'broadening model' is really adequate only for a Lorentzian line shape (like that observed for the uniform phase), while the 'line' in our case is a well resolved doublet for the range $\lambda \geq 0.4\mu\text{s}^{-1}$ as seen below, the maximum $\lambda \approx 0.9\mu\,\text{s}^{-1}$ gives a qualitatively correct idea of the splitting, $\Delta\nu \approx \lambda/\pi \approx 0.3$ MHz, equivalent to $\Delta B \approx \Delta\nu/(\gamma_\mu/2\pi) = 22$ G. The exact value is ≈ 29 G, so that the interval $B_1 < H < B_2$, where the domains exist, is *more than one third* of the period ΔH.

The Fourier transform of $\tilde{N}(t)$ for the central peak region of Fig. 9 is seen in Fig. 10. The field H for the upper spectrum is above H_m, at which λ is

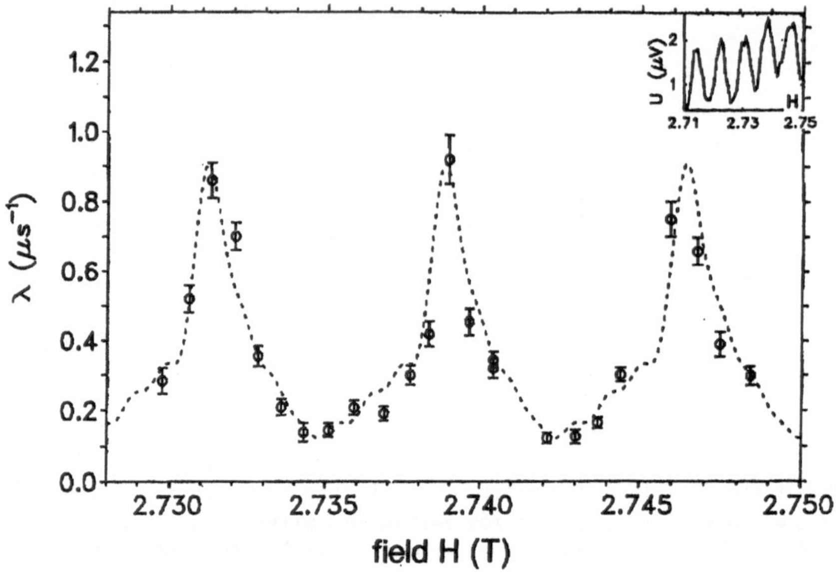

Fig. 9. Exponential damping rate λ of the precessing μ^+ polarization, as a function of H. The periodic sharp rise of λ (i.e. of the linewidth $\Delta\omega = 2\lambda$ at the frequency $\omega = \gamma_\mu B$) marks the onset of line splitting due to the domain phase. For $\lambda > 0.4\mu s^{-1}$ the 'broadened' line turns out to be a well resolved doublet (Fig. 10). The dashed line is a best fit to the data by a (truncated) Fourier series with a period of $\Delta H = 76$ G. This period agrees, within the present accuracy, with the dHvA period $\Delta H = 78.2$ G at $H_0 = 2.74$ T. The *in situ* measured oscillating thermopower is shown as insert [25].

maximum in Fig. 9, the spectrum in the middle is at H_m, and the lower one is for $H < H_m$. The *position* of the doublet lines is stable as H sweeps downwards, and the 'transfer' of intensity into the lower line shows how the diamagnetic domains grow while the paramagnetic ones decrease, Eq. (19).

The use of Eq. (28), allowing for *two* ω_j's there, where the splitting is resolved, leads to the result in Fig. 11. This shows the μ^+ frequency or frequencies, together with the normalized asymmetries $\overline{a^{(j)}} = a^{(j)}/(a^{(1)} + a^{(2)})$. Since $\nu = (\gamma_\mu/2\pi)B$, the ordinate is the induction B in MHz units, giving $\Delta B = 28.8 \pm 1.4$ G at this $H_0 = 2.740$ T. According to Eq. (18), the value $\Delta B/\Delta H = 0.37$ corresponds to $a = 4\pi\kappa_0 = 1.27$. The normalized asymmetry $\overline{a}^{(1)}$ is just the relative intensity of line 1, equal to the volume fraction $\alpha_{||}$ of Eq. (19); $\overline{a}^{(1)}$ and $\overline{a}^{(2)} = 1 - \alpha_{||}$ are seen to vary linearly with H, as predicted.

The beats in κ_0, a special property of the dHvA effect in beryllium for **H**||[0001] shown in Fig. 5, makes it possible to 'scan' also the interaction parameter $a = 4\pi\kappa_0$, and see the variation of $\Delta B/\Delta H$ (Eq. (18)) in a small H-interval. The doublet at $H_0 = 26412$ G, situated by ≈ 13 dHvA periods lower than previously, is shown in Fig. 12a. The splitting of ≈ 0.6 MHz gives $\Delta B \approx 40$ G, larger

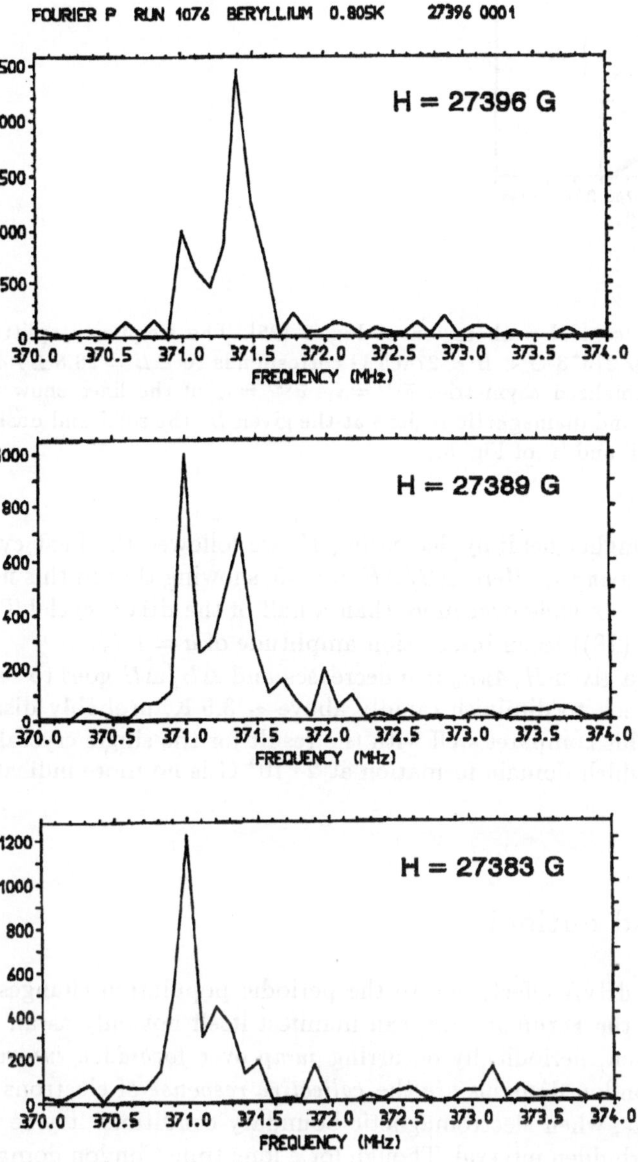

Fig. 10. Frequency spectrum data showing the splitted μSR line in the central peak region of Fig. 9 [25]. The applied field H decreases downwards in the figure. The position of the doublet lines ($\Delta\nu = 0.39$ MHz) is stable, the amplitude of the higher frequency line decreases and that of the lower increases, as the diamagnetic domains grow at the expense of the paramagnetic ones.

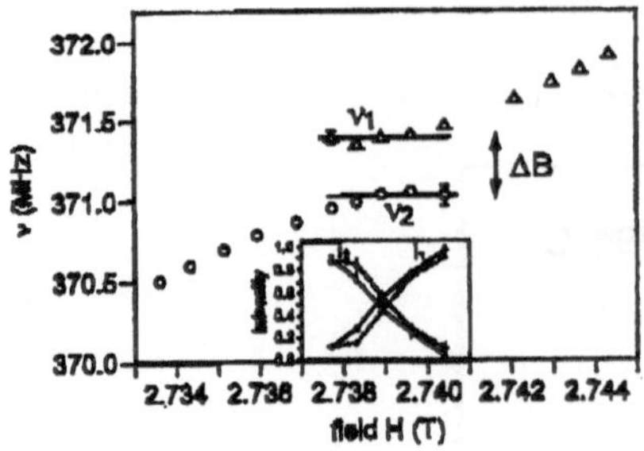

Fig. 11. Analysis of the central peak region of Fig. 9 [25]. The frequency splitting $\Delta\nu = 0.39 \pm 0.02$ MHz for 27378 G $< H <$ 27405 G corresponds to $\Delta B = 28.8 \pm 1.4$ G in the domains. The normalized asymetries $\overline{a}^{(1)} = i_1$, $\overline{a}^{(2)} = i_2$ of the lines show the relative volumes of para- and diamagnetic regions at the given H; the solid and dashed curves are for detectors 'l' and 'r' of Fig. 8.

than for the previous, higher field: by decreasing H, we followed the beat cycle in the direction of *increasing* κ_0. Here $\Delta B/\Delta H = 0.55$, showing that in this field range the domain phase extends over more than a half of the dHvA cycle! This value corresponds (Eq. (18)) to an interaction amplitude of $a = 1.75$.

By increasing T for a given H, $4\pi\kappa_0 = a$ decreases and $\Delta B/\Delta H$ goes to zero. In Fig. 12b, $\gamma_\mu \Delta B$ is seen to diminish rapidly above $\sim$ 3.5 K, probably disappearing for $T > 4$ K. This compares well with the result for the single crystal in ref.[17] in Fig. 5c, for which domain formation at $2 \cdot 10^4$ G is no more indicated above $\approx$3 K.

4 Conclusions and outlook

It was shown how the dHvA effect, due to the periodic population changes in Landau levels crossing the Fermi surface, can manifest itself not only as an *oscillation* but as a sudden, periodically occurring *jump* over *forbidden intervals* of the induction B. Condon domains are the *collective response* of electrons on Landau cyclotron orbits, when electromagnetic boundary conditions 'try to enforce' B to enter the forbidden interval. Though for a long time Condon domains had been observed solely in silver, and only very recently also in beryllium, it turns out that the condition $4\pi\kappa > 1$ for domain formation has *a priori* nothing exceptional: it is predicted to hold for sufficiently pure single crystals of *most* metals at low temperatures. The scarcity of data is due to experimental difficulties: 30 μm thin domains in a typical, 1 mm thick slab have to be seen in highly homogeneous fields of $10^4 - 10^5$ G, at $T \approx 1$ K.

That is why our knowledge of the Condon phase, even after the spectroscopic evidences in Ag and Be, is still rather *qualitative*. We know that the domains

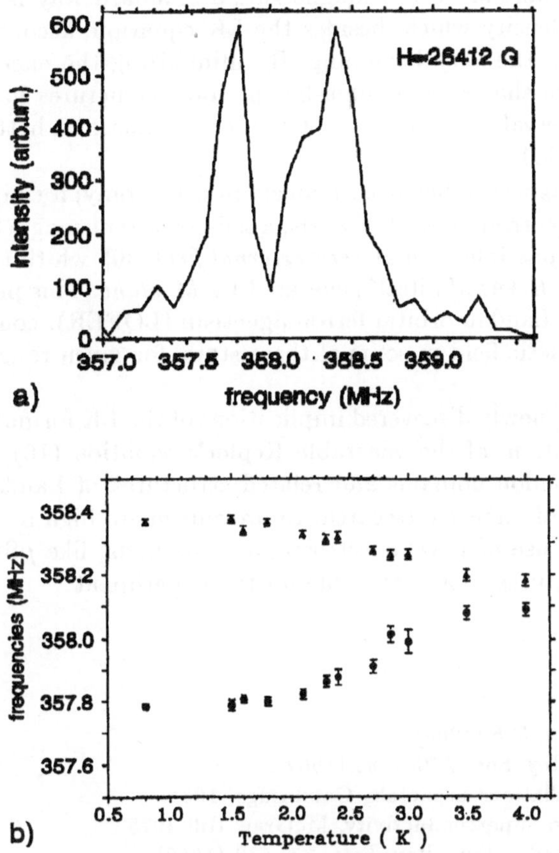

Fig. 12. **(a)** Line splitting at an applied field $H = 26412$ G, by ≈ 13 dHvA periods lower than in Fig. 10 [26]. The splitting in the doublet is ~ 0.6 MHz, corresponding to $\Delta B = (\omega_2 - \omega_1)/\gamma_\mu = 40$ G. **(b)** Temperature dependence of ΔB at the same field H as in (a).

form, as predicted, in phase with the dHvA cycles, and the sharp spectral lines in both metals confirmed that $w << b$, the wall regions with inhomogeneous B are indeed much thinner than the domains themselves. But how thin: is the assumption $w \approx r_c$ correct? What is the *form* of $\mathbf{M}(\mathbf{r})$ in the wall region, as function of the parameter $a(T,B) = 4\pi\kappa_0(T,H)$? Other questions concern the phase transition: how does the diamagnetic domain *nucleate* near $H = B_2$? What kind of a singularity is at $a(T,B) = 1$? And more generally, is the phase composed of essentially *homogeneous*, oppositely magnetized strips the most favourable configuration based on the LK equation?

The Condon domains, discussed in some length in these notes, correspond to the simplest self-consistent solution of the LK equation. More sophisticated solutions, like magnetization-density-waves (MDW) may also exist [27, 28], though not yet found experimentally. To predict quite generally the stable, periodic magnetic structures $\mathbf{M}(\mathbf{r})$ of Landau orbital magnetism, a standard way is to start [29] from a free energy density which, besides the LK equation, accounts also for the 'wall energy' from the very beginning. By minimizing the energy functional, a variety of domain shapes or domain-like periodic structures have been predicted for different intervals of T and H; to test these predictions, better experimental resolution is needed.

A new aspect of orbital magnetism has been realized in astronomy: for high electron densities, like those occurring in white dwarfs, a self-consistent magnetic solution of the LK equation is possible even at *zero external field* [30], whith the interaction amplitude $a(T, B = H + 4\pi M)$ itself generated by M *alone*. This permanent magnetic state, called Landau orbital ferromagnetism (LOFER), could account for the observed magnetic field in some of these stars, for given ranges of density and temperature.

In view of all the surprising, newly discovered implications of the LK formula, this quantum mechanical 'relation' of the venerable Kepler's equation (16), it is easy to understand that Condon domains and related structures of Landau orbital magnetism are subject of intensive research. Improvements in high magnetic field technology and the use of novel spectroscopical methods, like μSR, will make these proposed structures more accessible for the experiment.

References

1. J.H. Condon, Phys. Rev. **145**, 526 (1966).
2. D. Shoenberg, Phil. Trans. Roy. Soc. **A255**, 85 (1962).
3. D. Shoenberg, Magnetic oscillations in metals, Cambridge, 1984.
4. M. Tinkham, Introduction to Superconductivity, McGraw Hill 1975.
5. J.H. Condon and R.E. Walstedt, Phys. Rev. Lett. **21**, 612 (1968).
6. J.H. van Vleck, Theory of Electric and Magnetic Susceptibilities, Clarendon 1932.
7. L.D. Landau and E.M Lifshitz, Statistical Physics, part 1, Pergamon 1980.
8. L. Onsager, Phil. Mag. **43**, 1006, (1952) and I.M. Lifshitz 1954, cited in [3].
9. I.M. Lifshitz and A.M. Kosevich, Sov. Phys. JETP **2**, 636, (1955).
10. L.D. Landau and E.M Lifshitz, Electrodynamics of Continuous Media, Pergamon 1984.
11. A.B. Pippard, Proc. Phys. Soc. (London) **A272**, 192 (1963).
12. J.L. Lagrange, Oeuvres, ed. M.J.-A. Serret, vol.3, Gauthier-Villars, Paris 1869, pp.113-138.
13. D.H. Menzel, Fundamental Formulas of Physics, Dover 1960.
14. L.D. Landau and E.M Lifshitz, Mechanics, Pergamon 1976.
15. A.A. Abrikosov, Fundamentals of the Theory of Metals, North-Holland 1988.
16. W.A. Read and J.H. Condon, Phys. Rev. **B1**, 3504 (1970).
17. L.R. Testardi and J.H. Condon, Phys. Rev. **B1**, 3928 (1970).
18. V.S. Egorov, Sov. Phys. JETP **45**, 1161 (1977) and Sov.Phys. Solid State **30**, 1253 (1988).
19. A. Schenck, Muon Spin Rotation Spectroscopy, 1986 Hilger, Bristol.
20. S.F.J. Cox, J. Phys. **C20**, 3187, (1987).
21. E. Karlsson, Solid State Phenomena as Seen by Muons, Protons and Excited Nuclei, Oxford 1995.

22. R.L. Garwin, L.M. Lederman, M. Weinrich, Phys. Rev. **105**, 1415, (1957), and J.L. Friedman, V.L. Telegdi, Phys. Rev. **106**, 1290, (1957).
23. A. Abragam, Principles of Nuclear Magnetism, Oxford 1986.
24. Yu. M. Belousov and V.P. Smilga, Sov. Phys. Solid State **21**, 1417 (1979).
25. G.Solt, C. Baines, V.S. Egorov, D. Herlach, E. Krasnoperov, U. Zimmermann, Phys. Rev. Lett. **76**, 2575, (1996).
26. G.Solt, C. Baines, V.S. Egorov, D. Herlach, E. Krasnoperov, U. Zimmermann, Hyp. Inter. 1996, to appear.
27. J.J. Quinn, Phys. Rev. Lett. **16**, 731, (1966).
28. M.Ya. Azbel, Sov. Phys. JETP **26**, 1033, (1969).
29. M.A. Itskovski, G.F. Kventsel, T. Maniv, Phys. Rev. **B50**, 6779, (1994).
30. H.J Lee, V. Canuto, H.Y. Chiu, C. Chiuderi, Phys. Rev. Lett. **23**, 390, (1969).

Growth Instabilities in M.B.E.

Christophe Duport [2] *Paolo Politi* [1] *and Jacques Villain* [2]

[1] CEA, Département de Recherche Fondamentale sur la Matière Condensée SPMM/MP, 38054 Grenoble Cedex 9, France
[2] CEA, Département de Recherche Fondamentale sur la Matière Condensée, SPSMS 38054 Grenoble Cedex 9, France

Abstract: A growing crystal or any other object is said to have an instability when its surface does not remain planar but instead develops patterns such as dendrites. After a short outline of the instabilities in growth from the melt or solution or vapour, a detailed analysis is given of the instabilities in molecular beam epitaxy (MBE).

1 Growth instabilities

1.1 Definition

Consider a semi-infinite crystal initially limited by a plane perpendicular to the z-axis. Suppose that this crystal is growing under the effect of a temperature gradient parallel to z, or of an atomic beam of uniform intensity in space and parallel to z. If the surface, instead of remaining smooth, develops patterns of well-defined shape, we will say that there is a growth instability.

There is some difference between instabilities and roughness. We shall say that a surface is rough when there are small irregularities without characteristic length, while instabilities result in macroscopic features with a fairly well-defined length scale.

1.2 Examples

Snowflakes are familiar examples. They grow from a supersaturated vapour. The molecules of the vapour diffuse until they meet the solid. If there are protruding parts, diffusing molecules go first to the protruding parts (Fig. 1) because it is the shortest way to the growing solid. This is the mechanism of the instability. In the case of snow, the protruding parts initially form as a result of crystal symmetry and the first six branches develop. Then, irregularities arise on these branches because of stochastic fluctuations and then increase more or less deterministically, giving rise to dendrites.

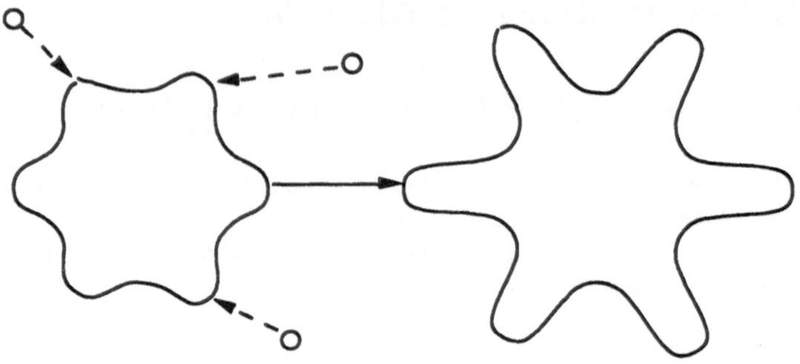

Fig. 1. Formation of a snow-flake from the undercooled vapour. In the absence of anisotropy, the initial shape would be a circle. The anisotropy first transforms the circle into a hexagon. Then, water molecules diffunding from the vapour go preferably to the corners of the hexagon, thus transforming them into dendrites. The formation of secondary dentrites has a similar origin.

This mechanism can be implemented by simulations. In the so-called DLA model (Diffusion limited aggregation), particles stop moving as soon as they meet the solid. This model gives rise to very irregular shapes. More regular shapes arise when diffusion at the solid-fluid interface takes place. Very often, theorists just neglect stochastic effects and write deterministic equations, more precisely the diffusion equation in the fluid, together with appropriate boundary conditions at the interface.

Similar instabilities arise in the growth from a supersaturated solution, or from the supercooled melt. Again, diffusion is an essential ingredient. In growth from the melt, the diffusing quantity is the impurity concentration or (if the melt is really very pure) the energy[1]. Generally, energy diffuses much more rapidly than impurities and does not play a great role.

1.3 Instabilities due to surface tension

It is often impossible to grow a material on another material because the adsorbed molecules do not like to be in contact with the substrate. In more scientific terms, the free energy per unit area $\sigma_{s\ell}$ (or interface tension) of the substrate-liquid or substrate-solid interface is too large. The adsorbate then forms droplets (Fig. 2). This type of growth is called Volmer-Weber[2, 3] type of growth, or Stranski-Krastanov if a few smooth atomic layers (one, two, three...) form before the drops. These types of growth thus correspond to the case when the adsorbate does not wet the substrate.

The quantitative condition for the instability to take place is of course that the substrate-vapour interface free energy σ_{sv} is less than the sum $\sigma_{s\ell} + \sigma_{v\ell}$

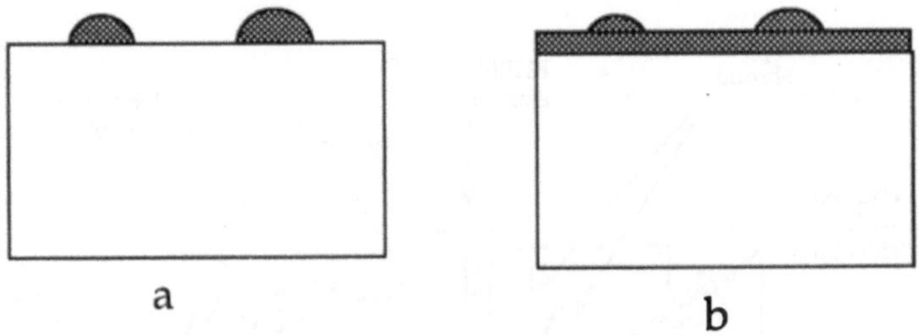

Fig. 2. Vollmer-Weber (a) and Stranski-Krastanov (b) growth

of the free energies of the substrate-condensed phase interface and the vapour-condensed phase interface, i.e.

$$\sigma_{sv} < \sigma_{s\ell} + \sigma_{v\ell}. \tag{1}$$

This condition can also be deduced from Young's formula which gives the contact angle of a drop with the substrate. Exercise: do that!

1.4 Surfactants

Surfactants are impurities which lower the surface tension. If one finds a surfactant which lowers $\sigma_{s\ell}$, then one can induce wetting.

We shall mainly be concerned by the case of a solid adsorbate. In that case, thermodynamic equilibrium is not always easy to reach... and this is good! Thus, it is possible to build objects which are very useful (multilayers, quantum wells, etc.) but metastable. Of course, this is only possible if the temperature is not too high. Thus, in addition to thermodynamic phenomena, we will be concerned with kinetic effects.

2 Molecular beam epitaxy

The remainder of this review is devoted to instabilities in molecular beam epitaxy. This method is used to grow complex crystals with controlled purity, especially multilayers.

Roughly speaking, molecular beam epitaxy (MBE) is defined as follows (Fig. 3). Under ultra-high vacuum conditions, atoms (single or in molecules) are sent onto the surface, where they diffuse until they meet a step where they are incorporated. These diffusing atoms are called *adatoms*. Free adatom diffusion clearly has a meaning only on a high symmetry terrace (between steps).

Imagine a crystal delimited by a high symmetry plane, (111) or (001). Suppose that this crystal grows by MBE. Atoms land onto the surface from the beam, are adsorbed and become adatoms. What do they do next?

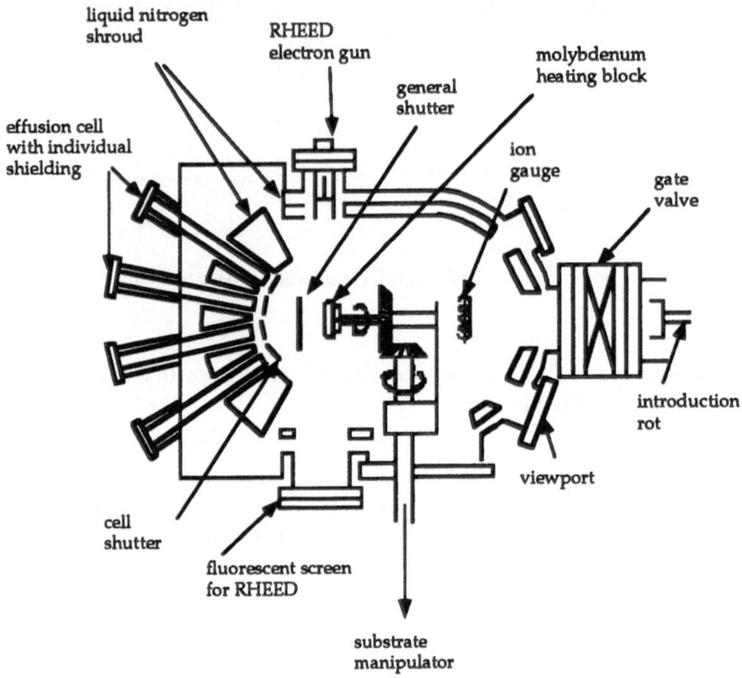

Fig. 3. Example of a MBE apparatus

We have taken care to choose a substrate temperature where surface diffusion is effective. Thus, each single adatom wanders about until it finds a good reason for halting.

The good reason might be another adatom. They then form a pair, or *dimer*, which is likely to be stable – that is, not willing to break up into two adatoms again – at low enough temperatures. Other adatoms will come to join the dimer, which constitutes thus the nucleus of a whole terrace. The same process is going on farther away, and terraces continue to form on the surface. When there are enough terraces, they will efficiently compete with each other for incoming adatoms, no new nucleation will take place, and terraces grow until they coalesce to yield a new complete layer.

If we now increase the substrate temperature, the dimers will no longer be stable enough to act as nuclei. It is likely that *trimers* will be, until the temperature becomes too high. At still higher temperature, critical nuclei are clusters of many atoms, and the distance between them increases. Eventually, growth no longer takes place through nucleation of terraces: rather, adatoms diffuse far enough to directly reach pre-existing steps. In fact, any real surface is atomically flat only up to a certain typical lengthscale, depending mostly on sample preparation condition. Steps are usually present due to an "imperfect" cut (*miscut*) away from the high symmetry direction. Their distance rarely exceeds a few microns, and is typically around 100 nanometers on a "nominally" high symmetry surface.

Also, steps are unavoidably present when screw dislocations cut the surface. Nowadays, macroscopic crystals of silicon or other semiconductors can be grown free of dislocations.

3 The Schwoebel effect

As seen above, atoms from the beam diffuse to a step where they are incorporated[4]. Now, a step has a "lower" side and an "upper" side (the surface is assumed horizontal, the solid being below). Do steps absorb atoms from the upper and the lower terrace with the same probability? There is no ground for this assumption[5, 6], which is in fact seen not to be true in several instances. Fig. 4 a shows the potential seen by an adatom[7]. The curve has a maximum which should be overcome by adatoms willing to go downstairs and prevents them from easily sticking to the down step. One can easily understand this maximum ("Schwoebel barrier"), since an adatom willing to step down should go through an uncomfortable position (Fig. 4 c) where it has few neighbours. This is the intuitive basis of the calculation of Bourdin et al. On the other hand, the energy barrier may be lower if the edge atom is pushed away by the incoming adatom (Fig. 4 d) which takes its place. Numerical calculations within the *effective medium theory* (see for instance the review by Stoltze[8]) show that on the (100) face of metals such an "exchange" mechanism has a lower energy than the over–edge "jump" for Au, but the opposite is true for Cu, Ag, Ni, Pd and Pt. On the (111) face the "jump" mechanism seems to be always favored. Both barriers are generally higher than for diffusion on a flat surface. However, in the only case where ab initio calculations have been performed, namely Aluminum, these calculations predict no Schwoebel barrier[9]).

4 The Schwoebel instability for a growing surface of high symmetry orientation

As seen in Section 2, if the growing surface has a high symmetry orientation, the growth results in the formation of "terraces" of atomic height [10, 3]. Atoms landing on these terraces cannot easily step down [5, 6] and therefore nucleate new terraces at the top of terraces, and finally towers or "mounds". Such mounds have been observed experimentally on Si [11], Cu [12] and Fe [13].

One might expect the instability to take place only for a strong Schwoebel effect. As a matter of fact, it takes place even with a weak Schwoebel effect. This can be seen from a continuous model in which the local surface current density $\boldsymbol{j}$ is assumed to be a function of the local slope $\boldsymbol{m} \equiv \nabla z$. The current density contains two terms.

i)The first term says that the adatoms do not like to go downstairs, which implies that the current is parallel to the slope, with the same sign. For weak

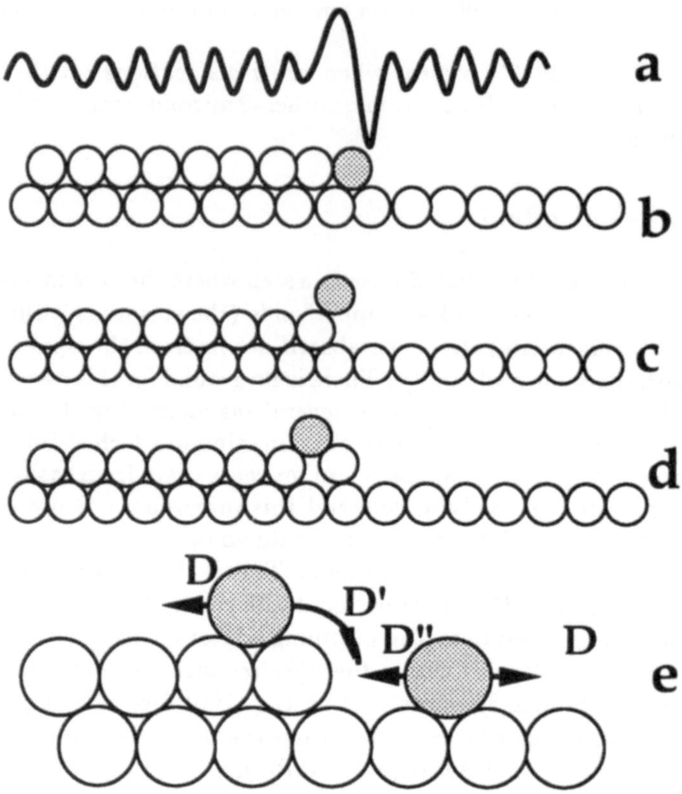

Fig. 4. a) Theoretical potential felt by an adatom. b) Intuitive reason of the minimum minimorum. c,d) Intuitive reason of the maximum maximorum. Scheme (c) corresponds to the mechanism considered by Bourdin et al, and scheme (d) to the exchange process (see Stoltze's review). e) Interpretation of jumping rates D, D', D''.

m, the current is proportional to m, but for larger slopes, there is a saturation effect and the current even turns out to decrease for large m.

ii) The second term is a healing term of the form $j = -\nabla\mu$, which expresses the fact that atoms try to decrease their free energy, i.e. their chemical potential μ. The latter is generally assumed to be proportional to the local curvature (this is a particular case of the Gibbs-Thomson formula). Finally, one obtains[14, 15, 16, 17]

$$j = \frac{F\ell_s\ell_c m}{2(1+|m|\ell_s)(1+|m|\ell_c)} + K\frac{\partial^2 m}{\partial x^2}. \tag{2}$$

This equation should be combined with the conservation law $\dot{z} = -\boldsymbol{\nabla}\cdot\boldsymbol{j}_s$. A linear stability analysis easily shows that the surface is stable with respect to fluctuations of wavelength shorter than the lower limit[17]

$$\lambda_c^{inf} = 2\pi\sqrt{\frac{2K}{F\ell_s\ell_c}}. \tag{3}$$

The surface is unstable with respect to fluctuations of longer wavelength. The instability appears after a time of order

$$t^* \approx \frac{K}{(F\ell_s\ell_c)^2}. \tag{4}$$

The equations of motion have "fixed points" which correspond to a periodic array of "mounds". These profiles are stable with respect to an amplitude variation, but unstable with respect to mound coalescence. The stability with respect to amplitude variations results from the following argument: on the right hand side of Eq. (2), the second term is stabilizing, while the first is destabilizing. The latter is less effective for large slopes, so that the slope has an upper limit. However, this limit increases with the wavelength of the modulation, because the stabilizing term of Eq. (2) becomes less effective. If large slopes are allowed, large amplitudes result, and indeed the amplitude turns out to be proportional to the square of the wavelength[16].

Now, if coalescence is not too fast, each mound has time to get close to the stationary profile corresponding to its radius. Numerical simulations performed on a one-dimensional model [18, 17] confirm this property. Therefore, the shape of the mounds becomes sharper when coarsening takes place[16, 17]. This is in agreement with some simulations and some experiments, but other simulations and other experiments suggest a "self-similar" shape, i.e. the mounds have a constant "aspect ratio" $h(t)/R(t)$ of height to radius.

Are there two different regimes? This question is still open. If there is a self-similar regime, it seems hard to describe it with the simple equation Eq. (2).

Simulations are often analyzed in terms of power laws and find $R(t) \sim t^{1/z}$, where z varies between 3 and 5. Such power laws are generally observed in self-similar structures and would therefore be compatible with a constant aspect ratio. If $h(t)/R(t)$ goes to infinity with time, a slower coarsening should be expected.

Of course the aspect ratio cannot really diverge. When it becomes of order unity, something has to occur. In that limit, the microscopic properties of the system or of the model become important. The expected universal-regime is $h(t)/R(t) \ll 1$, when the expected universal-equation Eq. (2) holds. Most simulations, and some experiments, are in this regime.

5 The Schwoebel effect for a vicinal surface: step bunching instability

A vicinal (001) surface, for instance, is a surface which is nearly a (001) surface. Vicinus means near in latin (as well as vicino in italian and vecino in spanish). Thus a vicinal (or "stepped") surface is a succession of high symmetry terraces separated by steps. These steps are generally more or less straight.

Let us assume that one terrace, A, is broader than the other ones (Fig. 5). When the crystal is growing by MBE (Fig. 5 a) the broad terrace A gathers more atoms, which preferably go to the upward step edge of B because of the Schwoebel effect. Consequently, the width of the broad terrace A decreases. Thus, the Schwoebel effect erases the irregularity.

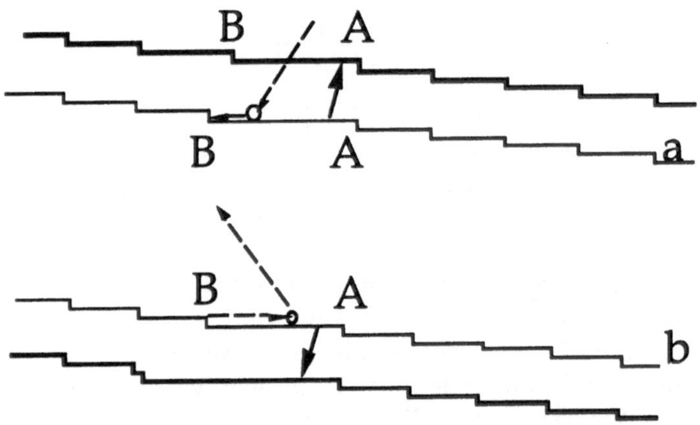

Fig. 5. Evolution of a broader terrace A when the Schwoebel effect is present. The thin line shows the initial profile and the thick line shows the profile at a later time. a) During growth, the terrace A gathers more atoms which preferably go to the upper ledge B. This ledge proceeds faster, so that the terrace A becomes narrower and the regular profile is stabilised. b) During evaporation, the terrace A evaporates more atoms which mostly come from the upper ledge B. This ledge recedes faster and catches up the next step, thus forming a bunch of two steps.

For an *evapourating* (instead of growing) stepped surface, the Schwoebel effect has more spectacular consequences. The broad terrace A now evapourates more atoms. These atoms predominantly come from the upward step edge of B, because the Schwoebel effect is difficult to cross in both directions. Therefore, this step recedes faster than the other ones and the broad terrace becomes even broader. *The Schwoebel effect is destabilizing during evapouration.* Eventually, steps form bunches. In contrast, the Schwoebel effect is stabilizing (with respect to step bunching) during growth.

Exceptionally, an inverse Schwoebel effect can occur: atoms prefer to go downstairs. It is easy to prove that this inverse Schwoebel effect is stabilizing with respect to step bunching during evapouration and destabilizing during growth.

One can wonder whether the instability of an evapourating, regular array of steps leads to a stable, less regular array. The step just above the broad terrace is caught by the upper step. However, if the broader terrace is initially just a little broader, its width will then cease to increase. But this step pairing is not the final stage, because all steps will pair[19, 20]. When all steps are paired, the system of double steps is again unstable as the system of single step was. Then the pairs form pairs, i.e. step quadruplets. Then the quadruplets form pairs of quadruplets, which are octuplets, etc. The step bunching process is not limited and does not lead to a steady state. However, the formation of large bunches may require a long time.

6 Step meandering on a growing vicinal surface

Let us consider more in detail the problem of a growing vicinal surface with Schwoebel effect. In the preceding sections, straight steps were assumed. But do they remain straight? In order to discuss the stability of a straight step, we can assume a weak perturbation to be present and ask whether it increases or decreases with time. The perturbation of the step may be a small bump (Fig. 6). Because of the Schwoebel effect, the adatoms mainly reach the step from the terrace ahead. Those which are just in front of the bump will of course mainly go to the bump itself, and those which are very far from the bump will not reach it. The important point is that the adatoms which are not just in front of the bump, but not too far, will preferentially diffuse to the nearest point on the step, which lies on the bump. Thus, the bump grows and the straight shape is unstable.

This mechanism is very similar to the formation of snowflakes, addressed in section 1, or to the production of fractal aggregates in DLA.

Thus, a growing vicinal surface with normal Schwoebel effect is unstable with respect to step meandering[21]. In the same way, it is easy to prove that

i) an evapourating vicinal surface with normal Schwoebel effect is stable with respect to step meandering;

ii) a growing vicinal surface with inverse Schwoebel effect is also stable with respect to step meandering;

iii) an evapourating vicinal surface with inverse Schwoebel effect is unstable with respect to step meandering.

7 Misfit dislocations in molecular beam heteroepitaxy

A layer of atoms adsorbed on a foreign substrate at equilibrium, is subjected to two competing effects, namely the adatom-adatom interaction (described by a Hamiltonian $\mathcal{H}_{aa}$) which favours an interatomic distance a equal to the natural distance a_0 in a free adatom layer and the adatom-substrate interaction $\mathcal{H}_{as}$,

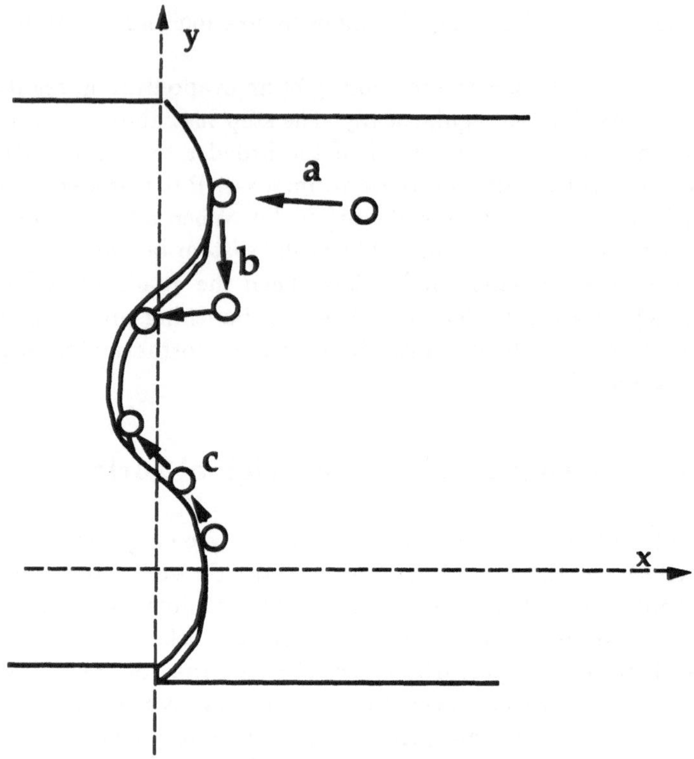

Fig. 6. A sinusoidal perturbation on a step: will the amplitude decrease or increase with time? The answer to this question is in the linear stability analysis. Adatoms go preferentially to the tips (a), and this effect is destabilizing. Then, in order to minimize the interface energy, they can diffuse to the holes after being reemitted (b) or slide along the step (c). Effects (b) and (c) tend to stabilize the straight step. They are neglected in the standard DLA model.

which, on the other hand, favours the commensurate structure where the atomic distance a of the adsorbate is equal to the lattice parameter of the substrate, b. It is nearly obvious that, for a given *lattice misfit* $(a_0 - b)/b$ the incommensurate structure ($a \neq b$) is stabilized by strong adatom-adatom couplings, compared to the adatom-substrate interaction; if the latter is the stronger, the commensurate ($a = b$) structure is stable. In this case, we say there is *epitaxy*. It is less obvious that the commensurate structure is stable even when $\mathcal{H}_{as}$ is weak, provided the misfit is small enough. In fact, the adsorbate-adsorbate coupling energy lost by letting the interatomic distance vary from a to b is proportional to $(a_0 - b)^2$, whilst the adsorbate-substrate energy gain is independent of $a_0 - b$.

What happens, at equilibrium, when the first monolayer is commensurate and one increases the number h of layers? Clearly, when h becomes very large the structure becomes incommensurate – at equilibrium. In fact, the energy gain in letting the adsorbate take its natural interatomic distance a is proportional to the layer volume, and eventually overcomes the energy loss due to giving up commensuration, which is only proportional to the interface area. If the first layers nonetheless stay commensurate, the appearance of the incommensurate structure requires the introduction of *misfit dislocations* (see Fig. 7). The critical thickness for the appearance of misfit dislocation at equilibrium has been investigated in many works, which are reviewed, for instance, by Jesser and van der Merwe [22].

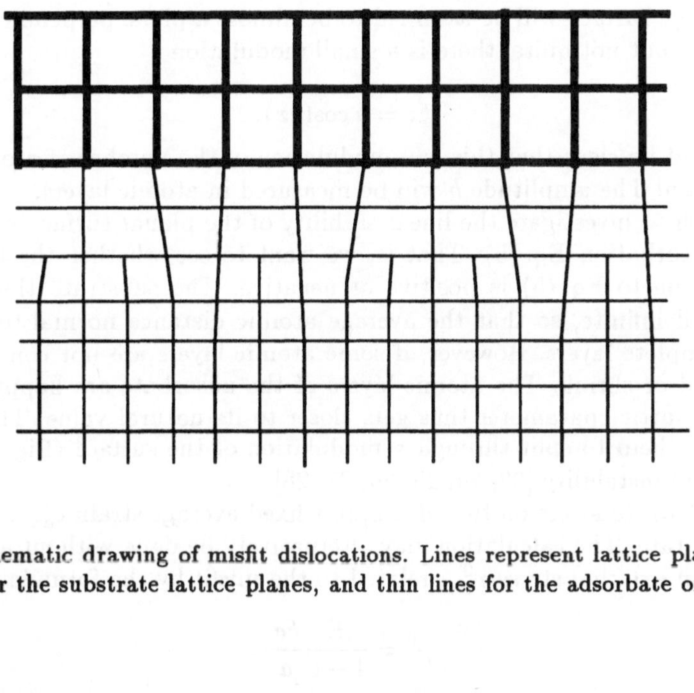

Fig. 7. Schematic drawing of misfit dislocations. Lines represent lattice planes. Thick lines are for the substrate lattice planes, and thin lines for the adsorbate ones.

Misfit dislocations are not really an instability in the sense of our definition, but can be the cause of an instability. Moreover, it is also of interest to worry about the stability of the commensurate structure, not only of the flat surface. For microelectronics, misfit dislocations are not recommended since it is difficult to control their regularity and, anyway, they spoil the regularity of the crystal.

The critical thickness calculated from equilibrium thermodynamics is not always in agreement with experiment, especially at low temperatures. This is a hint of the importance of kinetic effects, which will be seen in the following in other instances.

8 The Asaro-Tiller-Grinfeld instability in commensurate epitaxial films: thermodynamic analysis for a non-singular surface

We now investigate the stability of a commensurate adsorbate (i.e. without misfit dislocations) when its thickness is large.

We know that the commensurate state is thermodynamically unstable. However, misfit dislocations are difficult to form, because they require the movement of the atoms inside the solid, and these have a very low mobility. It is easier to move the atoms at the surface. As will be seen, surface diffusion of atoms does produce an instability, in which the crystal remains commensurate, but its surface does not remain planar. To see that, we assume a small deviation from the planar shape and we ask whether this deviation is amplified or decays at later times.

Thus, the surface will be assumed to be almost a plane perpendicular to the z direction, but not quite; there is a small modulation

$$\delta z = h \cos(qx). \tag{5}$$

It should be clear that this is a modulation of the number of atoms, not an elastic strain! The amplitude h will be measured in atomic layers.

We wish to investigate the linear stability of the planar surface with respect to the perturbation Eq. (5). That is, we want to see whether the free energy variation due to Eq. (5) is positive or negative. The substrate thickness will be assumed infinite, so that the average atomic distance normal to z is fixed within complete layers. However, if some atomic layers are not complete, they can expand or shrink. The atomic layers of the adsorbate are happy to do so, since their lattice parameter thus gets closer to its natural value. Therefore we can expect them to split through a modulation of the surface (Fig. 8). This is the Grinfeld instability [23, 24, 25, 26, 27, 28].

The substrate forces on the adsorbate a fixed average strain $\epsilon^0_{\alpha\gamma}$ with respect to its free state. The calculation may alternatively be done without a substrate, but with an external stress $p^0_{\alpha\gamma}$, related to the misfit by the formula

$$p^0 = \frac{E}{1-\zeta}\frac{\delta a}{a}.$$

A simplified calculation will be presented, introducing an average strain $\delta\epsilon$ (with respect to the flat surface $h = 0$) instead of the complete strain field. The free energy per unit area contains three contributions:

i) the increase of the surface area produces an increase of the "capillary" energy (due to chemical bonds which are broken when forming the surface). If the surface tension σ is orientation independent, the average capillary energy per unit projected area (perpendicular to the z axis) is increased by the modulation by a quantity

$$d\mathcal{F}_{cap}/d\mathcal{A} = \sigma h^2 q^2/2; \tag{6}$$

ii) the energy gained due to the relaxation in the modulated region is proportional to the height h of this region, to the average strain ϵ, and to the external stress p^0, viz.

$$d\mathcal{F}_{relax}/d\mathcal{A} \approx -hp^0\epsilon; \tag{7}$$

iii) this energy gain is partially compensated by the elastic energy paid to the inhomogeneity of the strain. This energy is mainly concentrated below the modulated region. Its three-dimensional density is proportional to ϵ^2 through an elastic constant C. In order to obtain the energy per unit area in the xy plane, one should multiply by the depth of the strained region. It is shown at the end of this section that this depth is of order $1/q$. The elastic energy per unit area is then

$$d\mathcal{F}_{el}/d\mathcal{A} \approx C\epsilon^2/(2q). \tag{8}$$

Minimizing the sum of contributions (ii) and (iii) with respect to ϵ yields

$$\epsilon \approx hqp^0/C$$

so that the variation of the total elastic free energy per unit area resulting from the modulation is the sum of the two contributions (ii) and (iii), namely

$$d\mathcal{F}_{el}^{tot}/d\mathcal{A} \approx -h^2p_0^2q/(2C). \tag{9}$$

If q is small enough, the positive term (i) due to surface tension is unable to compensate this negative term, so that increasing the modulation amplitude h lowers the free energy. There is an instability.

We close this section with the proof that the penetration depth of the strain (or the displacement which is its primitive) is $1/q$. The displacement $\boldsymbol{u}(x, y, z)$ satisfies the equation of elasticity, which we write in the case of a harmonic solid for simplicity, i.e.

$$(\lambda + \mu)\boldsymbol{\nabla}(div\boldsymbol{u}) + \mu\nabla^2\boldsymbol{u} = 0. \tag{10}$$

This equation must be combined with an appropriate equation at the surface, which will not be written. We shall only consider the (unphysical) case $\lambda + \mu = 0$. Then, the general solution of Eq. (10) with wave vector $(q, 0, 0)$ which vanishes for $z = -\infty$ is clearly $u(x, y, z) = u_0 \exp(iqx)\exp(qz)$. The physical case $\lambda + \mu \neq 0$ is a little more complicated and is left as an exercise to the reader.

In Eq. (9), the stress p^0 appears through its square, so that the Grinfeld instability is predicted to appear as an effect of either compression or extension. As a matter of fact, in heteroepitaxial films, it is rarely observed in expanded systems because shear motion is too easy, so that misfit dislocations, which are much more efficient in releasing strains, appear before the Grinfeld instability.

9 Thermodynamic instability and kinetic stabilisation in commensurate epitaxial films with a high symmetry orientation

9.1 Thermodynamic instability

Eq. (6) in the preceding section is strictly correct if the surface tension is isotropic. In a crystal, this condition is never fulfilled, but the capillary energy is still proportional to h^2q^2, in qualitative agreement with Eq. (6), in the neighbourhood of almost all orientations. The exceptions are called singular. Unfortunately, at usual temperatures, the singular orientations are the high symmetry orientations (001) and (111), and these are precisely the orientations commonly used in MBE growth.

In the neighbourhood of a singular surface, the excess capillary energy resulting from a misorientation θ is proportional to the number of steps, to the free energy γ of a step per unit length (line tension) and to the length L of the system. As a result, the average capillary free energy per unit projected area is

$$d\mathcal{F}_{cap}/d\mathcal{A} \approx \gamma|\theta|. \tag{11}$$

At a sufficiently high temperature, the line tension γ can vanish and the next order in the surface free energy should be used. This next order is just the non-singular formula Eq. (6). The temperature T_R at which γ vanishes and the capillary energy becomes analytic is called the roughening transition temperature.

For a sinusoidal modulation of amplitude h, the capillary contribution Eq. (11) dominates the elastic energy Eq. (9) for small h, and therefore a planar singular surface is stable with respect to small modulations, in contrast with a non-singular surface. However, it can be argued that, for large modulation amplitude, the calculation of the previous section is still applicable. In other words, there is still an instability, but now there is an activation energy. To find this energy, one should of course consider a perturbation of finite size instead of a sinusoidal modulation.

The calculation has been done by Tersoff and LeGoues [29] in the case of a truncated pyramid (Fig. 9). They obtained the energy barrier as

$$W_0 \approx \frac{\gamma^3(1-\zeta)^2}{E^2(1+\zeta)^2}\left(\frac{a}{\delta a}\right)^4 .$$

9.2 Kinetic stabilisation

The calculation of Tersoff and LeGoues is a thermodynamic one, where the free energy is minimized. Experimentally, bumps are not always observed, especially at low temperature[30].

This can be attributed to the fact that the adatoms are incorporated at the first step they meet. Let us assume that there are only two levels of steps. The

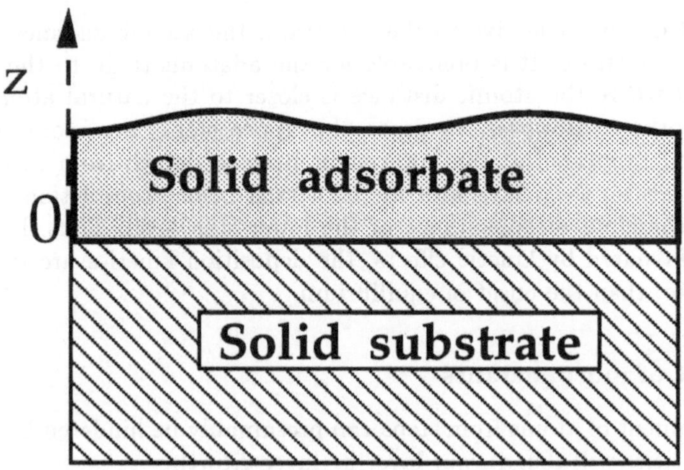

Fig. 8. The Grinfeld instability on a non-singular surface. A commensurate adsorbate gains energy by modulating its thickness, because at the top of the waves it can take an atomic distance closer to its "natural" one. Even for a vanishingly small modulation, there is an energy gain.

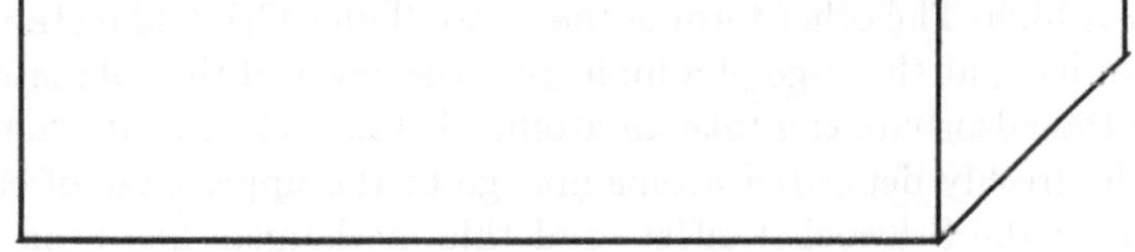

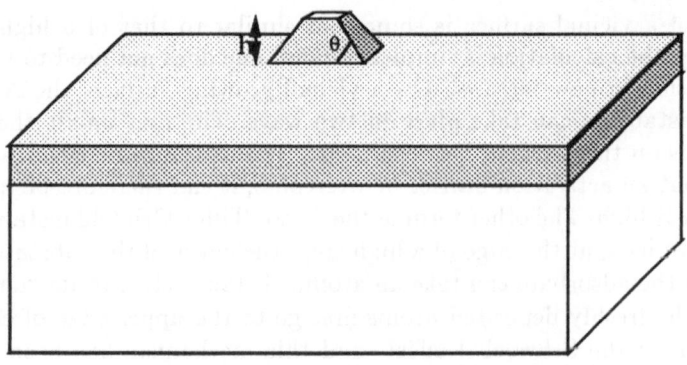

Fig. 9. Misfit instability on a singular surface. The commensurate adsorbate gains energy by forming pyramids as in figure 8, but now there is an activation energy corresponding to a minimal size of the pyramid.

lower level is more sensitive to the substrate, the atomic distance is closer to that of the substrate. It is preferable for the adatoms to go to the step of the upper level where the atomic distance is closer to the natural atomic distance of the adsorbate. However, to go to the upper steps, the atoms which have first been incorporated at the lower steps have to detach from those steps. At low temperature, they have no time to detach before completion of the lower layer and therefore disappearance of the lower step level. Thus, layer-by-layer growth is favoured by kinetic effects; the deposition kinetics are faster, at low temperature, than the onset of equilibrium.

9.3 Effect of morphactants

The detachment of atoms from steps can presumably be hindered by impurities. Such impurities will thus favour layer by layer growth.

There are indeed impurities which favour layer by layer growth. Preumably, they act according to this mechanism. These impurities have been called surfactants, in analogy with the effect of surfactants addressed in section I. However, the mechanism addressed here has nothing to do with surface tension, and the word "morphactant" proposed by Eaglesham is certainly preferable.

10 Step bunching instability and stabilisation in commensurate epitaxial films with a vicinal orientation

The case of a vicinal surface is somewhat similar to that of a high symmetry surface, but the calculation is simpler because one does not need to worry about the nucleation of new terraces. On a vicinal growing surface, the Asaro-Tiller-Grinfeld instability can take place in two ways. The first one is the formation of pyramids on the terraces between steps. This is the same effect as before. It requires that an activation barrier be overcome; it can therefore be quite slow if this barrier is high. The other form of the Asaro-Tiller-Grinfeld instability is step bunching. Indeed, at the edge of a high step, the effect of the substate is not too strong and the adsorbate can take an atomic distance close to its "natural" one. However, the freshly deposited atoms first go to the upper edge of each terrace as required by the Schwoebel effect, and this mechanism has been seen to be stabilizing. This kinetic stabilisation fails only if the atoms have enough time to detach from steps and to find a more favourable step before the incorporation of new atoms (Duport et al 1995).

11 Combination of dislocations and bumps

We have seen that two kinds of instabilities may arise in heteroepitaxy: the formation of misfit dislocations and the formation of bumps. Actually, they can appear simultaneously. The most usual scenario is that misfit dislocations appear first. If their distance is smaller than the thickness of the film deposited above the dislocations, the atomic distance at the surface is modulated. Far from the dislocations, it is closer to that of the substrate, and just above the dislocations it is closer to the natural distance of the adsorbate. Therefore, the new atomic layers are preferentially nucleated above the dislocations.

12 Spontaneous production of quantum dots

When the misfit between the atomic distances of the adsorbate and the substrate is large, it is sufficient to deposit very few atomic layers (say, two) in order to get an array of bumps or pyramids. These pyramids form a fairly regular array, more regular than expected from available theories. In particular, the dispersion of sizes is rather narrow. This kind of deposition has been proposed as a way to obtain quantum dots, i.e. semiconductor clusters with well-defined size, and therefore well-defined energy levels.

The failure of available theories is presumably that they underestimate the ability of atoms to find thermodynamically favoured configurations. For instance, they are assumed to form clusters which are not allowed to move. For a given density of clusters of a given size, the minimum of the elastic (free) energy is obtained when the clusters are equally spaced. This equal spacing clearly favours a narrow size distribution.

13 Conclusion

Homoepitaxial growth (Fe/Fe, Si/Si...) is, at low temperature, subject to a kinetic instability resulting from the difficulty of the atoms in reaching their most favourable position, which is a step edge. Heteroepitaxial growth, which is of course of much greater interest, is subject to the same instability, and in addition to an instability which arises from the different lattice constants. The latter instability is actually not specific to molecular beam epitaxy. It is a thermodynamic instability which appears earlier at higher temperature. If one wants to avoid the instability, one must therefore choose an appropriate temperature window. If the misfit is small, the instability does appear in principle, but only at long wavelengths because short wavelength fluctuations are stabilized by the surface tension. Moreover, for a high symmetry orientation, there is an activation barrier which can in practice provide a good metastability.

The present review is far from being exhaustive. Little has been said about the dynamics. Possible instabilities with respect to twinning, for instance, have not been considered. The development of instabilities beyond the initial, linear regime has not been addressed. Many of these questions are, in fact, still open.

References

1. Mullins, W.W., Sekerka, R.F. (1963) J. Appl. Phys. **34**, 323. and **35**, 444.
2. *Metallic Multilayers* Ed. A. Chamberod and J. Hillairet, Materials Science Forum **59-60** Transtech Publications, Aedermannsdorf (1990).
3. J. Villain, A. Pimpinelli, Physique de la Croissance Cristalline, Eyrolles-Alea-Saclay, Paris (1994) and english version, to be published.
4. BURTON, W.K., CABRERA, N., FRANK, F. (1951) Phil. Trans. Roy. Soc. **243**, 299.
5. G. Ehrlich, F.G. Hudda (1966) J. Chem. Phys. **44**, 1039.
6. R.L. Schwoebel and E.J. Shipsey, J. Appl. Phys. **37**,3682 (1966).
7. J.P. Bourdin, J.P. Ganachaud, J.P. Jardin, D. Spanjaard.
8. P. Stoltze, J. Phys.: Condens.Matter **6**, 9495 (1994).
9. R. Stumpf, M. Scheffler, Phys. Rev. Lett **72**, 254 (1994).
10. J. Villain, J. de Physique I **1**, 1, 19 (1991).
11. D.J. Eaglesham, H.J. Gossmann, M. Cerullo, Phys. Rev. Lett **65**, 1223 (1990).
12. H.J. Ernst, F. Fabre, R. Folkerts, J. Lapujoulade, Phys. Rev. Lett **72** 112 (1994)
13. J.A. Stroscio D.T. Pierce, M.D. Stiles, A. Zangwill, L.M. Sander, Phys. Rev. Lett **75** 4246 (1995).
14. M. Siegert and M. Plischke, Phys. Rev. Lett **73**, 1517 (1994).
15. M.D. Johnson et al., Phys. Rev. Lett **72**, 116 (1994).
16. A.W. Hunt, C. Orme, Williams, B. G. Orr, L.M. Sander, Europhys. Lett. **27**, 611 (1994).
17. Politi, P. and Villain, F. (1996) preprint.
18. I. Elkinani, J. Villain J. de Physique I **1**, 1991 (1994).
19. D. Kandel, J.D. Weeks, Phys. Rev. Lett **69**, 3785 (1992).
20. D. Kandel, J.D. Weeks, Phys. Rev. Lett **74**, 3632 (1995).
21. G.S. Bales and A. Zangwill, Phys. Rev. B **41**, 5500 (1990).
22. Jesser and van der Merwe [22](1989).
23. R.J. Asaro, W.A. Tiller, Metall.Trans. **3**, 1789 (1972).
24. M.Ya. Grinfeld, Dokl. Akad. Nauk SSSR **283**, 1139 (1985) [Sov. Phys. Dokl. **31**, 831 (1986)] and J. Nonlinear Sci. **3**, 35 (1993).
25. D.J. Srolovitz, Acta metall. **37**, 621 (1989).
26. P. Nozières, Lectures at the Beg Rohu Summer School (1989), in *Solids far from equilibrium*, C.Godrèche Ed. (Cambridge Univ. Press,1991) and J. de Physique I **3**, 681 (1993).
27. W.H. Yang and D.J. Srolovitz, Phys. Rev. Lett **71**, 1593 (1993).
28. K. Kassner and C. Misbah, Europhys. Lett. **28**, 245 (1994).
29. J. Tersoff and F.K. LeGoues, Phys. Rev. Lett **72**, 3570 (1994).
30. H. Zeng, G. Vidali, O. Biham, J. Vac. Sci. Tech. **A12**, 2058 (1994).

Scaling Behaviour in Submonolayer Film Growth - Beyond Mean Field Theory

J.A. Blackman and *P.A. Mulheran*

Department of Physics
University of Reading
Whiteknights, Reading RG6 6AF, UK.

1 Introduction

There has been a huge resurgence of interest in the mechanisms of thin film growth over the last few years. Electron microscopy has, for a long time, been the main tool for studying surfaces, but the development of scanning tunnelling microscopy (STM) [1, 2] and surface sensitive electron diffraction techniques [3]-[5] such as RHEED now allows one to probe surface details at sub-monolayer coverages. The theory of film growth [6, 7] was developed largely through the sixties and seventies, but the refinements in experiments have sparked a renewed interest in the corresponding theory [8].

Existing growth theory [6, 7] defines an initial 'transient' regime in which the density of monomers (isolated atoms on the surface) increases linearly with time. At the end of this period island nucleation is established. Following this, the density of islands becomes larger than that of monomers and one is into the aggregation regime which is dominated by the capture of diffusing monomers by existing islands; new nucleation occurs but more slowly than in the earlier period.

Scaling behaviour is observed in the island size distributions. This has been known for some time from computer simulations, but can now be observed experimentally for epitaxial growth using STM [2]. There are also predictions that averaged quantities like island density or mean island size show power law dependence in their growth both as functions of substrate coverage and also deposition rate. Again there is experimental access [5] to these quantities.

Following the aggregation regime at about 40% coverage one enters a coalescence phase dominated by the merging of growing islands. The growth properties are, of course, fundamentally different when one enters this new regime.

These lecture notes will review the current understanding of the theory of the scaling behaviour and the growth exponents. Rate equations, which are a form of mean field theory, have long been popular [6] as a way of studying the growth mechanisms. We shall find that they are remarkably good in their predictions of the growth exponents but fail dramatically to describe the distribution of island sizes. Attempts to go beyond mean field theory will be discussed.

2 Rate Equation Approach

Let us start by defining the parameters that characterize the growing islands. One has first of all the number density N_s. This is the number of islands on unit area of substrate that contain s atoms. It will depend on the coverage which is proportional to the deposition time and the rate of deposition F (assuming evaporation from the surface is negligible). Other quantities of interest are mean island size $\langle s \rangle$, total density of islands N and density of monomers N_1. Another important quantity is the critical island size, i; this is the size above which an island is stable and does not dissociate by the loss of monomers.

The basic rate equations that describe growth in the aggregation regime are

$$\frac{dN_1}{dt} = F - 2K_1N_1^2 - N_1 \sum_{s \geq 2} K_s N_s \tag{1}$$

$$\frac{dN_s}{dt} = N_1 \left(K_{s-1}N_{s-1} - K_s N_s\right) \tag{2}$$

F is the rate of deposition, and the capture kernels are conveniently written in terms of a common constant describing the monomer diffusion, $K_s = D\sigma_s$. We also define the substrate coverage $\Theta = Ft$ and the ratio $\mathfrak{R} = D/F$. The equations assume that $i = 1$; if $i > 1$ there are dissociation terms.

There has been some controversy about the form to use for σ_s. For some time it was assumed that this quantity is independent of s - the so called 'point island' model [9]. An alternative model [10] assumed a $s^{1/2}$ dependence. It appeared that the former predicted a $\mathfrak{R}$ dependence that agreed with computer simulations (and experiment) while the latter gave a better account of the Θ dependence. Subsequently it was shown [11] that using $\sigma_s \sim (Ns)^{1/2}$ resolves these difficulties and gives the observed dependence on both $\mathfrak{R}$ and Θ. Most recently [12] a fully self-consistent solution of the rate equations has shown almost exact agreement with computer simulations for N, N_1 and $\langle s \rangle$, but very poor agreement for the distribution function N_s itself.

2.1 A one-dimensional model

The breakdown of the theory for the distribution stems from the inadequacy of mean field theory. We will explore the origin of this breakdown more fully with the help of a one dimensional model. We imagine particles deposited on a line (rather than a surface); they are allowed to diffuse and if two meet they nucleate to form a trap which remains static, but can capture additional diffusing monomers. To avoid the complications of direct impact of particles onto the islands, they are not allowed to grow in size, but a record is kept of the number of monomers absorbed. We shall refer to this quantity as the trap size.

A modification of the existing mean field to the one- dimensional case consists (see below) in setting $\sigma_s = 4N$, independent of s as expected. The observed scaling behaviour in the island size distribution can be expressed by writing

$$N_s \sim \Theta \langle s \rangle^{-2} f\left(s / \langle s \rangle\right) \tag{3}$$

Using standard scaling theory techniques [10, 11, 13, 14] or asymptotic methods [15] together with the above expression for σ_s and restricting ourselves here to a critical island size i of 1, gives the following growth exponents

$$\mathrm{N} \sim \Theta^{1-z}\mathfrak{R}^{-\chi} \qquad \mathrm{N}_1 \sim \Theta^{-r}\mathfrak{R}^{-\omega} \qquad \langle s\rangle \sim \Theta^{z}\mathfrak{R}^{\chi} \tag{4}$$

where $r = 1/2$, $z = 3/4$, $\omega = 1/2$, $\chi = 1/4$.

The results of a Monte Carlo simulation [16] are shown in Fig. 1. The trap density N is shown as a function of $\Theta(= Ft)$, and the rescaled plot demonstrates the validity of the derived growth exponents.

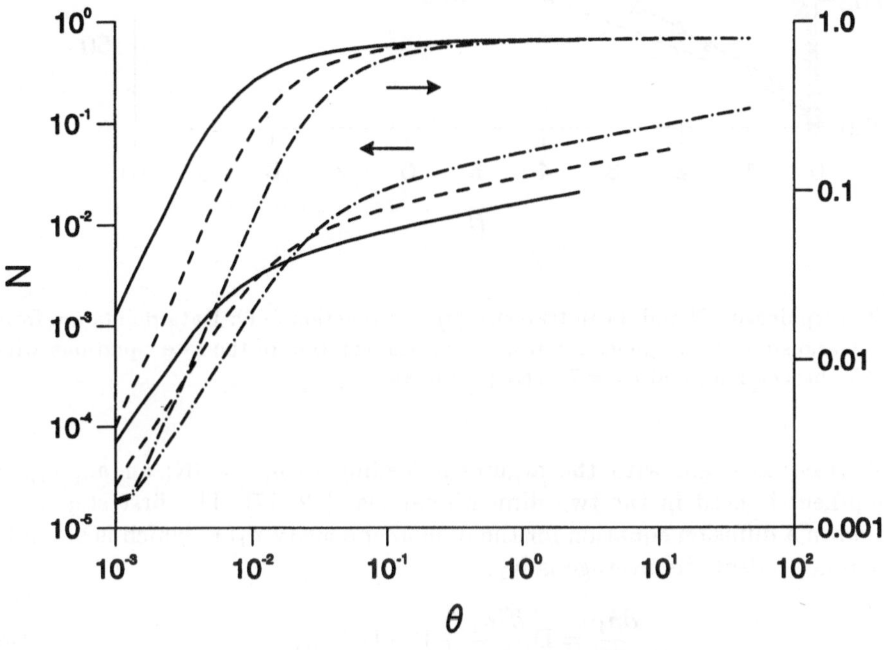

Fig. 1. Growth simulation in 1D. Trap density N as a function of ($\Theta = Ft$) for three values of $\mathfrak{R}$ (0.5×10^5 chain line; 0.5×10^6 broken line; 0.5×10^7 full line). Second group of plots shows the scaled trap density.

In Fig. 2 the broken lines represent a numerical solutions of the rate equations for both N and $\langle s\rangle$ with $\sigma_s = 4\mathrm{N}$ (the significance of the full lines will emerge later) and are to be compared with the data points which come from the Monte Carlo simulations. On a logarithmic scale the slopes of the two are the same which is why the exponents are given correctly. However, in terms of actual values, it is seen that there is some discrepancy and it is here that we begin to encounter the breakdown of mean field theory.

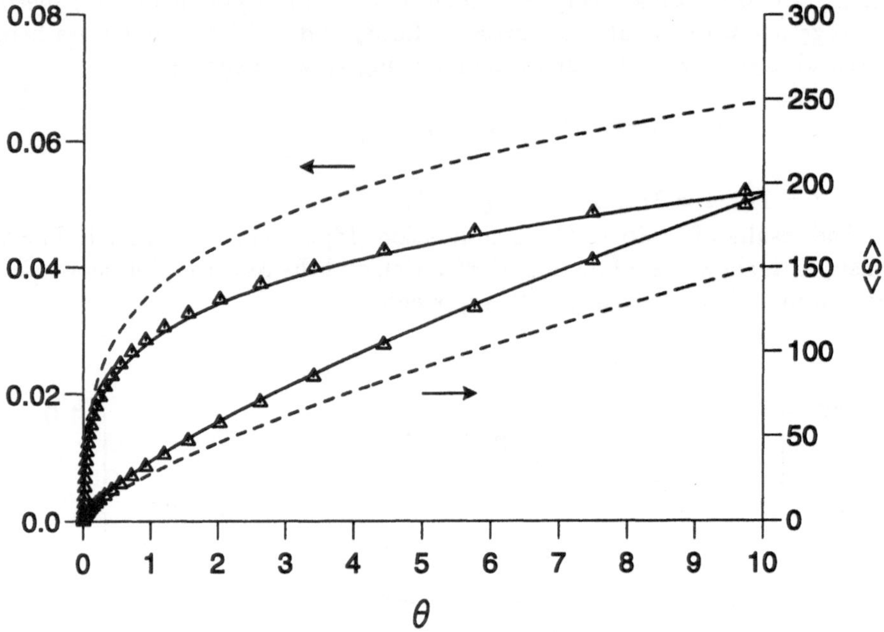

Fig. 2. Trap density N and mean trap size $\langle s \rangle$ as a function of Θ. Data points are from Monte Carlo simulations. Lines are from numerical solution of the rate equations with $\sigma_s = 4\mathrm{N}$ (broken line) and $\sigma_s = 7.717\mathrm{N}$ (full line).

We need to begin with the argument leading to $\sigma_s = 4\mathrm{N}$; an analogous development is used in the two dimensional case [12, 17]. The first step is to write down a diffusion equation for the monomer density $n_1(x)$ which is actually position dependent (its average is N_1)

$$\frac{\partial n_1}{\partial t} = \mathrm{D}\frac{\partial^2 n_1}{\partial x^2} + \mathrm{F} - \mathrm{D}\xi^{-2} n_1 \tag{5}$$

where ξ is the average distance a monomer travels before being captured by an island or another monomer. So that the equation is consistent with Eq. (1) after averaging, we make the identification

$$\xi^{-2} = \left[2\sigma_1 \mathrm{N}_1 + \sum_{s\geq 2} \sigma_s \mathrm{N}_s \right] \tag{6}$$

Subtracting Eq. (5) from Eq. (1), and assuming time variations in monomer density are small, we obtain for the monomer density as a function of distance x from an island, $n_1(x) = \mathrm{N}_1[1 - \exp(-x/\xi)]$.

The rate of capture of monomers is $2\mathrm{D}[dn_1/dx]_{x=0}$, where the factor of 2 is included because they arrive from both sides of the trap. Thus $\sigma_s = 2/\xi$

independent of s as expected and, using Eq. (6) and $N \gg N_1$ for the aggregation regime, we arrive at $\sigma_s = 4N$.

This is a mean field theory result because of the mean free path term in Eq. (5). We could alternatively omit that term from the equation and examine the monomer density *between a pair* of traps a distance y apart. Assuming quasi static behaviour, the density is given by $n_1(x) = \mathfrak{R}^{-1}x(y-x)/2$, and the total number of monomers in the gap is given by $y^3/12\mathfrak{R}$. If this is averaged over all gaps along the line, we can obtain the mean monomer density (the quantity denoted by N_1): $N_1 = \langle y^3 \rangle N/12\mathfrak{R}$. The mean gap size is given by $\langle y \rangle = N^{-1}$. If we introduce scaled variables, $Y = y/\langle y \rangle$, we obtain

$$12N^2N_1\mathfrak{R} = \langle Y^3 \rangle \tag{7}$$

The quantity $\langle Y^3 \rangle$ contains details about the spatial distribution of traps that is absent from mean field theory. This extra information will have an effect on the capture behaviour. The rate of capture of monomers by a trap at $x = 0$ is given by $2D[dn_1/dx]_{x=0} = D\mathfrak{R}^{-1}\langle y \rangle$, where we have averaged over all gap sizes. Identifying this with $D\sigma_s N_1$, noting that $N = \langle y \rangle^{-1}$, and using Eq. (7), we obtain an expression for σ_s

$$\sigma_s = 12N/\langle Y^3 \rangle \tag{8}$$

A best estimate of $\langle Y^3 \rangle$ is 1.555 (see next section). This yields $\sigma_s = 7.717N$. The rate equations are solved again using this value and the results are shown by the full line in Fig. 2. The excellent agreement has been obtained by going beyond mean field theory and including information about the spatial distribution of islands that is essentially the pair correlation.

2.2 Application to a coalescence process

Let us now turn to a model that focuses on growth by coalescence. The coalescence regime that follows aggregation is complicated by the presence of several processes. A simplified computer simulation of coalescence was introduced a few years ago by Family and Meakin [18]. The algorithm was stripped bare of all but the essentials; no diffusion was included in the model, and the islands remained circular throughout the simulation. Simply stated, the model consists of the deposition of D–dimensional hyperspherical droplets of radius r_0 onto a plane substrate. When a droplet falls onto an existing droplet, the two become one. If the enlarged droplet in turn overlaps with another then these two form a single island. Generally, when an island of radius r_1 merges with another of radius r_2, a new island is formed at the centre of mass of the two original ones, with a radius r given by $r = (r_1^D + r_2^D)^{1/D}$.

Family and Meakin [18] studied the scaling properties and growth properties of this homogeneous growth model. The purpose of this section is to examine a form of rate equation appropriate to their model and compare the predictions with the simulation.

The basic equations that we propose [19] to describe the homogeneous model are the following:

$$\frac{dN_1}{dt} = F - B_1 - \sum_{s\geq 1} B_s - \sum_{s\geq 1} A_{s1} - \frac{1}{2}\sum_{i,j} C_{ij1} \tag{9}$$

$$\frac{dN_s}{dt} = B_{s-1} - B_s + \frac{1}{2}\sum_{i,j} A_{ij}\delta(s-i-j) - \sum_{i\geq 1} A_{si} \tag{10}$$
$$+\frac{1}{2}\sum_{i,j,k} C_{ijk}\delta(s-i-j-k) - \frac{1}{2}\sum_{ij} C_{ijs} - \frac{1}{2}\sum_{i,j,k} C_{ijk}\delta(s-i-j)$$

where monomer now means the elementary droplet that falls on the surface, F is the rate of deposition of these droplets, and s is the number of elementary droplets in a composite droplet. The first terms, B_s, represent the probability of an incoming droplet making a direct impact on an existing island. We will specialise the discussion here to three dimensional islands, so that the impact term can be written:

$$B_s = s^{2/3} N_s \tag{11}$$

In the units used, purely numerical factors have been incorporated into the definition of the basic quantities.

The second type of term, describes the rate of coalescence of two islands of size i and j to form a new one of size $i+j$. The derivation of the expression for the process essentially follows the lines of Vincent [20] leading to:

$$A_{ij} = \frac{3}{4}\left(i^{1/3} + j^{1/3}\right) N_i N_j \tag{12}$$

Finally there are terms involving the coalescence of three or more islands. The simplest 3-island term arises, for example, when two large island coalesce and, in doing so, swallow up a smaller one in their vicinity. This term takes the following form:

$$C_{ijk} = A_{ij}(ij)^{1/3}\Phi(i/j)N_k \tag{13}$$

where Φ is a function of the ratio of the sizes of the two islands that are swallowing the third. $\Phi(i/j)$ is a maximum when i and j are equal.

Now let us invoke scaling [19] by assuming that Eq. (3) holds for the process described by Eqs. (9) and (10). There is no diffusion in this model so only the dynamic exponents are relevant. The most important for the scaling properties is z (where $\langle s\rangle \sim t^z$). Going through the standard procedure [19] yields $z = 3$, which agrees with the behaviour observed [18] in the simulations.

Now let us turn to the distribution function $N_s(t)$ for the coalescence model. For many coagulation models, a single peak distribution function appears. Family and Meakin [18] find instead, for the homogeneous model, a bimodal distribution with the second peak at the $s \to 0$ limit. What process produces this behaviour? To explore this matter, we have solved the rate equations (9), (10) numerically using a large but finite range of values of island sizes. In practice a hierarchy of 2000 equations was used. The results are shown in Figs. 3 and 4. For the first figure, terms B and A from Eqs. (11) and (12) are included but the C terms from Eq. (13) are excluded; in Fig. (4) all three terms are included. In can be seen immediately that the origin of the bimodal behaviour which is present only in the second figure are the three body terms which describe the swallowing of a third island that can result when the coalescence of a pair takes place.

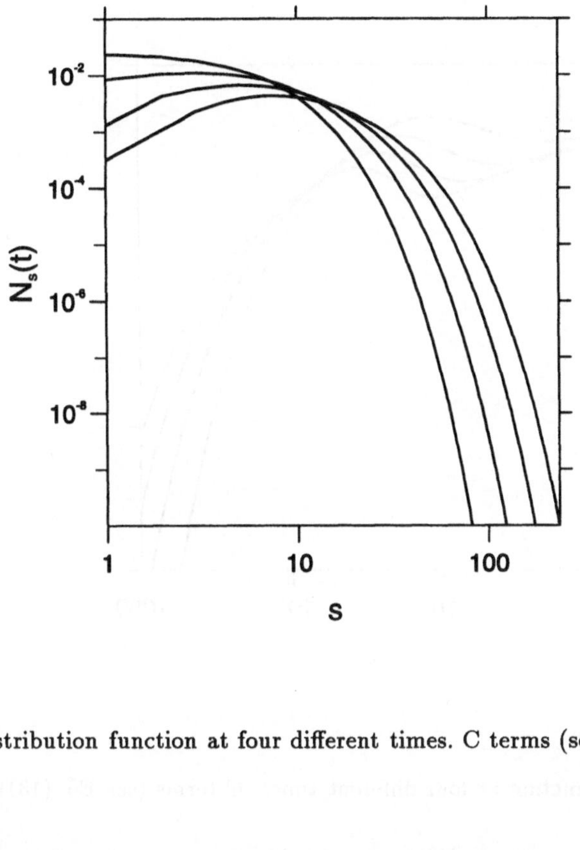

Fig. 3. Distribution function at four different times. C terms (see Eq. (13)) omitted.

The rate equation treatment has thrown light on this particular feature. The scaling properties of the distribution function also appear to be much better described by this method for the coalescence model [19] than they do for aggregation models [12, 16].

There are two key points to emphasise from our discussions about growth exponents in the two regimes (aggregation and coalescence): (i) the mean field equations do give a reliable prediction of growth exponents, (ii) the exponents are a good way of characterising particular processes [z is 3/4 (1 for a two dimensional substrate [10, 11, 13]) for the aggregation model and 3 for the coalescence model].

3 Distribution Functions for the Aggregation Regime

Let now us address the question of the island size distributions for the aggregation regime of growth. We have already seen in the discussion of the one-dimensional model that it is necessary to include some details of the local environment of the trap to describe the behaviour accurately. This becomes even more vital when it comes to treating the distribution function. The discussion can be developed for the one-dimensional model [16] but, for this lecture, we will focus on the two-dimensional case [21, 22].

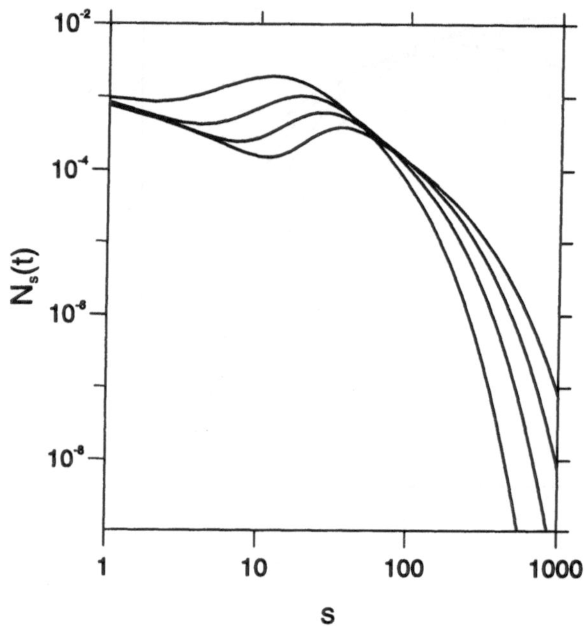

Fig. 4. Distribution function at four different times. C terms (see Eq. (13)) included.

One way of describing the local environment of an island on a two-dimensional substrate is via a Voronoi cell construction [21, 22]. This is done by forming the perpendicular bisectors of lines joining pairs of islands. The irregular polygons formed around each island are the Voronoi cells for those islands, and we postulate that a Voronoi cell is a good approximation to a capture zone. Monomers deposited within a particular capture zone will, on average, eventually be trapped by the associated island. The rate of capture will be proportional to the area of the capture zone.

Monte Carlo simulations for a homogeneous growth model were performed [21] for the deposition, diffusion, nucleation and capture of monomers on a two dimensional substrate. A typical snapshot at four different coverages is shown in Fig. 5 with the Voronoi network superposed on the pictures [21]. The critical island size was 1 in the simulations, and rearrangement of captured monomers on the islands were allowed thus maintaining a compact shape. Simulations that excluded rearrangements and resulted in the formation of dendritic islands were also performed [21, 22], and also models for heterogeneous growth [21] and for other critical island sizes [21, 22] were studied as well.

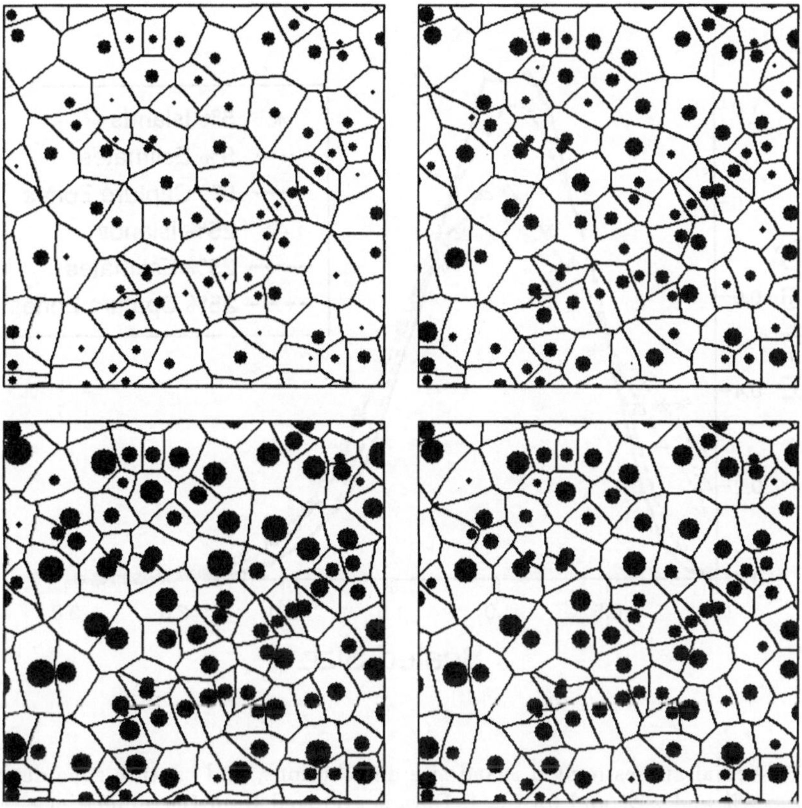

Fig. 5. Picture of the evolving microstructure in the i=1 circular island simulations. Coverages are $\Theta = 5, 10, 15, 20\%$ starting top left and moving clockwise.

During the simulations, the evolution of each island was followed. At any time, the area of its associated Voronoi cell was monitored thus giving an *estimate* of the rate at which monomers were being absorbed. From this estimate, an expected size could be calculated from the cumulative effects of its environmental history. The behaviour in the form of both actual and estimated distribution functions on scaled axes is shown at 5% and 25% coverages in Fig. 6.

The integrated frequency is normalised to 1, and the horizontal axis is the ratio of the island size to the mean size. The distribution functions for the capture zones are also shown and it is seen that they show reasonable scaling (certainly within the statistical noise). It is also seen that, not only do the island size distributions themselves show convincing scaling behaviour, but the actual size distributions and the estimated size distributions show excellent agreement. This demonstrates that, for any realistic theory of island growth that is to address size distributions, one really has to include the details and evolution of the islands' local environment.

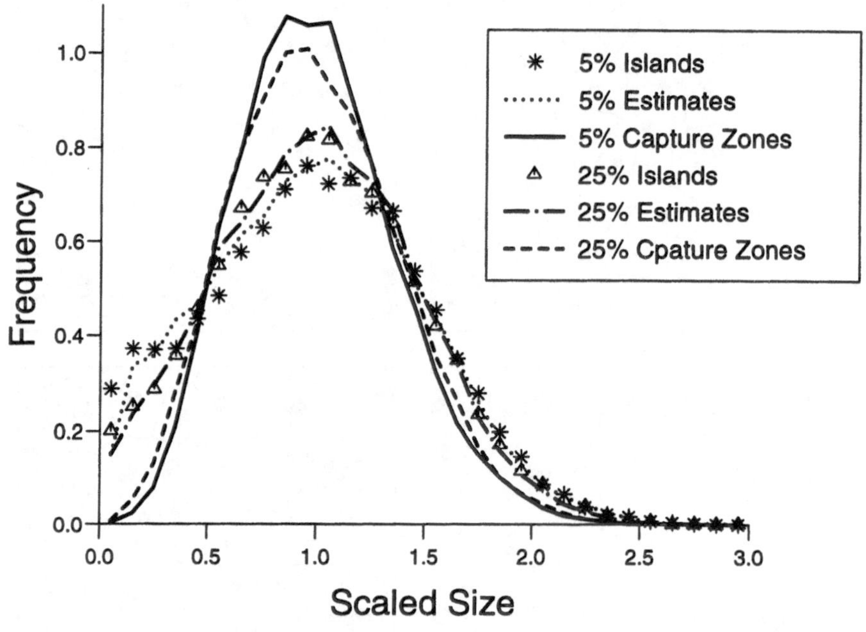

Fig. 6. Actual and estimated island size distributions, and capture zone distributions for Θ at 5% and 25% and for $i = 1$. Data is averaged from 100 simulations.

Similar plots are shown in Fig. 7 for simulations with a critical island size $i = 2$. The comments made for the $i = 1$ case apply, but it will be noticed that the distribution functions for the island sizes are closer to those of the Voronoi cells than previously. As the critical island size is increased, the nucleation is closer to completion before significant island growth occurs. This means that an island's environment has changed less significantly during its evolution than in the low i situation. An extreme case of this behaviour is heterogeneous growth [22], where there is an initial nucleation and negligible subsequent nucleation during growth. For this situation, the island and cell distributions are essentially an exact match [22].

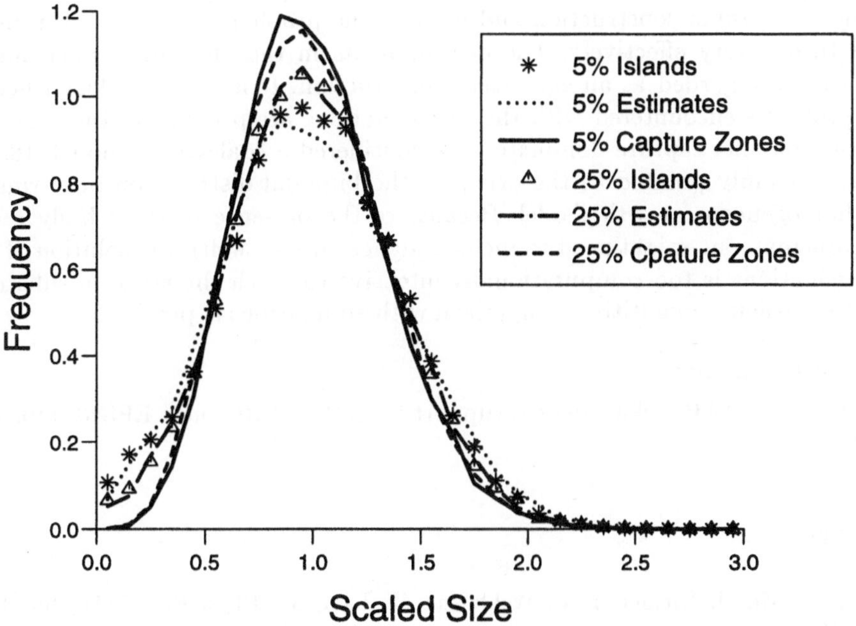

Fig. 7. As Fig. 6 except $i = 2$.

We conclude this section with a comment about the analogous situation in one-dimension. A similar discussion to the above about the distributions apply. Earlier in this paper, in Eq. (8), we related the capture kernel σ_s to the distribution in gap sizes, $\langle Y^3 \rangle$. Again this quantity is dependent on a monitoring of the local environment for its estimation. An evaluation of $\langle Y^3 \rangle$ was made during one-dimensional simulations analogous to those described above [16]. The result, quoted under Eq. (8), comes from those simulations.

4 Conclusions

The object of this paper is to show where mean field theory is applicable in growth models and where it breaks down - and when it does, why it fails. It is seen that the rate equations, despite neglecting all details about local environment, give an excellent account of the growth exponents which themselves provide a very effective flag to the occurrence of particular growth processes. The rate equations begin to show signs of unreliability when absolute values of quantities like N or $\langle s \rangle$ are required. This is most noticeable in one-dimension [16] and may well be much smaller at higher dimensionality [12].

It is with the size distribution functions that mean field theory becomes most unreliable. For the aggregation models, certainly, one has to include local environment effects. This is done in two dimensions [21, 22] in an elegant fashion using the Voronoi construction and resolves the problems encountered in mean field theory very effectively. The developments in the one dimensional model [16] can be regarded as an equivalent construction. It is still not clear whether the problems encountered with the rate equations are peculiar to the model in which monomer capture dominates. We considered a coalescence model [18, 19] which certainly pointed to the origin of the bimodal distribution occurring in the homogeneous growth model. Because of the presence of three body terms (in principal, one might need to include higher ones as well), the solution of the rate equations is too computationally intensive to reach the stage in substrate coverage when a quantitive comparison with simulations is possible.

Acknowledgments

One of us, JAB, acknowledges support from the SERC (now EPSRC) for this work.

References

1. Y.W. Mo, J. Kleiner, M.B. Webb and M. Langally, Phys. Rev. Lett., **66**, 1988 (1991).
2. J.A. Stroscio and D.T. Pierce, Phys. Rev. **B 49**, 8522 (1994).
3. H.-J. Ernst, F. Fabre and J. Lapujoulade, Phys. Rev. **B 46**, 1929 (1992).
4. W. Li and G. Vidali, Phys. Rev. **B 48**, 8336 (1993).
5. J.-K. Zuo, J.F. Wendelken, H. Dürr and C.-L. Liu, Phys. Rev. Lett., **72**, 3064 (1994).
6. J.A. Venables, G.D.T. Spiller and M. Hanbücken, Rep. Prog. Phys., **47**, 399 (1984).
7. S. Stoyanov and D. Kaschiev, in Current Topics in Materials Science, ed. E. Kaldis (North-Holland, 1981), vol. 7, p. 69.
8. J. Villain, A. Pimpinelli, L.-H. Tang and D. Wolf, J. Phys. I (France), **2**, 2107 (1992).
9. M. Bartelt and J.W. Evans, Phys. Rev., **B 46**, 12675 (1992); Surf. Sci., **298**, 421 (1993).
10. J.A. Blackman and A. Wilding, Europhys. Lett., **16**, 115 (1991).
11. C. Ratsch, A. Zangwill, P. Šmilauer and D.D. Vvedensky, Phys. Rev. Lett., **72**, 3194 (1994).
12. G.S. Bales and D.C. Chrzan, Phys. Rev. **B 50**, 6057 (1994).
13. J.G. Amar, F. Family and P.-M. Lam, in Mechanisms of Thin Film Evolution, eds. S.M. Yalisove, C.V. Thompson and D.J. Eaglesham (Materials Research Society, 1994), vol. 317, p. 167.
14. J.A. Blackman and A. Marshall, J. Phys. **A 27**, 725 (1994).
15. N.V. Brilliantov and P.L. Krapivsky, J. Phys. **A 24**, 4787 (1991).
16. J.A. Blackman and P.A. Mulheran, to be published.
17. J.A. Venables, Phil. Mag., **27**, 697 (1973).
18. F. Family and P. Meakin, Phys. Rev. **A 40**, 3836 (1989).
19. J.A. Blackman, Physica, **A 220**, 85 (1995).

20. R. Vincent, Proc. Roy. Soc. **A 321**, 53 (1971).
21. P.A. Mulheran and Blackman, Phys. Rev. **B53**,10261 (1996).
22. P.A. Mulheran and Blackman, Phil. Mag. Lett. **72**, 55 (1995).

Low-Dimensional Correlated Particle Systems: Exact Results for Groundstate and Thermodynamic Quantities

A. Klümper

Institut für Theoretische Physik, Universität zu Köln
Zülpicher Str. 77,
D-50937 Köln 41, Germany
Email: kluemper@thp.uni-koeln.de

Abstract: A one-dimensional fermion model with bond charge-interactions as well as Hubbard-type interactions is investigated exactly. The large distance asymptotics of the density-density and pair correlations are calculated. The system shows Luther-Emery liquid behaviour with a crossover from a density-density dominated regime to one with dominant pair correlations. Furthermore, the integrable tJ chain is studied at finite temperatures. Some concepts for this analysis are introduced, notably the Trotter-Suzuki mapping and the quantum transfer matrix. Finally, the specific heat is presented and its structure discussed.

PACS 71.28.+d,74.20.-z,75.10.Lp

1 Introduction

In this contribution to the Winterschool, I will give a short introduction into the physics of one-dimensional quantum chains. I will focus on the study of integrable systems which succumb to exact treatments, notably by the Bethe ansatz. Quite generally integrability imposes strong constraints on the interaction terms. Nevertheless, the physical properties of such systems appear to be generic and quite representative of other non-integrable systems.

The most famous examples of integrable models are the spin-1/2 Heisenberg chain and the 1d Hubbard model [1, 2, 3]. Here, however, I want to discuss fermionic systems with additional interactions different from the on-site Coulomb repulsion. In the course of the search for purely electronic mechanisms for high-T_c superconductivity a generalized Hubbard model was derived as a more detailed tight-binding description of the electronic system of solid matter

$$H = -\sum_{<i,j>} t_{ij}\,(c^{+}_{i\sigma}c_{j\sigma} + h.c.) + U\sum_{i} n_{i\uparrow}n_{i\downarrow} + V\sum_{<i,j>} n_i n_j, \qquad (1)$$

with U the on-site Hubbard term, V the nearest-neighbour Coulomb interaction and the hopping integral

$$t_{ij} = t_0 - \Delta t(n_{i,-\sigma} + n_{j,-\sigma}). \qquad (2)$$

Such correlated hopping terms (bond-charge interactions) arise from overlap integrals which are different for singly occupied orbitals than for multiply occupied orbitals.

For repulsive electrons, the model parameters in Eq. (1) should be positive t_0, U, $V > 0$. In particular $\Delta t > 0$ leads to a suppression of the hopping of a particle between two sites if one of them is already occupied by an electron (with opposite spin). However, the nature of all of these apparently repulsive interaction terms is different as revealed by the particle hole transformation and sublattice phase shift, $c_{i\sigma} \to \pm c_{i\sigma}^+$ [4, 5]. Neglecting a chemical potential term, the result on Eq. (1) is a transformation of

$$U, V, t_0 \to +U, +V, +(t_0 - 2\Delta t), \qquad \Delta t \to -\Delta t. \tag{3}$$

Thus, the Hubbard terms U and V are repulsive for holes as for electrons, the correlated hopping term, however, becomes attractive for holes. This observation was a cornerstone of the concept of hole superconductivity in Refs. [4, 5]. For mean-field and Lanczos studies of Eq. (1) the reader is referred to Ref. [6].

We are interested in exact results for Eq. (1) obtained in 1d. Unfortunately, even the dimensional restriction does not generally guarantee exact solutions. For Eq. (1), only the special case of the standard Hubbard model is integrable. On the other hand, we may add certain interaction terms which keep the physics of Eq. (1), but render the mathematics more tractable. If we consider a one-dimensional Hamiltonian with correlated hopping terms, Hubbard on-site interaction and pair hopping processes

$$\begin{aligned} H = &- \sum_{j,\sigma} (c_{j\sigma}^+ c_{j+1\sigma} + c_{j+1\sigma}^+ c_{j\sigma}) \exp\left[-\frac{1}{2}(\eta - \sigma\gamma) n_{j,-\sigma} - \frac{1}{2}(\eta + \sigma\gamma) n_{j+1,-\sigma}\right] \\ &+ \sum_j \left[U n_{j\uparrow} n_{j\downarrow} + t_p \left(c_{j\uparrow}^+ c_{j\downarrow}^+ c_{j+1\downarrow} c_{j+1\uparrow} + h.c.\right)\right], \end{aligned} \tag{4}$$

we have integrability under the condition [7]

$$t_p = U/2 = \pm \left[2e^{-\eta}(\cosh\eta - \cosh\gamma)\right]^{1/2}, \tag{5}$$

where we restrict ourselves to the physical positive sign in the following. As a criterion of integrability we may either use the condition of factorizing S-matrices [8] or the existence of a classical model satisfying the Yang-Baxter equation [3] and leading to Eq. (4) in the Hamiltonian limit.

The system Eq. (4) shows an apparent left-right asymmetry in the correlated hopping term, which however does not affect transport properties of the system. In fact, no charge or spin current is imposed by the asymmetry. The asymmetry seems to be irrelevant from the physical point of view. Mathematically, however, it is essential for integrability and the analytical study.

The second system I will consider is the one-dimensional tJ model describing hopping of electrons from singly occupied lattice sites to empty sites and a Heisenberg like spin exchange for nearest-neighbours

$$H = -t \sum_{j,\sigma} \mathcal{P}(c_{j,\sigma}^\dagger c_{j+1,\sigma} + c_{j+1,\sigma}^\dagger c_{j,\sigma})\mathcal{P} + J \sum_j (S_j S_{j+1} - n_j n_{j+1}/4). \tag{6}$$

Note the projector $\mathcal{P} = \prod_j (1 - n_{j\uparrow} n_{j\downarrow})$ which ensures that double occupancies of sites are forbidden. At the supersymmetric point $2t = J$, the system was shown to be integrable [9, 10].

In Section 2, the groundstate properties of Eq. (4) will be discussed on the basis of a Bethe ansatz for its eigenstates. The concept of Luttinger liquids will be introduced whose validity is much larger than the class of integrable one-dimensional systems. In Section 3, the finite temperature properties of Eq. (6) will be investigated via a combination of a Trotter-Suzuki mapping and an application of the so-called quantum transfer matrix.

2 Groundstate properties

The scattering processes for the system Eq. (4) factorize thereby permitting a systematic construction of eigenstates in the form of a Bethe ansatz, i.e. an appropriate superposition of plane waves in certain fundamental regions of phase space [8, 3]. There are two sets of rapidity variables, λ_j and Λ_α describing the charge and spin degrees of the system. For periodic boundary conditions they have to satisfy a set of coupled Bethe ansatz equations

$$\left[\frac{\sin(\lambda_j - ia)}{\sin(\lambda_j + ia)}\right]^L = \prod_{\alpha=1}^{N_\downarrow} \frac{\sin(\lambda_j - \Lambda_\alpha + i\gamma/2)}{\sin(\lambda_j - \Lambda_\alpha - i\gamma/2)},$$
$$\prod_{j=1}^{N} \frac{\sin(\Lambda_\alpha - \lambda_j + i\gamma/2)}{\sin(\Lambda_\alpha - \lambda_j - i\gamma/2)} = -\prod_{\beta=1}^{N_\downarrow} \frac{\sin(\Lambda_\alpha - \Lambda_\beta + i\gamma)}{\sin(\Lambda_\alpha - \Lambda_\beta - i\gamma)}, \tag{7}$$

where L is the number of lattice sites and a is a constant,

$$a = \frac{1}{4}\left\{\ln\left[\frac{\sinh\frac{1}{2}(\eta+\gamma)}{\sinh\frac{1}{2}(\eta-\gamma)}\right] - \gamma\right\}. \tag{8}$$

The energy of the corresponding state is given in terms of the particle rapidities λ_j as

$$E = 2\sum_{j=1}^{N}\left[\cosh 2a - \frac{\sinh^2 2a}{\cosh 2a - \cos 2\lambda_j}\right]. \tag{9}$$

For the groundstate the magnetization is zero, requiring $N_\downarrow = N/2$. Furthermore, the minimization of the total energy requires a symmetrical distribution of λ_j rapidities around zero as compact as possible. The $N/2$ rapidities Λ_α always fill up the total available space on the real axis from $-\pi$ to π. This structure of the groundstate distribution has important consequences for the excitations; see Fig. 1. For generic particle densities we have quasilinear energy-momentum excitations of the particle-hole type. The spin system does not allow for such excitations. The simplest possible excitation in this case is of the hole type which turns out to possess a gap; see Fig. 2.

The existence of a spin gap also implies the existence of a charge gap. Let us manipulate the groundstate, which is a perfect singlet for even particle number

N, by adding one particle. The cost in energy is given by the chemical potential plus the spin gap. If we add two particles we have to pay the chemical potential twice, but nothing else. We therefore find a positive binding energy of two particles

$$E_{binding} = 2[E_0(N+1) - E_0(N)] - [E_0(N+2) - E_0(N)] = 2\Delta_{spin} > 0, \quad (10)$$

which is twice the spin gap.

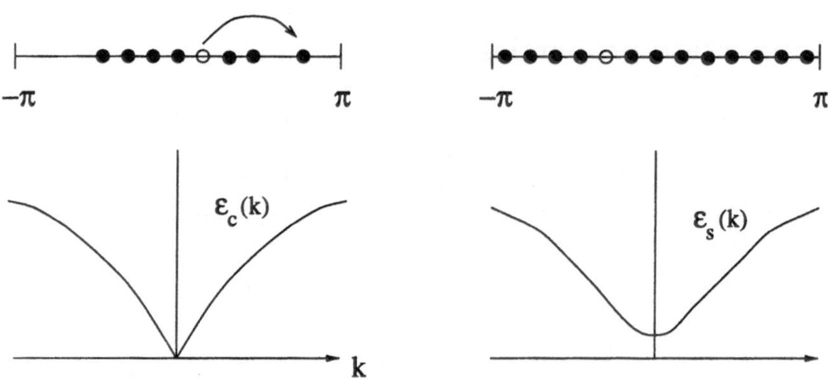

Fig. 1. Depiction of the distribution of a) charge and b) spin rapidities in the groundstate of Eq. (4). Also shown are the elementary charge and spin excitations of the system.

The interesting question to answer is whether or not this binding of particles leads to a coherent state with off-diagonal long-range order, which is the criterion for superconductivity. Strictly speaking we do not expect such a condensation phenomenon to take place in one dimension. Nevertheless, it makes sense to look whether we have quasi long-range order, i.e. algebraic decay of the pair-correlation functions and whether these correlations are longer ranged than the density-density correlations.

For the long-distance asymptotics of the density-density and pair correlations, we find

$$\langle \rho(r)\rho(0)\rangle \simeq \rho^2 + A_1 r^{-2} + A_2 r^{-\alpha}\cos(2k_F r), \qquad 2k_F = \pi\rho,$$

$$\langle c^+_{r\uparrow} c^+_{r\downarrow} c_{0\downarrow} c_{0\uparrow}\rangle \simeq B r^{-\beta}, \quad (11)$$

whereas the one particle functions decay exponentially

$$\langle c^+_{r\sigma} c_{0\sigma}\rangle \simeq \cos(k_F r) e^{-r/\Delta_{spin}}. \quad (12)$$

These properties, namely spin-charge separation and continuous exponents constitute a Luttinger liquid (i.e. a non-Fermi liquid). In fact, the spin excitations are massive, so the system belongs to the Emery-Luther universality class. For such systems the momentum distribution function n_k is absolutely smooth as it is the Fourier transform of the exponentially decaying correlation Eq. (12).

The concrete calculation of the scaling dimensions x, describing the algebraic decay of correlations $C_r \simeq r^{-2x}$, consists of a combination of conformal field theory and finite-size scaling of energy levels

$$E_x - E_0 = \frac{2\pi}{L} v(x + N^+ + N^-) \tag{13}$$

where L denotes the system size and v the velocity of the charge excitations, $N^{\pm}$ are the numbers of particle-hole excitations at the two Fermi points. The finite-size study can be done analytically and one finds an expression with so-called Gaussian dimensions

$$x = \frac{n^2}{2K} + \frac{K}{2} m^2, \tag{14}$$

where $n = \Delta N/2$ and ΔN is the change in particle number; m is the number of particle-hole excitations from one Fermi point to the other one. The parameter K characterizes the particular Emery-Luther theory and depends on the interaction strengths and the particle density ρ.

The parameter K as well as the groundstate energy are determined by various integral equations for the density functions of charge and spin rapidities in the thermodynamic limit. As the spin system is massive the corresponding degrees of freedom may be 'integrated out' leaving for the charge system

$$2\pi\rho(\lambda) = \frac{2\sinh 2a}{\cosh 2a - \cos 2\lambda} + \int_{-\lambda_0}^{\lambda_0} \varphi(\lambda - \mu)\rho(\mu)d\mu, \tag{15}$$

where $\phi(\lambda) = 1 + 4\sum_{n=1}^{\infty} \cos 2n\lambda/(1 + \exp(2n\gamma))$ [7]. The parameter λ_0 is determined by the subsidiary condition for the total density $\rho = N/L$ of the electrons

$$\int_{-\lambda_0}^{\lambda_0} \rho(\lambda)d\lambda = \rho. \tag{16}$$

Finally, the parameter K is related to the so-called dressed charge $\xi(\lambda)$ by

$$K = \xi(\lambda_0)^2/2. \tag{17}$$

The dressed charge function in turn has to satisfy the integral equation

$$2\pi\xi(\lambda) = 2\pi + \int_{-\lambda_0}^{\lambda_0} \varphi(\lambda - \mu)\xi(\mu)d\mu. \tag{18}$$

This equation can be solved numerically from which the exponents follow via Eq. (14). For the correlations in Eq. (11) we obtain

$$\alpha = 1/\beta = K. \tag{19}$$

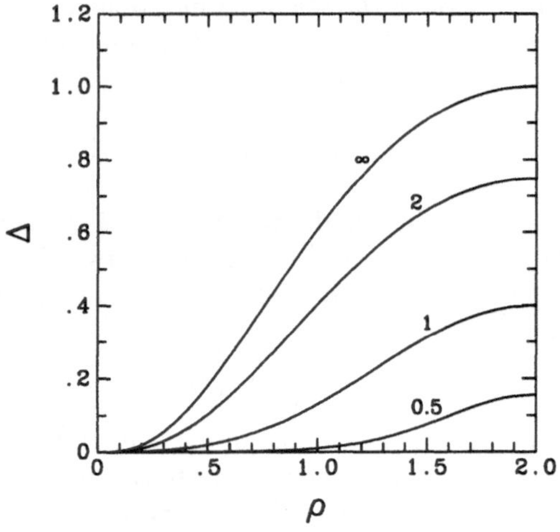

Fig. 2. The spin gap of the model for various particle densities and interaction parameters η (note that the special case with maximum correlated hopping interaction $\gamma = \eta$ is considered).

The model does not show finite off-diagonal long-range order. However, we observe a longer range of the pair correlations in comparison to density-density correlations for certain regimes of interactions and particle densities; see Fig. 3. For all values of the interaction parameters η and γ, there is a crossover from a regime with dominant density-density correlations ($1 < \beta < 2$) to a regime with dominant pair correlations ($\beta < 1$) at a "critical density" ρ_c.

Finally, we should like to comment on the relation of the parameter K characterizing the Emery-Luther liquid with macroscopic quantities such as the compressibility and charge stiffness. By combining Eqs. (13) and (14), we obtain an explicit expression for the second derivative of the groundstate energy with respect to the density of particles

$$\frac{\partial^2 e}{\partial \rho^2} = \frac{\pi v}{2K}, \tag{20}$$

which is nothing but $(\kappa \rho^2)^{-1}$ where κ denotes the compressibility. A little more involved is the derivation of the charge stiffness resp. the Drude peak D of the electrical conductivity at zero frequency

$$D = \frac{Kv}{\pi}. \tag{21}$$

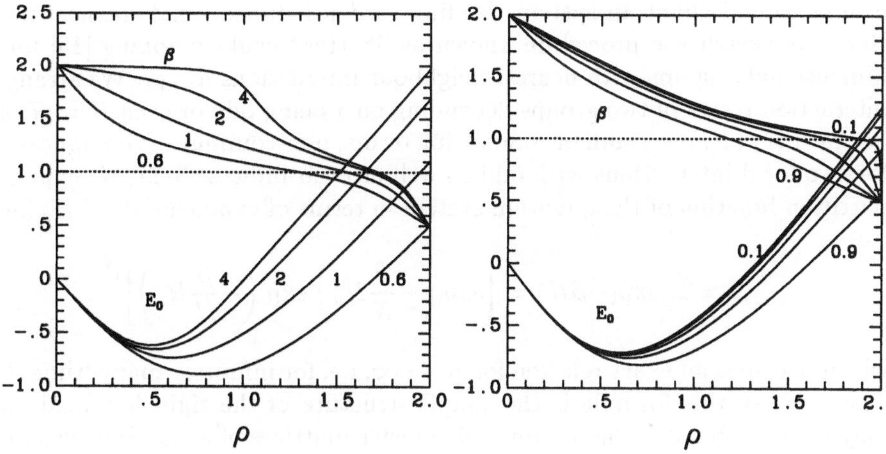

Fig. 3. a) Depiction of the density dependence of the groundstate energy per site E_0 and the exponent β of the pair correlation function for values $\gamma = 0.5$, and $\eta = 0.6$, 1, 2, 4. b) Similar to a) for $\eta = 1$, and $\gamma = 0.1, 0.3, 0.5, 0.7, 0.9$.

For technical details we refer to Ref. [11].

3 Thermodynamics of integrable systems

Next we study integrable quantum chains at finite temperature. At first glance, the knowledge of the Bethe ansatz equations for the energy eigenstates seems to provide the necessary information for calculating the partition function. However, only low-lying energy states are easily accessible, i.e. the elementary excitations of the system. A superposition of these excitations even for the low-temperature limit will fail as we have to study states where the number of elementary excitations is comparable to the number of sites. In this case the elementary excitations cease to be independent. The traditional way to deal with the residual interaction between the elementary excitations consists of a description of the rapidity distribution in terms of density functions. The interaction of the elementary excitations is due to the Bethe ansatz equations and takes the form of non-linear integral equations at finite temperature. In general, however, there are infinitely many such equations to be solved [12]. This poses quite a numerical problem. In addition, the traditional approach does not provide information for the correlations of the model.

In order to face the mentioned problems a different route is taken here [13, 14]. Instead of tackling the quantum Hamiltonian at finite temperature directly, the system is first mapped to a classical model. Quite generally this is possible for d-dimensional quantum systems leading to $d+1$-dimensional classical models. Here, we sketch the procedure known as Trotter-Suzuki mapping [15] for one-dimensional systems with nearest-neighbour interactions $h_{j,j+1}$. We arrange the interaction terms in two groups depending on j being odd or even $H = H_o + H_e$. So, each H_o (H_e) is a sum of commuting terms, non-commutativity appears only for two local interactions with odd as well as even indices. Next, we express the partition function of the quantum system in terms of exponentials of H_o and H_e

$$Z = \operatorname{Tr} \exp(-\beta H) = \left[\exp\left(-\frac{\beta}{N} H_o\right) \exp\left(-\frac{\beta}{N} H_e\right)\right]^N, \tag{22}$$

which becomes an exact relation for $N \to \infty$, i.e. for infinitely many 'time slices'. The merit of this formula is the simple structure of the right hand side which may be interpreted as the product of transfer matrices of a classical system on a square lattice with local interactions of spins a, b, c, d around lattice faces giving rise to Boltzmann weights

$$w(a,b,c,d) = \langle a, b \left| \exp\left(-\frac{\beta}{N} h\right) \right| c, d \rangle, \tag{23}$$

where h is the local quantum mechanical interaction.

The further analytic study of the system is greatly simplified by the right choice of an adequate transfer matrix. The first idea of taking the row-to-row transfer matrix proves to be a dead end as this matrix has an excitation gap of order $1/N$ and we have to take the limit $N \to \infty$. Therefore, all eigenvalues of this matrix would have to be taken into account for the calculation of the partition function. The column-to-column transfer matrix (also known as the quantum transfer matrix) enjoys quite different and more convenient properties [16, 17]. First, this transfer matrix shows a gap persisting in the limit $N \to \infty$. Second, the corresponding transfer direction is identical to the space direction. Hence, only the largest eigenvalue of the column-to-column transfer matrix is needed for calculating the partition function; the next-largest eigenvalues give the correlation lengths of the static correlation functions.

For the supersymmetric tJ model the quantum transfer matrix can be diagonalized by an algebraic Bethe ansatz [14]. I skip the technical details of the derivation, but should like to comment on some aspects of the method. The Bethe ansatz equations in the traditional form are difficult to treat in the limit $N \to \infty$. However, they can be transformed into a set of two non-linear integral equations which, first, permit the limit $N \to \infty$ to be taken analytically and, secondly, admit a straightforward numerical solution.

Now we consider various thermodynamic quantities at arbitrary temperatures as obtained numerically from the integral equations [14]. Fig. 4 shows the specific heat as a function of T for different fixed particle densities. First of all, we note a linear temperature dependence at low T. According to conformal field theory, the coefficient is given by $\pi(1/v_s + 1/v_c)/3$, where v_s and v_c are the velocities of the elementary spin and charge excitations. Our numerical data are consistent with this expression. Furthermore, we observe two maxima with changing dominance for increasing particle density n. The nature of this structure can be understood from the elementary excitations of the system. In the groundstate the particles are bound in singlet pairs with binding energies varying from 0 to some density dependent value. There are two types of excitations. First, there are charge excitations due to energy-momentum transfer onto individual pairs. Second, there are excitations due to the breaking of pairs. The latter excitation is of a spin type at lower excitation energies, but changes character at higher (density dependent) energies to a charge type as it describes the motion of single particles [18]. Therefore, the first and second maximum at lower densities (Fig. 4a) are caused by charge excitations due to pairs and single particles, respectively. At higher densities (Fig. 4b), the maximum at lower temperatures is dominated by excitations of pairs whereas the second one at higher temperatures is caused by spin excitations. For increasing concentration the spin contribution becomes dominant as the charge excitations freeze out. This is in accordance with the limiting case $n = 1$ leading to the spin-1/2 Heisenberg chain.

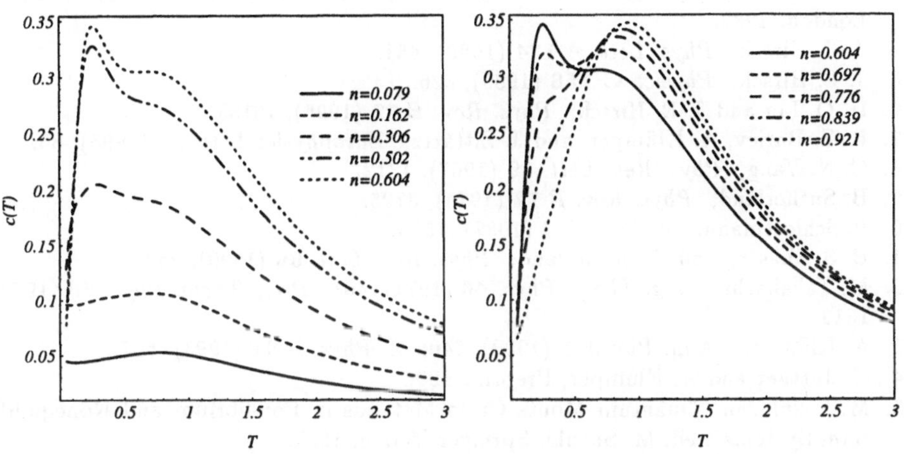

Fig. 4. a) Specific heat as function of T for different particle densities n with $n \leq 0.6$. b) Similar to a) with $n \geq 0.6$.

4 Conclusion

I have presented analytical results for one-dimensional correlated fermion systems. The low-lying excitations and correlation functions of a model with bond-charge and Hubbard-type interactions have been presented. The properties of this system constitute a Luther-Emery liquid with a crossover from a regime dominated by density-density correlations to a regime with dominant pair correlations. Finally, methods for the treatment of the thermodynamical aspects of integrable systems have been presented, pemitting a test of the Luttinger liquid picture at finite temperatures. So far, the concept of the quantum transfer matrix was used for the thermodynamical potential of the integrable tJ chain, but it is potentially more powerful as it also provides information on the correlations at finite temperatures [13].

Acknowledgments

The author acknowledges financial support by the *Deutsche Forschungsgemeinschaft* under grant No. Kl 645/3-1 and useful discussions with G. Jüttner and J. Suzuki. The work was performed within the research program of the Sonderforschungsbereich 341, Köln-Aachen-Jülich.

References

1. H. A. Bethe, *Z. Phys.* **71** (1931), 205.
2. E. H. Lieb and F. Y. Wu, *Phys. Rev. Lett.* **20** (1968), 1445.
3. R. J. Baxter, "Exactly Solved Models in Statistical Mechanics", Academic Press, London, 1982.
4. J. E. Hirsch, *Phys. Lett. A* **134** (1989), 451.
5. J. E. Hirsch, *Physica* **C 158** (1989), 326.
6. H. Q. Lin and J. E. Hirsch, *Phys. Rev. B* **52** (1995), 16155.
7. R. Z. Bariev, A. Klümper, and J. Zittartz, *Europhysics Lett.* **32** (1995), 85.
8. C. N. Yang, *Phys. Rev. Lett.* **19** (1967), 1312.
9. B. Sutherland, *Phys. Rev. B* **12** (1975), 3795.
10. P. Schlottmann, *J. Phys. C* **4** (1987), 7565.
11. B. S. Shastry and B. Sutherland, *Phys. Rev. Lett.* **66** (1990), 243.
12. M. Takahashi, *Prog. Theor. Phys.* **46** (1971), 401; *Prog. Theor. Phys.* **50** (1973), 1519.
13. A. Klümper, *Ann. Physik* **1** (1992), 540; *Z. Phys. B* **91** (1993), 507.
14. G. Jüttner and A. Klümper, Preprint 1996
15. M. Suzuki, in "Quantum Monte Carlo Methods in Equilibrium and Nonequilibrium Systems", ed. M. Suzuki, Springer Verlag, 1987.
16. M. Suzuki, *Phys. Rev. B* **31** (1985), 2957.
17. M. Suzuki and M. Inoue, *Prog. Theor. Phys.* **78** (1987), 787.
18. P. A. Bares, G. Blatter and M. Ogata, *Phys. Rev. B* **44** (1991), 130.

Critical Behavior of Weakly-Disordered Anisotropic Systems in Two Dimensions

G. Jug [1] *and* *B.N. Shalaev* [2]

[1] INFM–Istituto di Scienze Matematiche, Fisiche e Chimiche
Università di Milano a Como, Via Lucini 3, 22100 Como (Italy)
and INFN–Sezione di Pavia, 27100 Pavia (Italy)

[2] A.F. Ioffe Physical & Technical Institute, Russian Academy of Sciences,
194021 St. Petersburg (Russia)

Abstract: The critical behaviour of two-dimensional (2D) anisotropic systems with weak quenched disorder described by an Ising model (IM) with random bonds, the N-colour Ashkin-Teller model (ATM) and some of its generalizations is studied. In the critical region, these models are shown to be described by a multifermion field theory similar to the Gross-Neveu model with a few independent quartic coupling constants. Renormalization group calculations are used to obtain the temperature dependence near the critical point of some thermodynamic quantities and the large distance behaviour of the two-spin correlation function. The equation of state at criticality is also obtained within this framework. We find that the random models under consideration belong to the same universality class as that of the two-dimensional IM. The critical behaviour of the 3- and 4-state random-bond Potts models is also briefly discussed.
PACS numbers: 05.50.+q, 05.70.Jk, 75.10.Hk, 75.40.Cx

1 Introduction

The critical properties of two-dimensional random spin systems have been extensively studied in the past few years [1],[2],[3]. Two-dimensional (2D) systems are particularly interesting for the following reasons: first, there are numerous examples of layered crystals undergoing continuous antiferromagnetic and structural phase transitions [4], [5]. Perfect crystals are the exception rather than the rule, quenched disorder always existing in different degrees. Even weak disorder may drastically affect the critical behaviour. Secondly, the conventional field-theoretic renormalization group (RG) approach based on the standard ϕ^4 theory in $(4-\epsilon)$-dimensions, as applied to study properties of disordered systems by Harris and Lubensky [6] and Khmelnitskii [7], does not work in 2D due to the hard restriction $\epsilon \ll 1$. Some early exact results concerning the 2D random-bond IM with a special type of disorder (where only the vertical bonds are allowed to acquire random values, the horizontal bond couplings being fixed) have been obtained by McCoy and Wu [8]. This type of 1D quenched disorder without frustration was shown to smooth out the logarithmic singularity of the specific heat (the frustrated case was considered by Shankar and Murthy [9]). Many years ago Dotsenko and Dotsenko [1] initiated some considerable progress in the study of 2D random-bond IMs by exploiting the remarkable equivalence between this model and the N=0 Gross-Neveu model. For weak dilution, the new temperature dependence of the specific heat was found to be $\ln \ln \tau$, $\tau = \frac{T-T_c}{T_c}$ being the reduced deviation from the critical temperature. However, their results concerning the two-spin correlation function at the Curie point were later questioned by Shalaev [10], Shankar [11], and Ludwig [12], [13]. Using the RG approach and the bosonization technique these authors showed that the large-distance behaviour of this function at criticality was the same as in the pure case. Some arguments in favour of the pure IM fixed point governing the critical behaviour of the 2D IM with impurities were given earlier by Jug [36]. Recently, a good number of papers devoted to Monte-Carlo simulations of the critical behaviour of the impure IM have been published [15]. Most of the Monte-Carlo data are in good agreement with the analytical results obtained in [10],[11],[12].

Now, there is some scope to extend the previous analysis for the 2D random IM to other issues. An interesting possibility is to study critical phenomena in 2D dilute anisotropic systems with many-component order parameters. The analysis of the critical behaviour of such systems in $(4-\epsilon)$-dimensions was developed in great detail many years ago [16] but cannot be directly applied to the 2D case. Therefore it would be interesting and important to study these 2D models. The key ingredient of our treatment is that, following Shankar, we map the initial Landau Hamiltonian, written in terms of scalar fields, onto a multifermion field theory of the Gross-Neveu type with a few independent quartic couplings [17] (see also Refs. [18], [19]). This transformation can be done for Hamiltonians containing only even powers of each order parameter component, the fourth-order term being an invariant of the hypercubic symmetry group. It should be stressed that this method is quite general and may be extended to other systems.

The work presented in this paper is organized as follows. In Section II, we consider the 2D IM with random bonds. The transfer matrix formalism is set up and the corresponding equations are written down. In Section III, we give a description of the critical behaviour of the pure N-colour Ashkin-Teller model, as well as the critical properties of two interacting N- and M-colour quenched disordered ATM's. The RG method is used to obtain the exact temperature dependence of the correlation length, specific heat, susceptibility and spontaneous magnetization near criticality and the equation of state at the critical point. The computation of the two-spin correlation function for pure and impure models at criticality is also reviewed. Section IV contains a brief discussion of the critical behaviour of random minimal conformal field theory models and some concluding remarks.

2 The Two-Dimensional Ising Model with Random Bonds

The classical Hamiltonian of the 2D Ising model with random bonds defined on a square lattice with periodic boundary conditions is

$$H = - \sum_{i=1,j=1}^{N} [J_1(i,j)s_{ij}s_{ij+1} + J_2(i,j)s_{ij}s_{i+1j}], \tag{2.1}$$

where i, j label the sites of the square lattice, $s_{ij} = \pm 1$ are the spin variables, $J_1(i,j)$ and $J_2(i,j)$ are the horizontal and vertical random independent couplings, each having the same probability distribution,

$$P(x) = (1-p)\delta(x-J) + p\delta(x-J'), \tag{2.2}$$

with p being the concentration of impurity bonds, and both J and J', assumed positive so that the Hamiltonian favours aligned spins. Notice that both antiferromagnetic couplings (creating frustration) and broken bonds ($J' = 0$) lead to ambiguities in the transfer matrix and must be excluded in this treatment. Let us now consider the calculation of the partition function of the model under discussion, namely

$$Z = \sum \exp(-\frac{H}{T}), \tag{2.3}$$

where H is defined in Eq. (2.1) and the sum runs over all 2^{N^2} possible spin configurations. The partition function is represented as the trace of the product of the row-to-row transfer matrices $\hat{T}_i$ [20], [21], [22], i.e.

$$Z = Tr \prod_{i=1}^{N} \hat{T}_i. \tag{2.4}$$

The Hermitian $2^N \times 2^N$ matrix $\hat{T}_i$ rewritten in terms of spin variables [2], [20], [21], [22] is

$$\hat{T}_i = \exp(\frac{1}{T}\sum_{j=1}^{N} J_1(i,j)\sigma_3(j)\sigma_3(j+1)) \exp(\frac{1}{T}\sum_{l=1}^{N} J_2^*(i,l)\sigma_1(l)), \tag{2.5}$$

where σ_α, $\alpha = 1,2,3$ are the Pauli spin matrices; here J_2 and J_2^* are related by the Kramers-Wannier duality relation [20], [21], [22], viz.

$$\tanh(\frac{J_2^*}{T}) = \exp(-\frac{2J_2}{T}). \tag{2.6}$$

In Eq. (2.5) we have set an irrelevant factor to unity. Since the non-averaged operator $\hat{T}_i$ in Eq. (2.7) is random, the representation in Eq. (2.4) is in fact inappropriate for the computation of the partition function. In order to get a convenient starting point for further calculations, we apply the replica trick. We introduce n identical "replicas" of the original model labelled by the index α, $\alpha = 1, \ldots, n$ and use

$$\bar{F} = -T\overline{\ln Z} = -T \lim_{n\to 0} \frac{1}{n}\overline{(Z^n - 1)}, \tag{2.7}$$

the well-known identity for the averaged free energy. Substituting Eq. (2.4) into Eq. (2.7), one obtains

$$\overline{F} = -T \lim_{n\to 0}\{Tr \overline{\prod_{\alpha=1}^{n}\prod_{i=1}^{N} \hat{T}_i^\alpha - 1}\}\frac{1}{n}. \tag{2.8}$$

In contrast to the case of random-site disorder, for the random-bond problem, the two matrices $\hat{T}_i^\alpha$ and $\hat{T}_j^\beta$ with different row indices $i \neq j$ depend on two different sets of random coupling constants and commute with each other for any α, β. This allows us to average these two operators independently. After some algebra one arrives at

$$\overline{Z^n} = Tr\hat{T}^N, \tag{2.9}$$

where the transfer matrix $\hat{T}$ of the 2D random-bond IM [2] is given by

$$\begin{aligned}\hat{T} &= \overline{\prod_{\alpha=1}^{n} \hat{T}_i^\alpha} \\ &= \exp\left\{\sum_{j=1}^{N} \log[(1-p)\exp(\frac{J}{T}\sum_{\alpha=1}^{n}\sigma_3^\alpha(j))\sigma_3^\alpha(j+1) + \right. \\ &+ p\exp(\frac{J'}{T}\sum_{\alpha=1}^{n}\sigma_3^\alpha(j)\sigma_3^\alpha(j+1))]\Bigg\} \times \\ &\times \exp\{\sum_{j=1}^{N}\log[(1-p)\exp(\frac{J^*}{T}\sum_{\alpha=1}^{n}\sigma_1^\alpha(j)) + p\exp(\frac{J^{*\prime}}{T}\sum_{\alpha=1}^{n}\sigma_1^\alpha(j))]\}\end{aligned} \tag{2.10}$$

Setting p to zero (or $J = J'$) one is indeed led to the well known expression for the T-operator of the pure IM [20], viz.

$$\hat{T} = \exp\{\frac{J}{T}\sum_{j=1}^{N}\sigma_3(j)\sigma_3(j+1)\}\exp\{\frac{J^*}{T}\sum_{j=1}^{N}\sigma_1(j)\}. \tag{2.11}$$

The T-matrix is known to possess the Kramers-Wannier dual symmetry. In the language of spin variables this nonlocal mapping [21],[22] is

$$\tau_1(k) = \sigma_3(k)\sigma_3(k+1), \qquad \tau_2(k) = i\sigma_1(k)\sigma_3(k), \qquad \tau_3(k) = \prod_{m<k}\sigma_1(m), \tag{2.12}$$

where the operators $\tau_\alpha(k)$ satisfy the very same algebra as the Pauli spin matrices $\sigma_\alpha(n)$. It is easy to see that if $p = 0, 0.5, 1$, the T-matrix given by Eq. (2.11) is invariant under the dual transformation. The plausible assumption that there is a single critical point yields the equation for the Curie temperature T_c, obtained by Fisch [23],

$$\exp(-\frac{2J'}{T_c}) = \tanh(\frac{J}{T_c}). \tag{2.13}$$

Notice that the point $p = 0.5$ is not the percolation threshold, because the coupling constants J and J' are assumed to take nonzero values with ferromagnetic sign. Writing $\hat{T}$ in the exponential form

$$\hat{T} = \exp(-\hat{H}), \tag{2.14}$$

one obtains the partition function

$$Z = Tr \exp(-N\hat{H}), \tag{2.15}$$

where by definition $\hat{H}$ is just the logarithm of the transfer matrix $\hat{T}$ (the "quantum" Hamiltonian), which is not a simple local operator. The crucial simplification occurs after taking the y-continuum limit with $a_y \to 0$ (the lattice spacing along the y-axis). After setting a_y to zero, the logarithmic derivative of $\hat{T}$ with respect to a_y, the "quantum" Hamiltonian, takes the simple form (for details see [2])

$$\hat{H} = \frac{d\ln\hat{T}}{da_y}|_{(a_y=0)} = -\sum_{j=1}^{N}\{K_1\sigma_3^\alpha(j)\sigma_3^\alpha(j+1) + K_2\sum_{\alpha=1}^{n}\sigma_1^\alpha(j)$$
$$+ K_4'(\sigma_3^\alpha(j)\sigma_3^\alpha(j+1))^2 + K_4''(\sum_{\alpha=1}^{n}\sigma_1^\alpha(j))^2\}. \tag{2.16}$$

The higher-order terms in the spin operators are known to be irrelevant in the critical region, so that they can be dropped in Eq. (2.16). The replicated Hamiltonian, Eq. (2.16), may be converted into the fermionic one by means of the Jordan-Wigner transformation [21],[22], i.e.

$$c^\alpha(m) = \sigma_-^\alpha(m)\prod_{j=1}^{m-1}\sigma_1^\alpha(j)Q^\alpha,$$
$$c^{\alpha^+}(m) = \sigma_+^\alpha(m)\prod_{j=1}^{m-1}\sigma_1^\alpha(j)Q^\alpha, \tag{2.17}$$
$$\sigma_\pm = \frac{1}{2}(\sigma_3 \pm i\sigma_2), \qquad Q^\alpha = \prod_{\beta=1}^{\alpha-1}\prod_{j=1}^{N}\sigma_1^\beta(j), \qquad \alpha = 1, ..., n,$$

where $c^\alpha(m)$ and $c^{\alpha^+}(m)$ are the standard annihilation and creation fermionic operators that satisfy the canonical anticommutation relations

$$\{c^\alpha(m), c^{\beta^+}(n)\} = \delta^{\alpha\beta}\delta_{mn}, \qquad \{c^\alpha(m), c^\beta(n)\} = 0. \tag{2.18}$$

After making the different species anticommute, the Klein factors Q^α drop out of H. For each species it is convenient to introduce a two-component Hermitean Majorana spinor field [24]

$$\begin{aligned}\psi_1^\alpha(n) &= \frac{1}{\sqrt{2a_x}}[c^\alpha(n)\exp(-i\frac{\pi}{4}) + c^{\alpha^+}(n)\exp(i\frac{\pi}{4})],\\ \psi_2^\alpha(n) &= \frac{1}{\sqrt{2a_x}}[c^\alpha(n)\exp(i\frac{\pi}{4}) + c^{\alpha^+}(n)\exp(-i\frac{\pi}{4})]\end{aligned} \tag{2.19}$$

with standard anticommutation rules

$$\{\psi_c^\alpha(n), \psi_b^\beta(m)\} = \frac{1}{a_x}\delta^{\alpha\beta}\delta_{bc}\delta_{mn} \qquad c, b = 1, 2 \tag{2.20}$$

where a_x is the lattice spacing along the x-axis. Now let us note that in the vicinity of T_c, the correlation length ξ goes to infinity and the system "forgets" the discrete nature of the lattice. For this reason we can simplify the Hamiltonian by taking the continuum limit $a_x \to 0$. Performing simple but cumbersome calculations, we arrive at the $0(n)$-symmetric Lagrangian of the Gross-Neveu model ([1])

$$L = \int d^2x[i\bar{\psi}_a\hat{\partial}\psi_a + m_0\bar{\psi}_a\psi_a + u_0(\bar{\psi}_a\psi_a)^2], \tag{2.21}$$

where $\gamma_\mu = \sigma_\mu, \hat{\partial} = \gamma_\mu\partial_\mu, \mu = 1, 2, \bar{\psi} = \psi^T\gamma_0$ and

$$m_0 \sim K_1 - K_2 \sim \tau = \frac{T - T_c}{T_c}, \qquad u_o \sim K_4' + K_4''. \tag{2.22}$$

Here m_0 and u_0 are the bare mass of the fermions and the quartic coupling constant, respectively. Note that if $p \ll 1$, then $u_o \sim p$. Provided $p = 0.5$ and $T = T_c$, we have $u_o \sim (J - J')^2$.

3 The n-Colour Ashkin-Teller Model

The N-colour ATM (introduced by Grest and Widom [27]) is a system of N 2D IM models coupled together as in the conventional 2-colour model. The lattice Hamiltonian of the isotropic N-colour ATM is

$$H = -\sum_{<nn>}\{J\sum_{a=1}^{N} s_i^a s_j^a + J_4[\sum_{a=1}^{N} s_i^a s_j^a]^2\}, \tag{3.1}$$

where $s^a = \pm 1$, $a = 1, ..., N$, $<>$ indicates that the summation is over all the nearest-neighbour sites and J_4 is a coupling constant between Ising planes.

This model was shown to be the lattice version of a model with hypercubic anisotropy, describing a set of magnetic and structural phase transitions in variety of solids [16], [28]. In particular, in the replica limit the Hamiltonian, Eq. (3.1), describes the random-bond Ising model (for $J_4 < 0$).

By exploiting the operator product expansion (OPE) approach, Grest and Widom obtained the one-loop β-function for J_4. If $J_4 < 0$ and $N > 2$, the phase transition was shown to be continuous and the critical behaviour to belong to the 2D IM universality class [27].

The exact solution of the multi-colour ATM in the large N-limit was found by Fradkin [29], who showed that a second order phase transition with IM critical exponents occurs if $J_4 < 0$. As was shown by Aharony [30], the model with hypercubic anisotropy, Eq. (3.1), in the large N-limit is appropriate for the description of the critical behaviour of the annealed random IM in accordance with the exact solution obtained by Lushnikov [31].

In the critical region the N-colour ATM was shown to be equivalent to the $0(N)$-symmetric Gross-Neveu model, Eq. (2.21), where we must set $N = n$ [17] (see also [18] and [19]). Note that the discrete hypercubic symmetry of the N-colour ATM evolves into a continuous $0(N)$ symmetry, hidden when the system approaches the critical point.

The one-loop RG equations and its initial conditions are given by

$$\frac{du}{dt} = \beta(u) = -\frac{(N-2)u^2}{\pi}, \qquad \frac{d\ln F}{dt} = -\gamma_{\bar{\psi}\psi}(u) = \frac{(1-N)u}{\pi},$$
$$u(t=0) = u_0, \qquad F(t=0) = 1, \tag{3.2}$$

where u is a quartic fermionic coupling constant, $\beta(u)$ is the beta-function, $\gamma_{\bar{\psi}\psi}(u)$ is the anomalous dimension of the composite operator $\bar{\psi}_a\psi_a$, $t = \ln\frac{1}{ma}$ and a and m are the lattice spacing and renormalized mass, respectively. F is the Green's function

$$F = \frac{dm}{d\tau} = \int d^2x d^2y < \bar{\psi}(x)\psi(y)\bar{\psi}(0)\psi(0) > \tag{3.3}$$

at zero external momenta. The solution of these equations gives the temperature dependence of the correlation length ξ and specific heat C in the asymptotic region $t \to \infty$ [19], i.e.

$$u = \frac{\pi}{(N-2)t}, \qquad F \sim t^{-\frac{N-1}{N-2}},$$
$$\xi = m^{-1} \sim \tau^{-1}[\ln\frac{1}{\tau}]^{\frac{N-1}{N-2}},$$
$$C \sim \int dt F(t)^2 \sim [\ln(\frac{1}{\tau})]^{\frac{N}{2-N}}. \tag{3.4}$$

In order to complete the calculation of the temperature dependence of the other thermodynamic quantities, we have to compute the large-distance asymptotic behaviour of the two-spin correlation function at criticality. The most effective way of calculating correlation functions for the 2D IM is to use bosonization. Below we shall give a brief description of this procedure, exploiting simple physical arguments.

Let us now begin with the action

$$L = \int d^2x \{ i\bar{\psi}\hat{\partial}\psi + [m_0 + \tau(x)]\bar{\psi}\psi \}, \tag{3.5}$$

where ψ is a Majorana spinor and $\tau(x)$ is a Gaussian distributed field. In fact, the action Eq. (3.5) describes free fermions moving in the random potential $\tau(x)$ responsible for spatial fluctuations of a local $T_c(x)$ introduced by weak disorder. After applying the replica trick and averaging over "all" the possible configurations of $\tau(x)$, one arrives at the very same Gross-Neveu Lagrangian given by Eq. (2.21). The representation of the square of the two-spin correlation function of the pure 2D IM in terms of the path integral over the real bosonic field ϕ of quantum sine-Gordon model was found by Zuber and Itzykson [24] (see also [25] and [26]) and is

$$\begin{aligned} G(x-y)^2 &= Z^{-1}\frac{1}{2\pi^2 a^2}\int D\phi \sin(\sqrt{4\pi}\phi(x))\sin(\sqrt{4\pi}\phi(y))\exp\{-S\}, \\ S &= \frac{1}{2}\int d^2x \{(\partial_\mu\phi)^2 + \frac{2m_0}{\pi a}\cos(\sqrt{4\pi}\phi)\}, \\ Z &= \int D\phi \exp\{-S\}. \end{aligned} \tag{3.6}$$

At criticality, $m_0 = 0$, the path integral is Gaussian and the result of its evaluation is easily shown to be

$$G(x-y) \sim |x-y|^{-\frac{1}{4}}. \tag{3.7}$$

The representation of the two-spin correlation function may be extended to the dilute system by replacing in Eq. (3.7) the bare mass $m_0 \sim \tau$ with the random one $m_0 + \tau(x)$. The averaged correlation function $\overline{G(x-y)}$ at the Curie point may be computed (even without using the replica trick) in two stages: (i) first, the square root of $G(x,y)^2$ is formally evaluated by means of an expansion in a power series in $\tau(x)$; (ii) secondly, the resulting expression is integrated with respect to $\tau(x)$ (for technical details of the calculations, see [2], [26]). The conventional RG equation for the renormalized averaged correlation function is

$$\{\mu\frac{\partial}{\partial\mu} + \beta(u)\frac{\partial}{\partial u} + \eta(u)\}\overline{G_R(p,u,\mu)} = 0, \tag{3.8}$$

with μ and $\eta(u)$ being a renormalization momentum and the anomalous spin dimension, respectively, and

$$\eta(u) = \beta(u)\frac{d\ln Z_\sigma(u)}{du}. \tag{3.9}$$

Here the spin renormalization constant $Z_\sigma(u)$ and the renormalized correlation function are defined in the standard way, viz.

$$\overline{G(p)} = Z_\sigma(u)\overline{G_R(p, u, \mu)}. \tag{3.10}$$

The Kramers-Wannier symmetry was shown to apply to some vanishing terms linear in u in the expansions for $\eta(u)$ and $Z_\sigma(u)$ [2], that is

$$Z_\sigma(u) = 1 + 0(u^2), \qquad \eta(u) = \frac{7}{4} + 0(u^2). \tag{3.11}$$

Given $\beta(u)$ and $\eta(u)$, the solution of the Ovsyannikov-Callan-Symanzik equation for the two-spin correlation function is quite simple, namely

$$G(p) \sim p^{-\frac{7}{4}}, \qquad G(R) \sim R^{-\frac{1}{4}}. \tag{3.12}$$

So the Fisher critical exponent has the very same value $\eta = \frac{1}{4}$ as in the 2D IM, irrespective of the value of N. From Eq. (3.12), it follows that the temperature dependence of the homogeneous susceptibility and the spontaneous magnetization are described by power-law functions of the correlation length ξ (without logarithmic corrections like $\ln \xi$), viz.

$$\chi \sim \xi^{2-\eta} \sim \tau^{-\frac{7}{4}}[\ln \frac{1}{\tau}]^{\frac{7(N-1)}{4(N-2)}},$$
$$M \sim \xi^{\frac{\eta}{2}} \sim (-\tau)^{\frac{1}{8}}[\ln \frac{1}{(-\tau)}]^{\frac{(N-1)}{(N-2)}}. \tag{3.13}$$

The equation of state at the Curie point may be obtained from the usual scaling relation

$$H \sim M^{\frac{4+\eta}{\eta}} \sim M^{15}. \tag{3.14}$$

Notice that these results are valid only for $N > 2$ and $J_4 < 0$. If $J_4 > 0$, the discrete γ_5 symmetry $\psi \rightarrow \gamma_5\psi$, $\bar{\psi}\psi \rightarrow -\bar{\psi}\psi$ is spontaneously broken. From the γ_5-symmetry-breaking, it follows that $< \bar{\psi}\psi > \neq 0$. This means that we have a finite correlation length, or, in other words, a first-order phase transition [17], [27].

Thus equations Eq. (2.22) reproduce the well-known results for particular cases, namely $N = 0, 1$, and ∞ corresponding to the random-bond problem, Onsager problem, and IM with equilibrium impurities, respectively.

The symmetric eight-vertex model (or the Baxter model) is known to be isomorphic to the $N =$ 2-colour ATM in the vicinity of the critical line. The phase diagram of the 2-colour ATM was shown to contain the ferromagnetic phase transition line beginning at the IM critical point and ending at the point corresponding to the 4-state Potts model. Along this line the model exhibits nonuniversal critical behaviour, the critical exponents varying continuously. In this case the above results obviously show the pathology at $N = 2$ due to the factor $\frac{1}{N-2}$. The system under discussion is described by the 0(2)-symmetric

Gross-Neveu model, or, equivalently, by the massive Thirring model, presenting nonuniversal critical exponents [17]. Since the $N = 3$-colour ATM is equivalent to the 0(3)-symmetric Gross-Neveu model which is known to be supersymmetric ([32]), this model should possess a hidden supersymmetry (see for details [17], [18]).

The main conclusion of this Section is that the critical behaviour of the 2D N-colour ATM and the random bond IM is governed by the pure IM fixed point. This implies that all of the critical exponents of these systems are the very same as the 2D IM. Randomness gives rise to the self-interaction of the spinor field which leads to logarithmic corrections to the power laws.

We may also extend our study of the N-colour ATM to two interacting M- and N-colour quenched disordered ATM, giving a generalized Ashkin-Teller model (the particular case $N = M = 1$ was studied in [34]). The cumbersome RG calculations show that, in contrast to a 2D IM with random bonds, the weak quenched disorder is here irrelevant near T_c. Moreover, in the critical region the decoupling of the two interacting multi-colour ATM's was found to occur even in the presence of quenched disorder. The temperature dependence of the main thermodynamic quantities near the critical point, the two-spin correlation function and equation of the state at criticality are given by Eqs. (3.4), (3.13) and (3.14) (see for details [33]).

4 Conclusions

It has been shown that the critical behaviour of a good number of 2D anisotropic systems controlled by the IM fixed point is stable in the presence of quenched disorder. This statement was found to hold quite generally for the 2D IM, the multi-colour ATM, and some of its generalizations for which randomness is marginally relevant. In the case of the 2-colour ATM, or the Baxter model, disorder drastically changes the nonuniversal critical behaviour inherent in these models over to the Ising-type critical behaviour. Although some of these models exhibit a breakdown of the Harris criterion, this does not affect, in general, the stability of the IM fixed point. It is commonly believed that the type of randomness (random-bond, or site-disorder) does not play a role, despite the fact that random-site disorder has not been studied in great detail as yet.

Basically Monte Carlo simulation results are in a good agreement with the analytical results based on RG calculations [15]. For instance, the high-accuracy MC simulation results for a 1024×1024 Ising lattice with ferromagnetic impurity bonds recently obtained by Schur and Talapov [15], show that the two-spin correlation function at criticality is numerically very close to that of the pure model. On the other hand, numerical results obtained by Wiseman and Domany [15] somewhat contradict the theoretical predictions. These results for a 256×256 lattice favour a log-type behaviour of the specific heat near T_c for the disordered 2-colour ATM and 4-state Potts model, and a double-log behaviour of the specific heat for the random-bond IM. As was established by Dotsenko and Dotsenko

[34], the specific heat of the impure 2-colour ATM should exhibit a double-log divergence at the critical point.

A very interesting problem which remains to be solved is the critical behaviour of the minimal models of a conformal field theory with $c < 1$ perturbed by a small number of impurities. In accordance with the Harris criterion, weak quenched disorder is expected to be strongly relevant near criticality since the critical exponent α of these models is always positive and given by $\alpha = \frac{2(m-3)}{3(m-1)}, m = 3, 4, \ldots$ [35]. In particular, for the 3- and 4-state Potts model, we have $\alpha = \frac{1}{3}, (m = 5)$, and $\alpha = \frac{2}{3}, (m = \infty)$, respectively. The first results in this field were obtained in the pioneering papers by Ludwig [36], Ludwig and Cardy [37] and Dotsenko, Picco and Pujol [38]. They succeeded in developing an approach essentially based on the powerful conformal field theory technique. These authors suggested a special kind of ϵ-expansion, with $\epsilon = c - \frac{1}{2}$, to compute the critical exponents. Here c is the central charge of the minimal models without randomness and $\frac{1}{2}$ is the conformal anomaly of the 2D IM. The main result of their considerations is that the $\beta(u)$ and $\gamma_{\bar{\psi}\psi}(u)$ functions coincide with the corresponding functions for the 0(N)-symmetric Gross-Neveu model obtained in the framework of the minimal substraction scheme combined with dimensional regularization. The distinguished feature of this scheme is that these functions do not depend on ϵ except for the first term in the β-function. The three-loop results for the critical exponent ν for random 3- and 4-state Potts models obtained in [33] are $\nu(5) = 1.018$ and $\nu(\infty) = 1.081$. The critical exponent $\eta(u)$ in the three-loop approximation was obtained in [38]. It turns out that the numerical values of η are in the close vicinity of the IM value of $\eta = 0.25$. Thus the critical behaviour of the random 3- and 4-state Potts model was shown to be described by a new impure fixed point which does not coincide with the IM one [38],[36], but the numerical values of critical exponents of weakly-disordered minimal models are very close to the critical exponents of the 2D IM. It is interesting to compare the estimate for the critical exponent of the 4-state Potts model based on the 3-loop approximation with the known numerical results. For the Baxter-Wu model (equivalent to the 4-state Potts model), Novotny and Landau [39] obtained $\nu = 1.00(7)$. The result of Andelman and Berker [40] is given $\nu = 1.19$. Finally, the recent result obtained by Schwenger, Budde, Voges, and Pfnur [41] is $\nu = 1.03(8)$.

Thus, the introduction of disorder leads to critical behaviour as characterised by the IM fixed point, for the minimal models of conformal field theory this Ising behaviour, conjectured in [42] for the 2D Potts models, is actually only approximate. The accuracy with which the Ising values of the exponents is observed, however, justifies the use of the term "IM superuniversality" for all these discrete-symmetry models when disordered.

References

1. Vl.S. Dotsenko and Vik.S. Dotsenko, *Adv. in Phys.* **32**, 129 (1983).
2. B.N. Shalaev, *Phys. Rep.* **237**, 129 (1994).

3. Vik.S. Dotsenko, *Sov. Phys.-Uspekhi* **38**, 310 (1995).
4. K. Hirakawa and H. Ikeda, in *Magnetic Properties of Layered Transition Metal Compounds*, L.J. de Jongh (Ed.) (Kluwer Academic Publishers, New York 1990); and related articles in this book.
5. I.F. Lyuksyutov, A.G. Naumovets and V.L. Pokrovsky, *Two-Dimensional Crystals* (Academic Press, London 1992).
6. A.B. Harris and T.C. Lubensky, *Phys. Rev. Lett.* **33**, 1540 (1974).
7. D.E. Khmelnitskii, *Soviet Phys. JETP.* **68**, 1960 (1975).
8. B.M. McCoy and T.T. Wu, *Phys. Rev.* **176**, 631 (1968).
9. R. Shankar and G. Murthy, *Phys. Rev.* **B 36**, 536 (1987).
10. B.N. Shalaev, *Soviet Physics-Solid State* **26**, 1811 (1984).
11. R. Shankar, *Phys. Rev. Lett.* **58**, 2466 (1987).
12. A.W.W. Ludwig, *Phys. Rev. Lett.* **61**, 2388 (1988).
13. A.W.W. Ludwig, *Nucl. Phys.* **B 330**, 639 (1990).
14. G. Jug, *Phys. Rev.* **B 27**, 609 (1983).
15. J.-S. Wang, W. Selke, Vl.S. Dotsenko and V.B. Andreichenko, *Nucl. Phys.* **B 344**, 531 (1990); J.-S. Wang, W. Selke, VI.S. Dotsenko and V.B. Andreichenko, *Physica* **A 164**, 221 (1990); J.-S. Wang, W. Selke, VI.S. Dotsenko and V.B. Andreichenko, *Europhys. Lett.* **11**, 301 (1990); H.-O. Heuer H.-O, *Phys. Rev.* **B 45**, 5691 (1992); J.-K. Kim, *Phys. Rev. Lett.* **70**, 1735 (1993); L.N. Schur and A.L. Talapov, *Europhys. Lett.* **27**, 193 (1994); S. Wiseman and E. Domany, *Phys. Rev.* **E 51**, 3074 (1995).
16. P. Bak, *Phys. Rev.* **B 14**, 3980 (1976); P. Bak and D. Mukamel, *Phys. Rev.* **B 13**, 5086 (1976); D. Mukamel and S. Krinsky, *Phys. Rev.* **B 13**, 5065, 5078 (1976); S.A. Brazovskii, I.E. Dzyaloshinskii and B.G. Kukharenko, *Sov. Phys.-JETP* **43**, 1178 (1976); P. Bak, S. Krinsky and D. Mukamel, *Phys. Rev. Lett.* **36**, 52 (1976).
17. R. Shankar, *Phys. Rev. Lett.* **55**, 453 (1985).
18. Y.Y. Goldschmidt, *Phys. Rev. Lett.* **56**, 1627 (1986).
19. B.N. Shalaev, *Sov. Phys.-Solid State* **31**, 51 (1989).
20. R.J. Baxter, *Exactly Solved Models in Statistical Mechanics* (Academic Press, London 1982).
21. J.B. Kogut, *Rev. Mod. Phys.* **51**, 659 (1979).
22. E. Fradkin and L. Susskind, *Phys. Rev.* **D 17**, 2637 (1978).
23. R. Fisch, *J. Stat. Phys.* **18**, 111 (1978).
24. J.-B. Zuber and C. Itzykson, *Phys. Rev.* **D 15**, 2875 (1977).
25. P. Di Francesco, H. Saleur and J.-B. Zuber, *Nucl. Phys., B 290*, [FS20], 527 (1987).
26. C. Itzykson and J.-M. Drouffe, *Statistical field theory* **vol.2** (Cambridge Univ. Press, Cambridge 1989).
27. G. Grest and M. Widom, *Phys. Rev.* **B 24**, 6508 (1981).
28. J.-C. Toledano, L. Michel, P. Toledano and E. Brezin, *Phys. Rev.* **B 31**, 7171 (1985).
29. E. Fradkin, *Phys. Rev. Lett.* **53**, 1967 (1984).
30. A. Aharony, *Phys. Rev. Lett.* **31**, 1494 (1973).
31. A.A. Lushnikov, *Sov. Phys.-JETP* **56**, 215 (1969).
32. E. Witten, *Nucl. Phys.* **B 142**, 285 (1978).
33. G. Jug and B.N. Shalaev, submitted to *Phys. Rev.* **B** .
34. VI.S. Dotsenko and Vik.S. Dotsenko, *J. Phys.* **A 17**, L301 (1984); Vik.S. Dotsenko, *J. Phys.* **A 18**, L241 (1985).
35. Vl.S. Dotsenko and V.A. Fateev, *Nucl. Phys.* **B 240**, 312 (1984).
36. A.W.W. Ludwig, *Nucl. Phys.* **B 285**, 97 (1987).

37. A.W.W. Ludwig and J. Cardy, *Nucl. Phys.* **B 285**, [FS19], 687 (1987).
38. VI.S. Dotsenko, M. Picco and P. Pujol, *Phys. Lett.* **B 377**, 113 (1995).
39. M.A. Novotny and D.P. Landau, *Phys. Rev.* **B 24**, 1468 (1981).
40. D. Andelman and A.N. Berker, *Phys. Rev.* **B 29**, 2630 (1984).
41. L. Schwenger, K. Budde, C. Voges and H. Pfnur, *Phys. Rev. Lett.* **73** 296 (1994).
42. M. Kardar, A. Stella, G. Sartoni and B. Derrida, *Phys. Rev.* **E 52**, R1269 (1995); Cardy J.L. cond–mat@xxx.lanl.gov. No. 9511112.

Critical Behaviour in Non-Integer Dimension

Yurij Holovatch

Institute for Condensed Matter Physics
of the Ukrainian Academy of Sciences
1 Svientsitskii St.,
UA-290011 Lviv, Ukraine

Abstract: A method for studying critical behaviour in non-integer space dimensions is discussed. The critical exponents of several models commonly used in the theory of phase transitions are calculated for the case of non-integer space dimension. The calculations are performed using a fixed-dimension field theoretical approach. The renormalization group functions in the Callan--Symanzik scheme are considered directly in non-integer dimensions. Perturbation theory expansions are resummed with the use of Padé-Borel transformation.

1 Introduction

The notion of non-integer space dimension is now common in the theory of critical phenomena. There exist different reasons for introducing this concept: on the one hand, the treating of the space dimension (or its deviation from some definite fixed value) as a continuous variable and a perturbation theory series expansion parameter makes it possible to obtain results for integer d as well. Here one should mention not only the famous $\varepsilon = 4 - d$ expansion [1] whose application to the theory of critical phenomena led to the calculation of reliable values of the critical exponents for a whole range of $3d$ models (see [2] - [4]) but also $\varepsilon = d - 1$ expansion, introduced for the near-planar interface [5] - [7] and droplet [8] models, and the $\sqrt{\varepsilon}$-expansion for the weakly dilute Ising model [9, 10], etc.

On the other hand, continuous variations of the space dimension d by means of analytic continuation of hypercubic lattices to non-integer d are used to link the results obtained for certain fixed (integer) d to exact ones (if they exist for some value of d) or to results of other calculational methods. Further, there exist models in which new phenomena appear at some (non-integer) space dimension and the problem of the definition of this marginal dimension arises. Thus the task of studying the critical behaviour of some model directly at non-integer d can occur. In the case of the Ising model such studies have been made using various methods [11] - [20].

In the majority of the above-mentioned papers, the purely formal character of the analytic continuation in terms of dimension d has resulted in an interest in the comparison of the critical exponents obtained in this manner with the values of the critical exponents of the corresponding spin systems placed on the sites of self-similar fractal lattices (see [21, 22], where the non-integer fractal dimension can be treated as a purely geometric property. Unfortunately, the description of the fractal involves several factors that can vary independently of one another. In addition to the fractal dimension they are: the topological dimension, ramification, connectivity and lacunarity. It appears that the critical exponents strongly depend on these parameters [14] - [16], [23] and only in the limit of zero lacunarity is it now believed that the results for spin systems on fractal lattices may be interpolated by analytic continuation to non-integer d. The question of the correspondence of the critical behaviour on fractal lattices to the critical behaviour on interpolated hypercubic lattices is still open [23] - [25].

This article reviews some of our recent calculations of the universal characteristics of the critical behaviour of model spin systems on interpolated hypercubic lattices in non-integer d. Our work involves the fixed dimension renormalization group approach. Following Parisi [26] who performed calculations directly in 3 (or 2) dimensions, we consider the renormalization group functions directly in an arbitrary non-integer dimension d. This allows us to avoid the ε-expansion when studying the critical behaviour in non-integer d. As will be seen, for some models we can complete the existing data whereas for certain cases the approach is the only one way to obtain reliable values of critical exponents in non-integer dimension d.

2 Models and the renormalization procedure

We concentrate here on three models commonly used in the theory of phase transitions: the Ising model, m-vector ($O(m)$-symmetrical) model and m-vector model in the presence of weak quenched disorder. This choice enables us both to demonstrate the main peculiarities of critical behaviour in non-integer d and to show how the method under consideration works in different cases. The Ising model has a Hamiltonian which, in the absence of an external magnetic field, is given by:

$$H = J \sum_{i,j} S_i S_j, \qquad (2.1)$$

where the spins S_i take the values ± 1, and the summation is over nearest-neighbour sites on the lattice. As is well known [3, 27], one can describe the long-distance properties of such a model in the neighbourhood of a second-order phase transition in the terms of a continuous Euclidian field theory with the Lagrangian:

$$\mathcal{L}(\phi) = \int d^d R \Big\{ \frac{1}{2} \left[|\nabla \phi|^2 + m_0^2 \phi^2 \right] + \frac{u_0}{4!} \phi^4 \Big\}, \qquad (2.2)$$

where m_0^2 is a linear function of the temperature, u_0 is the bare coupling, $\phi = \phi(R)$ is an one-component field.

This model can be generalized by introducing into Eq. (2.2) a multiplet of m fields forming a representation of the group $O(m)$. In this case the Lagrangian reads:

$$\mathcal{L}(\phi) = \int d^dR\Big\{\frac{1}{2}\left[|\nabla\boldsymbol{\phi}|^2 + m_0^2|\boldsymbol{\phi}|^2\right] + \frac{u_0}{4!}|\boldsymbol{\phi}|^4\Big\}, \tag{2.3}$$

where $\boldsymbol{\phi} = \boldsymbol{\phi}(R)$ is the vector field $\boldsymbol{\phi} = (\phi^1, \phi^2, \ldots, \phi^m)$. The corresponding spin Hamiltonian is the scalar product of m-component vectors $\boldsymbol{S}$, i.e.

$$H = J\sum_{i,j} \boldsymbol{S}_i\boldsymbol{S}_j. \tag{2.4}$$

The spin system described by the Hamiltonian Eq. (2.4) in the case of weak quenched disorder, in which a portion of the sites is empty and the occupied sites and randomly distributed are fixed in certain positions, is the so-called quenched m-vector model. At small dilution (i.e. far from the percolation threshold), the critical behaviour is governed by the Lagrangian,

$$\mathcal{L}(\phi) = \int d^dR\Big\{\frac{1}{2}\sum_{\alpha=1}^{n}\left[|\nabla\boldsymbol{\phi}^\alpha|^2 + m_0^2|\boldsymbol{\phi}^\alpha|^2\right] + \frac{v_0}{4!}\left(\sum_{\alpha=1}^{n}|\boldsymbol{\phi}^\alpha|^2\right)^2 + \frac{u_0}{4!}\sum_{\alpha=1}^{n}\left(|\boldsymbol{\phi}^\alpha|^2\right)^2\Big\}, \tag{2.5}$$

where, in the replica limit, $n \to 0$ [10]. Again each component $\boldsymbol{\phi}^\alpha$ is a vector $\boldsymbol{\phi}^\alpha = (\phi^{\alpha,1}, \phi^{\alpha,2}, \ldots, \phi^{\alpha,m})$ and $u_0 > 0, v_0 < 0$ are the bare couplings.

In order to study the critical properties of the field theories Eqs. (2.2), (2.3) and (2.5) in a general space dimension d, we use the standard procedure of renormalizing the one-particle irreducible vertex function[1]

$$\Gamma^{(L,N)}(p_1, .., p_L; k_1, .., k_N; m_0^2, \{u_0\}, d)$$

at zero external momenta $\{p_j; k_j\}$ and nonzero mass (see [3, 27] for example). Asymptotically close to the critical point, the renormalized vertex-functions

$$\Gamma_R^{(N)}(\{k_j\}; m^2, \{u\}; d)$$

satisfy the homogeneous Callan-Symanzik equation [3, 27], i.e.

$$\Big\{m\frac{\partial}{\partial m} + \sum_i \beta_{u_i}(\{u\})\frac{\partial}{\partial u_i} - \frac{N}{2}\gamma_\phi(\{u\})\Big\}\Gamma_R^{(N)}(\{k_j\}; m^2, \{u\}; d) = 0, \tag{2.6}$$

where $\{u\}$ and m are the renormalized conventionally defined coupling constants and mass, respectively. At the stable fixed point $\{u^*\}$, its coordinates are determined by the zero of the β-functions and γ_ϕ gives the value of the pair correlation function critical exponents η. The correlation length critical exponent ν can be

[1] Here and below $\{u_0\}$, $\{u\}$ stands for the set of couplings; e.g. for the Lagrangian, Eq. (2.2), $\{u\} = u$, whereas for Eq. (2.5) $\{u\} = u, v$.

calculated upon consideration of the two-point vertex function with a ϕ^2 insertion and gives one more γ-function $\bar{\gamma}_{\phi^2}(\{u\})$ which at the fixed point gives the value of the combination $2 - \nu^{-1} - \eta$ (ν being correlation length critical exponent). The other critical exponents can be obtained on the basis of ν and η using the familiar scaling relations.

3 Perturbation theory in non-integer *d*

Equation (2.6) may be studied, in principle, for arbitrary non-integer space dimension d [29, 30]. Imposing the zero-momentum renormalization conditions for the conventionally defined 2-point and 4-point single-particle irreducible vertex functions $\Gamma_R^{(2)}(k, -k; m_0^2, \{u_0\}; d)$, $\Gamma_R^{(4)}(\{k_i\}; m_0^2, \{u_0\}; d)$ one obtains the expressions, shown in Fig. 1 in the three loop approximation. For every internal line

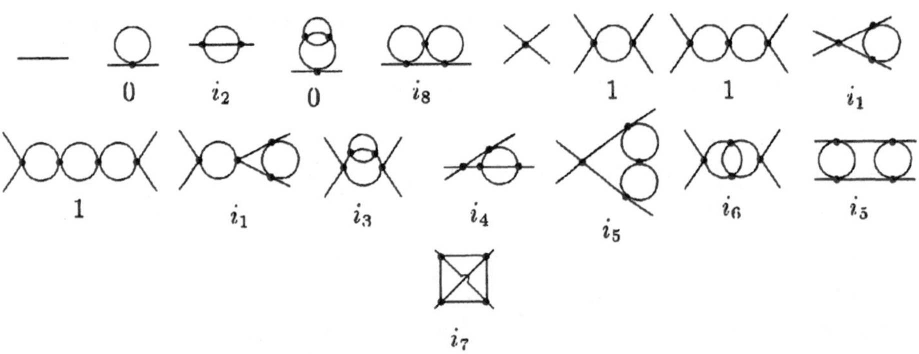

Fig. 1. Graphs of the vertex functions $\Gamma^{(2)}$, $\Gamma^{(4)}$ in the three-loop approximation and the normalized values of the corresponding loop integrals (for the function $\Gamma^{(2)}(k)$ the value of the derivative $\partial/\partial k^2 \,|_{k^2=0}$ is given).

i there corresponds a propagator $(1 + k_i^2)$, integration over internal momenta is imposed and a momentum conservation law is carried out at every point. The correspondence between the graphs for the vertex functions $\Gamma^{(2)}, \Gamma^{(4)}$ and the numerical values of loop integrals follows from Fig. 1. Expanding the integrals i_1 through i_8 in $\varepsilon = 4 - d$ one passes to the ε-expansion technique. Alternatively one can consider them directly in a space of fixed dimension (for $d = 2$ and $d = 3$, the corresponding values are given in [28]). The simplest way to obtain the expressions for the numerical procedure for evaluating the integrals i_1 through i_8 at non-integer d is to make use of the Feynman parameters method (see [27], for example). These expressions, in which the dimension of space is a parameter, were evaluated numerically for a general value of d and their values are listed in [31]. The dependence of the loop integrals on the space dimension d for continuous change of d is shown in Figs. 2 and 3.

Upon the basis of expansions of the renormalized Γ-functions, one can obtain expressions for the β and γ-functions of the field theories Eqs. (2.2), (2.3) and

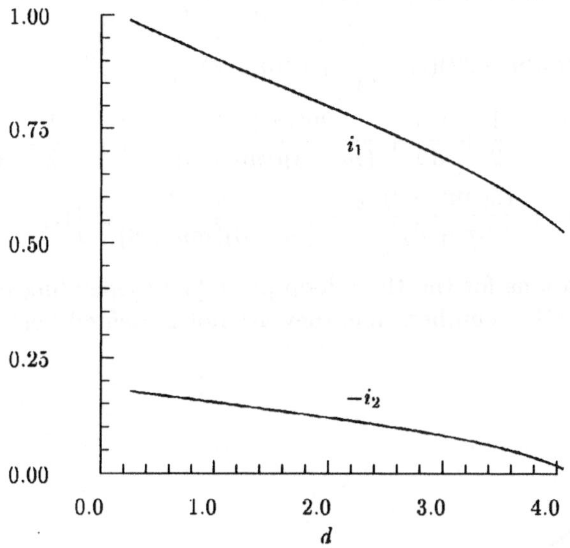

Fig. 2. Two-loop integrals as a function of the space dimension d.

(2.5) in the three-loop approximation. In the case of the Lagrangian Eq. (2.5), they are[2]

$$\beta_u(u,v) = -(4-d)\Big\{u - u^2 - \frac{12}{mn+8}uv + \beta_u^{(2)} + \beta_u^{(3)} + \dots\Big\}, \quad (3.1)$$

$$\beta_v(u,v) = -(4-d)\Big\{v - v^2 - \frac{2(m+2)}{m+8}uv + \beta_v^{(2)} + \beta_v^{(3)} + \dots\Big\}, \quad (3.2)$$

$$\gamma_\phi(u,v) = -4(4-d)\Big\{\Big[\frac{m+2}{(m+8)^2}u^2 + \frac{mn+2}{(mn+8)^2}v^2 + \frac{2(m+2)}{(m+8)(mn+8)}uv\Big]i_2 + \gamma_\phi^{(3)} + \dots\Big\}, \quad (3.3)$$

$$\bar\gamma_{\phi^2}(u,v) = (4-d)\Big\{\frac{m+2}{m+8}u + \frac{mn+2}{mn+8}v + \bar\gamma_{\phi^2}^{(2)} + \bar\gamma_{\phi^2}^{(3)} + \dots\Big\}, \quad (3.4)$$

where the indices $^{(2)}$ and $^{(3)}$ refer to the two- and three-loop parts of the corresponding functions:

$$\beta_u^{(2)} = \frac{8}{(m+8)^2}u^3\Big[(5m+22)(i_1-\frac{1}{2}) + (m+2)i_2\Big] + \frac{96}{(m+8)(mn+8)}u^2v\Big[(m+5)(i_1-\frac{1}{2}) + \frac{m+2}{6}i_2\Big] + \quad (3.5)$$

[2] The convenient numerical scale, in which the first coefficients of the β-functions are -1 and 1, (see [29]) for details) is used.

$$\frac{24}{(mn+8)^2}uv^2\Big[(mn+14)(i_1-\frac{1}{2})+\frac{mn+2}{3}i_2\Big],$$

$$\beta_v^{(2)}=\frac{8}{(mn+8)^2}v^3\Big[(5mn+22)(i_1-\frac{1}{2})+(mn+2)i_2\Big]+$$

$$\frac{24(m+2)}{(m+8)^2}u^2v\Big[i_1-\frac{1}{2}+\frac{i_2}{3}\Big]+\frac{96(m+2)}{(m+8)(mn+8)}uv^2\Big[i_1-\frac{1}{2}+\frac{i_2}{6}\Big],$$

$$\bar{\gamma}_{\phi^2}^{(2)}=-\Big[\frac{12(m+2)}{(m+8)^2}u^2+\frac{12(mn+2)}{(mn+8)^2}v^2+\frac{24(m+2)}{(m+8)(mn+8)}uv\Big](i_1-\frac{1}{2}).$$

As the explicit expressions for the three loop parts [32] (depending on integrals i_1 through i_8) are rather cumbersome, they are not presented here.

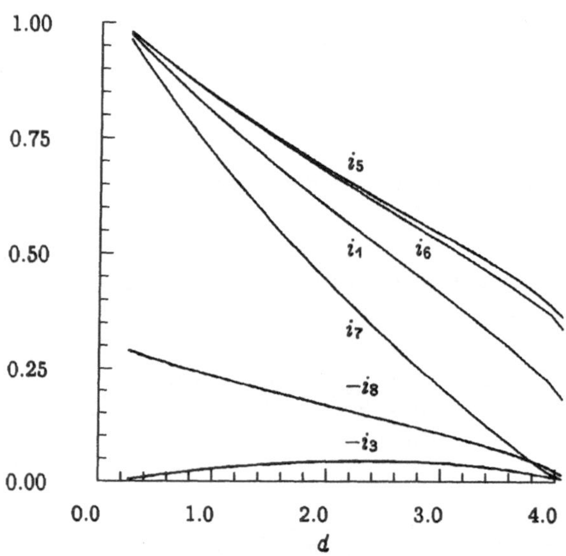

Fig. 3. Three-loop integrals as a function of the space dimension d.

Setting $v=0$ in Eqs. (3.1)-(3.4), one obtains the corresponding data for the $O(m)$-symmetrical model (Eq. 2.3) [30] while substituting $v=0,\ m=1$ gives the case of the Ising model [33]. Expressions (3.1)-(3.4), the main result of this study, are analysed in the next section.

4 Results and discussion

Substituting the values of loop integrals i_1 through i_8 as continuous functions of d into Eqs. (3.1)-(3.4), one can extract the dimensional dependence of the critical exponents governing the second-order phase transition. These series are known to be asymptotic (see e.g. [34] - [38]). To make them convergent we have used several resummation techniques and present the results below.

a) The Ising model

In order to study the dimensional dependence of the expressions Eq. (3.1)-(3.4) in the case $v = 0$, $m = 1$ we have applied [30] a simple Padé-Borel resummation technique [35] and then, to improve the results obtained at low dimensions we imposed Eqs. (3.1)-(3.4) to yield the values of the critical exponents of the Ising model for the cases where results are known exactly, i.e. for $d = 1$ ($\nu = \infty$, $\eta = 1$, see e.g. [39]) and for $d = 2$ ($\nu = 1, \eta = \frac{1}{4}$, [40]). This can be done by choosing the highest-order term of the series under consideration as a fitting parameter and setting it so as to obtain the known results. In the case of the ε-expansion such a procedure was considered in [41]. The results presented in this subsection were obtained using the three loop approximation with the fourth order term as the fitting parameter.

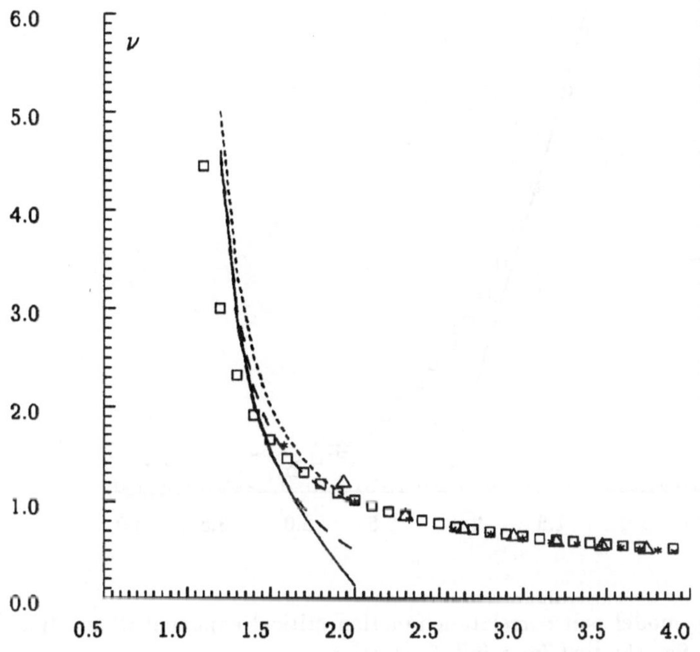

Fig. 4. The Ising model correlation-length critical-exponent ν as a function of the dimension d. See the text for a full description.

In Fig. 4 we compare our results for the critical exponent ν (given by squares) with the results of a $\varepsilon' = d-1$ expansion for the near-planar interface model[3] (for low values of d) and with data, obtained on the basis of a Kadanoff lower-bond

[3] Strong arguments have been given for the correspondence of the critical behaviour of this model to that of the Ising model [5] - [7].

renormalization transformation [11] (asterisks) and by the study of the physical branch of the exact renormalization-group equation solution [42, 43] (triangles). The value of the critical exponent ν for the near-planar interface model is shown to first, second, third and fourth orders (dotted, dot-dashed, dashed and solid lines) [5] - [7].

Fig. 5 shows the comparison of our data for the critical exponent η (squares) with results obtained from a Kadanoff lower-bond renormalization transformation [11] (asterisks), from the exact renormalization group equation [42, 43] (triangles) and with the value of η for the droplet model [8] (solid line).

b) $O(m)$-symmetrical model

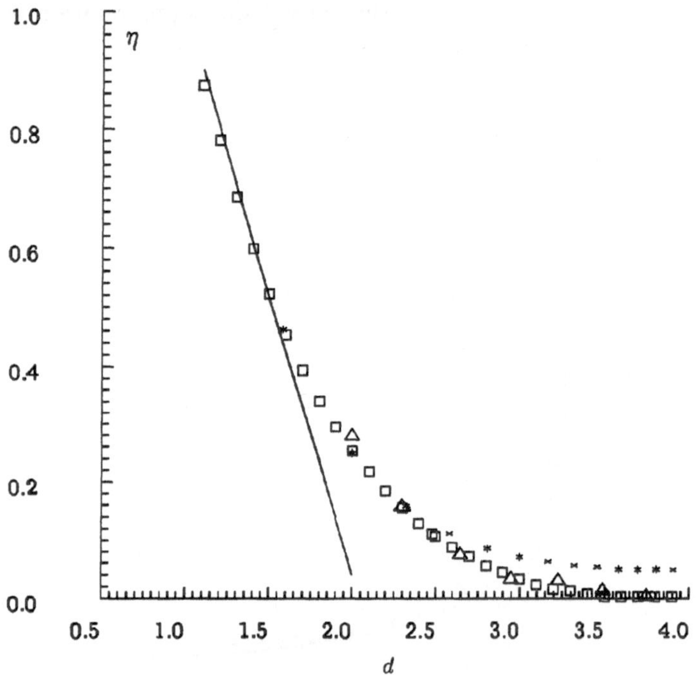

Fig. 5. The Ising model pair correlation function critical exponent η as a function of the dimension d. See the text for a full description.

In contrast to the Ising model, the $O(m)$-symmetric model has not been intensively studied for the case of arbitrary space dimension d. Another reason for considering the case $m > 1$ is that, in accordance with the Mermin-Wagner-Hohenberg theorem,[4] [44, 45] a continuous symmetry can be spontaneously broken only if the space dimension is larger than two,[5] whereas for the Ising model

[4] Let us mention here the generalization of the Mermin-Wagner-Hohenberg to the case of fractals of continuous symmetry [46].

[5] Exact solutions for the $2d$ classical Heisenberg model for $m = 4$ and $m = 3$ also

($m = 1$) the lower critical dimension is $d = 1$. So by means of a continuous change of the space dimension d one can try to study the mechanism of disappearance of a phase transition near a lower critical dimension.

Starting from the expressions for the β- and γ-functions, Eqs. (3.1) - (3.4) for the case of $v = 0$, $m = 2, 3, 4$, and performing a resummation procedure analogous to the one applied to the Ising model in the previous subsection (the additional parameters in the resummation procedure were chosen on the assumption that the lower critical dimension is two) one can obtain information about the dimensional dependence of the critical exponents. Table 1 contains the three-loop results for the stable fixed point value u^*, the critical exponent ν and the magnetic susceptibility critical exponent γ. Note that while in [48] the singular behaviour of the critical exponents in the point $d = 2$ was found, one of the results of [42, 43] shows that the critical exponents at this point remain analytic. The correspondence of this fact to the Mermin-Wagner-Hohenberg theorem is discussed there and a conjecture is made that at $d = 2$ the ordered state in a system with continuous symmetry exists only inside the critical region. Our results, involving information about the location of singularity at $d = 2$, are in accordance with [48] for the case of d close to $d = 2$ and for $d \simeq 3$ agree with those obtained in [42, 43]. In the case $m = \infty$ we recover the exact result for the spherical model [49].

Table 1. The fixed point coordinate u^* and critical exponents ν and γ of the $O(m)$-symmetric model as a function of d for m = 2, 3, 4.

	$m=2$			$m=3$			$m=4$		
d	u^*	ν	γ	u^*	ν	γ	u^*	ν	γ
2.0	-	∞	∞	-	∞	∞	-	∞	∞
2.2	2.0296	1.329	2.540	1.9895	1.512	2.884	1.9464	1.687	3.219
2.4	1.7907	.971	1.878	1.7604	1.072	2.072	1.7275	1.166	2.253
2.6	1.6285	.816	1.593	1.6043	.882	1.721	1.5779	.941	1.837
2.8	1.5041	.725	1.425	1.4841	.770	1.513	1.4623	.810	1.592
3.0	1.4024	.662	1.310	1.3854	.694	1.371	1.3671	.721	1.426
3.2	1.3349	.618	1.226	1.3210	.638	1.267	1.3061	.656	1.303
3.4	1.2644	.580	1.156	1.2534	.593	1.182	1.2419	.604	1.204
3.6	1.1888	.549	1.095	1.1808	.556	1.110	1.1726	.562	1.123
3.8	1.1069	.522	1.044	1.1022	.525	1.050	1.0975	.528	1.056

c) $O(m)$-symmetrical model in the presence of quenched disorder

The critical behaviour of dilute m-vector model is governed by the Lagrangian Eq. (2.5) in the replica limit $n \to 0$. In this case, the series for the renormalization group functions Eq. (3.1) - (3.4) are a series of two variables and we have used [29,

demonstrate the absence of magnetic ordering (see [47] and references therein).

51] rational approximants of two variables (Chisholm approximants [50]) to make an analytical continuation of their Borel transforms. In another approach we applied a Padé-Borel resummation to the resolvent series (see [52]) constructed on the basis of Eq. (3.1)-(3.4.)

Table 2. The fixed point coordinates u^* and v^* and critical exponents ν and γ of the dilute Ising model as a function of d.

d	u^*	v^*	ν	γ
2.00	2.0228	-0.20878	0.974	1.845
2.20	2.0350	-0.29577	0.887	1.704
2.40	2.0599	-0.38728	0.817	1.587
2.60	2.1009	-0.48628	0.760	1.488
2.80	2.1633	-0.59736	0.712	1.402
3.00	2.2569	-0.72816	0.670	1.327
3.20	2.4004	-0.89255	0.634	1.260
3.40	2.6345	-1.11960	0.601	1.200
3.60	3.0689	-1.48464	0.572	1.143
3.80	4.0882	-2.23115	0.543	1.087

For sufficiently large values of m, $m > m_c$, the pure m-vector ($u^* \neq 0$, $v^* = 0$) fixed point is stable (this is in accord with the Harris criterion [53, 54] which predicts new critical behaviour for the dilute system only if the specific heat critical exponent α^{pure} of the pure system is negative). When the number of order parameter components decreases, beginning at the marginal value m_c, the pure fixed-point becomes unstable and a crossover to a mixed fixed-point ($u^* \neq 0$, $v^* \neq 0$) occurs. The Harris criterion at $d = 2$ together with the exact value of $\alpha^{pure} = 0$ for the Ising model (see [55] as well) means that $m_c(d = 2) = 1$. The best theoretical value for m_c in the $3d$ case was obtained in [56] where $m_c(d = 3) = 1.195 \pm 0.002$. For this reason we are mainly interested in the dilute Ising model ($m = 1$) where new critical behaviour is observed for $2 < d < 4$. The d-dependence of the critical exponent ν obtained in the two- [29] and three-loop approximations [51] is plotted in Fig. 6 using dashed and solid lines, respectively. The resummed three-loop results for the fixed point coordinates u^* v^* and critical exponents ν, γ are given in Table 2. We compare our results obtained on the basis of Chisholm-Borel resummation technique with the available data obtained by scaling field method for dimensionalities $2.8 \leq d \leq 4$ [57] (shown by stars). The straightforward extrapolation of these results to $d = 2$ as well as the comparison with the most accurate theoretical value ν ($d = 3$) $= 0.6701$ [37, 38] (shown by a square in Fig. 6) suggests that the results obtained by a fixed-dimension renormalization group approach are preferable in the case of non-integer d.

At present there are no reliable values of the critical exponents of dilute

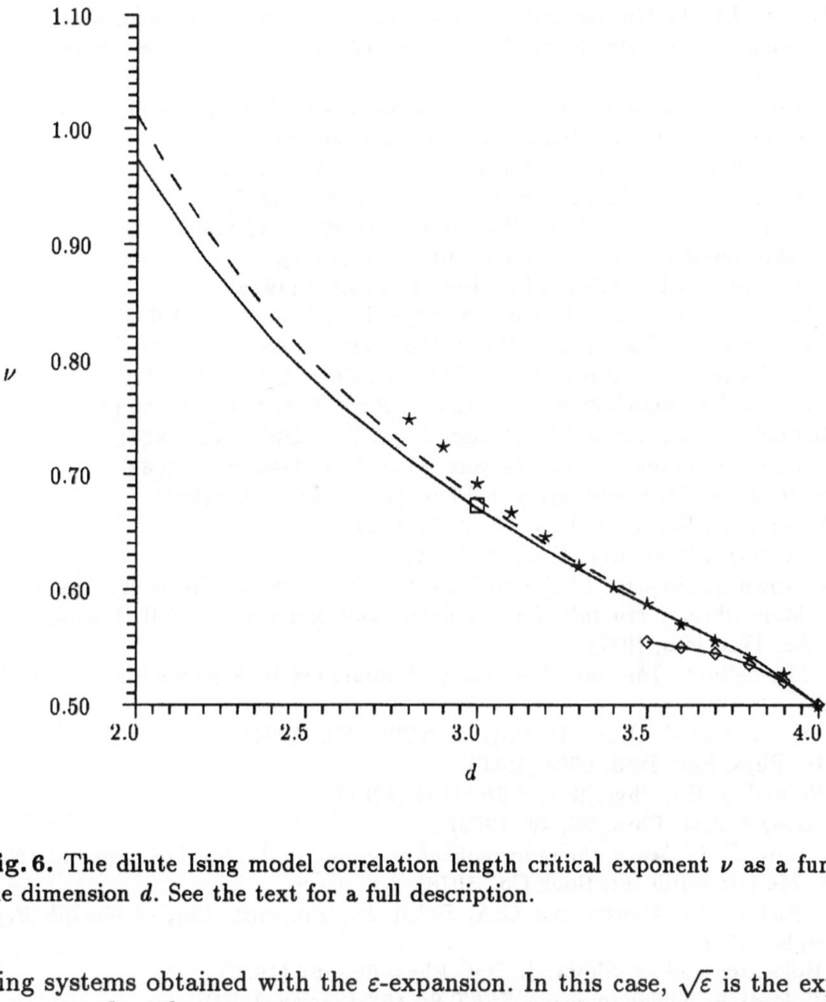

Fig. 6. The dilute Ising model correlation length critical exponent ν as a function of the dimension d. See the text for a full description.

Ising systems obtained with the ε-expansion. In this case, $\sqrt{\varepsilon}$ is the expansion parameter [9, 10] and only two terms of expansion have been calculated [58, 59]. The results of the application of the Padé-Borel resummation technique to the $\sqrt{\varepsilon}$ expansion for critical exponent ν are plotted in Fig. 6 with diamonds. They are reasonable only close to $d = 4$.

Thus the fixed-dimension renormalization group approach [26] directly applied at arbitrary d may be a useful tool for studying critical behaviour in non-integer space dimension. As the above analysis has shown, for certain models it gives results in reasonable agreement with those obtained by the other methods; further there exist cases where it seems to be the only one way to obtain reliable values of the critical exponents in the context of the field-theoretical approach.

References

1. K.G. Wilson and M.E. Fisher, Phys. Rev. Lett. **28**, 240 (1972).
2. K.G. Wilson and J. Kogut, Phys. Rep. **C12**, 75 (1974).

3. E. Brezin, J.C. Le Guillou and J. Zinn-Justin, in: *Phase Transitions and Critical Phenomena*, ed. C. Domb and M.S. Green (Academic Press, New York, 1975), Vol. 6, p.125
4. S.-K. Ma, *Modern theory of critical phenomena.* (Benjamin, Reading, MA, 1976).
5. D.J. Wallace and R.K.P. Zia, Phys. Rev. Lett. **43**, 808 (1979).
6. D. Forster and A. Gabriunas, Phys. Rev. **A23**, 2627 (1981).
7. D. Forster and A. Gabriunas, Phys. Rev. **A24**, 598 (1981).
8. A.D. Bruce and D.J. Wallace, Phys. Rev. Lett. **47**, 1743 (1981).
9. D.E. Khmelnitskii, Sov. Phys. JETF **68**, 1960 (1975).
10. G. Grinstein and L. Luther, Phys. Rev. **B13**, 1329 (1976).
11. S.L. Katz, M. Droz and J.D. Gunton, Phys. Rev. **B15**, 1597 (1977).
12. J.C. Le Guillou and J. Zinn-Justin, J. Phys. (Paris) **48**, 19 (1987).
13. S.G. Gorishny, S.A. Larin and F.V. Tkachov, Phys. Lett. **A101** 120 (1984).
14. Y. Gefen and B. Mandelbrot, A. Aharony, Phys. Rev. Lett. **45** 855 (1980).
15. B. Bonnier, Y. Leroyer and C. Meyers, Phys. Rev. **B37** 5205 (1988).
16. B. Bonnier, Y. Leroyer and C. Meyers, Phys. Rev. **B40** 8961 (1989).
17. B. Bonnier and M. Hontebeyrie, J. Phys. (Paris) **I1**, 5205 (1991).
18. M.A. Novotny, Europhys. Lett. **17**, 297 (1992)
19. M.A. Novotny, Phys. Rev. **B46**, 2939 (1992).
20. M.A. Novotny, *Preprint FSU-SCRI-92-111.* (Florida State University, 1992).
21. B.B. Mandelbrot , *Fractals: form, chance, and dimension.* (W.H. Freeman and Co., San Francisco, 1977).
22. B.B. Mandelbrot, *The fractal geometry of nature* (W.H. Freeman and Co., N.-Y., 1983).
23. P. Tomczak and W. Jezewski, Physica **A209**, 275 (1994).
24. B. Hu, Phys. Rev. **B33**, 6503 (1986).
25. Y. Wu and B. Hu, Phys. Rev. **A35**, 1404 (1987).
26. G. Parisi, J. Stat. Phys. **23**, 49 (1980).
27. D.J. Amit, *Field theory, the renormalization group, and critical phenomena.* (New York: Mc Graw-Hill Int. Book Co., 1978).
28. B.G. Nickel, D.I. Meiron and G.A. Baker Jr., *Preprint, Univ.of Guelph Rep.*, (Guelph, 1977).
29. Yu. Holovatch and M. Shpot, J. Stat. Phys. **66**, 867 (1992).
30. Yu. Holovatch, *Saclay preprint, SPhT-92-123*, (Saclay, 1992); Int. J. Mod. Phys. **A8**, 5329 (1993).
31. Yu. Holovatch and T. Krokhmalskii, J. Math. Phys. **35**, 3866 (1994).
32. Yu. Holovatch, unpublished.
33. Yu. Holovatch, Theor. Math. Phys. (Moscow) **96**, 482 (1993).
34. J.C. Le Guillou and J. Zinn-Justin, Phys. Rev. **B21**, 3976 (1980)
35. G.A. Baker Jr., B.G. Nickel and D.J. Meiron, Phys. Rev. **B17**, 1365 (1978).
36. G. Jug, Phys.Rev. **B27**, 609 (1983).
37. I.O. Mayer, A.I. Sokolov and B.N. Shalaev, Ferroelectrics, **95**, 93 (1989).
38. I.O. Mayer, J. Phys. **A22**, 2815 (1989).
39. H. Stanley, *Introduction to phase transitions and critical phenomena.* (Clarendon press, Oxford, 1971).
40. L. Onsager, Phys. Rev. **65**, 117 (1944)
41. L. Schäfer, C. von Ferber, U. Lehr and B. Duplantier, Nucl. Phys. **B374**, 473 (1992).
42. A.E. Filippov and A.V. Radievskii, Sov. Phys. JETF **102**, 1899 (1992).
43. S.A. Breus and A.E. Filippov, Physica **A192**, 486 (1992).

44. N.D. Mermin and P. Wagner, Phys. Rev. Lett. **17**, 1133 (1966).
45. P. Hohenberg, Phys. Rev. **158**, 383 (1967).
46. D. Cassi, Phys. Rev. Lett. **68**, 3631 (1992).
47. Yu.A. Izyumov and Yu.N. Skriabin, *Statistical mechanics of magnetically ordered systems*, (New York, Consultants Bureau 1988).
48. J.L. Cardy and H.W. Hamber, Phys. Rev. Lett. **45**, 499 (1980).
49. M. Kac, T.N. Berlin, Phys. Rev. **86**, 821 (1952).
50. J.S.R. Chisholm, Math. Comp. **27**, 841 (1973)
51. Yu. Holovatch and T. Yavors'kii, unpublished
52. G.A. Baker Jr and P. Graves-Morris, *Padé approximants* (Addison-Wesley Publ. Co., Reading, Mass., 1981).
53. A.B. Harris, J. Phys. **C7**, 1671 (1974)
54. see chap. X in [4].
55. B.N. Shalaev, Fiz. Tverd. Tela. **30**, 895 (1988); ibid. **31**, 93 (1989).
56. C. Bervillier, Phys. Rev. **B34**, 8141 (1986).
57. K.E. Newman and E.K. Riedel, Phys. Rev. **B25**, 264 (1982).
58. C. Jayaprakash and H.J. Katz, Phys. Rev. **B66**, 3987 (1987).
59. B.N. Shalaev, Sov. Phys. JETF **73**, 2301 (1977).

The Peierls Instability and the Flux Phase Problem

Nicolas Macris

Institut de Physique Théorique
Ecole Polytechnique Fédérale de Lausanne
CH-1015 Lausanne, Switzerland
e-mail: macris@eldp.epfl.ch

Abstract: The Peierls instability and the flux phase problem are treated in a unified way for certain models of strong electronic correlations. The treatment relies on an adaptation of the reflection positivity technique valid for certain models of itinerant fermions. We discuss three applications: the dia- or paramagnetic behavior of annulenes, the instability in a two dimensional Peierls-Hubbard model and some properties of coupled polyacetylene chains.

1 Introduction

We will discuss two apparently different problems, which are in fact closely related. The first one is the Peierls instability and the second the so called "flux phase problem". We will make the point that both problems can be viewed and treated in a unified way. The Peierls instability is a familiar phenomenon in solid state physics which was discovered independently by R. Peierls [1] and H. Fröhlich [2] in 1954. It seems that the original motivation of Fröhlich was a theory of one dimensional superconductivity, while that of Peierls was directly related to crystalline structure. The fact that certain metals and alloys are more complicated than the closed packed or body centered lattices led Peierls to study the distorsions of a linear chain of atoms. He pointed out that in one dimension the instability would be quite generic, but at that time one dimensional structures could not be easily studied experimentally. In recent years however observations on real materials have shown that a connection between the mathematical treatment and reality exists. See [3] for a nice discussion. Amusingly the flux phase problem, which is much more recent, has come into the scene via (gauge field) theories of strong electronic correlations, thought to be relevant for high temperature superconductivity. To my knowledge the problem was first discussed in [4], [5] but many variants exist. For the moment the flux phase problem constitutes a piece of mathematical and rather speculative treatments of models of strong electronic correlations. It is not yet clear that it has any connection with reality. Nevertheless the problem is interesting in its own right because as we will see it is intimately related to the Peierls instability, dia- or para-magnetism of molecules and Hückel versus anti-Hückel behavior of conjugated polymers, orbital magnetism in mesoscopic rings. Moreover there is an underlying non trivial mathematical structure that we will explain: "reflection positivity for fermions" [14],[15]. We inform the interested reader that the flux phase problem is also related to Kasteleyn's theorem in statistical mechanics (see [6] where the first mathematical study appeared).

This note is organised as follows. In paragraphs 1.1 and 1.2 we review briefly the basics of the Peierls instability and the flux phase problem. There we set up the context and the various model hamiltonians that are used. In section 2 the unified formulation of the problems is presented, together with a general theorem which gives a partial solution. The underlying mathematical structure is explained in section 3 and three applications are presented in section 4.

1.1 The Peierls instability

Let us first recall what is the Peierls instability. Consider a one dimensional chain of atoms with a uniform spacing a between atoms. We suppose that this is the equilibrium configuration of the potential energy of the chain. If each atom can accommodate, say one orbital associated to itinerant electrons, the system is half filled (because of electron spin). The states with $|k| < \pi/2a = k_f$ are occupied and the states with $|k| > k_f$ are empty. Suppose that we displace all the even atoms to the left by a small distance $\delta << a$. Then the new configuration has a period $2a$, and the one-electron spectrum now has a gap above the fermi points $\pm k_f$. This time the lower band is full and the upper band is empty therefore the system is an insulator. By second order perturbation theory one can prove that the opening of the gap (by the distorsion) lowers the electronic energy by an amount of the order $-\delta^2 |\ln \delta|$ [2]. On the other hand the distorsion costs some elastic energy which can be assumed to be proportional to δ^2 (since $\delta << a$). The logarithmic term implies that the distorsion is favourable. It is generally believed that in one dimension the phenomenon is generic and in particular it is not limited to half filling. Indeed one can always find a distorsion with wave vector $2k_f$ which will open a gap in the one electron spectrum just above the Fermi points $\pm k_f$. In higher dimensions this coincidence is not the general rule (except at special densities and for special geometries of the fermi surface) and the instability may or may not be favorable.

The arguments given above disregard at least three essential points. First the quantum fluctuations of the lattice are not taken into account. We will not discuss this aspect here (except in paragraph 4.1). Second the Coulombic interactions between the electrons have been neglected. We will include their effect at the level of simple Hubbard interaction terms. Finally the configuration space of the chain has not been fully explored. In fact only two periodic configurations have been considered and it is conceivable that the true minimum occurs for some other configuration. Therefore the argument tells us that the period a configuration must be unstable, and that the period $2a$ configuration has less energy, but it does not give the true minimum. For some special models one can determine the true minimum and confirm the general intuition. However one can also construct counterexamples.

A famous physical system displaying the instability is the (trans-) polyacetylene chain. A reasonable model for the chain is as follows. The carbon-hydrogen groups (CH) are bonded together by σ bonds which form the squeleton of the chain $(CH)_x$. They are modelled by an elastic energy term of the form $\sum_l V(u_{l+1} - u_l)$, where u_l is the displacement of the l -th C atom with respect to its equilibrium position (for the potential energy Vone can have in mind a harmonic form but we will not limit ourselves to this case). The π electrons coming from the delocalisation of the carbon p_z orbitals are itinerant. Since there is one such orbital per C the system is half filled. The kinetic energy of the itinerant electrons is given by a hopping matrix $t_{l,l+1}(u_{l+1} - u_l)$ which is a function of the distance between adjacent C atoms. The resulting hamiltonian corresponds to the Hückel model

$$H_{HUC} = \sum_{l,\sigma} t_{l,l+1}(u_{l+1} - u_l)(c^\dagger_{l+1,\sigma} c_{l,\sigma} + c^\dagger_{l,\sigma} c_{l+1,\sigma}) + \sum_l V(u_{l+1} - u_l) \quad (1.1)$$

If $V(u) = \frac{1}{2}\kappa^2 u^2$ and $t(u) = t_0 - \alpha u$, where κ, t_0 and α are constants (1.1) is known as the Su, Schrieffer, Heeger hamiltonian (SSH)[7] who argued that at half filling the ground state is dimerized (see figure 1a). This means that the optimal configuration of C atoms has period two, $u_l^{(0)} = (-1)^l \delta$, where δ is a function of t_0,α and κ. A rigorous examination of the SSH model has been performed by Kennedy and Lieb [8] who showed that the true minimum indeed has period two. One can also define a bond order parameter by $p_{l,l+1} = < c^\dagger_{l+1,\sigma} c_{l,\sigma} + c^\dagger_{l,\sigma} c_{l+1,\sigma} >$ and show that it also has period two. One says that the ground state is a bond order wave.

Other systems described by Hückel type of hamiltonians are finite conjugated polymers. These are finite molecules constituted of CH groups, such as Benzene, Anthracene, Naphtalene (see figure 1b). A relevant question is whether or not the bond lengths are distorted and if yes, what is the pattern. Longuet-Higgins and Salem [9] where the first chemists to show that a dimerised structure would certainly occur in large annulenes (these are like polyacetylene but instead of being lincar they have the shape of a ring). In the chemistry literature one speaks of "bond alternation" (of short and long bonds) which is more natural in view of the finite size effects which may occur. Another relevant question is whether the molecule is dia- or para-magnetic. This issue was resolved by London in the framework of the Hückel model for annulenes. The result is that when the length of the molecule is equal to 0 mod 4 it is paramagnetic and if it is equal to 2 mod 4 it is diamagnetic. It will become clear that this is related to the flux phase problem.

The effects of Coulomb interactions are important in conjugated polymers and have been much studied. The simplest model for the interaction is an on-site interaction $U \sum_l n_{l,\uparrow} n_{l,\downarrow}$. This was first considered by Pariser, Parr, and Pople [10], in the chemistry literature. The resulting hamiltonian is commonly called the Peierls-Hubbard hamiltonian in the community of physicist

$$H_{PH} = \sum_{l,\sigma} t_{l,l+1}(u_{l+1} - u_l)(c^\dagger_{l+1,\sigma} c_{l,\sigma} + c^\dagger_{l,\sigma} c_{l+1,\sigma}) + \sum_l V(u_{l+1} - u_l)$$
$$+ U \sum_l (n_{l,\uparrow} - \frac{1}{2})(n_{l,\downarrow} - \frac{1}{2}) \tag{1.2}$$

Here the interaction is written in a form convenient to exhibit the symmetries of H at half-filling and corresponds to a suitable adjustment of the chemical potential. For more extensive information on these models the reader can consult the review article [11].

1.2 The flux phase problem

As said before the flux phase problem appeared as an ingredient of "mean field" theories of strongly correlated electronic systems. Within these theories it appears that charges moving in a two dimensional magnetic environment, created by fluctuations of the spin degrees of freedom, acquire quantum mechanical phases. The flux phase is a state corresponding to a special arrangement of these phases so that the ground state energy is minimal. We do not attempt here to reproduce the mean field treatment but we refer the reader to [4],[5] and [12]. In the treatment proposed in [5], which is the simplest, one is led from the Heisenberg hamiltonian (which becomes the $t-J$ model upon doping) to the mean field model

$$H_{MF} = \sum_{<xy>,\sigma=\uparrow,\downarrow} \chi_{xy} c^\dagger_{x,\sigma} c_{y,\sigma} + J \sum_{<xy>} |\chi_{xy}|^2 \tag{1.3}$$

where $< xy >$ are bonds of the two dimensional lattice Λ, and χ_{xy} is a mean field associated to the bonds of the lattice

$$\chi_{xy} = < c^\dagger_{x,\uparrow} c_{y,\uparrow} + c^\dagger_{x,\downarrow} c_{y,\downarrow} > \tag{1.4}$$

The mean field configuration $\{\chi_{xy}\}$ has to be determined by minimising the ground state energy of (1.3).

Let us interpret (1.3)-(1.4). The mean field hamiltonian possesses a gauge invariance. Indeed setting $\chi_{xy} = |\chi_{xy}| \exp(i\phi_{xy})$ we see that the local transformations $c^\dagger_{x,\sigma} \to \exp(i\theta_x) c^\dagger_{x,\sigma}$, $c_{y,\sigma} \to \exp(-i\theta_y) c_{y,\sigma}$, $\phi_{xy} \to \phi_{xy} - (\theta_x - \theta_y)$ leave the hamiltonian invariant. Therefore it is natural to interpret the phases ϕ_{xy} as those produced by a (fictitious) magnetic field with a flux Φ_F through the faces F of Λ

$$\Phi_F = \sum_{<xy>\in F} \phi_{xy} = \arg\Big(\prod_{<xy>\in F} \chi_{xy} \Big) \tag{1.5}$$

We emphasize that here the magnetic field is generated by the system itself and is a model for taking into account the effects of magnetic fluctuations on the motion of charges. The system also generates a "distorsion" of the hopping amplitude $|\chi_{xy}|$. This costs a certain energy given by the quadratic term in (1.3).

It is interesting to note that there is no energy associated to the magnetic field itself since the "elastic energy" in (1.3) depends only on the modulus of χ_{xy}.

Because of gauge invariance the ground state energy is only a function of $\{|\chi_{xy}|\}$ and $\{\Phi_F\}$, $E_0(\{|\chi_{xy}|\}, \{\Phi_F\})$. Essentially two types of mean field solutions have been studied in the literature [12].

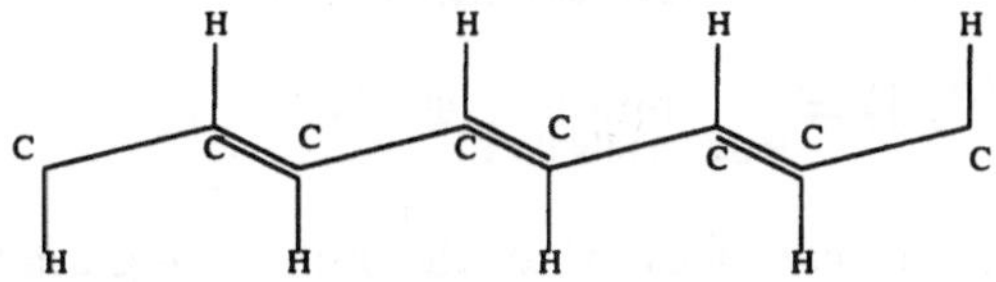

(a)

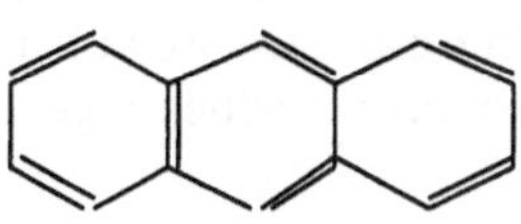

(b)

Fig. 1. (a) A dimerised ground state for the polyacetylene linear chain. (b) Anthracene molecule.

The first one is the flux phase. One takes $|\chi_{xy}| = \chi$ uniform and minimises only over the configuration of fluxes $\{\Phi_F\}$. A minimum with $\Phi_F \neq 0$ is analogous to a Peierls instability, with the important difference that there is no energy cost associated to the appearance of the non zero flux. In general time reversal invariance can be broken unless $\Phi_F = 0$ or π because then there exist a gauge in which the hamiltonian is a real matrix.

The second type of mean field solution are the valence bond cristals. There one sets $\Phi_F = 0$ (i.e. $\phi_{xy} = 0$) and minimises E_0 over the configurations of hopping amplitudes $\{|\chi_{xy}|\}$. A comparison with the Hückel hamiltonian will convince the reader that this is exactly the Peierls problem in higher dimensions. If the optimal configuration of hopping amplitudes is non uniform there is a Peierls instability, although the modulus of χ_{xy} is not necessarily related to the lattice spacing in this context.

2 A Unified Formulation

Consider a finite two dimensional graph consisting of a set of vertices Λ and a set of bonds $\{< xy >\}$. the hopping matrix associated to the graph has matrix elements $t_{xy} = \overline{t_{yx}}$ with $t_{xy} \neq 0$ if and only if $< xy >$ belongs to the graph. In general we have $t_{xy} = |t_{xy}|\exp(i\phi_{xy})$. Our general hamiltonian is

$$H = \sum_{x,y\in\Lambda,\sigma=\uparrow,\downarrow} t_{xy} c^{\dagger}_{x,\sigma} c_{y,\sigma} + \sum_{x,y\in\Lambda} F(|t_{xy}|) + U \sum_{x\in\Lambda} (n_{x,\uparrow} - \frac{1}{2})(n_{x,\downarrow} - \frac{1}{2}) \quad (2.1)$$

The ground state energy depends on the configurations of $\{|t_{xy}|\}$ and $\{\Phi_F\}$ (due to gauge invariance) where here F denote the faces of Λ

$$E_0(\{|t_{xy}|\}, \{\Phi_F\}) = \min_{\Psi\in\mathcal{F}(\Lambda),||\Psi||=1} (\Psi, H\Psi) \quad (2.2)$$

Using particle-hole symmetries one can show that the minimising state Ψ in the Fock space of electrons $\mathcal{F}(\Lambda)$ belongs to the subspace with total number of electrons equal to $|\Lambda|$. Therefore when we study (2.2) we are automatically in the half filled band.

The main problem that we consider is that of minimising E_0 in the space of all configurations of $\{|t_{xy}|\}$ and $\{\Phi_F\}$. In some cases we are able to solve partially this variational problem. The following theorem describes a general solution for a square lattice.

Theorem 1

Let Λ be a square lattice with periodic boundary conditions (so it is a torus) and with an even number of sites N_1, N_2 in the two coordinate directions 1, 2. Assume F is bounded below, continuous and increasing at infinity. Then for any U the minimum of E_0 is attained for the following configurations of $\{|t^{(0)}_{xy}|\}$ and $\{\Phi^{(0)}_F\}$:

(i) $\Phi^{(0)}_F = \pi$ through all square plaquettes F and $\Phi^{(0)}_i = \pi(\frac{N_i}{2} - 1)$, $i = 1,2$ through the two non trivial loops of the torus.

(ii) Let $x = (x_1, x_2)$ in coordinate form. We have

$$|t^{(0)}_{(x_1,x_2)(x_1+1,x_2)}| = |t^{(0)}_{(x_1+2,x_2)(x_1+3,x_2)}|\,,$$

$$|t^{(0)}_{(x_1,x_2)(x_1,x_2+1)}| = |t^{(0)}_{(x_1,x_2+2)(x_1,x_2+3)}|\,,$$

$$|t^{(0)}_{(x_1,x_2)(x_1+1,x_2)}| = |t^{(0)}_{(x_1,x_2+1)(x_1+1,x_2+1)}|\,,$$

$$|t^{(0)}_{(x_1,x_2)(x_1,x_2+1)}| = |t^{(0)}_{(x_1+1,x_2)(x_1+1,x_2+1)}|\,.$$

The same result holds for the maximum of $\mathrm{Tr}\exp(-\beta H)$. The theorem tells us what is the optimal flux configuration, but not what is the precise value of $|t^{(0)}_{xy}|$. However it does restrict the search of the true minimum to the configurations satisfying the above constraints. The hopping amplitude can take only four values t_1, t_2, t_3, t_4, geometrically arranged like in figure 2. It is not known if there exist other global minima outside of the class described here. On the other hand one easily shows that there exist many other local extrema. For a discussion of this aspect see [13]

Such a theorem was first obtained by Lieb [14] and then an improved proof along the lines explained in section 3 was given in [15]. In fact, in these references the minimisation is carried on only over the fluxes and the moduli of the hopping terms are fixed. The extension presented here is straightforward. Similar results can be proven for other periodic lattices or finite graphs with appropriate symmetries, in two or three dimensions (see [15]).

3 Reflection Positivity for Fermions

In this section we present the mathematical structure that is hidden behind theorem 1. Here we will limit ourselves to present a proof of this theorem, but the same ideas are useful in other situations.

3.1 Reflection positivity: abstract set-up

Reflection positivity is a notion first introduced in quantum field theory [16] and later applied to statistical mechanics in the context of spin systems [17,18,19]. While in quantum field theory it is very natural, it is less so in statistical mechanics where its applicability depends on the particular hamiltonian. For this reason here we take the point of view to consider it as a property of the hamiltonian itself.

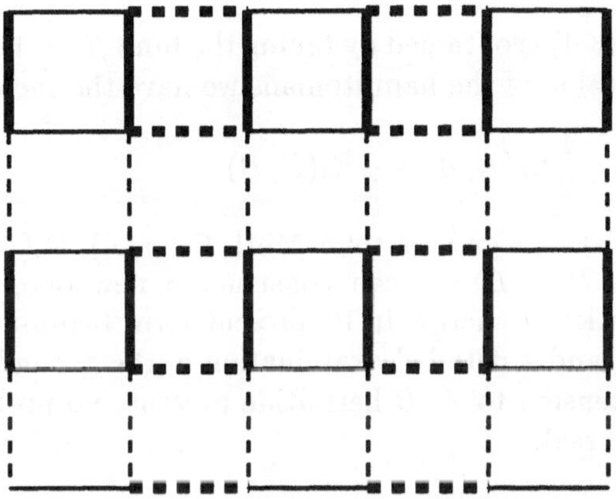

Fig. 2. The allowed configurations of $|t_{xy}|$ in Theorem 1. The values of t_1, t_2, t_3, t_4 are associated to the normal, thick, dashed, thick dashed, lines.

Suppose that the Hilbert space $\mathcal{H}$ can be decomposed as $\mathcal{H}_L \otimes \mathcal{H}_R$ where $\mathcal{H}_{L,R}$ are two copies of a d−dimensional space. Then $\mathcal{H}$ is d^2 dimensional. Let A, B be $d \times d$ hermitian matrices and $D^{(i)}$, $i = 1, ..., n$ real $d \times d$ matrices. The $d \times d$ identity is $\mathbf{1}$. We set

$$A_L = A \otimes \mathbf{1}, \qquad B_R = \mathbf{1} \otimes B, \qquad D_L^{(i)} = D^{(i)} \otimes \mathbf{1}, \qquad D_R^{(i)} = \mathbf{1} \otimes D^{(i)} \tag{3.1}$$

The hamiltonian (acting on $\mathcal{H}$) is said to be reflection positive if it has the form

$$H(A,B) = A_L + \overline{B_R} + \sum_{i=1}^{n} \gamma_i (D_L^{(i)} - D_R^{(i)})^2 \tag{3.2}$$

with $\gamma_i \geq 0$ for all $i = 1, ..., n$ and the sum over $i = 1, ..., n$ is a symmetric matrix in $\mathcal{H}$. In (3.2) $\overline{B_R}$ denotes the matrix obtained by complex conjugation of all the matrix elements. A celebrated example of a hamiltonian of the form (3.2) is the Heisenberg antiferromagnet.

Such hamiltonians have two basic properties summarised in the two Lemmas

***Lemma 1** (positivity of correlation functions)*

Let $\mathcal{O}$ be a $d \times d$ hermitian matrix. We have that

$$< \mathcal{O}_L \mathcal{O}_R >= \frac{Tr \mathcal{O}_L \mathcal{O}_R \exp(-\beta H(A,A))}{Tr \exp(-\beta H(A,A))} \tag{3.3}$$

is non negative

***Lemma 2** (Schwartz inequality)*

$$Tr \exp(-\beta H(A,B)) \leq \Big(Tr \exp(-\beta H(A,A))\Big)^{1/2} \Big(Tr \exp(-\beta H(B,B))\Big)^{1/2} \tag{3.4}$$

The ground state version of (3.4) is obtained by taking the limit $\beta \to +\infty$ and states that for the lowest eigenvalue of the hamiltonians we have the inequality

$$E_0(A,B) \geq \frac{1}{2} E_0(A,A) + \frac{1}{2} E_0(B,B) \tag{3.5}$$

The right hand side of (3.5) is always greater than $\mathrm{Min}\{(E_0(A,A), E_0(B,B)\}$. Therefore given a hamiltonian $H(A,B)$ we can construct a new one, either $H(A,A)$ or $H(B,B)$, which has lower energy. In its present form Lemma 2 was proved in [18] for A and B real and a detailed examination of the ground state version appears in [20]. The extension to A, B hermitian presents no problems, but it is important to keep $D^{(i)}$ real.

3.2 Application to fermions

In their standard form the Peierls-Hubbard hamiltonians (1.2) or (2.1) are not reflection positive. However there is a definite procedure which transforms the hamiltonian into a reflection positive form. Here we outline this procedure which was introduced in [15].

The first task is to decide what is $\mathcal{H}_L$ and $\mathcal{H}_R$. We separate the square lattice Λ in two equal halves by a plane P, not containing any site of Λ, called the reflection plane (see figure 3). The left (L) and right (R) parts of the lattice $\Lambda = L \cup R$ can be mapped onto each other by geometric reflections through P. We let $\mathcal{H}_L = \mathcal{F}(L)$ and $\mathcal{H}_R = \mathcal{F}(R)$ be the Fock spaces associated to each side of the lattice. Since the hamiltonian acts on $\mathcal{F}(\Lambda)$ we first perform the transformation

$$d^{\dagger}_{x,\sigma} = c^{\dagger}_{x,\sigma}\exp(i\pi N_L), \qquad \text{all} \qquad x \in \Lambda, \qquad N_L = \sum_{x\in L,\sigma=\uparrow,\downarrow} c^{\dagger}_{x\sigma}c_{x,\sigma} \tag{3.6}$$

One easily checks that the commutation relations are preserved for $d^{\#}_x$, $d^{\#}_y$, if $(x,y) \in L$ or $(x,y) \in R$, but the operators commute if $x \in L$, $y \in R$ or vice vera. Moreover in terms of the $d^{\#}$'s the hamiltonian has the same form. Now we identify the algebras $(d^{\#}_x, x \in L)$ with $\mathcal{F}(L)$ and $(d^{\#}_x, x \in R)$ with $\mathcal{F}(R)$, and consider the hamiltonian as a matrix acting on $\mathcal{F}(L) \otimes \mathcal{F}(R)$.

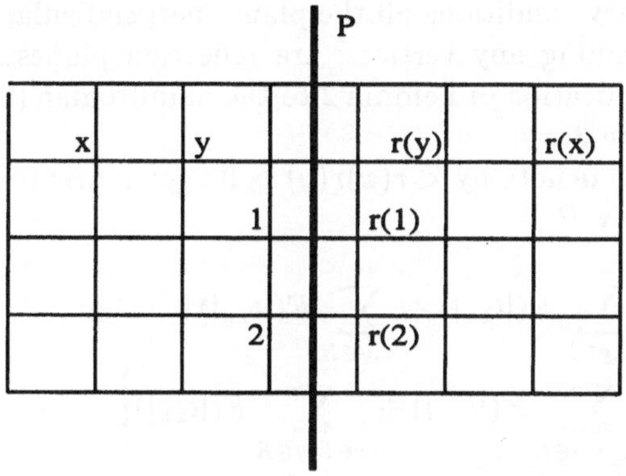

Fig. 3. The setting used in paragraph 3.3

The second transformation consists of a particle-hole transformation $d^{\dagger}_{x,\sigma} \to d_{x,\sigma}$, $d_{x,\sigma} \to d^{\dagger}_{x,\sigma}$ for all $x \in R$. These two transformations put the hamiltonian into the form (3.2) except that the coefficient corresponding to γ_i might not have the correct sign yet.

For this reason we perform a third transformation for sites $y \in R$ such that there exist a bond $< xy >$ with $x \in L$,

$$d^{\dagger}_{y,\sigma} \to \exp(i\theta_y)d^{\dagger}_{y,\sigma} \tag{3.7}$$

with $\{\theta_y\}$ such that

$$t_{xy} = |t_{xy}|\exp(i\phi_{xy}) \to t'_{xy} = -|t_{xy}| = -|t'_{xy}|, \qquad < xy > \cap P \neq \emptyset \tag{3.8}$$

The final result is a hamiltonian of the form (3.2) with

$$A_L = \sum_{x,y\in L} t'_{xy}d^{\dagger}_{x,\sigma}d_{y,\sigma} + U\sum_{x\in L}(n_{x,\uparrow} - \frac{1}{2})(n_{x,\downarrow} - \frac{1}{2}) \tag{3.9}$$

$$B_R = \sum_{x,y\in R} (-\overline{t'_{xy}}) d^\dagger_{x,\sigma} d_{y,\sigma} + U \sum_{x\in R} (n_{x,\uparrow} - \frac{1}{2})(n_{x,\downarrow} - \frac{1}{2}) \tag{3.10}$$

$$\sum_{i=1}^{n} \gamma_i (D_L^{(i)} - D_R^{(i)})^2 = \frac{1}{2} \sum_{x\in L, y\in R} |t'_{xy}|(d^\dagger_{x,\sigma} - d^\dagger_{y,\sigma})^2 + \frac{1}{2} \sum_{x\in R, y\in L} |t'_{xy}|(d_{x,\sigma} - d_{y,\sigma})^2 \tag{3.11}$$

One must also add to these three terms the "elastic" energy (a c-number) $\sum F(\{|t_{xy}|\})$.

3.3 Sketch of the Proof of Theorem 1

Because of the periodic boundary conditions all the planes perpendicular to the coordinate axis, and not containing any vertices, are reflection planes P. the proof consists of a repeated application of Lemma 2 to the hamiltonian (3.9-11). The setting is depicted on figure 3.

Given a bond $< xy > \in L$ we denote by $< r(x)r(y) >$ its geometric reflection through P. We note that for any P

$$\begin{aligned} \textstyle\sum_{x,y\in\Lambda} F(|t_{xy}|) = \frac{1}{2}\Big[&\sum_{x,y\in L} F(|t_{xy}|) + \sum_{x,y\in R} F(|t_{xy}|) \\ &+ \sum_{x\in R, y\in L} F(|t_{xy}|) + \sum_{x\in L, y\in R} F(|t_{xy}|)\Big] \\ &+ \frac{1}{2}\Big[\mathrm{L} \longleftrightarrow \mathrm{R}\Big] \end{aligned} \tag{3.12}$$

By applying (3.12) and (3.5) with respect to P one gets a new hamiltonian, either $H(A,A)$ or $H(B,B)$, with a lower ground state energy. In $H(A,A)$ we include the term in the first bracket on the right hand side of (3.12) and in $H(B,B)$ we include the second. For the new hamiltonian we necessarily have

$$t'_{xy} = -t'_{r(x)r(y)} \qquad < xy > \cap P = \emptyset \tag{3.13}$$

Therefore $|t'_{xy}|$ is invariant under a reflection through P. Since the energy is lowered by repeated application of Lemma 2 with respect to all other planes P, the configuration $\{|t'_{xy}| = |t_{xy}|\}$ of the ground state must be invariant under reflections through all planes P. This yields the configurations of figure 2.

Let us examine the consequence of (3.13) on the flux through a plaquette intersected by P. If the vertices of the plaquette are denoted by $(1, 2, r(1), r(2))$ we have

$$\begin{aligned} \Phi_F &= \arg(t_{1r(1)} t_{r(1)r(2)} t_{r(2)2} t_{21}) = \arg(t'_{1r(1)} t'_{r(1)r(2)} t'_{r(2)2} t'_{21}) \\ &= \arg(|t'_{1r(1)}||t'_{r(1)r(2)}||t'_{r(2)2}|(-t'_{r(2)r(1)})) = \arg(-1) = \pi \end{aligned} \tag{3.14}$$

This will hold for all plaquettes since all reflection planes are used. The flux through the nontrivial loops of the torus is determined similarly.

Finally we remark that the conditions that $\Phi_F = \pi$ for all plaquettes and $\{|t_{xy}|\}$ invariant under reflections are preserved under further application of Lemma 2. If there is another configuration not satisfying these conditions which has minimal energy then it must necessarily be degenerate.

3.4 Remarks

On a square lattice if $|t_{xy}|$ is invariant under reflections and $\Phi_F = \pi$ Lemma 1 implies

$$< \prod_{x\in S} d^{\#}_{x,\sigma} \prod_{x\in r(S)} \widetilde{d^{\#}_{x\sigma}} >\geq 0 \tag{3.15}$$

where $\tilde{d^{\#}}$ is the particle hole transform of $d^{\#}$. By translating back to the original operators $c^{\#}_{x,\sigma}$ we get useful inequalities for certain correlation functions. For example

$$(-1)^{|x|+|r(x)|} < S^{(3)}_x S^{(3)}_{r(x)} >\geq 0 \tag{3.16}$$

for all β, U and all x, $r(x)$ obtained with any reflection plane. This indicates that the system has a tendency towards antiferromagnetic order. We emphasize that it does not imply LRO, in fact in one dimension (3.16) holds and there is no LRO.

More generally inequalities of the type (3.15) are interesting because they give constraints that variational wave functions must satisfy.

Finally we point out that a system of hard core boson on a square lattice with flux zero per plaquette satisfies the same "positivity inequalities" than the same fermion system with flux π.

4 Applications

In this section we give three applications to specific problems. The results are rigorous, but the detail of the proofs are not given since they consist of minor modification of the method of section 3. Refs. [15] and [21] discuss other applications, namely the existence of LRO in the $t-V$ model at low temperatures, and the ground states of Falicov-Kimball type models.

4.1 Orbital magnetism of annulenes

Annulenes are ring shaped molecules belonging to the class of conjugated polymers. Relevant hamiltonians are of the Hückel or Peierls-Hubbard type. Here we also add the quantum fluctuations of the u_l variable, and apply an external magnetic field with flux Φ through the ring. Thus the hamiltonian for a molecule of length L is

$$\begin{aligned} H = &\sum_{l=1,\sigma=\uparrow,\downarrow}^{L-1} t_{l,l+1}(u_{l+1}-u_l)e^{i\phi_{l,l+1}}(c^{\dagger}_{l+1,\sigma}c_{l,\sigma} + c^{\dagger}_{l,\sigma}c_{l+1,\sigma}) \\ &+U\sum_{l=1}^{L}(n_{l,\uparrow}-\frac{1}{2})(n_{l+1,\downarrow}-\frac{1}{2}) - \frac{1}{2M}\sum_{l=1}^{L}\frac{\partial^2}{\partial u_l^2} + \sum_{l=1}^{L-1} V(u_l - u_{l+1}) \end{aligned} \tag{4.1}$$

with $\Phi = \sum_{l=1}^{L-1} \phi_{l,l+1}$

For the case $U = 0$, $M = +\infty$ and periodic boundary conditions, London understood as early a 1930 that if $L = 4n + 2$ the molecule is diamagnetic

$$E_0(\Phi) \geq E_0(\Phi = 0) \tag{4.2}$$

and for $L = 4n$ it is paramagnetic, i.e. the inequality is reversed. With the method of section 3 these facts can be justified rigorously and also generalised to the full hamiltonian (4.1) for even L.

Theorem 2

For the hamiltonian (4.1) with periodic boundary conditions, we have $E_0(\Phi) \geq E_0(\Phi = 0)$ if $L = 2$ mod 4, and $E_0(\Phi) \geq E_0(\Phi = \pi)$ if $L = 0$ mod 4.

Since the method of proof is not restricted to one dimension the theorem can be generalised to more complicated molecules (or graphs) that are constituted of many adjacent rings.

The size of the largest annulenes is of the order of $20A$ and so a flux equal to π through the annulene corresponds to a magnetic field strength of the order of $10^8 G$, which is not achievable in laboratory. Therefore it could be thought that the second inequality is not physically relevant. However there are at least two situations where it could be relevant.

A flux equal to π is equivalent to a change of boundary conditions from periodic to antiperiodic ones. Chemically this could in principle be realised by replacing a p_z orbital of one carbon atom by a d orbital. The overlaps of the d orbital with the two neighbouring p_z orbitals have opposite signs and therefore there is a total effective flux of π. Such molecules, sometimes called anti-Hückel systems, have not been synthetised but the possibility has been discussed in the chemical literature,see [22].

Mesoscopic rings have a size much larger than annulenes and therefore a flux π corresponds to more realistic magnetic field strengths. Models of Hubbard type with spin orbit interactions have been proposed to explain some features related to the magnetic susceptibility of these systems. The kinetic energy term is of the form

$$\sum_{l=1,\alpha,\beta=\uparrow,\downarrow}^{L} e^{i\phi_{l,l+1}} (U_{l,l+1})_{\alpha,\beta} (c_{l,\alpha}^{\dagger} c_{l+1,\beta} + c_{l+1,\alpha}^{\dagger} c_{l,\beta}) \tag{4.3}$$

where $U_{l,l+1}$ are $SU(2)$ matrices representing the spin orbit coupling (see [23] and references therein). Under some conditions on the choice of these matrices inequalities like in theorem 2 can be proven.

4.2 Rigorous results for the Peierls Hubbard model at half filling

Kennedy and Lieb [6] showed for the SSH hamiltonian that the ground state is dimerised at half filling. Their proof uses the fact that one can represent the ground state energy as a convex functional. Later this result was extended to $U \neq 0$, but $M = +\infty$ for (4.1) [20]. If $\Phi = 0$ and $L = 4n+2$ or $\Phi = \pi$ and $L = 4n$

the ground state is of the form $u_l = (-1)^l \delta$. If δ is non zero then the ground state has period two, otherwise it is uniform. The method used in [20] is to map the system on a spin chain by using a standard Jordan-Wigner transformation and is therefore limited to one dimension.

All of these results are recovered by the method of section 3 applied to a lattice Λ which is a ring of length L. Theorem 1 also solves a two dimensional problem on a square lattice considered by Tang and Hirsch [24] in the case where a flux π per plaquette is fixed. These authors study the hamiltonian (2.1) in two dimensions with the choice

$$|t_{xy}| = t|1 - \alpha \hat{e}_{xy} \cdot (\boldsymbol{u}_x - \boldsymbol{u}_y)| \tag{4.4}$$

and

$$V(\{|t_{xy}|) = \frac{\kappa}{2}|\boldsymbol{u}_x - \boldsymbol{u}_y|^2 \tag{4.5}$$

where $\hat{e}_{xy}$ is a unit vector along the bond $< xy >$, and $\boldsymbol{u}$ is the displacement with respect to the equilibrium position. Theorem 1 reduces the search for the optimal configuration of the displacements to those compatible with figure 1.

4.3 Coupled polyacetylene chains

Usually polyacetylene chains are coupled between themselves through inter-coulombic interactions or interhopping terms. Let us take two infinite chains with hamiltonians H_1 and H_2, coupled by a hopping term $t_l = t + (-1)^l t'$. The total hamiltonian is

$$H = H_1 + H_2 + \sum_{l,\sigma} t_l \left(c_{l,\sigma}^{(1)\dagger} c_{l,\sigma}^{(2)} + c_{l,\sigma}^{(2)\dagger} c_{l,\sigma}^{(1)} \right) \tag{4.6}$$

with H_i, $i = 1, 2$ equal to (1.2) with the $c^\#$'s replaced by $c^{\#(i)}$ and $t_{l,l+1}(u_l - u_{l+1})$ by $t_{l,l+1}^{(i)}(u_l^{(i)} - u_{l+1}^{(i)})$.

If $t' > t$ the sign of t_l alternates and the flux through a square plaquette is equal to π. On the other hand if $t' < t$ the sign of t_l is constant and the flux is zero. Here the flux is not related to an external magnetic field but rather is an effective flux coming from chemical bonding. Motivated by some experimental observations on the structure of coupled chains, Baeriswyl and Maki [25] argued that in some situations the chains lie in different planes making an angle of almost ninety degrees. Then it turns out that the p_z orbitals of carbon atoms have a different overlap whether l is odd or even. This may lead to alternating signs for t_l and therefore to effective flux equal to π.

In this case the methods of section three apply and one can prove that the optimal configuration of bond lengths of the two chains are uniform or alternate "out of phase" (see figure 4a). More precisely for (4.6) with $t' > t$ the optimal configuration must satisfy

$$t_{l,l+1}^{(1)} = t_{l+1,l+2}^{(2)}, \qquad t_{l+1,l+2}^{(1)} = t_{l,l+1}^{(2)} \tag{4.7}$$

For $t' < t$ and H_i the SSH hamiltonians one can show by explicit computations that an out of phase configuration of alternating bond lengths has a higher energy than the "in phase" configuration (see figure 4b). In this case the general methods developed here do not give any rigorous information.

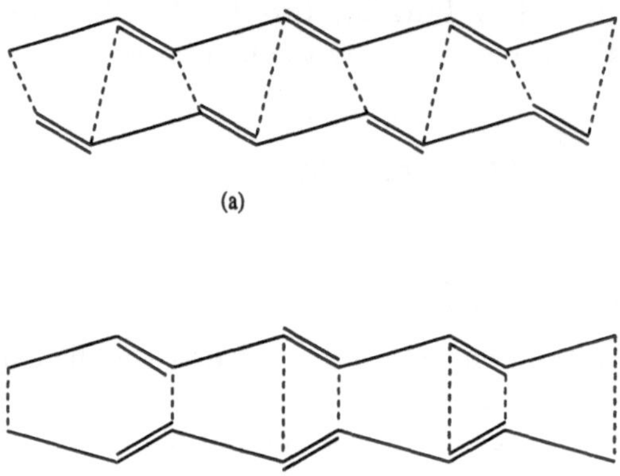

(a)

(b)

Fig. 4. (a) out of phase configuration of two coupled chains. (b) in phase configuration.

Physical consequences of these configurational properties are discussed in [11],[25].

Acknowledgments

The author is grateful to the organisers of the 32 nd winter school in Karpacz, Poland, for the invitation to present this work. The analysis presented in section 3 was developed in collaboration with Bruno Nachtergaele.

References

1. R.E. Peierls, *Quantum theory of solids*, Clarendon, Oxford 1955, p 108.
2. H. Fröhlich, Proc. R. Soc. London Ser A **223**, 296 (1954).
3. R.E. Peierls, *More surprises in theoretical physics*, Princeton University Press.
4. P. Wiegman, Physica **153** C, 103 (1988).
5. I. Affleck and J.B. Marston, Phys. Rev. B **37**, 3774 (1988).
6. E.H. Lieb and M. Loss, Duke Math. J. **71**, 337 (1993).
7. W.P. Su, J.R. Schrieffer and A.J. Heeger, Phys. Rev. B **22**, 2099 (1980).
8. T. Kennedy and E.H. Lieb, Phys. Rev. Lett. **59**, 1309 (1987).
9. H.C. Longuet-Higgins and L. Salem, Proc. R. Soc. London Ser A **251**, 172 (1959).
10. R. Pariser and R.G. Parr, J. Chem. Phys **21**, 466 (1953); Pople, Trans. Faraday. Soc **49**, 1375 (1953).
11. D. Baeriswyl, D.K. Campbell and S. Mazumdar, in *Conjugated conducting polymers*, ed. H. Kiess, Springer Series in Solid State Sciences vol 102 (Springer, New York) p 7.

12. E. Fradkin, *Field theories of condensed matter systems*, Addison Wesley 1991.
13. N. Macris and J. Ruiz, "On the orbital magnetism of itinerant electrons", preprint (1995).
14. E.H. Lieb, Phys. Rev. Lett. **73**, 2158 (1994).
15. N. Macris and B. Nachtergaele, "On the flux phase conjecture at half filling: an improved proof" preprint (1995).
16. J. Glimm and A. Jaffe, *Quantum Physics a functional integral point of view*, Springer Verlag, New York, 1981, p88.
17. J. Fröhlich, B. Simon and T. Spencer, Comm. Math. Phys **50**, 79 (1976).
18. F.J. Dyson, E.H. Lieb and B. Simon, J. Stat. Phys **18**, 335 (1978).
19. J. Fröhlich, R. Israel, E.H. Lieb and B. Simon, Comm. Math. phys **62**, 1 (1978); J. Stat. Phys **22**, 297 (1980).
20. E.H. Lieb and B. Nachtergaele, Phys. Rev. B **51** 4777 (1995).
21. N. Macris, "Periodic ground states in simple models of itinerant fermions interacting with classical fields" Proceeding of the VI Max Born Symposium, Poland (1995), to appear in Physica A.
22. Dewar, Dougherty, *The PMO theory of organic chemistry*, Plenum Press, New York, 1975, p106.
23. S. Fujimoto and N. Kawakami, Phys. Rev. B **48**, 17406 (1993).
24. S. Tang and J.E. Hirsch, Phys. Rev. B **16**, 9546 (1988).
25. D. Baeriswyl and K. Maki, Phys. Rev. B **38**, 8135 (1988).

Fully and Partially Dressed States in Quantum Field Theory and in Solid State Physics

G. Compagno [1], *R. Passante*[2], *F. Persico*[1]

[1] Istituto Nazionale di Fisica della Materia and Istituto di Fisica dell'Universita', Via Archirafi 36, I-90123 Palermo, Italy
[2] Istituto per le Applicazioni Interdisciplinari della Fisica, Consiglio Nazionale delle Ricerche, Via Archirafi 36, I-90123 Palermo, Italy

1 Introduction

The importance of the cultural links between solid state physics and quantum field theory can hardly be overemphasized. They have led to an impressive cross-fertilization between these two main traditional branches of physics, which dates back to the end of the 50s, when a lively exchange of new ideas was established [1] between quantum field theory and condensed matter physics.

It is easy to identify some of these ideas: those which come first to our mind are symmetry breaking, the renormalisation group and elementary excitations [2], although of course the influence of the theory of superconductivity is such that nowadays it deserves a separate chapter in the most modern books on quantum field theory [3]. The present review is embedded in the cultural ground common to condensed matter physics and quantum field theory, and it is aimed at promoting the continued exchange of fundamental concepts. We shall try to obtain this result by concentrating on the concept of elementary excitations and by pointing out the similarity of some aspects of their dynamics in solid state physics and in quantum field theory. In particular, we shall consider the interactions of electrons and excitons with lattice phonons, and we shall discuss their counterparts in relativistic quantum field theory and in QED. Indeed as an exciton or an excited electron moves through a semiconductor, it carries along a lattice distortion which changes its physical properties. Such a lattice distortion can be described as a cloud of virtual phonons surrounding the mobile charged particles [4, 5] by virtue of continuous emission and reabsorption processes. Thus one speaks of a dressed electron (or polaron) and a dressed exciton (or polaronic exciton). We intend to insist on the analogy of these objects with the dressed nucleon [6], which is a nucleon surrounded by a cloud of virtual mesons, as well as the dressed atom [7], that is an atom endowed with a cloud of virtual photons. We shall argue that the similarity of the dressing processes may lead to a deeper understanding of the dressing phenomenon and may suggest new experiments in solid state physics, capable of shedding new light on yet only partially explored fundamental processes of quantum field theory and in particular of QED.

Part I: Fully Dressed States

2 Virtual quanta in ground-state source-field systems

We shall begin by showing that a nonvanishing number of radiation quanta are usually present in the ground state of a source which is coupled with a radiation field.

The Fröhlich polaron

In a semiconductor crystal it is possible to excite an electron from the valence to the conduction band [8]. Disregarding for the moment the Coulomb attraction of the excited electron with the simultaneously created hole in the valence band, if the crystal possesses a sufficiently ionic character the electron is expected to interact mainly with the longitudinal optical phonons, which are capable of generating large electric polarization fields. Moreover, if the electron created is not too energetic, one can approximate the rigid lattice band as parabolic. In these conditions the system can be modelled by the Hamiltonian [9]

$$H = H_0 + V; \;\; H_0 = \frac{1}{2m}\sum_{\boldsymbol{p}} p^2 c_{\boldsymbol{p}}^\dagger c_{\boldsymbol{p}} + \hbar\omega \sum_{\boldsymbol{k}} a_{\boldsymbol{k}}^\dagger a_{\boldsymbol{k}};$$
$$V = \sum_{\boldsymbol{pk}} \left(V_k a_{\boldsymbol{k}} c_{\boldsymbol{p}+\hbar\boldsymbol{k}}^\dagger c_{\boldsymbol{p}} + V_k^* a_{\boldsymbol{k}}^\dagger c_{\boldsymbol{p}}^\dagger c_{\boldsymbol{p}+\hbar\boldsymbol{k}} \right) \tag{1}$$

where m is the effective mass of the slow electron in the rigid lattice, i.e. in the absence of phonons. The sum on $\boldsymbol{p}$ runs over the electronic band states of wavevector $\boldsymbol{p}/\hbar$, $c_{\boldsymbol{p}}^\dagger(c_{\boldsymbol{p}})$ are Fermi operators which create (annihilate) a spinless electron of charge $-e$ in the band state $\boldsymbol{p}$. The frequency common to all longitudinal optical modes is ω and $a_{\boldsymbol{k}}^\dagger(a_{\boldsymbol{k}})$ creates (annihilates) a longitudinal optical phonon with wavevector $\boldsymbol{k}$. The electron-phonon coupling constant is [10]

$$V_k = i\frac{\hbar\omega}{k}\left(\frac{\hbar}{2m\omega}\right)^{1/4}\left(\frac{4\pi\alpha_p}{V}\right)^{1/2}; \;\; \alpha_p = \frac{e^2}{\hbar\bar{\epsilon}}\left(\frac{m}{2\hbar\omega}\right)^{1/2}; \;\; \bar{\epsilon} = \frac{\epsilon(0)\epsilon(\infty)}{\epsilon(0)-\epsilon(\infty)} \tag{2}$$

where $\epsilon(0)$ and $\epsilon(\infty)$ are, respectively, the static and the high-frequency dielectric constants of the semiconductor.

In the absence of V_k, the set of eigenstates $\mid \boldsymbol{p}, \{0_{\boldsymbol{k}}\}\rangle$ of H_0 describe an electron of momentum $\boldsymbol{p}$ in the crystal with no phonons. The presence of V_k introduces changes in these eigenstates, which can be evaluated approximately by perturbation theory, if V_k is not too large and provided the kinetic energy of the electron is small compared to $\hbar\omega$. At first order in V_k the new eigenstates are [11]

$$\mid \boldsymbol{p} = \hbar\boldsymbol{K}, \{0_{\boldsymbol{k}}\}\rangle' = \mid \boldsymbol{p} = \hbar\boldsymbol{K}, \{0_{\boldsymbol{k}}\}\rangle + \sum_{\boldsymbol{K}} \frac{V_k^*}{\Delta_{\boldsymbol{k}}} \mid \boldsymbol{p}' = \hbar(\boldsymbol{K}-\boldsymbol{k}), 1_{\boldsymbol{k}}\rangle \tag{3}$$

where $\hbar \boldsymbol{K}$ is the total (electron+phonons) momentum, $\Delta_{\boldsymbol{k}} = (p^2 - p'^2)/2m - \hbar\omega$ and $1_{\boldsymbol{k}}$ indicates the presence of one phonon with momentum $\boldsymbol{k}$. A simple evaluation of the quantum average of the operator $n = \sum_{\boldsymbol{k}} a^\dagger_{\boldsymbol{k}} a_{\boldsymbol{k}}$, corresponding to the total number of phonons, over the perturbed state Eq. (3) yields

$$\langle n \rangle \simeq \frac{\alpha_p}{2} \tag{4}$$

which is different from zero. We remark that, strictly speaking, the state Eq. (3) is not the ground state of H; neverthless is quite near to it if $\boldsymbol{K}$ is small.

The two-level atom in QED

Within the electric dipole approximation [12], the interaction of a two-level atom (at the origin of the reference frame) with the electromagnetic field is described by [7]

$$H = H_0 + V_1 + lV_2 \; ; \; H_0 = \hbar\omega_0 S_z + \sum_{\boldsymbol{k}j} \hbar\omega_k a^\dagger_{\boldsymbol{k}j} a_{\boldsymbol{k}j} \, ;$$

$$V_1 = \sum_{\boldsymbol{k}j} \left(\epsilon_{\boldsymbol{k}j} a_{\boldsymbol{k}j} S_+ + \epsilon^*_{\boldsymbol{k}j} a^\dagger_{\boldsymbol{k}j} S_- \right) \; ; \; V_2 = \sum_{\boldsymbol{k}j} \left(\epsilon_{\boldsymbol{k}j} a^\dagger_{\boldsymbol{k}j} S_+ + \epsilon^*_{\boldsymbol{k}j} a_{\boldsymbol{k}j} S_- \right) \tag{5}$$

where ω_0 is the natural atomic frequency, $S_+ (S_-)$ is the atomic operator which raises (lowers) the atom to the upper (lower) level, ω_k is the frequency of the $\boldsymbol{k}j$ mode of the field whose creation (annihilation) operator is $a^\dagger_{\boldsymbol{k}j} (a_{\boldsymbol{k}j})$, and $\epsilon_{\boldsymbol{k}j}$ is the atom-photon coupling constant. Moreover l is a constant of modulus 1 which we keep in the calculations for diagnostic purposes, since it appears as a factor of V_2 whose contributions we wish to be able to follow. We label the eigenstates of H_0 by $\mid \{n_{\boldsymbol{k}j}\}, \uparrow(\downarrow)\rangle$. This indicates a photon distribution $\{n_{\boldsymbol{k}j}\}$ in the field and the atom in the excited $\uparrow$ (ground $\downarrow$) state. Consider in particular the ground state $\mid \{0_{\boldsymbol{k}j}\}, \downarrow\rangle$ of H_0, with no photons and the atom in its ground state. The normalized expression of the perturbed ground state can be shown to be [13]

$$\begin{aligned} &\mid \{0_{\boldsymbol{k}j}\}, \downarrow\rangle' = \\ &\left(1 + l\frac{1}{E_0 - H_0}(1 - P_0)\, V_2 + l\frac{1}{E_0 - H_0}(1 - P_0)\, V_1 \frac{1}{E_0 - H_0}(1 - P_0)\, V_2 \right. \\ &\left. - \frac{1}{2} l^2 \langle \{0_{\boldsymbol{k}j}\}, \downarrow \mid V_2 \frac{1}{(E_0 - H_0)^2}(1 - P_0)\, V_2 \mid \{0_{\boldsymbol{k}j}\}, \downarrow\rangle \right) \mid \{0_{\boldsymbol{k}j}\}, \downarrow\rangle \end{aligned} \tag{6}$$

where $E_0 = -\hbar\omega_0/2$ and P_0 is the projection operator onto $\mid \{0_{\boldsymbol{k}j}\}, \downarrow\rangle$. Using this expression it is easy to see that the quantum average of the total number of photons $n = \sum_{\boldsymbol{k}j} a^\dagger_{\boldsymbol{k}j} a_{\boldsymbol{k}j}$ over the ground state Eq. (6) of H is

$$\langle n \rangle = \frac{l^2}{\hbar^2} \sum_{\boldsymbol{k}j} \frac{\mid \epsilon_{\boldsymbol{k}j} \mid^2}{(\omega_0 + \omega_k)^2} . \tag{7}$$

The essential role played by V_2 in this result should be noted, since $\langle n \rangle$ vanishes for $l = 0$.

The polaronic exciton

If the Coulomb attraction between the excited electron and the hole left behind in the valence band of a semiconductor cannot be neglected, one has to take into account the existence of levels within the gap separating the conduction and the valence band. These levels correspond to bound electron-hole states, analogous to the hydrogenic states of a positronium atom [14]. The composite electron-hole system is called an exciton. Since both the constituent particles (electron and hole) interact in a polaron-like fashion with the lattice, so does the exciton leading to the formation of a polaronic exciton [15]. For a two-band crystal model, the standard Hamiltonian for a Frenkel exciton can be written as [16]

$$H = \hbar\omega_0 \sum_m \left(S_z^{(m)} + \frac{1}{2} \right) + \sum_{mn} J_{mn} S_+^{(m)} S_-^{(n)} + \sum_k \hbar\omega_k a_k^\dagger a_k + \sum_{mk} \chi_k^{(m)} \hbar\omega_k \left(a_k^\dagger + a_{-k} \right) \left(S_z^{(m)} + \frac{1}{2} \right), \tag{8}$$

where the superscript m labels the lattice sites, $\hbar\omega_0$ is the rigid-lattice ground state exciton energy, J_{mn} is the matrix element for transfer from site m to site n and $\chi_k^{(m)}$ is the (generally complex) exciton-phonon coupling constant. The phonons are described by a_k operators, where k indicates both wavevector and branch index. Realistic polaronic exciton Hamiltonians are more complicated and capable of taking into account excitonic states other than the ground state [17], but Eq. (8) will suffice for the present, limited purposes. In fact exact diagonalization of H is possible only if we put $J_{mn} = 0$, which describes the case of a localized exciton. In this case the ground state of the polaronic exciton is given by [16]

$$\mid \{0_k\}, \uparrow_n\rangle' = U \mid \{0_k\}, \uparrow_n\rangle \tag{9}$$

where $\mid \{0_k\}, \uparrow_n\rangle$ describes an exciton at site n and no phonons, and where U is an appropriate unitary operator. The quantum average of the total number of phonons over this state turns out to be

$$\langle n\rangle = \sum_k \mid \chi_k^{(n)} \mid^2 . \tag{10}$$

The static nucleon

The simplest possible model for a nucleon interacting with a neutral meson field is a static source linearly coupled with a scalar field [6, 18]. In this model, which can be solved exactly, the source is fixed at the origin of the reference frame and is described by a source density $\rho(\boldsymbol{x})$ which is unchanged by the interaction with the meson field. The latter consists of a real Klein-Gordon field whose Hamiltonian is [19]

$$H_F = \sum_{\boldsymbol{k}} \hbar\omega_k a_{\boldsymbol{k}}^\dagger a_{\boldsymbol{k}}; \quad \omega_k = c\sqrt{k^2 + \kappa^2} \tag{11}$$

where κ is the Compton wavevector of the meson field. The source-field interaction is taken of the form

$$H_{SF} = -g\sum_{\boldsymbol{k}}\sqrt{\frac{\hbar}{2V\omega_k}}\left(\rho_{\boldsymbol{k}}^* a_{\boldsymbol{k}} + \rho_{\boldsymbol{k}} a_{\boldsymbol{k}}^\dagger\right); \quad \rho_{\boldsymbol{k}} = \int \rho(\boldsymbol{x}) e^{i\boldsymbol{k}\cdot\boldsymbol{x}} d^3\boldsymbol{x}, \tag{12}$$

where g is the strong-force nucleon-meson coupling constant and V is the field quantization volume. The total Hamiltonian $H = H_F + H_{SF}$ can be diagonalized with the help of a set of unitary operators

$$D_{\boldsymbol{k}} = \exp\left\{-g[2\hbar V{\omega_k}^3]^{-1/2}\left(\rho_{\boldsymbol{k}}^* a_{\boldsymbol{k}} - \rho_{\boldsymbol{k}} a_{\boldsymbol{k}}^\dagger\right)\right\}, \tag{13}$$

and the ground state of the nucleon-meson system is given by

$$\mid 0\rangle' = \prod_{\boldsymbol{k}} D_{\boldsymbol{k}} \mid \{0_{\boldsymbol{k}}\}\rangle, \tag{14}$$

where $\mid \{0_{\boldsymbol{k}}\}\rangle$ represents the vacuum of the meson field, which corresponds to the ground state of H_F. A simple calculation yields the quantum average of the total number of mesons $n = \sum_{\boldsymbol{k}} a_{\boldsymbol{k}}^\dagger a_{\boldsymbol{k}}$ on $\mid 0\rangle'$ as

$$\langle n\rangle = \frac{g^2}{2\hbar V}\sum_{\boldsymbol{k}}\frac{1}{{\omega_k}^3}\mid \rho_{\boldsymbol{k}}\mid^2 . \tag{15}$$

All the examples considered above lead us to conclude that the occurrence of a nonvanishing number of radiation quanta in the ground state of a source-field system is far from uncommon. In the next section we shall discuss the physical nature of this virtual cloud.

3 The physical nature of the ground-state quanta

The localization, in the ground state of the system, of quanta in a region near to the source can be understood on the basis of energy considerations. In fact at zero temperature the quanta of the field which appear in the course of a quantum fluctuation, lead to an increase of the energy of the source-field system, if the source-field interaction energy is not taken into account. In other words, the bare energy E of the source-field system is not necessarily conserved during such a fluctuation, and states having a bare energy $E' \neq E$ can be excited. The magnitude of the energy unbalance $\delta E = \mid E' - E\mid$, however, is constrained by the Heisenberg uncertainty principle

$$\delta E \sim \hbar/\tau \tag{16}$$

where τ is the duration of the fluctuation. In this time τ, the energy balances again and the extra quanta of the field are reabsorbed. It is important to realize that here we are discussing fluctuations of a purely quantum nature, which take place also at zero temperature. Since these fluctuations take place continuously, the source can be described as surrounded by a steady-state cloud of virtual quanta continuously emitted into, and reabsorbed from, the field. In contrast with the behaviour of a real quantum, which is emitted during an energy-conserving process with $E' = E$, and which in the absence of boundaries can abandon the source for ever, a virtual quantum can only attain a finite distance from the source, roughly given by

$$r \sim c\tau \sim \frac{\hbar c}{\delta E} \tag{17}$$

where c is a velocity scale for the quantum. Consequently we should describe the source as surrounded by a cloud of virtual quanta and we should expect the linear dimensions of the virtual cloud surrounding the source to coincide approximately with r given by Eq. (17) for virtual transitions characterized by an energy unbalance δE [20].

We have thus reached the conclusion, in agreement with an early suggestion by Van Hove [21], that a quantum mechanical source at zero temperature can be surrounded by a permanent cloud of virtual quanta which constitute a kind of dress for the bare source. In this sense we speak here of fully dressed sources. The question arises whether these virtual quanta are related to the ground-state quanta we have shown to exist in the fundamental state of several source-field systems. In order to answer this question we have to look more closely into the physical models of the previous section and we have to apply to these models the qualitative ideas about dressed sources developed in the present section.

Considering first the case of a two-level atom, we can attribute the energy unbalance δE to the fluctuations induced by the "counterrotating" term V_2 in Eq. (5). These fluctuations lead the atom from the ground to the excited state and also contribute to the number of virtual photons of mode $\boldsymbol{k}j$ according to Eq. (7). For a given fluctuation, the energy unbalance is then $\delta E = \hbar(\omega_0 + \omega_k)$ while c in Eq. (17) coincides with the velocity of light. Considering long-wavelength photons with $\omega_k \ll \omega_0$, Eq. (17) yields $r \sim c/\omega_0 = l_0/2\pi$, where l_0 is the wavelength of the atomic frequency involved in the process. Thus for this kind of fluctuations, and disregarding the possibility of two-photon processes, one obtains typical dimensions of the dressed atom which are of the order of $10^{-4} \div 10^{-5}$ cm. On the other hand, in the elementary model of the nucleon we have considered, the source is rigid. Thus the minimum energy required to create a meson from the vacuum is equal to mc^2, while the velocity scale for the bare field is chosen again to be the velocity of light. Introduction of these values in Eq. (17) yields $r \sim 1.4 \cdot 10^{-13} cm$, which is of the right order of magnitude for the experimental mean square radius of the proton and of the neutron [22]. In the case of the Fröhlich polaron, the electron should be considered much lighter than an atom or a nucleon, and in practice the only mobile particle in view of the small group velocity of the optical phonons assumed in Eq. (1). Consequently its recoil cannot be neglected, unlike the two previous cases. Thus the appropriate value for the energy fluctuations to be substituted in Eq. (17) is $\delta E = \hbar\omega + p^2/2m$. Moreover, one should expect that the strongest interaction is with phonons of wavelength l such that in a time $2\pi/\omega$ the electron travels over a distance corresponding to l. In fact this permits the electron to feel a phonon electric field which is always of the same sign. This yields $2\pi p/m\omega \sim l$, or $p \sim m\omega/k$. Moreover, assuming the velocity acquired by the electron during a fluctuation to originate only from overall wavevector conservation in the process of emission of the phonon in a one-phonon process (small coupling assumption), one has $p = \hbar k$, which together with $p \sim m\omega/k$ gives $p^2/2m \sim \hbar\omega/2$. This implies that we are dealing with low-frequency electrons, in agreement with the assumptions of the previous section, and in addition yields $\delta E \sim 3\hbar\omega/2$ for the amplitude of energy fluctuations. The small group velocity of the optical photons suggests that for c, appearing in Eq. (17), we should choose the velocity of the electron $p/m \sim \sqrt{\hbar\omega/m}$. Substitution in Eq. (17) of this value for c and of the value obtained for δE gives $r \sim \frac{2}{3}\sqrt{\hbar/\omega m}$. Using the value of the free electron mass for m and $\omega \sim 10^{14}$ Hz we obtain $r \sim 10^{-7}$ cm, in good order of magnitude agreement with the polaron radius evaluated from experimental data [23]. For the polaronic exciton, we remark that in the Hamiltonian Eq. (8) with $J_{mn} = 0$, the exciton is rigid, in the sense that the interaction with phonons does not change its internal state. In this sense one might be tempted to apply the same qualitative considerations developed for the nucleon, although the small mass of the exciton should be taken into account. As mentioned, however, internal states of the exciton possessing higher energy than the ground state do obviously exist. In this case processes whereby phonons are exchanged with the lattice simultaneously with changes of exciton internal energy might play a role. These processes would be described, in the linear approximation, by terms of the same form as V_1 and V_2 in Eq. (5) [24], and their contribution to the virtual cloud of the polaronic exciton should resemble that of the dressed atom. Regrettably, however, a consistent treatment of the virtual cloud of the polaronic exciton does not seem to exist yet.

Finally, we must spend a few words to dispel an ambiguity in the language concerning dressed states as referred to atoms or molecules in QED. In fact, the same term is in use in quantum optics to describe the atom-field correlated states arising when an atom interacts with a real external quantum field such as that generated by a laser [25]. In these lectures we will always use the term "dressed atom" in the sense discussed in the present section.

4 The shape of the ground-state virtual cloud

The semiquantitative considerations of the previous section suggest the presence of a cloud of virtual quanta surrounding the sources of a field at zero temperature. This raises the following questions:

- Is it possible to identify the ground-state quanta obtained in Section 2 with those which belong to the cloud?

- Can we find a quantitative description of the virtual cloud in terms of a space-dependent distribution of fields around the source?

- Does the presence of this cloud give rise to measurable effects, which depend on its detailed shape in space around the source?

The first two questions can be answered together, since we shall show in this Section that the shape of the virtual cloud can be obtained from the explicit form of the ground state of the total source-field system, which we have evaluated in Section 2 and which we have shown to be at the origin of the finite number of ground-state quanta. We shall henceforth call this state the "dressed ground state". The third question will be discussed in the next section.

Clearly the information on the shape of the virtual cloud, as a function of the distance from the source, will be available given an operator functional of the field distribution around the source, whose quantum average on the ground state of the coupled source-field system can be taken as a reliable measure of the form of the virtual cloud. Thus it is likely that more than one operator of this kind exists, and that any sensible choice should give equivalent results.

In the case of the static nucleon coupled to a scalar meson field, an obvious choice is the amplitude of the scalar field, defined as [26]

$$\phi(\boldsymbol{x}) = \sum_{\boldsymbol{k}} \sqrt{\frac{\hbar}{2V\omega_k}} \left(a_{\boldsymbol{k}} e^{i\boldsymbol{k}\cdot\boldsymbol{x}} + a_{\boldsymbol{k}}^{\dagger} e^{-i\boldsymbol{k}\cdot\boldsymbol{x}} \right) . \tag{18}$$

From Eq. (14) we easily find

$$'\langle 0 \mid \phi(\boldsymbol{x}) \mid 0\rangle' = \frac{1}{2V} g \sum_{\boldsymbol{k}} \frac{1}{\omega_k{}^2} \left(\rho_{\boldsymbol{k}} e^{i\boldsymbol{k}\cdot\boldsymbol{x}} + \rho_{\boldsymbol{k}}^{*} e^{-i\boldsymbol{k}\cdot\boldsymbol{x}} \right) . \tag{19}$$

For a point-like source $\rho(\boldsymbol{x}) = \delta(\boldsymbol{x})$ and $\rho_{\boldsymbol{k}} = 1$. Thus after some algebra it can be shown that

$$'\langle 0 \mid \phi(\boldsymbol{x}) \mid o\rangle' = \frac{g}{4\pi c^2} \frac{e^{-\kappa x}}{x} . \tag{20}$$

We see that the field decays in a Yukawa-like fashion [27] far from the source. The divergence at $x = 0$ is unphysical and depends on the assumption of a point-like source, but we note that the virtual cloud is concentrated within one Compton wavelength of the nucleon, in agreement with the semiquantitative considerations of the previous section. A cloud of virtual mesons is also an essential constituent of more sophisticated models of the nucleon, such as the cloudy bag model [28].

For the Fröhlich polaron, the operator corresponding to the electric polarization created in the lattice by the presence of the phonons is [9]

$$\phi(\boldsymbol{x}) = i \left(\frac{2\pi\hbar\omega}{\bar{\epsilon}V} \right)^{1/2} \sum_{\boldsymbol{k}} \frac{1}{k} \left(a_{\boldsymbol{k}}^{\dagger} e^{-i\boldsymbol{k}\cdot\boldsymbol{x}} - a_{\boldsymbol{k}} e^{i\boldsymbol{k}\cdot\boldsymbol{x}} \right) . \tag{21}$$

Starting from this, and using Eq. (3) for the dressed ground state, we obtain the following result for the quantum average $\phi(\boldsymbol{x}, \boldsymbol{x}_e)$ of the potential created at $\boldsymbol{x}$ by the presence of an electron at $\boldsymbol{x}_e$

$$\begin{aligned} \phi(\boldsymbol{x}, \boldsymbol{x}_e) &= -\frac{4\pi e}{\bar{\epsilon}V} \hbar\omega \sum_{\boldsymbol{k}} \frac{\cos \boldsymbol{k} \cdot (\boldsymbol{x}_e - \boldsymbol{x})}{k^2 \Delta_{\boldsymbol{k}}} \\ &= -\frac{e}{\bar{\epsilon} \mid \boldsymbol{x}_e - \boldsymbol{x} \mid} \left(1 - \exp\left\{ - \left[\frac{2m\omega}{\hbar} \right]^{1/2} \mid \boldsymbol{x}_e - \boldsymbol{x} \mid \right\} \right) . \end{aligned} \tag{22}$$

Thus at distances from the electron larger than $\sqrt{\hbar/2m\omega}$ the polarization potential is of the normal screened Coulomb type, whereas at smaller distances the presence of the virtual phonon cloud eliminates the Coulomb singularity [11]. More refined calculations yield results in agreement with Eq. (22) [29]. We remark that this result is also in accord with the conclusions of the previous section.

Turning to the two-level atom, quadratic expressions [30, 31] present the advantage of possessing a direct physical meaning. For example, $E^2(\boldsymbol{x})$ is proportional to the electric energy density operator [32]

$$\mathcal{H}_{\text{el}}(\boldsymbol{x}) = \frac{1}{8\pi}\boldsymbol{E}^2(\boldsymbol{x}) \tag{23}$$

A rather nontrivial calculation of the quantum average of $\mathcal{H}_{\text{el}}(\boldsymbol{x})$ on the dressed ground state Eq. (6) yields a complicated result which, however, can be easily understood in the two limiting cases of short distances x from the atom (near zone) and of large distances (far zone). The near zone is defined for $x < c/\omega_0$. The lifetime of virtual photons with $\omega_k > \omega_0$ is shorter than $\omega_0{}^{-1}$, so that these photons during their lifetime cannot reach distances greater than c/ω_0 from the atom. Thus in this region the virtual cloud is dominated by energetic high-frequency virtual photons. On the contrary, low-energy virtual photons with $\omega_k < \omega_0$ have a longer lifetime, and consequently they can reach the far zone, where they dominate the energy density due to the scarcity of high-frequency photons [33]. In each of the two zones we have approximately [13]

$$'\langle\{0_{\boldsymbol{k}j}\},\downarrow|\,\mathcal{H}_{\text{el}}(\boldsymbol{x})\,|\{0_{\boldsymbol{k}j}\},\downarrow\rangle' - \frac{1}{2V}\sum_{\boldsymbol{k}j}\frac{\hbar\omega_k}{2}$$
$$= \begin{cases} \frac{1}{8\pi}\mu_{21}^2\left(1+3\cos^2\theta\right)x^{-6} & (x < c/\omega_0) \\ \frac{1}{32\pi^2}\hbar c\alpha_{mn}\left(13\delta_{mn}+7\hat{x}_m\hat{x}_n\right)x^{-7} & (x > c/\omega_0) \end{cases} \tag{24}$$

where V is the quantization volume as usual, $\hat{\boldsymbol{x}}$ is the unit vector along $\boldsymbol{x}$, $\boldsymbol{\mu}_{21}$ is the electric dipole matrix element between the two atomic states, θ is the angle between $\boldsymbol{\mu}_{21}$ and $\hat{\boldsymbol{x}}$ and

$$\alpha_{mn} = \frac{2\,(\boldsymbol{\mu}_{21})_m\,(\boldsymbol{\mu}_{21})_n}{\hbar\omega_0} \tag{25}$$

is the static polarizability tensor of the two-level atom [34]. The negative contribution to the LHS of Eq. (24) is the infinite, spatially uniform energy density due to the zero-point electric field of the vacuum modes, which of course is of no interest here. The fact that at the approximate boundary of the two zones, for $x \sim c/\omega_0$, Eq. (24) presents a decrease in the slope (from x^{-6} to x^{-7}) is in qualitative agreement with the considerations of the previous section. The agreement cannot be complete because the calculations described in this section take into account two-photon processes, which were neglected in the previous section.

To the best of our knowledge, no realistic evaluation of the shape of the virtual phonon cloud has been proposed up to now for the case of the polaronic exciton in 3 dimensions. Some qualitative considerations, however, can be found in review papers [35].

5 Observable effects related to the shape of the virtual cloud

We are now in a position to answer the third question raised at the beginning of the previous section, whether the shape of the ground-state virtual cloud can be related to observable effects.

To begin with, it is evident that the terms present in the source-field Hamiltonians discussed hitherto and responsible for the existence of the virtual cloud in different systems, are also responsible for energy shifts of a single source coupled to its own field. Consider first the nucleon-meson system whose Hamiltonian is given in Section 2. It is not difficult to show that all the eigenstates of the complete source-field Hamiltonian $H_F + H_{SF}$ are shifted by the common amount

$$\Delta = -\frac{1}{2}g^2\frac{1}{V}\sum_{\boldsymbol{k}}\frac{1}{{\omega_k}^2} \mid \rho_{\boldsymbol{k}} \mid^2 = -\frac{1}{2}g^2\frac{1}{(2\pi)^3}\int\frac{1}{{\omega_k}^2} \mid \rho_{\boldsymbol{k}} \mid^2 d^3\boldsymbol{k} \qquad (26)$$

with respect to the free Hamiltonian H_F [6]. This shift can be regarded as an energy renormalization, which could be identified with a mass renormalization of the source if we had dealt with a mobile, rather than a fixed, nucleon. As for the Fröhlich polaron, a simple perturbative calculation yields the electron energy in the form $\hbar^2 K^2/2m^*$ where [10]

$$m^* = \frac{m}{1-\alpha_p/6}. \qquad (27)$$

This evidently indicates a mass renormalization of the electron, similar in nature to the mass renormalization of an electric charge in QED [36]. Finally, for a two-level atom, second-order perturbation theory yields the ground-state shift

$$\Delta = -\lambda^2\sum_{\boldsymbol{k}j}\frac{\mid \epsilon_{\boldsymbol{k}j} \mid^2}{\hbar(\omega_0+\omega_k)} \qquad (28)$$

which can be shown to contribute part of the two-level atom Lamb shift [7]. In spite of the fact, however, that all the energy shifts above are certainly related to the virtual photons, it is clear that they are quantities of a global nature, that is they do not depend in an immediate way on the shape of the virtual cloud. In what follows we set out to show that the shape of the virtual cloud can actually be detected, at least in principle, by the simple trick of using sources in pairs rather than one at a time.

Consider in fact two static meson sources at points 0 and $\boldsymbol{R}$. The Hamiltonian of the system is the same as in Section 2, provided we take $\rho(\boldsymbol{x}) = \rho_1(\boldsymbol{x}) + \rho_2(\boldsymbol{x})$. Here $\rho_1(\boldsymbol{x})$ and $\rho_2(\boldsymbol{x})$ are assumed to be similarly shaped, sharp and nonoverlapping peaks centred at 0 and $\boldsymbol{R}$ respectively. Thus one can reliably use the approximation

$$\rho_{\boldsymbol{k}} \sim 1 + e^{-i\boldsymbol{k}\cdot\boldsymbol{R}} \tag{29}$$

Substitution of this into the shift Eq. (26) yields, after some algebra,

$$\Delta = 2\Delta_0 - g^2 \frac{1}{(2\pi)^3} \int \frac{\cos \boldsymbol{k}\cdot\boldsymbol{R}}{{\omega_k}^2} d^3\boldsymbol{k} = 2\Delta_0 - \frac{1}{4\pi c^2} g^2 \frac{1}{R} e^{-\kappa R} \tag{30}$$

In this expression Δ_0 is the shift caused by each of the two sources as if the other were absent, and it is evidently independent of $\boldsymbol{R}$. The total shift, however, is R-dependent and for this reason yields a Yukawa potential [6, 27] and its related Yukawa force which can certainly be measured. Moreover, comparison with Eq. (20) shows that

$$\Delta = 2\Delta_0 - g' \langle 0 \mid \phi(\boldsymbol{R}) \mid 0\rangle' \tag{31}$$

Thus the shape of the virtual meson cloud Eq. (20) can be obtained by measuring the attraction of the two sources as function of the distance $\boldsymbol{R}$.

The same principle can be used for the virtual cloud of the Fröhlich polaron [37]. In fact second-order perturbation theory yields, for two infinitely massive model electrons at 0 and $\boldsymbol{R}$, the ground-state shift [38]

$$\Delta = 2\Delta_0 - \frac{2}{\hbar\omega} \sum_{\boldsymbol{k}} \mid V_k \mid^2 \cos \boldsymbol{k}\cdot\boldsymbol{R} = 2\Delta_0 - \frac{4\pi e^2}{\bar{\epsilon} V} \sum_{\boldsymbol{k}} \frac{1}{k^2} \cos \boldsymbol{k}\cdot\boldsymbol{R} \tag{32}$$

where V is the volume of the crystal and Δ_0 has the same meaning as in Eq. (30). On the other hand, for massive electrons we can put $\Delta_{\boldsymbol{k}} = \hbar\omega$ in Eq. (22), and we get

$$\phi(\boldsymbol{x}, \boldsymbol{x}_e) = -\frac{4\pi e}{\bar{\epsilon} V} \sum_{\boldsymbol{k}} \frac{1}{k^2} \cos \boldsymbol{k}\cdot(\boldsymbol{x}_e - \boldsymbol{x}) \tag{33}$$

from which we obtain

$$\Delta = 2\Delta_0 + e\phi(\boldsymbol{x} = \boldsymbol{x}_e - \boldsymbol{R}, \boldsymbol{x}_e) \tag{34}$$

which is the counterpart of Eq. (31). We see that, at least in our simple model, we obtain the shape of the virtual cloud in the polaron case by measuring the mutual force acting between two polarons.

For the atomic case, we start from the well-known result that the general van der Waals potential between two ground state, isotropic and electrically polarizable sources T and S at distance $\boldsymbol{R}$ is given by [39]

$$V(\boldsymbol{R}) = -\frac{4}{9\pi\hbar c R^2} \sum_{ts} \mid \boldsymbol{\mu}_{t0}^T \mid^2 \mid \boldsymbol{\mu}_{s0}^S \mid^2 \int_0^\infty \frac{k_t k_s u^4 e^{-2uR}}{(k_t^2 + u^2)(k_s^2 + u^2)}$$
$$\times \left(1 + \frac{2}{uR} + \frac{5}{u^2R^2} + \frac{6}{u^3R^3} + \frac{3}{u^4R^4}\right) du \qquad \left(k_{t,s} = \frac{\omega_{t,s}}{c}\right). \tag{35}$$

In this expression $\boldsymbol{\mu}_{t0}^T$ $(\boldsymbol{\mu}_{s0}^S)$ is the electric dipole matrix element between the ground state of source T (S) and its t (s) excited state. Moreover, the sum over

t (s) runs over all excited states of source T (S), and ω_t (ω_s) is the frequency of the $t \leftrightarrow 0$ ($s \leftrightarrow 0$) transition of the source $t(s)$. The following properties can be proved, starting from Eq. (35):

i) If each source is in the far zone of the other, $V(\boldsymbol{R})$ reduces to

$$V(\boldsymbol{R}) = -\frac{23}{4\pi}\hbar c \alpha_S \alpha_T \frac{1}{R^7} \tag{36}$$

where

$$\alpha_T = \frac{2}{3}\sum_t \frac{1}{\hbar\omega_t} \mid \boldsymbol{\mu}_{t0}^T \mid^2; \quad \alpha_S = \frac{2}{3}\sum_s \frac{1}{\hbar\omega_s} \mid \boldsymbol{\mu}_{s0}^S \mid^2 \tag{37}$$

are the isotropic static ground-state polarizabilities of the sources. If we smear the anisotropy out of the far-zone result Eq. (24) for a two-level atom by taking $\alpha_{mn} = \alpha\delta_{mn}$, we obtain

$$'\langle\{0_{\boldsymbol{k}j}\},\downarrow \mid \mathcal{H}_{\text{el}}(\boldsymbol{x}) \mid \{0_{\boldsymbol{k}j}\},\downarrow\rangle' - \frac{1}{2V}\sum_{\boldsymbol{k}j}\frac{\hbar\omega_k}{2} = \frac{23}{12\pi^2}\hbar c\alpha\frac{1}{x^7} \qquad (x > c/\omega_0). \tag{38}$$

It is thus evident that, at least for a pair of two-level atoms, $V(\boldsymbol{R})$ in Eq. (36) is proportional to the excess energy density created in the far zone by each source at the position of the other. Hence the force between the two sources is related in a simple way to the quantity we have taken as a measure of the virtual photon cloud.

ii) If source S is in the far zone of T and T is in the near zone of S, the general expression Eq. (35) can be shown to yield [40]

$$V(R) = -\alpha^T \langle \phi_0^S \mid (\boldsymbol{\mu}^S)^2 \mid \phi_0^S \rangle \frac{1}{R^6} \tag{39}$$

where $\mid \phi_0^S\rangle$ is the ground state of source S. If we take into account that $\mu_{21}^2 = \langle\downarrow \mid \mu^2 \mid \downarrow\rangle$ and furthermore if we smear the anisotropy out of the near-zone two-level atom result Eq. (24) by taking $\cos^2\theta = 1/3$, we obtain

$$'\langle\{0_{\boldsymbol{k}j}\},\downarrow \mid \mathcal{H}_{\text{el}}(\boldsymbol{x}) \mid \{0_{\boldsymbol{k}j}\},\downarrow\rangle' - \frac{1}{2V}\sum_{\boldsymbol{k}j}\frac{\hbar\omega_k}{2} = \frac{1}{4\pi}\langle\downarrow \mid \mu^2 \mid \downarrow\rangle\frac{1}{x^6} \qquad \left(x < \frac{c}{\omega_0}\right). \tag{40}$$

It is thus evident that $V(R)$ in Eq. (39), at least for a two-level atom, is proportional to the excess energy density created in the near zone by source S at the position of source T. Consequently also in the near zone the energy density of the virtual cloud can be related to a measurable interparticle force. We briefly mention that another way of detecting the shape of the virtual cloud has been suggested recently [41, 42, 43], which is based on the fact that the rate of spontaneous emission of real photons by an excited atom is influenced by the presence of the virtual cloud.

We conclude that in all cases considered (dressed nucleon, Fröhlich polaron, dressed atom) the details of the space dependence of the virtual cloud can be measured, at least in principle. This is rather reassuring, because it shows that the concept of virtual cloud is a physical one. In Part II we will extend the measurability analysis to other physical quantities which are directly or indirectly influenced by the presence of the virtual cloud.

Part II: Partially Dressed States

6 Measurements on dressed atoms and excitons

Up to this point, the time variable has not played any significant role in our treatment. This is to be expected since we have defined the dressed ground state as an eigenstate of the total Hamiltonian of the source-field system, and there is no time dependence in any of these eigenstates. The role of time, however, becomes evident as soon as we start discussing the quantum theory of measurement of a dressed source.

To see why this is the case, we must introduce briefly the main ideas of the theory of measurements of finite duration in the version given by Peres and coworkers [44]l although an earlier version was used for the measurement of the electromagnetic field [45]. According to this theory, before the measurement process starts the system to be measured is in a state $\mid \varphi\rangle$, the measuring apparatus (occasionally called pointer) is in a state $\mid \psi\rangle$, the state of the total system being in the uncorrelated state $\mid \Psi\rangle =\mid \psi\rangle \mid \varphi\rangle$. States $\mid \psi\rangle$ and $\mid \varphi\rangle$ are defined in two different Hilbert spaces pertaining to two different Hamiltonians H_M and H_R respectively. The eigenstates of H_M are denoted by $\mid \psi_k\rangle$ and those of H_R by $\mid \varphi_k\rangle$. It is assumed that one can establish a one-to-one correspondence between a given $\mid \psi_k\rangle$ and a given $\mid \varphi_k\rangle$.

The measurement process starts at $t = 0$ and is described in terms of an appropriate time-dependent Hamiltonian $H_{MR}(t)$ containing object as well as apparatus variables. Thus in this theory the measurement is a quantum mechanical process describable in terms of a unitary operator $U(t)$ such that during the period of measurement the total state $\mid \Psi(t)\rangle$ is given by

$$\mid \Psi(t)\rangle = U(t) \mid \Psi(0)\rangle. \tag{41}$$

The interaction Hamiltonian H_{MR} must be such as to yield correlations between the object being measured and the apparatus, in such a way that

$$\mid \Psi(t)\rangle = \sum_k c_k \mid \psi_k\rangle \mid \varphi_k\rangle \tag{42}$$

where c_k are c-numbers. The measurement is taken to end abruptly at time τ_m, when an observation takes place which induces a collapse of Eq. (42) into one of the possible states $\mid \psi_k\rangle \mid \varphi_k\rangle$. This observation is a nonunitary process. The main difference between the conventional theory of measurement generally presented in textbooks and the theory of measurements of finite duration is the distinction which is made in the latter between measurement and observation. Thus this theory seems to provide a reasonable generalization of the conventional quantum theory of measurement. This generalization is achieved by eliminating an objectionable feature of the latter, namely the assumption of instantaneous experiments. The preliminary concepts for such a generalization can be found in the book by von Neumann [46]; whereas the criteria for an apropriate choice of H_{MR} have been discussed more recently [47].

In preparation for the application of these ideas to the theory of dressed atoms and of polaronic excitons, we assume that the object to be measured is an isolated two-level atom at a fixed space-point, described by the Hamiltonian

$$H_R = \hbar\omega_0 S_z. \tag{43}$$

We model the pointer as a body of large mass M and momentum p, free to move in one dimension such that

$$H_M = \frac{1}{2M}p^2. \tag{44}$$

For the atom-pointer Hamiltonian we take the form

$$H_{MR} = g(t)pS_z; \qquad g(t) = \frac{1}{\tau_m}\left[\theta(t) - \theta(t-\tau_m)\right]; \qquad \int_0^\infty g(t)dt = g_0 \tag{45}$$

where $g(t)$ is a gate-function. We now proceed to show that H_{MR} is of the appropriate form to measure S_z. The latter operator is directly related to the bare atomic population, since the average values of $\frac{1}{2} \pm S_z$ yield directly the occupation numbers of either atomic or excitonic levels. On the basis of the general ideas of the theory of measurement of finite duration, one expects that in the interval $0, \tau_m$ the pointer gets correlated with the atom, and one postulates that the wavefunction collapses at $t = \tau_m$ as a result of an observation [44]. This bare two-level atom problem can be solved exactly. The total Hamiltonian is

$$H = H_M + H_R + H_{MR} = \frac{1}{2M}p^2 + \hbar\omega_0 S_z + g(t)pS_z \tag{46}$$

and the equations of motion of the dynamical variables are

$$\dot{q} = -\frac{i}{\hbar}[q, H] = \frac{1}{M}p + g(t)S_z; \qquad \dot{p} = 0; \qquad \dot{S}_z = 0 \tag{47}$$

where q is the position of the pointer. Thus p and S_z are constants of motion, and for $t \geq \tau_m$ integration of Eq. (47) gives

$$\begin{aligned} q(t) &= g_0 S_z(0) + \frac{1}{M}p(0)t + q(0); \\ q^2(t) &= \frac{1}{4}g_0^2 + 2g_0 S_z(0)\left[\frac{1}{M}p(0)t + q(0)\right] + \left[\frac{1}{M}p(0)t + q(0)\right]^2. \end{aligned} \tag{48}$$

Taking quantum averages on a state of the system in which $\mid \Psi\rangle$ is such that $\langle q(0)\rangle = \langle p(0)\rangle = 0$, and exploiting the assumption that the mass of the pointer is large, we have

$$\langle q(\tau_m)\rangle = g_0\langle S_z(0)\rangle;$$
$$\langle q^2(\tau_m)\rangle - \langle q(\tau_m)\rangle^2 = g_0^2\left(\frac{1}{4} - \langle S_z(0)\rangle^2\right). \qquad (49)$$

These two quantities are the average value and the variance of the function $P(q)$, defined as the probability distribution of the pointer positions at the observation time τ_m. Since it is in general true that $\mid \varphi\rangle = a \mid \uparrow\rangle + b \mid \downarrow\rangle$, we have

$$\langle q(\tau_m)\rangle = g_0\left(\mid a \mid^2 - \frac{1}{2}\right);$$
$$\langle q^2(\tau_m)\rangle - \langle q(\tau_m)\rangle^2 = g_0^2 \mid a \mid^2 (1- \mid a \mid^2). \qquad (50)$$

Moreover, it can be shown [48] that at time τ_m there is a perfect correlation of q and S_z, in the sense that if in a particular measurement the pointer is found at position $+\frac{1}{2}g_0$ $(-\frac{1}{2}g_0)$, the atom is certainly in the upper (lower) level. Thus in the general case with $\mid a \mid^2 \neq 1$, the expressions in Eq. (50) are only compatible with a two-peaked $P(q)$: one of the two sharp peaks, of intensity $\mid a \mid^2$, is located at $q = +g_0/2$ and the other, of intensity $1- \mid a \mid^2$, is located at $q = -g_0/2$. Evidently one can deduce the value of $\mid a \mid^2$ from the experiment, and consequently one is able to determine the populations in the upper and lower state as

$$\frac{1}{2} \pm \langle S_z(0)\rangle = \frac{1}{2} \pm \left(\mid a \mid^2 - \frac{1}{2}\right).$$

Thus the gedanken experiment modelled by Hamiltonian Eq. (46) is capable of measuring the bare atomic population distribution.

We now perform the same measurement on the atom dressed by the zero-point fluctuations of a single-mode field of frequency ω. Thus the atom-field Hamiltonian, which is derived from Eq. (5) with $l = -1$, is

$$H_R = \hbar\omega_0 S_z + \hbar\omega a^\dagger a + \epsilon a S_+ + \epsilon^* a^\dagger S_- - \epsilon a^\dagger S_+ - \epsilon^* a S_- \qquad (51)$$

and the total Hamiltonian is

$$H = \frac{1}{2M}p^2 + g(t)S_z p + H_R \qquad (52)$$

Proceeding as previously, we obtain

$$q(t) = \int_0^t S_z(t')g(t')dt' + \frac{1}{M}p(0)t + q(0) \qquad (53)$$

where S_z is not a constant of motion and cannot be taken out of the integral. The expression for $q^2(t)$ can be obtained directly by squaring Eq. (53). For $\mid \psi\rangle$ we take the same state as previously, and for $\mid \varphi\rangle$ we choose the dressed ground state of the atom, which we find by specializing Eq. (6) to the single-mode case. After some algebra, the average value and the variance of $P(q)$ at time τ_m are found to be [49]

$$\langle q(\tau_m)\rangle = g_0\langle S_z(0)\rangle;$$
$$\langle q^2(\tau_m)\rangle - \langle q(\tau_m)\rangle^2 = g_0^2\left(\frac{1}{4} - \langle S_z(0)\rangle^2\right)\frac{2\left[1-\cos(\omega_0+\omega)\tau_m\right]}{(\omega_0+\omega)^2\tau_m^2}. \tag{54}$$

This result for the fully dressed atom should be compared with Eq. (49) for a bare atom. In the limit of a "short" measurement, such that $\tau_m \ll (\omega_0+\omega)^{-1}$, the variance of $P(q)$ given by Eq. (54) coincides with the two-peaked variance given by Eq. (49) for a bare atom. Thus the measuring apparatus perceives the atom as essentially bare and the measurement is not influenced much by the coupling with the single-mode vacuum fluctuations. On the contrary, in the limit $\tau_m \gg (\omega_0+\omega)^{-1}$, the variance of the dressed atom vanishes. This indicates the presence of a single peak in $P(q)$, rather than of two peaks as in the general bare-atom case. This shows that "long" measurements, such that $\tau_m \gg (\omega_0+\omega)^{-1}$, detect a new object, namely the dressed atom. This also shows that long measurements are strongly influenced by the coupling with vacuum fluctuations. We remark that at this point time has made its appearance in the theory of dressed states, albeit in the form of the parameter τ_m which brings about a time scale, the duration of the measurement.

The single-mode model just discussed can be generalized to the many-mode case described by the Hamiltonian, stemming from Eq. (5) with $l=-1$,

$$\begin{aligned} H = {} & \frac{1}{2M}p^2 + \hbar\omega_0 S_z + g(t)S_z p + \sum_k \hbar\omega_k a_k^\dagger a_k \\ & + \sum_k\left(\epsilon_k a_k S_+ + \epsilon_k^* a_k^\dagger S_- - \epsilon_k a_k^\dagger S_+ - \epsilon_k^* a_k S_-\right) \end{aligned} \tag{55}$$

where subscript k stands for $\boldsymbol{k}j$. We shall report here only the final result, where the quantum average is taken on dressed ground state Eq. (6) [48, 50]

$$\langle q(\tau_m)\rangle = g_0\left(-\frac{1}{2} + \sum_k \frac{|\epsilon_k|^2}{\hbar^2(\omega_0+\omega_k)^2}\right);$$
$$\langle q^2(\tau_m)\rangle - \langle q(\tau_m)\rangle^2 = g_0^2\sum_k \frac{|\epsilon_k|^2}{\hbar^2(\omega_0+\omega_k)^2}\frac{2\left[1-\cos(\omega_0+\omega_k)\tau_m\right]}{(\omega_0+\omega_k)^2\tau_m^2}. \tag{56}$$

From the form of the variance one deduces that a measurement of duration τ_m detects the atom as if it were dressed only by virtual photons of high frequency such that $\omega_0+\omega_k > \tau_m^{-1}$, since these photons do not contribute to the variance of $P(q)$. The low-frequency photons, of frequency such that $\omega_0+\omega_k < \tau_m^{-1}$, are not perceived by the measuring apparatus. Thus it is as if the apparatus perceived an atom dressed by the high-frequency photons only or, equivalently, as if the virtual cloud around the atom were deficient of the low-frequency components. Since the low-frequency virtual photons are the main contributors to the energy density in the far zone, the atom is perceived as dressed by an incomplete virtual cloud, whose external part is missing. The question of the extent at which this model can be used to represent the physics of measurement on the polaronic exciton will be postponed until the end of Section 9. However the above discussion leads us quite naturally to the theory of partially dressed states.

7 The physical nature of partially dressed states

The concept of a partially dressed source was proposed in clear terms by Feinberg in 1966 [51], but the fundamental idea was probably contained in Ginzburg's earlier work [52] cited by Feinberg and summarized by Ginzburg in a more recent publication [53]. Feinberg's original QED argument can be understood even without recourse to the theory of a free electron dressed by virtual photons [54]. Indeed a dressed, but otherwise free, electron of momentum $\boldsymbol{p}$ is well known to create an electromagnetic field in space whose quantum average, as in classical electrodynamics, depends on the angle θ between $\boldsymbol{p}$ and the observation point $\boldsymbol{x}$ [55]. This leads to a dependence on θ of the energy density, although in the latter case interesting differences between the classical and the quantum case are found [54]. Thus if the direction of $\boldsymbol{p}$ changes abruptly, such as during a scattering event, the energy density after this change is different from that which would normally pertain to the new value of θ. In order to attain the new equilibrium virtual cloud, real photons must be exchanged with the field in the form of Cherenkov radiation [56] and a finite time for the rearrangement of the energy density around the electron is necessary.

It is not difficult to imagine how to make a partially dressed source in the context of atomic or molecular physics, although experiments of this kind have not yet been performed. Consider, for example, an atom in a local static electric field which can be switched on and off. The static electric field influences the atomic energy levels, and consequently the structure of the virtual cloud, via the Stark shift [57]. Thus the virtual cloud found when the static electric field is present, becomes out of equilibrium when the static electric field is turned off. Also, it is clear that when an electron is excited from the valence to the conduction band in a semiconductor, some time must elapse before the phonon virtual cloud is formed around it. During this time we are dealing with an object which should be described as a partially dressed electron, or as an incomplete polaron.

In all these examples, the obvious questions to ask are:

- How does reconstruction of the virtual cloud take place after a more or less traumatic event leading to the formation of a partially dressed source?
- Is a partially dressed source experimentally observable?
- Which are its physical properties?

If we are to answer these questions and to describe the phenomenon of cloud regeneration in quantitative terms, it is necessary to limit the scope of the discussion and to adopt a model which can be treated mathematically, since it is evident that a great variety of partially dressed states exist even for the same source-field system, corresponding to the infinitely many ways in which a virtual cloud can be distorted from its equilibrium configuration.

In order to simplify our task and to make the model amenable to mathematical treatment, we shall find it useful to define an abstract entity which we call the "bare source". Generally speaking this entity should be thought of as belonging to the same class of abstract concepts such as point mass and point charge. The latter are abstractions obtained by a conceptual limiting process of masses and charges of smaller and smaller dimensions, but they do not necessarily exist in nature, in the sense that they may never be attained in practice. Neverthless these abstractions have proved themselves extremely useful, for example, in the domain of classical physics.

Consider now a ground-state hydrogen atom at the centre of a perfect cavity. The QED of atoms in a cavity has been well-known for a long time [58, 59] and we will not dwell here on the details of the model. We shall rather assume that the cavity is spherical of radius L, to conform to the spherical symmetry of a central spinless hydrogen atom, and that it is grounded to zero potential to ensure that its external surface does not bear charges. Then the cloud of virtual photons of the atom is confined inside the cavity and its linear dimension is also L. Thus the system, with respect to a free-space dressed hydrogen atom [60], is deprived essentially of the low-frequency part of the spectrum of virtual photons with wavelength larger than L which, as we have discussed, could reach out to distances $x > L$ from the atom. We now imagine that the cavity disappears suddenly at $t = 0$. We will not be concerned with the details of the way this event takes place, provided the disappearance time is shorter than all the inverse frequency scales of the problem [61]. In view of this we will assume that disappearance of the cavity walls at $t = 0$ leaves a well-defined atomic structure, a virtual cloud which vanishes for $x \geq L$ and a spectrum of virtual photons which is missing of its equilibrium low-frequency part. This is really a partially dressed atom, but it may possibly be considered as effectively bare if viewed from a distance $x \gg L$. Such a distance in fact could be attained only by the low-frequency photons which are missing.

It should be noted that in solid state physics all the subtle problems connected with the definition of a bare source, that make life difficult in QED, do not seem to arise. This is the case for essentially two reasons. First, if we consider dressing by virtual phonons, as for the polaronic exciton, the electromagnetic force holding the source together is relatively independent of the elastic force responsible for the virtual cloud. Thus we can in principle let the exciton-photon coupling vanish without destroying the internal structure of the exciton. Second, consider, for example, an exciton created by a laser inside the semiconductor gap at $T = 0K$. Since the frequency of the laser photons is in the optical range ($\sim 10^{15} Hz$), the characteristic time for creating the exciton can be taken of the order of $10^{-15} s$. Due to the exciton-phonon interaction, the lattice will tend to readjust around the exciton. Such a readjustment, however, cannot take place in a time shorter than the inverse of the highest frequency of the lattice normal modes, which is the Debye frequency ω_D, typically of the order of $10^{13} Hz$. Consequently, although the final result of the readjustment is a polaronic exciton, this final state is attained in a time which may be orders of magnitude larger than the time necessary to create the bare exciton. Thus it is certainly legitimate to regard the exciton as bare at the beginning of this time interval, and as getting partially dressed by a virtual phonon cloud in the course of time.

8 The virtual cloud of partially dressed states

In this section we shall investigate the reconstruction of the virtual cloud starting from the vacuum for different physical models. In particular we shall assume, in agreement with the qualitative considerations of Section 7, that we deal with a source which is bare at $t = 0$, and we shall follow the dynamics of the virtual cloud which develops from the vacuum around the source in the course of time.

Polaronic exciton

Here we assume that the Hamiltonian of the system is Eq. (8). The operator representing the displacement of the nth lattice site from its equilibrium configuration, in a one-dimensional model is

$$u_n(t) = \sum_k \sqrt{\frac{\hbar}{2NM\omega_k}} e^{-ikn} \left[a_k^\dagger(t) + a_{-k}(t)\right] \tag{57}$$

where M is the mass of the site and N the total number of them. The exact solutions of the Heisenberg equations for the phonon operators with $J_{mn} = 0$ are

$$a_k(t) = a_k(0)e^{-i\omega_k t} - \left(1 - e^{-i\omega_k t}\right) \sum_m \chi_{-k}^{(m)} \left(S_z^{(m)}(0) + \frac{1}{2}\right). \tag{58}$$

For the initial state we take an exciton localized at $n = 0$ and a quiescent undistorted lattice

$$\mid \phi(0)\rangle = \mid \{0_{kj}\}, \uparrow_0\rangle. \tag{59}$$

The quantum average of $u_n(t)$ on this state gives [16]

$$\langle \phi(0) \mid u_n(t) \mid \phi(0)\rangle =$$

$$-\sqrt{\frac{2\hbar}{NM}} \sum_k \frac{1}{\sqrt{\omega_k}} e^{-ikn} \chi_{-k}^{(0)} \langle \phi(0) \mid \left(S_z^{(0)} + \frac{1}{2}\right) \mid \phi(0)\rangle \left(1 - \cos\omega_k t\right). \tag{60}$$

This can be shown to be equivalent to a static distortion around the exciton at $n = 0$, plus two elastic pulses moving away from the exciton site with opposite velocity $\pm c$, where c is the velocity of sound. Thus switching the source on at $t = 0$ in this one-dimensional example yields a dressed localized polaronic exciton, in addition to real elastic pulses moving away from the location of the dressed source.

Static nucleon

The energy density of a real Klein-Gordon field is [62]

$$\mathcal{H}(\boldsymbol{x}) = \frac{1}{2}\left(\dot{\phi}^2 + c^2 (\nabla\phi)^2 + \kappa^2 c^2 \phi^2\right) \tag{61}$$

where $\phi(x)$ is the field amplitude defined in Eq. (18). In the presence of a static nucleon, the Hamiltonian is given by Eq. (11) and Eq. (12). The quantum average of the energy density on a state which is initially devoid of mesons is [38]

$$\langle\{0_{\boldsymbol{k}}\} \mid \mathcal{H}(\boldsymbol{x},t) \mid \{0_{\boldsymbol{k}}\}\rangle = \frac{1}{32\pi^2 c^2} g^2 \frac{1}{x^2} \left\{ \left(\frac{\partial \mathcal{F}}{\partial t}\right)^2 + c^2 \left(\frac{\partial}{\partial x}\frac{1}{x}\frac{\partial}{\partial x}\int_0^t \mathcal{F} dt'\right)^2 + \kappa^2 c^2 \left(\frac{\partial}{\partial x}\int_0^t \mathcal{F} dt'\right)^2 \right\}. \tag{62}$$

Since

$$\mathcal{F} = \mathcal{F}(\boldsymbol{x},t) = J_0\left(\kappa c\sqrt{t^2 - x^2/c^2}\right)\theta(t - x/c) \tag{63}$$

where J_0 is the zero-order Bessel function, it is evident that the quantum average of the energy density vanishes outside a sphere of radius ct centered on the source, which is sometimes called the causality sphere. As $t \to \infty$ expression Eq. (62) tends to

$$\langle\{0_{\boldsymbol{k}}\} \mid \mathcal{H}(\boldsymbol{x}) \mid \{0_{\boldsymbol{k}}\}\rangle = \frac{1}{32\pi^2 c^2} g^2 \frac{1}{x^4}\left(2\kappa^2 x^2 + 2\kappa x + 1\right) e^{-2\kappa x} \tag{64}$$

which coincides with its stationary dressed ground state value, obtainable by the methods mentioned in Section 4. Thus reconstruction of the virtual cloud take place causally. The disappearance at $t = 0$ of a fully dressed static nucleon has also been investigated [63]. In this case the quantum average of the energy density outside the causality sphere is unchanged, whereas it decays to zero in quite a short time for $x < ct$.

Two-level atom

For a two-level atom coupled to the quantum fluctuations of the electromagnetic field, we adopt Hamiltonian Eq. (5). As discussed in Section 4, we describe the virtual cloud by means of the electric energy density of the field $\mathcal{H}_{\mathrm{el}}(\boldsymbol{x},t)$, given by expression Eq. (23). Using the multipolar scheme [64, 65] amounts to choosing $l = -1$ and [13]

$$\epsilon_{\boldsymbol{k}j} = -i\sqrt{\frac{2\pi\hbar\omega_k}{V}}\boldsymbol{\mu}_{21}\cdot\hat{\boldsymbol{b}}_{\boldsymbol{k}j} \tag{65}$$

where the polarization vector $\hat{\boldsymbol{b}}_{\boldsymbol{k}j}$ is taken as real for semplicity. Then the total electric field outside the atom coincides with the transverse displacement field [31]

$$D_{\perp}(\boldsymbol{x}) = i\sum_{\boldsymbol{k}j}\sqrt{\frac{2\pi\hbar\omega_k}{V}}\hat{e}_{\boldsymbol{k}j}\left(a_{\boldsymbol{k}j}e^{i\boldsymbol{k}\cdot\boldsymbol{r}} - a^{\dagger}_{\boldsymbol{k}j}e^{-i\boldsymbol{k}\cdot\boldsymbol{r}}\right). \tag{66}$$

Thus we need explicit expressions for $a_{\boldsymbol{k}j}(t)$ and $a^{\dagger}_{\boldsymbol{k}j}(t)$. The coupled Heisenberg equations for field and atom operators can be obtained from Eq. (5) as

$$\dot{a}_{\boldsymbol{k}j} = -i\omega_k a_{\boldsymbol{k}j} - \frac{i}{\hbar}\epsilon^*_{\boldsymbol{k}j} S_- - \frac{i}{\hbar} l \epsilon_{\boldsymbol{k}j} S_+$$

$$\dot{S}_+ = i\omega_0 S_+ - 2\frac{i}{\hbar}\sum_{\boldsymbol{k}j}\epsilon^*_{\boldsymbol{k}j} a^{\dagger}_{\boldsymbol{k}j} S_z - 2\frac{i}{\hbar} l \sum_{\boldsymbol{k}j}\epsilon^*_{\boldsymbol{k}j} a_{\boldsymbol{k}j} S_z . \tag{67}$$

These equations are solved keeping terms up to order of ϵ^2 and evaluating the quantum averages of all operators quadratic in $a(t)$ and $a^{\dagger}(t)$ on the initial state $| \{0_{\boldsymbol{k}j}\}, \downarrow\rangle$ (i.e. the bare ground state with the atom unexcited and no photons). The results are then combined to obtain $\langle\{0_{\boldsymbol{k}j}\}, \downarrow | \mathcal{H}_{\rm el}(\boldsymbol{x}, t) | \{0_{\boldsymbol{k}j}\}, \downarrow\rangle$. This procedure yields [66, 67]

$$\langle\{0_{\boldsymbol{k}j}\}, \downarrow | \mathcal{H}_{\rm el}(\boldsymbol{x}, t) | \{0_{\boldsymbol{k}j}\}, \downarrow\rangle = \frac{1}{2V}\sum_{\boldsymbol{k}j}\frac{1}{2}\hbar\omega_k - \frac{l^2}{V}\sum_{\boldsymbol{k}j} f_{mn}(\boldsymbol{k}j)\frac{k}{k_0 + k}$$

$$\mathrm{Re}\left(e^{i\boldsymbol{k}\cdot\boldsymbol{x}} D^x_{mn}\left\{\left[e^{-ikx} - e^{i(k_0+k)ct} e^{-ik_0 x}\right]\theta(ct - x)\right\}\right) \tag{68}$$

where $k_0 = \omega_0/c$, $f_{mn}(\boldsymbol{k}j) = (\hat{\boldsymbol{b}}_{\boldsymbol{k}j})_\ell(\hat{\boldsymbol{b}}_{\boldsymbol{k}j})_m(\boldsymbol{\mu}_{21})_\ell(\boldsymbol{\mu}_{21})_n$ and where we have introduced the differential operator

$$D^x_{mn} = \frac{1}{x}\left\{(\delta_{mn} - \hat{x}_m\hat{x}_n)\frac{\partial^2}{\partial x^2} + (\delta_{mn} - 3\hat{x}_m\hat{x}_n)\left(\frac{1}{x^2} - \frac{1}{x}\frac{\partial}{\partial x}\right)\right\} . \tag{69}$$

The first contribution on the RHS of Eq. (68) is the usual infinite, space- and time-independent zero-point electric energy density of the field, which is of no interest here. The l^2 term represents the virtual cloud which at $t = 0$ vanishes everywhere, as expected of an initially bare atom. The virtual cloud develops causally, as indicated by the presence of the θ-function. In fact it vanishes outside the causality sphere of radius ct. Furthermore, the sums in the l^2 term can be evaluated for $t \to \infty$ both in the near ($k_o x \ll 1$) and in the far zone ($k_0 x \gg 1$), with results in agreement with the static case Eq. (24).

We conclude that in all the cases considered in this section (exciton, static nucleon, two-level atom), the reconstruction of the virtual cloud, starting from an initially bare source, takes place within the causality sphere of radius ct centred on the source. The virtual cloud inside this sphere tends more or less rapidly to assume the shape of the ground-state virtual cloud discussed in Section 4.

9 Observable effects related to the reconstruction of the virtual cloud

In this section we take up the question whether it is possible to detect experimentally the reconstruction of the virtual cloud.

Static nucleon

We consider two localized static meson sources S and T, with S fixed at the origin and T mobile at point R. In our scheme T plays the role of a field detector. The idea here is that S is bare and T fully dressed at $t = 0$, and that we wish to investigate if the motion of T at $t > 0$ is influenced by the reconstruction of the virtual cloud around S. The basic Hamiltonian is the same as that used in section 5. We shall not give here the details of the calculations [68] and we only mention that the quantum average of the force acting on T is given by

$$\langle F(t)\rangle = -\frac{1}{4\pi c^2} g^2 \frac{\partial}{\partial R}\frac{1}{R}\frac{\partial}{\partial R}\int_0^t \mathcal{F}(R,t')dt' \tag{70}$$

where $\mathcal{F}$ is the same as in Eq. (63). Thus, in view of the θ-function in $\mathcal{F}$, although the initially bare source S starts the self-dressing process immediately after $t = 0$, the force Eq. (70) vanishes until $t = R/c$. After this time $\langle F(t)\rangle$ has a rather complicated behaviour, but it is possible to show that asymptotically it tends to

$$\langle F(\infty)\rangle = \frac{\partial}{\partial R}\left(\frac{1}{4\pi c^2} g^2 \frac{1}{R} e^{-\kappa R}\right) \tag{71}$$

in agreement with Eq. (30). Thus the motion of T changes upon arrival of the front of the virtual cloud of S at $t = R/c$, and by virtue of this fact T can be used as a detector of the recostruction of the virtual cloud of S.

Two-level atom

Here S and T are two two-level atoms. The geometry of the system as well as the role of S and T in the gedanken experiment are the same as in the preceeding example. We assume, however, that each atom is in the far zone of the other. In this case the coupling of each source i localized at $\boldsymbol{x}_i$ with the field can be described by the Craig-Power Hamiltonian [69]

$$H_{int}^{(i)} = -\frac{1}{2}\alpha_{mn}^{i} D_{\perp m}(\boldsymbol{x}_i) D_{\perp n}(\boldsymbol{x}_i) \tag{72}$$

where $D_\perp$ is the transverse part of the electric displacement field, as usual. This form of the interaction can be obtained from Eq. (5) by a unitary transformation which eliminates the atom-field interaction at first order in the electron charge $-e$, and α_{mn} is given by Eq. (25). Only the low-frequency modes of the field are adequately taken into account by Eq. (72) [7], but this is sufficient for our purposes, given the far-zone assumption. As previously, we take S as bare and T as fully dressed at $t = 0$. The quantum average of the force acting on T at time t turns out to be [68, 70]

$$\langle F(t)\rangle = \frac{1}{4\pi}\hbar c \alpha^S \alpha^T \left\{-\frac{161}{R^8}[1-\theta(R-ct)] + \mathrm{SC}\right\}. \tag{73}$$

In this expression α^i are spherical averages of α_{mn}^i, taken by using $\alpha_{mn}^i = \alpha^i \delta_{mn}$, and SC stands for a complicate singular contribution which can be expressed, for a point-like source, in terms of $\delta(R - ct)$ and of its derivatives up to the fifth. This shows that the force on T vanishes up to $t = R/c$. After the front of the virtual cloud of S reaches T, the nonvanishing force Eq. (73) sets T in motion. This motion can thus be used to detect the reconstruction of the virtual cloud of S. Moreover as $t \to \infty$, Eq. (73) tends to the usual van der Waals force, which can be obtained from Eq. (36) by straightforward differentiation. It should be noted, however, that in any realistic experimental setup that can be implemented with present technology, the effects of the force on the test body T are likely to be undetectably small [13, 38]. Thus we shall discuss another possible, albeit less direct, way of observing the reconstruction of the photon virtual cloud. We consider the atomic dynamics during the self-dressing process. Thus the starting point is the Hamiltonian Eq. (5) and the initial state of the system is the bare state $\mid \{0_{kj}\}, \downarrow\rangle$ with the atom in its bare ground state and no photons in the field. From the Heisenberg equations, at second order in e we get [38]

$$\langle \{0_{kj}\}, \downarrow \mid S_z(t) \mid \{0_{kj}\}, \downarrow\rangle = -\frac{1}{2} + 2\frac{l^2}{\hbar^2} \sum_{kj} \mid \epsilon_{kj} \mid^2 \frac{1 - \cos(\omega_0 + \omega_k)t}{(\omega_0 + \omega_k)^2}. \tag{74}$$

We evaluate thus this in the minimal coupling scheme [12, 39, 64] with $l = 1$ and [7]

$$\epsilon_{kj} = -i\sqrt{\frac{2\pi\hbar\omega_0{}^2}{V\omega_k}}\boldsymbol{\mu}_{21} \cdot \hat{\boldsymbol{e}}_{kj}. \tag{75}$$

In this scheme Eq. (74) yields, for $\omega_0 t > 1$,

$$\langle \{0_{kj}\}, \downarrow \mid S_z(t) \mid \{0_{kj}\}, \downarrow\rangle = -\frac{1}{2} + l^2 \frac{\gamma}{\pi\omega_0} \left(\ln \frac{\omega_M}{\omega_0} - 1 - \frac{\cos \omega_0 t}{(\omega_0 t)^2} \right). \tag{76}$$

In this expression ω_M is a high-frequency cutoff, and terms oscillating at frequency ω_M have been neglected. These terms, in fact, are considered to be an artifact of the model and indeed disappear when the electric dipole approximation is not performed and the cutoff at ω_M is not introduced [71]. Furthermore γ is the usual spontaneous emission rate [72]

$$\gamma = \frac{4 \mid \mu_{21} \mid^2 \omega_0{}^3}{3\hbar c^3}. \tag{77}$$

For $t \to \infty$, the l^2 term in Eq. (76) is proportional to the Lamb shift [73]. We draw attention, however, to the term which oscillates at frequency ω_0. This term is similar to those obtained in the theory of spontaneous decay [74] and is related to the reconstruction of the virtual cloud. Moreover, as we have seen in Section 6, S_z is simply related to the population distribution among atomic levels. Thus, if one could detect these population oscillations, one could diagnose the presence of self-dressing.

Polaronic exciton

Direct detection of the reconstruction of the cloud of virtual phonons starting from an initially bare exciton has not been discussed so far, to the best of our knowledge. However the processes leading to the formation of a polaronic exciton in a semiconductor have been studied both theoretically [75] and experimentally [76]. The model of Section 2, as well as its more recent versions [77], are not capable of describing hybridization of different bare excitonic states due to the exciton-phonon interaction. Indeed we see by inspection that, for a localized exciton with $J_{mn} = 0$, the Hamiltonian Eq. (8) yields $S_z^{(m)} = const$ and consequently cannot give rise to internal dynamics for the exciton of the same sort as that leading to Eq. (74) for an atom coupled to the electromagnetic field. Thus on the basis of Hamiltonian Eq. (8) one would not expect that measurements of the population distribution of a partially dressed exciton bear traces of the reconstruction of the virtual phonon cloud. Neverthless hybridization of internal polaronic exciton states has recently been shown experimentally to exist [78] and should also be expected on theoretical grounds [79]. Thus we are led to suggest that a more realistic exciton-phonon Hamiltonian should include terms yielding transitions between different internal excitonic states, accompanied by absorption and emission of phonons. We now know that these terms, present in the atom-photon Hamiltonian Eq. (5), should lead to an internal dynamics of the excitonic population similar to that obtained in Eq. (76) for a two-level atom. This opens up the perspective that, by following the internal dynamics of the exciton, one might obtain information on the reconstruction of the virtual phonon cloud. Although no theoretical or experimental result is yet available, the possibility of ultrafast measurements of exciton population on the femtosecond scale seems already within experimental reach [80]. Moreover it has been shown [48] that it is possible in principle to measure the population distribution of a two-level system during the reconstruction of its virtual cloud and, as discussed at the end of Section 7, the reconstruction of the virtual cloud of an initially bare exciton takes a sizeable time and it should thus be amenable to experimental observation. Thus observations of the reconstruction of the virtual photon cloud in polaronic excitons seems within reach of solid state experimentalists. Naturally some uncertainty remains, insofar as it is not clear whether the atom-photon analogy is valid for the excitonic polaron. We wish to point out, however, the importance of measurements of this kind, which would constitute a valuable guidance since they would simulate experiments on self-dressing in quantum field theory for which direct or indirect measurements are hindered by technical difficulties as yet unsurmounted.

10 Summary and conclusions

In this set of lectures we have exploited the important cultural tradition of cross-fertilization between solid state physics and quantum field theory. The main idea common to both branches of physics which we have discussed is that of a source interacting with a radiation field. In solid state physics we have assumed that the source is implemented as an elementary excitation such as the polaron or the polaronic exciton. In Part I we have shown that the radiation-source interaction leads to a ground state of the source-field system in which the source is surrounded by a cloud of virtual quanta of the field which can be described as continuously emitted and reabsorbed by the source. This gives rise to a complex object which we have called the dressed source. We have shown that the virtual cloud around the source gives rise to observable effects which depend on its detailed structure. We have also shown that this is true for a variety of different sources in solid state physics, in hadronic physics and in QED.

The quantum theory of measurement summarized at the beginning of Part II has been the vehicle leading from fully dressed to partially dressed sources. We have shown that for measurements of appropriately short duration, even a fully dressed source can be detected as dressed by an incomplete virtual cloud, and we have defined the quantum mechanical states corresponding to this situation as partially dressed states. The quantitative analysis of the properties and the time-dependence of these states have led us to suggest that also partially dressed sources should be experimentally observable. At this point the importance of solid state physics becomes evident. In fact observation of the controlled reconstruction of the virtual cloud in hadronic physics or in QED seems unattainable by the presently available experimental techniques, whereas the same phenomenon is probably already within reach of solid state experimentalists. We hope that this paper will stimulate the necessary theoretical as well as experimental steps necessary for a complete investigation of partially dressed states in solid state physics. We emphasize that if such a programme were successful, it would constitute an important contribution to the study of fundamental processes in other branches of physics, as well as to the cultural cross-fertilization which is at the basis of our present efforts.

Acknowledgments

We are pleased to acknowledge partial financial support by the Comitato Regionale per le Ricerche nucleari e di Struttura della Materia and by the Ministero dell'Universita' e della Ricerca Scientifica e Tecnologica. We are also indebted to F. Bassani for interesting conversations on the subject of this paper.

References

1. S.S. Schweber, Physics Today, Nov. 1993.
2. P.W. Anderson, *Elementary Excitations*, Benjamin 1963.
3. S. Weinberg, *The Quantum Theory of Fields*, Cambridge University Press 1995.
4. C.G. Kuper, G.D. Whitfield eds., *Polarons and Excitons*, Oliver & Boyd 1963.
5. F. Bassani, G. Iadonisi, in *Quantum Field Theory*, F. Mancini ed., North-Holland 1986.
6. E.M. Henley, W. Thirring, *Elementary Quantum Field Theory*, McGraw-Hill 1962.
7. G. Compagno, R. Passante, F. Persico, *Atom-Field Interactions and Dressed Atoms*, Cambridge University Press 1995.
8. see e.g. N.W. Ashcroft, N.D. Mermin, *Solid State Physics*, Holt, Reinhart and Winston 1976.
9. H. Fröhlich, in ref. [4], p. 1.
10. D. Pines, in ref. [4], p. 63.
11. T.D. Lee, F. Low, D. Pines, Phys. Rev. **90**, 297 (1953).

12. E.A. Power, *Introductory Quantum Electrodynamics*, Longmans, Green and Co. 1964.
13. G. Compagno, G.M. Palma, R. Passante, F. Persico, J. Phys. B **28**, 1105 (1995).
14. H. Haken, *Quantum Field Theory of Solids*, North-Holland 1988.
15. J. Pollmann, H. Büttner, Phys. Rev. B **16**, 4480 (1977).
16. D.W. Brown, K. Lindenberg, B.J. West, J. Chem. Phys. **84**, 1574 (1986).
17. G. Iadonisi, F. Bassani, Nuovo Cimento D **9**, 703 (1987).
18. M. Bolsterli. J. Math. Phys. **32**, 254 (1991).
19. see e.g. F. Gross, *Relativistic Quantum Mechanics and Field Theory*, John Wiley & Sons 1993.
20. F. Persico, G. Compagno, R. Passante, in *Quantum Optics IV*, J.D. Harvey and D.F. Walls eds., Springer-Verlag 1986, p. 172.
21. L. Van Hove, Physica **21**, 901 (1955); **22**, 343 (1956).
22. See e.g. I.R. Kenyon, *Elementary Particle Physics*, Routledge and Kegan Paul 1987.
23. E. Kartheuser, in *Polarons in Ionic Crystals and Polar Semiconductors*, J.T. Devreese ed., North-Holland 1972, p. 715.
24. M. Menšik, J. Phys. Condens. Matter **7**, 7349 (1995).
25. C. Cohen-Tannoudji, J. Dupont-Roc, G. Grynberg, *Atom-Photon Interactions*, John Wiley & Sons 1992.
26. S.S. Schweber, *An introduction to Relativistic Quantum Field Theory*, Weatherhill 1964.
27. G.E. Brown, A.D. Jackson, *The Nucleon-Nucleon Interaction*, North-Holland 1976.
28. S. Theberge, A.W. Thomas, G.A. Miller, Phys. Rev. D **22**, 2838 (1980); **24**, 216 (1981).
29. F.M. Peeters, J.D. Devreese, Phys. Rev. B **31**, 1985 (1985).
30. E.A. Power, T. Thirunamachandran, in *Coherence and Quantum Optics V*, L. Mandel and E. Wolf eds., Plenum Press 1984, p. 549.
31. E.A. Power, T. Thirunamachandran, Phys. Rev. A **28**, 2663 (1983).
32. J.D. Jackson, *Classical Electrodynamics*, John Wiley & Sons 1962.
33. R. Passante, G. Compagno, F. Persico, Phys. Rev. A **31**, 2827 (1985).
34. see e.g. B.H. Bransden, C.J. Joachain, *Physics of Atoms and Molecules*, Longman 1986.
35. G. Iadonisi, Riv. Nuovo Cim. **7**,1 (1984).
36. W. Heitler, *The Quantum Theory of Radiation*, Oxford University Press 1960.
37. C. Kittel, *Quantum Theory of Solids*, John Wiley & Sons 1963.
38. G. Compagno, G.M. Palma, R. Passante, F. Persico, in *Quantum Electrodynamics and Quantum Optics*, A.O. Barut ed., Plenum 1990, p. 129.
39. D.P. Craig, T. Thirunamachandran, *Molecular Quantum Electrodynamics*, Academic Press 1984.
40. G. Compagno, R. Passante, F. Persico, Phys. Lett. A **121**, 19 (1987).
41. H. Khosravi, R. Loudon, Proc. R. Soc. London A **433**, 337 (1991).
42. F. De Martini, Phys. Scr. T **21**, 58 (1988).
43. F. De Martini, M. Marrocco, P. Mataloni, P. Murra, R. Loudon, J. Opt. Soc. Am. B **10**, 360 (1993).
44. A. Peres, W.K. Wootters, Phys. Rev. D **32**, 1968 (1985)
    A. Peres, Phys. Rev. D **39**, 2943 (1989)
    A. Peres, *Quantum Theory: Concepts and Methods*, Kluwer 1993.

45. N. Bohr, L. Rosenfeld, Mat.-fys. Medd. Dan. Vid. Selsk. **12**, no. 8 (1933), translated and reprinted in *Quantum Theory and Measurement*, J.A. Wheeler and W.H. Zurek eds., Princeton University Press 1983, p. 479.
46. J. Von Neumann, *Mathematical Foundations of Quantum Mechanics*, Princeton University Press 1955 .
47. M. Grabowski, Ann. der Phys. **47**, 391 (1990).
48. G. Compagno, R. Passante, F. Persico, Phys. Rev. A **44**, 1956 (1991).
49. G. Compagno, R. Passante, F. Persico, Europhys. Lett. **12**, 301 (1990).
50. G. Compagno, R. Passante, F. Persico, Acta Phys. Slov. **45**, 231 (1995).
51. E.L. Feinberg, Zh. Eks. Teor. Fiz. **50**, 202 (1966) [Sov. Phys. JETP **23**, 132 (1966)].
52. V.L. Ginzburg, Dokl. Akad. Nauk. **23**, 131, 775 (1939).
53. V.L. Ginzburg, *Applications of Electrodynamics in Theoretical Physics and Astrophysics*, Gordon & Breach 1989.
54. G. Compagno, G.M. Salamone, Phys. Rev. A **44**, 5390 (1991).
55. see e.g. R. Becker, *Electromagnetic Fields and Interactions*, Dover 1982.
56. E.L. Feinberg, Usp. Fiz. Nauk. **132**, 225 (1980) [Sov. Phys. Usp. **23**, 629 (1980)].
57. G. Compagno, G.M. Palma, Phys. Rev.A **37**, 2979 (1988).
58. E.A. Power, T. Thirunamachandran, Phys. Rev. A **25**, 2473 (1982).
59. P. Dobiasch, H. Walther, Ann. Phys. **10**, 825 (1987).
60. R. Passante, E.A. Power, Phys. Rev. A **35**, 188 (1987).
61. Yu.E. Lozovik, V.G. Tvetus, E.A. Vinogradov, Phys. Scripta **52**, 184 (1995).
62. C. Itzykson, J.B. Zuber, *Quantum Field Theory*, McGraw-Hill 1985.
63. F. Persico, E.A. Power, Phys. Rev. A **36**, 475 (1987).
64. E.A. Power, S. Zienau, Phil. Trans. R. Soc. London A **251**, 427 (1959).
65. E.A. Power, T. Thirunamachandran, Phys. Rev. A **28**, 2649 (1983).
66. G. Compagno, G.M. Palma, R. Passante, F. Persico, Europhys. Lett. **9**, 215 (1989).
67. G. Compagno, G.M. Palma, R. Passante, F. Persico, in *Coherence and Quantum Optics VI*, J. Eberly, L. Mandel and E. Wolf eds., Plenum 1990, p.191.
68. G. Compagno, R. Passante, F. Persico, Phys. Rev. A **38**, 600 (1988).
69. D.P. Craig, E.A. Power, Int. J. Quantum Chem. **3**, 903 (1969).
70. G. Compagno, R. Passante, F. Persico, in *Vacuum in Non-Relativistic Matter-Radiation Systems*, F. Persico and E.A. Power eds. (Physica Scripta T**21** 1988), p. 40.
71. L. Davidovich, PhD Thesis, Rochester 1975.
72. See e.g. R. Loudon, *The Quantum Theory of Light*, Oxford University Press 1981.
73. H.A. Bethe, Phys. Rev. **72**, 339 (1947).
74. K. Wodkiewicz, J.H. Eberly, Ann. Phys. **101**, 574 (1976)
P.L. Knight, P.W. Milonni, Phys. Lett. A **56**, 275 (1976)
G. Compagno, R. Passante, F. Persico, J. Mod. Opt. **37**, 1377 (1990).
75. I.G. Lang, S.T. Pavlov, A.V. Prokhorov, Sov. Phys. Sol. State **32**, 528 (1990).
76. G.S. Kanner, X. Wei, B.C. Hess, L.R. Chen, Z.V. Vardeny, Phys. Rev. Lett. **69**, 538 (1992).
77. J. Singh, Chem. Phys. **198**, 63 (1995).
78. T. Itoh, M. Nishijima, A.I. Ekimov, C. Gourdon, Al.L. Efros, M. Rosen, Phys. Rev. Lett. **74**, 1645 (1995).
79. F. Bassani, private communication.
80. A.P. Heberle, J.J. Baumberg, K. Köhler, Phys. Rev. Lett. **75**, 2598 (1995).

Density Functional Theory and Density Matrices

A. Holas

Institute of Physical Chemistry of the Polish Academy of Sciences
44/52 Kasprzaka, 01-224 Warsaw, Poland

Abstract: Recent investigations of the exchange-correlation potential of the Kohn-Sham (KS) scheme, making use of three equations satisfied by density matrices, are summarized and systematized. They lead to three exact expressions for the potential in terms of low-order density matrices of the interacting system and the KS system, and three approximations for the exchange-only potential in terms of the KS matrices. The application of the perturbation theory of Görling and Levy permits the formulation of a computational scheme in which the exact exchange potential and consecutive terms of the expanded correlation potential can be obtained within an extended KS approach.

1 Introduction

Density functional theory (DFT) is known to be a powerful computational method applicable in several branches of physics and chemistry: solids, liquids, surfaces, molecules, atoms, nuclei etc. Its basic aim is the determination of the ground-state energy and density of many-particle systems, but extensions to excited states, time-dependent phenomena, magnetic and superconducting states have appeared. In a number of recent monographs on DFT [1]–[8], the reader can find both introductory material and extended elaborations of the various aspects of DFT. A practical computational scheme offered by DFT, the Kohn-Sham equations, requires the knowledge of the exchange-correlation energy E_{xc} and potential $v_{\mathrm{xc}}(\boldsymbol{r})$ as functionals of the particle density $n(\boldsymbol{r})$. Various approximate forms have been developed for them, starting from the local density approximation and ending with generalized gradient approximations. Since these approximations happen to be the main sources of errors in DFT calculations, the improvement of existing or construction of new approximations is a prerequisite for further progress of the theory. The present paper summarizes some recent developments in this direction, connected with the essential use of density matrices.

2 Many-electron system

2.1 Hamiltonian

Our object of interest is a finite many-electron system, such as an atom, a molecule or a cluster, having, for simplicity, a nondegenerate ground state (GS). Its N electrons move in a common external potential $v(\boldsymbol{r})$ (e.g. due to nuclei in fixed positions) and are mutually interacting with a pair potential $u(\boldsymbol{r}_1, \boldsymbol{r}_2)$ (e.g. Coulomb $1/|\boldsymbol{r}_1 - \boldsymbol{r}_2|$). So the Hamiltonian of the system is given by

$$\hat{\mathcal{H}} = \hat{\mathcal{T}} + \hat{\mathcal{U}} + \hat{\mathcal{V}}, \tag{1}$$

with the kinetic energy operator

$$\hat{\mathcal{T}} = \sum_{i=1}^{N} \hat{t}(\boldsymbol{r}_i); \quad \hat{t}(\boldsymbol{r}_i) = -\tfrac{1}{2}\nabla_i^2, \tag{2}$$

the electron-electron repulsion operator

$$\hat{\mathcal{U}} = \sum_{i=1}^{N-1} \sum_{j=i}^{N} u(\boldsymbol{r}_i, \boldsymbol{r}_j); \quad u(\boldsymbol{r}_j, \boldsymbol{r}_i) = u(\boldsymbol{r}_i, \boldsymbol{r}_j), \tag{3}$$

and the electron-nuclei attraction (external energy) operator

$$\hat{\mathcal{V}} = \sum_{i=1}^{N} v(\boldsymbol{r}_i). \tag{4}$$

It will be convenient also to have a notation for a bare one-electron Hamiltonian

$$\hat{h}(\boldsymbol{r}_i) = \hat{t}(\boldsymbol{r}_i) + v(\boldsymbol{r}_i). \tag{5}$$

Atomic units are used throughout.

2.2 Density matrices

Let $\Psi(\boldsymbol{x}_1, \boldsymbol{x}_2, \ldots, \boldsymbol{x}_N)$ be a normalized, antisymmetric wave function. The notation $\boldsymbol{x}_i = (\boldsymbol{r}_i, s_i)$ for ith space and spin coordinate is adopted, abbreviated further to $\mathbf{12}\ldots$ for $\boldsymbol{x}_1, \boldsymbol{x}_2, \ldots$. The Nth order density matrix (N-DM) generated by Ψ is just the product

$$\gamma_N(\mathbf{12}\ldots\boldsymbol{N};\mathbf{1}'\mathbf{2}'\ldots\boldsymbol{N}') = \Psi(\mathbf{12}\ldots\boldsymbol{N})\,\Psi^*(\mathbf{1}'\mathbf{2}'\ldots\boldsymbol{N}), \tag{6}$$

while the pth order reduced DM (p-DM), for $p < N$, is obtained from γ_N by integrating $(N-p)$ coordinates, i.e.

$$\gamma_p(\mathbf{12}\ldots\boldsymbol{p};\mathbf{1}'\mathbf{2}'\ldots\boldsymbol{p}') = \binom{N}{p} \int d(\boldsymbol{p}+\mathbf{1})\ldots d\boldsymbol{N}\,\gamma_N. \tag{7}$$

Here the abbreviation $\int di$ to $\int d^4x_i$ means integration $\int d^3r_i$ and summation over s_i together with the replacement of x_i' by x_i in the integrand. For many applications, spinless DMs are sufficient, namely

$$\rho_p(12\ldots p;1'2'\ldots p') = \sum_{s_1,\ldots,s_p} \gamma_p(\mathbf{12}\ldots\mathbf{p};\mathbf{1'2'}\ldots\mathbf{p'})|_{s_i'=s_i} . \tag{8}$$

The abbreviation $12\ldots$ means $\boldsymbol{r}_1, \boldsymbol{r}_2, \ldots$. The diagonal elements of the DMs are denoted as

$$\xi_p(\mathbf{12}\ldots\mathbf{p}) = \gamma_p(\mathbf{12}\ldots\mathbf{p};\mathbf{12}\ldots\mathbf{p}) , \tag{9}$$

$$n_p(12\ldots p) = \rho_p(12\ldots p;12\ldots p) . \tag{10}$$

From the definition, Eq. (7), the reduction property follows:

$$\int d\boldsymbol{x}_p\, \gamma_p = \frac{N+1-p}{p}\gamma_{p-1} . \tag{11}$$

The subscript '1' will be often omitted. The basic quantity of DFT — the electron number density — is thus

$$n(\boldsymbol{r}) = \rho(\boldsymbol{r};\boldsymbol{r}) = \rho_1(\boldsymbol{r};\boldsymbol{r}) . \tag{12}$$

3 Equations satisfied by low-order density matrices

3.1 Energy equation for the 1-DM

We are going to establish the equations satisfied by the DMs generated from any eigenfunction Ψ of the system Hamiltonian $\widehat{\mathcal{H}}$, Eqs. (1)–(4):

$$\widehat{\mathcal{H}}\Psi(\mathbf{1}\ldots\boldsymbol{N}) = E\Psi(\mathbf{1}\ldots\boldsymbol{N}) . \tag{13}$$

After multiplying Eq. (13) by $\Psi^*(\mathbf{1}'\ldots\boldsymbol{N}')$ and integrating over all coordinates except $\mathbf{1}$, the following equation

$$\begin{aligned} E\,\gamma_1(\mathbf{1};\mathbf{1}') = \hat{h}(1)\,\gamma_1(\mathbf{1};\mathbf{1}') + 2\int d\mathbf{2}\,\{\hat{h}(2) + u(12)\}\,\gamma_2(\mathbf{12};\mathbf{1'2'}) \\ + 3\int d\mathbf{2}\,d\mathbf{3}\,u(23)\,\gamma_3(\mathbf{123};\mathbf{1'23}) , \end{aligned} \tag{14}$$

is obtained, which is known as the energy equation for the 1-DM (see [9] or [10]). This equation provides the basis for our further investigations.

To find the system energy in terms of the DM, we integrate Eq. (14) over $d\mathbf{1}$ and then use the reduction property Eq. (11) to obtain

$$EN = \left(1 + 2\,\frac{N-1}{2}\right)\int d\mathbf{1}\,\hat{h}(1)\,\gamma_1(\mathbf{1};\mathbf{1}') + \left(2 + 3\,\frac{N-2}{3}\right)\int d\mathbf{1}\,d\mathbf{2}\,u(12)\,\xi_2(\mathbf{12}) . \tag{15}$$

This equation, rewritten in terms of spinless DMs, gives the well known expression for the total energy as a sum of the kinetic, external potential and interaction energy, viz.

$$E = \int d1\,\hat{t}(1)\,\rho_1(1;1')_{|1'=1} + \int d1\,v(1)\,n(1) + \int d1\,d2\,u(12)\,n_2(12)\,. \qquad (16)$$

3.2 Integro-differential equation for the 1-DM

One eliminates the energy E from Eq. (14) by substituting in its place Eq. (15) divided by N (see Ref. [11]) and finds

$$\begin{aligned} 0 &= \{\hat{t}(1) + v(1)\}\,\gamma_1(\mathbf{1};\mathbf{1}') \\ &+ 2\int d\mathbf{2}\,\{[\hat{t}(2) + v(2)][\gamma_2(\mathbf{12};\mathbf{1}'\mathbf{2}') - \tfrac{1}{2}\gamma_1(\mathbf{1};\mathbf{1}')\,\gamma_1(\mathbf{2};\mathbf{2}')] + u(12)\,\gamma_2(\mathbf{12};\mathbf{1}'\mathbf{2})\} \\ &+ 3\int d\mathbf{2}\,d\mathbf{3}\,u(23)\,\{\gamma_3(\mathbf{123};\mathbf{1}'\mathbf{23}) - \tfrac{1}{3}\gamma_1(\mathbf{1};\mathbf{1}')\,\gamma_2(\mathbf{23};\mathbf{23})\}\,. \end{aligned} \qquad (17)$$

This is an exact relation at coordinate pair $\mathbf{11}'$, involving the potentials v and u, and the 1-DM, 2-DM and 3-DM, generated from any eigenfunction of $\widehat{\mathcal{H}}$, which is expressed in terms of these potentials, Eqs. (1)–(4). The 1-DM enters this equation both in the integrand, and in a "free" term in which the differential operator $\hat{t}$ acts on it — this justifies naming Eq. (17) the integro-differential equation (IDE) for the 1-DM.

3.3 Equation of motion for the 1-DM

After subtracting the hermitian conjugate of Eq. (14) from the original Eq. (14), we arrive at (see, e.g., Refs. [12], [13])

$$\{\,[\hat{t}(1)+v(1)]-[\hat{t}(1')+v(1')]\,\}\,\gamma_1(\mathbf{1};\mathbf{1}')+2\int d\mathbf{2}\,\{u(12)-u(1'2)\}\,\gamma_2(\mathbf{12};\mathbf{1}'\mathbf{2}) = 0\,. \qquad (18)$$

We observe the remarkable cancellation of terms involving E, $\hat{h}(2)$ and γ_3. Eq. (18) is an exact relation at $\mathbf{11}'$, involving the potentials, 1-DM and 2-DM. This is, in fact, an equation of motion (EOM) for the 1-DM, which is coupled with the 2-DM due to electron-electron interaction.

3.4 Differential virial equation

The two-point $\mathbf{11}'$ dependence in the EOM Eq. (18) can be reduced to a one-point dependence by applying to it the operator $\frac{1}{2}(\boldsymbol{\nabla}_1 - \boldsymbol{\nabla}_{1'})$ first and taking the limit $\boldsymbol{r}_1' \to \boldsymbol{r}$ next (see Ref. [14]), viz.

$$\boldsymbol{z}(\boldsymbol{r}_1;[\rho_1]) - \tfrac{1}{4}\boldsymbol{\nabla}_1\nabla_1^2 n(\boldsymbol{r}_1) + n(\boldsymbol{r}_1)\boldsymbol{\nabla}_1 v(\boldsymbol{r}_1) + 2\int d^3r_2\, n_2(\boldsymbol{r}_1,\boldsymbol{r}_2)\boldsymbol{\nabla}_1 u(\boldsymbol{r}_1,\boldsymbol{r}_2) = 0\,, \qquad (19)$$

(the spinless version). Here the vector field $\boldsymbol{z}$ is defined in terms of the kinetic energy density tensor

$$t_{\alpha\beta}(\boldsymbol{r};[\rho_1]) = \frac{1}{4}\left(\frac{\partial^2}{\partial r'_\alpha \partial r''_\beta} + \frac{\partial^2}{\partial r'_\beta \partial r''_\alpha}\right)\rho_1(\boldsymbol{r}+\boldsymbol{r}';\boldsymbol{r}+\boldsymbol{r}'')\big|_{\boldsymbol{r}'=\boldsymbol{r}''=0}, \tag{20}$$

as

$$z_\alpha(\boldsymbol{r};[\rho_1]) = 2\sum_\beta \frac{\partial}{\partial r_\beta} t_{\alpha\beta}(\boldsymbol{r};[\rho_1]). \tag{21}$$

We note that the global kinetic energy — the integrated kinetic energy density (which is the trace of $t_{\alpha\beta}$) —

$$T = \int d^3r \sum_\alpha t_{\alpha\alpha}(\boldsymbol{r};[\rho_1]), \tag{22}$$

is equivalent to that given by the first term of the total energy Eq. (16).

Eq. (19) is called the differential virial equation (DVE) because, after acting on it with the operator $\int d^3r\, \boldsymbol{r}_1$, the familiar virial equation is obtained [14] [here for Coulomb $u(12)$], i.e.

$$2T + E_{\text{ee}} = \int d^3r\, n(\boldsymbol{r})\, \boldsymbol{r}\cdot\nabla v(\boldsymbol{r}), \tag{23}$$

where the interaction energy E_{ee} is given by the third term of the total energy Eq. (16). The DVE Eq. (19) is an exact, local (at $\boldsymbol{r}_1$) relation between gradients of the potentials and the DMs n, ρ_1, n_2.

4 Equivalent Kohn-Sham system

As pointed out by Kohn and Sham (KS) [15], for each N-interacting-electron system, described in Sec. 2.1, it is convenient, for the purposes of DFT, also to introduce an N-noninteracting-electron system, equivalent to the original one in that it has the same GS energy E and density $n(\boldsymbol{r})$ (all quantities specific to this new system will be distinguished with traditional index 's' or 'KS'). So, in analogy to Eqs. (1)–(4), we have

$$\widehat{\mathcal{H}}_{\text{s}} = \widehat{\mathcal{T}} + \widehat{\mathcal{V}}_{\text{s}}, \tag{24}$$

with $\widehat{\mathcal{T}}$ given by Eq. (2), $\widehat{\mathcal{U}}$ absent, and

$$\widehat{\mathcal{V}}_{\text{s}} = \sum_{i=1}^{N} v_{\text{s}}(\boldsymbol{r}_i), \tag{25}$$

where the single-particle (effective KS) potential is the sum

$$v_{\text{s}}(\boldsymbol{r}) \equiv v_{\text{KS}}(\boldsymbol{r}) = v(\boldsymbol{r}) + v_{\text{int}}(\boldsymbol{r}), \tag{26}$$

in which $v_{\rm int}(\boldsymbol{r})$ accounts for all effects due to interactions among electrons in the original system, i.e.

$$v_{\rm int}(\boldsymbol{r}) = v_{\rm es}(\boldsymbol{r}) + v_{\rm xc}(\boldsymbol{r}) \tag{27}$$

with

$$v_{\rm es}(\boldsymbol{r}_1) = \frac{\delta E_{\rm es}[n]}{\delta n(\boldsymbol{r}_1)} = \int d^3r_2\, u(\boldsymbol{r}_1,\boldsymbol{r}_2)\, n(\boldsymbol{r}_2)\,, \tag{28}$$

and

$$v_{\rm xc}(\boldsymbol{r}) = \frac{\delta E_{\rm xc}[n]}{\delta n(\boldsymbol{r})}\,. \tag{29}$$

The GS eigenfunction $\Phi_{\rm s}$ of the Schrödinger equation

$$\hat{\mathcal{H}}_{\rm s}\Phi_{\rm s} = E_{\rm s}\Phi_{\rm s} \tag{30}$$

is a Slater determinant of N lowest-energy KS orbitals $\phi_j(\boldsymbol{x})$ — the solutions of the KS equation

$$\{\,\hat{t}(\boldsymbol{r}) + v_{\rm KS}(\boldsymbol{r})\,\}\,\phi_j(\boldsymbol{x}) = \epsilon_j\,\phi_j(\boldsymbol{x})\,. \tag{31}$$

Due to determinantal character of $\Phi_{\rm s}$, all of the DMs, $\overset{\rm KS}{\gamma_p}$, generated from it, can be obtained (see, e.g., [1] or [11]) in terms of the 1-DM

$$\overset{\rm KS}{\gamma_1}(\boldsymbol{x}_1;\boldsymbol{x}_1') = \sum_{j=1}^{N} \phi_j(\boldsymbol{x}_1)\,\phi_j^*(\boldsymbol{x}_1')\,. \tag{32}$$

In particular, the noninteracting electron density, identical to the interacting electron density, is given by

$$n(\boldsymbol{r}) = \sum_s \overset{\rm KS}{\gamma_1}(\boldsymbol{r}s;\boldsymbol{r}s) = \sum_s\sum_{j=1}^{N} |\phi_j(\boldsymbol{r}s)|^2, \tag{33}$$

while the 2-DM is

$$\overset{\rm KS}{\gamma_2}(\mathbf{12};\mathbf{1'2'}) = \tfrac{1}{2}\{\,\overset{\rm KS}{\gamma_1}(\mathbf{1};\mathbf{1'})\,\overset{\rm KS}{\gamma_1}(\mathbf{2};\mathbf{2'}) - \overset{\rm KS}{\gamma_1}(\mathbf{1};\mathbf{2'})\,\overset{\rm KS}{\gamma_1}(\mathbf{2};\mathbf{1'})\,\}\,. \tag{34}$$

The exchange-correlation energy $E_{\rm xc}$, involved in Eq. (29), is traditionally split into its exchange and correlation contributions

$$E_{\rm xc} = E_{\rm x} + E_{\rm c}\,, \tag{35}$$

defined by

$$E_{\rm x} = \int d1\, d2\, u(12)\,\{\,\overset{\rm KS}{n_2}(12) - \tfrac{1}{2}n(1)\,n(2)\,\}\,, \tag{36}$$

and

$$E_{\rm c} = \int d1 \sum_\alpha t_{\alpha\alpha}(1;[\rho_1 - \overset{\rm KS}{\rho_1}]) + \int d1\, d2\, u(12)\,\{\,n_2(12) - \overset{\rm KS}{n_2}(12)\,\}\,, \tag{37}$$

which induces a corresponding splitting of the exchange-correlation potential in Eq. (29), viz.

$$v_{\mathrm{xc}}(\boldsymbol{r}) = v_{\mathrm{x}}(\boldsymbol{r}) + v_{\mathrm{c}}(\boldsymbol{r}) = \frac{\delta E_{\mathrm{x}}[n]}{\delta n(\boldsymbol{r})} + \frac{\delta E_{\mathrm{c}}[n]}{\delta n(\boldsymbol{r})}\,. \tag{38}$$

The equations obtained in Sec. 3 can be easily adapted to the noninteracting system — here we give their spinless versions: for IDE [the analog of Eq. (17)]

$$\begin{aligned}&[\hat{t}(1) + v_{\mathrm{KS}}(1)]\,\overset{\mathrm{KS}}{\rho_1}(1;1')\\ &+2\int d2\,[\hat{t}(2) + v_{\mathrm{KS}}(2)][\overset{\mathrm{KS}}{\rho_2}(12;1'2') - \tfrac{1}{2}\,\overset{\mathrm{KS}}{\rho_1}(1;1')\,\overset{\mathrm{KS}}{\rho_1}(2;2')] = 0\,,\end{aligned} \tag{39}$$

for the EOM [the analog of Eq. (18)]

$$\{\,[\hat{t}(1) + v_{\mathrm{KS}}(1)] - [\hat{t}(1') + v_{\mathrm{KS}}(1')]\,\}\,\overset{\mathrm{KS}}{\rho_1}(1;1') = 0\,, \tag{40}$$

and for the DVE [the analog of Eq. (19)]

$$\boldsymbol{z}(\boldsymbol{r}_1;[\overset{\mathrm{KS}}{\rho_1}]) - \tfrac{1}{4}\boldsymbol{\nabla}_1\nabla_1^2 n(\boldsymbol{r}_1) + n(\boldsymbol{r}_1)\boldsymbol{\nabla}_1 v_{\mathrm{KS}}(\boldsymbol{r}_1) = 0\,. \tag{41}$$

5 Exchange-correlation potential in terms of density matrices

5.1 Results using DVE

We solve Eq. (41) for $\boldsymbol{\nabla}_1 v_{\mathrm{KS}}(\boldsymbol{r}_1)$, Eq. (19) for $\boldsymbol{\nabla}_1 v(\boldsymbol{r}_1)$, and next take the difference of these results [taking into account Eq. (26)] to obtain

$$\begin{aligned}&\boldsymbol{\nabla}_1 v_{\mathrm{KS}}(\boldsymbol{r}_1) - \boldsymbol{\nabla}_1 v(\boldsymbol{r}_1) = \boldsymbol{\nabla}_1 v_{\mathrm{int}}(\boldsymbol{r}_1) =\\ &\{n(\boldsymbol{r}_1)\}^{-1}\{\,\boldsymbol{z}(\boldsymbol{r}_1;[\rho_1 - \overset{\mathrm{KS}}{\rho_1}]) + 2\int d^3r_2\, n_2(\boldsymbol{r}_1,\boldsymbol{r}_2)\boldsymbol{\nabla}_1 u(\boldsymbol{r}_1,\boldsymbol{r}_2)\,\}\,.\end{aligned} \tag{42}$$

The definitions Eqs. (27) and (28) allow us to rewrite Eq. (42) as

$$\boldsymbol{\nabla}_1 v_{\mathrm{xc}}(\boldsymbol{r}_1) = -\boldsymbol{f}_{\mathrm{xc}}(\boldsymbol{r}_1;[u, n, \overset{\mathrm{KS}}{\rho_1}, \rho_1, n_2]) \tag{43}$$

with

$$\begin{aligned}\boldsymbol{f}_{\mathrm{xc}}(\boldsymbol{r}_1) = -\ &\{n(\boldsymbol{r}_1)\}^{-1}\Big\{\,\boldsymbol{z}(\boldsymbol{r}_1;[\rho_1 - \overset{\mathrm{KS}}{\rho_1}])\\ &+ 2\int d^3r_2\,[n_2(\boldsymbol{r}_1,\boldsymbol{r}_2) - \tfrac{1}{2}\,n(\boldsymbol{r}_1)n(\boldsymbol{r}_2)]\boldsymbol{\nabla}_1 u(\boldsymbol{r}_1,\boldsymbol{r}_2)\Big\}\,.\end{aligned} \tag{44}$$

Equation (43) may be viewed as a differential equation for the potential $v_{\mathrm{xc}}(\boldsymbol{r})$. The other way around it demonstrates that the force field $\boldsymbol{f}_{\mathrm{xc}}(\boldsymbol{r}_1)$ is conservative. Therefore the potential can be calculated as the work in bringing an electron from a reference point (say at infinity) to a given point $\boldsymbol{r}$ against this force field (see Holas and March [14]), i.e.

$$v_{\rm xc}(\boldsymbol{r}) = \int_{\boldsymbol{r}}^{\infty} d\boldsymbol{r}_1 \cdot \boldsymbol{f}_{\rm xc}(\boldsymbol{r}_1;[u,n,\overset{\rm KS}{\rho_1},\rho_1,n_2])\,. \tag{45}$$

Because of the conservative character of $\boldsymbol{f}_{\rm xc}(\boldsymbol{r}_1)$, the evaluated integral is independent of the particular path chosen for the integration in Eq. (45). The arbitrary constant of integration has been fixed by the requirement of the standard gauge $v_{\rm xc}(\infty) = 0$.

Equation (45) represents an exact expression for $v_{\rm xc}(\boldsymbol{r})$ as a line integral involving u and the DMs n, $\overset{\rm KS}{\rho_1}$, ρ_1, n_2.

5.2 Results using IDE

It is convenient to have the diagonal $\boldsymbol{r}'_1 = \boldsymbol{r}_1$ of Eq. (39), namely

$$\begin{aligned} &\widehat{t}(1)\,\overset{\rm KS}{\rho_1}(1;1')_{|1'=1} + v_{\rm KS}(1)\,n(1) \\ &+ 2\int d2\,[\widehat{t}(2) + v_{\rm KS}(2)][\overset{\rm KS}{\rho_2}(12;12') - \tfrac{1}{2}\,n(1)\,\overset{\rm KS}{\rho_1}(2;2')] = 0\,, \end{aligned} \tag{46}$$

and of Eq. (17) (its spinless variant)

$$\begin{aligned} &\widehat{t}(1)\,\rho_1(1;1')_{|1'=1} + v(1)\,n(1) \\ &+ 2\int d2\,\{[\widehat{t}(2) + v(2)][\rho_2(12;12') - \tfrac{1}{2}\,n(1)\,\rho_1(2;2')] + u(12)\,n_2(12)\} \\ &+ 3\int d2\,d3\,u(23)\,\{n_3(123) - \tfrac{1}{3}\,n(1)\,n_2(23)\} = 0\,. \end{aligned} \tag{47}$$

After subtracting Eq. (47) from Eq. (46) and taking into account Eqs. (26)–(28), the resulting equation can be written as

$$\begin{aligned} &n(1)\,v_{\rm xc}(1) = \\ &-2\int d2\,\{\overset{\rm KS}{n_2}(12) - \tfrac{1}{2}\,n(1)\,n(2)\}\,v_{\rm xc}(2) + w_{\rm xc}(1;[u,v,n,\overset{\rm KS}{\rho_1},\rho_1,\overset{\rm KS}{\rho_2},\rho_2,n_3]) \end{aligned} \tag{48}$$

(see Holas and March [11] for details concerning $w_{\rm xc}$). This integral equation for $v_{\rm xc}(\boldsymbol{r}_1)$ is in a form suitable for iterative solution. It is gauge invariant: if $v_{\rm xc}(1)$ is a solution, then $\{v_{\rm xc}(1) + C\}$, with arbitrary constant C, is also a solution, as follows from the reduction property Eq. (11).

A solution $v_{\rm xc}(1)$ of Eq. (48) gives the exact result for this potential obtained from the potentials u and v, and the DMs n, $\overset{\rm KS}{\rho_1}$, ρ_1, $\overset{\rm KS}{\rho_2}$, ρ_2, n_3.

5.3 Results using EOM

The KS potential can be easily obtained from Eq. (40) as

$$v_{\mathrm{KS}}(1) = v_{\mathrm{KS}}(1') - \{\rho_1^{\mathrm{KS}}(1;1')\}^{-1}[\hat{t}(1) - \hat{t}(1')]\,\rho_1^{\mathrm{KS}}(1;1')\,, \tag{49}$$

and, similarly, the external potential from the spinless version of Eq. (18), viz.

$$\begin{aligned} v(1) = v(1') - \{\rho_1(1;1')\}^{-1}\Big\{&[\hat{t}(1) - \hat{t}(1')]\,\rho_1(1;1') \\ &+ 2\int d2\,[u(12) - u(1'2)]\,\rho_2(12;1'2)\Big\}. \end{aligned} \tag{50}$$

By subtracting Eq. (50) from Eq. (49) and using Eqs. (26)–(28), we get (see Holas and March [13])

$$\begin{aligned} v_{\mathrm{XC}}(1) = v_{\mathrm{XC}}(1') &+ \frac{[\hat{t}(1) - \hat{t}(1')]\,\rho_1(1;1')}{\rho_1(1;1')} - \frac{[\hat{t}(1) - \hat{t}(1')]\,\rho_1^{\mathrm{KS}}(1;1')}{\rho_1^{\mathrm{KS}}(1;1')} \\ &+ \int d2\,[u(12) - u(1'2)]\left\{\frac{2\rho_2(12;1'2)}{\rho_1(1;1')} - n(2)\right\}. \end{aligned} \tag{51}$$

In order to have the potentials defined uniquely at point $\boldsymbol{r}_1$ in Eqs. (49)–(51), one should choose and fix a reference point $\boldsymbol{r}_1'$ (e.g., at infinity) and a value of the potential there (say 0). This freedom reflects the fact that potentials can be known only up to an additive constant.

Thus Eq. (51) provides an exact expression for $v_{\mathrm{XC}}(\boldsymbol{r}_1)$ in terms of u and the noninteracting and interacting DMs n, ρ_1^{KS}, ρ_1, ρ_2.

5.4 Density functional theory view of the results

A common feature of all three equations giving v_{XC} — Eqs. (45), (48), (51) — is appearance of low-order DMs. But this fact situates these results outside the scope of DFT. The only way to return to DFT is to express these DMs as functionals of the density. While such expressions are (and apparently will remain) unknown in general, some approximations to them may be constructed. Then the discussed equations allow for a direct generation of the approximate v_{XC} (connected with a particular approximation for the DMs) avoiding the functional differentiation in Eq. (29). The last route is used in the case when E_{XC} is known as a direct (although approximate) functional of the density. It should be mentioned that, despite the high accuracy of E_{XC} achieved by modern approximations, after functional differentiation of such E_{XC}, as a rule, significant loss of accuracy of the resulting v_{XC} is observed.

5.5 Approximate exchange-only potential

As seen in Eq. (37), the correlation energy of the KS approach arises due to differences between the interacting and noninteracting DMs, $(\rho_1 - \rho_1^{\rm KS})$ and $(n_2 - n_2^{\rm KS})$. Therefore one may expect that by replacing the interacting DMs with noninteracting ones in Eqs. (45), (48) and (51), the correlation effects are removed, leaving thus equations for the approximate exchange-only potentials, say $\tilde{v}_{\rm x}^{\rm DVE}$, $\tilde{v}_{\rm x}^{\rm IDE}$, $\tilde{v}_{\rm x}^{\rm EOM}$ (see Holas and March in [14], [11], [13], respectively):

$$\tilde{v}_{\rm x}^{\rm DVE}(\boldsymbol{r}) = \int_{\boldsymbol{r}}^{\infty} d\boldsymbol{r}_1 \cdot \tilde{\boldsymbol{f}}_{\rm x}^{\rm DVE}(\boldsymbol{r}_1;[u,n,n_2^{\rm KS}]) \tag{52}$$

with

$$\begin{aligned}\tilde{\boldsymbol{f}}_{\rm x}^{\rm DVE}(\boldsymbol{r}_1;[u,n,n_2^{\rm KS}]) &= \boldsymbol{f}_{\rm xc}(\boldsymbol{r}_1;[u,n,\rho_1^{\rm KS},\rho_1^{\rm KS},n_2^{\rm KS}]) \\ &= -\int d^3r_2 \left(\frac{2n_2^{\rm KS}(\boldsymbol{r}_1,\boldsymbol{r}_2)}{n(\boldsymbol{r}_1)} - n(\boldsymbol{r}_2) \right) \boldsymbol{\nabla}_1 u(\boldsymbol{r}_1,\boldsymbol{r}_2)\,,\end{aligned} \tag{53}$$

$$\begin{aligned}\tilde{v}_{\rm x}^{\rm IDE}(\boldsymbol{r}_1) = &- \int d^3r_2 \left(\frac{2n_2^{\rm KS}(\boldsymbol{r}_1,\boldsymbol{r}_2)}{n(\boldsymbol{r}_1)} - n(\boldsymbol{r}_2) \right) \tilde{v}_{\rm x}^{\rm IDE}(\boldsymbol{r}_2) \\ &+ \frac{\tilde{w}_{\rm x}^{\rm IDE}(\boldsymbol{r}_1;[u,n,n_2^{\rm KS},n_3^{\rm KS}])}{n(\boldsymbol{r}_1)}\end{aligned} \tag{54}$$

with

$$\tilde{w}_{\rm x}^{\rm IDE}(\boldsymbol{r}_1;[u,n,n_2^{\rm KS},n_3^{\rm KS}]) = w_{\rm xc}(\boldsymbol{r}_1;[u,v,n,\rho_1^{\rm KS},\rho_1^{\rm KS},\rho_2^{\rm KS},\rho_2^{\rm KS},n_3^{\rm KS}])\,, \tag{55}$$

which happen to be independent of the external potential v (see [11] for the explicit form of $\tilde{w}_{\rm x}^{\rm IDE}$), and

$$\begin{aligned}\tilde{v}_{\rm x}^{\rm EOM}(\boldsymbol{r}_1) = &\ \tilde{v}_{\rm x}^{\rm EOM}(\boldsymbol{r}_1') + \\ &\int d^3r_2 \left(\frac{2\rho_2^{\rm KS}(\boldsymbol{r}_1,\boldsymbol{r}_2;\boldsymbol{r}_1',\boldsymbol{r}_2)}{\rho_1^{\rm KS}(\boldsymbol{r}_1;\boldsymbol{r}_1')} - n(\boldsymbol{r}_2) \right) [u(\boldsymbol{r}_1,\boldsymbol{r}_2) - u(\boldsymbol{r}_1',\boldsymbol{r}_2)]\,.\end{aligned} \tag{56}$$

It is worth noting that a combination of $n_2^{\rm KS}$ and n, occurring in Eqs. (53) and (54), is known as the exchange hole at $\boldsymbol{r}_2$ of an electron at $\boldsymbol{r}_1$:

$$\rho_{\rm x}(\boldsymbol{r}_1,\boldsymbol{r}_2) = \frac{2n_2^{\rm KS}(\boldsymbol{r}_1,\boldsymbol{r}_2)}{n(\boldsymbol{r}_1)} - n(\boldsymbol{r}_2)\,. \tag{57}$$

The form of the approximate exchange-force field $\tilde{\boldsymbol{f}}_{\rm x}^{\rm DVE}$ obtained from Eq. (53) by rewriting it in terms of the exchange hole $\rho_{\rm x}$ and Coulomb u,

$$\tilde{\boldsymbol{f}}_{\rm x}^{\rm DVE}(\boldsymbol{r}_1) = \int d^3r_2\, \rho_{\rm x}(\boldsymbol{r}_1,\boldsymbol{r}_2) \frac{(\boldsymbol{r}_1 - \boldsymbol{r}_2)}{|\boldsymbol{r}_1 - \boldsymbol{r}_2|^3}\,, \tag{58}$$

happens to be identical with the force field proposed by Harbola and Sahni Ref. [16] in their work formalism (for the latest review, see Sahni in [8]). Due to the approximate character of this force, the line integral Eq. (52) may be path dependent.

While all three approximations to the exchange potential are different, there is no criterion at present to indicate the most accurate one. However, in Sec. 7, by applying perturbation theory, the correction terms will be established, restoring the exact value of v_x for $\widetilde{v}_x^{\rm DVE}$, $\widetilde{v}_x^{\rm IDE}$, $\widetilde{v}_x^{\rm EOM}$, separately.

Since all of the DMs occurring in Eqs. (52)–(56) can be written in terms of the KS orbitals $\phi_j(\boldsymbol{x})$ [see examples in Eqs. (32)–(34)], the approximate $\widetilde{v}_x$ generated by them can be directly used (possibly together with some available modern approximation for v_c) within the KS scheme, the explicit dependence of $\widetilde{v}_x(\boldsymbol{r})$ on $n(\boldsymbol{r})$ being bypassed.

6 Perturbation theory expansion for density matrices

Görling and Levy [17] proposed to link the interacting system and the equivalent noninteracting KS system by a family of intermediate systems having electron-electron potential scaled as $u^\alpha(12) = \alpha u(12)$, $0 \le \alpha \le 1$. Then the properties of fully interacting system (at $\alpha = 1$) can be viewed as derived from unperturbed ($\alpha = 0$) KS system by means of a perturbation theory with respect to the coupling parameter α. An important constraint is imposed on these systems, namely that their GS density be independent of α, and thus equal to the sought after density of interacting system. It is assumed, for simplicity, that all these systems are nondegenerate in their GS.

Each system is described by the Hamiltonian [cf. Eqs. (1)–(4), (24), (25)]:

$$\widehat{\mathcal{H}}^\alpha = \widehat{\mathcal{T}} + \alpha\widehat{\mathcal{U}} + \widehat{\mathcal{V}}^\alpha , \tag{59}$$

with

$$\widehat{\mathcal{V}}^\alpha = \sum_{i=1}^{N} v^\alpha(\boldsymbol{r}_i) , \tag{60}$$

where the effective "external" potential depending on α is

$$v^\alpha(\boldsymbol{r}) = v_{\rm KS}(\boldsymbol{r}) - v_{\rm int}^\alpha(\boldsymbol{r}) . \tag{61}$$

Obviously, in order to fit the limiting situations, the identities

$$v_{\rm int}^0(\boldsymbol{r}) = 0 , \tag{62}$$

and

$$v_{\rm int}^1(\boldsymbol{r}) = v_{\rm int}(\boldsymbol{r}) = v_{\rm es}(\boldsymbol{r}) + v_{\rm xc}(\boldsymbol{r}) \tag{63}$$

must hold [see Eqs. (25), (26), (4)]. For arbitrary α, $v_{\rm int}^\alpha$ is given [17] by the series

$$v_{\rm int}^\alpha(\boldsymbol{r}) = \sum_{j=1}^{\infty} \alpha^j \, v_{{\rm int}/j}(\boldsymbol{r}) , \tag{64}$$

with

$$v_{\mathrm{int}/1}(\boldsymbol{r}) = v_{\mathrm{es}}(\boldsymbol{r}) + v_{\mathrm{x}}(\boldsymbol{r}) \tag{65}$$

[see Eq. (38) for definition of v_{x}]. This allows us to rewrite Eqs. (62)–(65) as

$$v^{\alpha}_{\mathrm{int}}(\boldsymbol{r}) = \alpha\, v_{\mathrm{esx}}(\boldsymbol{r}) + v^{\alpha}_{\mathrm{c}}(\boldsymbol{r}), \tag{66}$$

i.e. in terms of the electrostatic-plus-exchange potential $v_{\mathrm{esx}}(\boldsymbol{r}) = v_{\mathrm{es}}(\boldsymbol{r}) + v_{\mathrm{x}}(\boldsymbol{r})$ and the α-dependent correlation potential, commencing at the second order [17]

$$v^{\alpha}_{\mathrm{c}}(\boldsymbol{r}) = \sum_{j=2}^{\infty} \alpha^j\, v_{\mathrm{c}/j}(\boldsymbol{r}) = \sum_{j=2}^{\infty} \alpha^j\, v_{\mathrm{int}/j}(\boldsymbol{r}), \tag{67}$$

being just the correlation potential, Eq. (38), at full coupling strength $v^{1}_{\mathrm{c}}(\boldsymbol{r}) = v_{\mathrm{c}}(\boldsymbol{r})$.

In order to find the GS solution of the intermediate system Schrödinger equation

$$(\widehat{\mathcal{H}}^{\alpha} - \mathcal{E}^{\alpha})\Psi^{\alpha} = 0 \tag{68}$$

by means of perturbation theory, the leading term $\widehat{\mathcal{H}}_{/0}$ of the expanded Hamiltonian

$$\widehat{\mathcal{H}}^{\alpha} = \sum_{j=0}^{\infty} \alpha^j\, \widehat{\mathcal{H}}_{/j} \tag{69}$$

is considered to be the unperturbed Hamiltonian,

$$\widehat{\mathcal{H}}_{/0} = \sum_{i=1}^{N} \hat{h}_{\mathrm{KS}}(\boldsymbol{r}_i) = \sum_{i=1}^{N} \{\hat{t}(\boldsymbol{r}_i) + v_{\mathrm{KS}}(\boldsymbol{r}_i)\}, \tag{70}$$

while the sum of remaining terms in Eq. (69) is treated as a perturbation. Assuming $\mathcal{E}^{\alpha}$ and Ψ^{α} are analytic functions of α in the range $[0,1]$, we expand these quantities in a manner similar to the expansion of $\widehat{\mathcal{H}}^{\alpha}$ in Eq. (69), i.e.

$$\mathcal{E}^{\alpha} = \sum_{j=0}^{\infty} \alpha^j\, \mathcal{E}_{/j}, \tag{71}$$

$$\Psi^{\alpha} = \sum_{j=0}^{\infty} \alpha^j\, \Psi_{/j}, \tag{72}$$

and insert them into Eq. (68). Thus a set of equations (for each power of α) is obtained, viz.

$$\sum_{\ell=0}^{j} (\widehat{\mathcal{H}}_{/\ell} - \mathcal{E}_{/\ell})\Psi_{/j-\ell} = 0; \qquad j = 0, 1, \ldots. \tag{73}$$

This set of equations (together with the imposed normalization) can be solved in a similar way as in the Rayleigh-Schrödinger perturbation theory (which must be extended here to include terms of the perturbing Hamiltonian higher than linear order in α).

The resulting partial wave functions $\Psi_{/j}$ are expanded in a complete set $\{\Phi_A\}$ of eigenfunctions of the unperturbed (i.e. KS) Hamiltonian Eq. (70). In this way the results for $\Psi_{/j}$ and $\mathcal{E}_{/j}$ are obtained in terms of the KS eigenenergies ϵ_i and orbitals $\phi_i(\boldsymbol{x})$ (both occupied, $i \leq N$, and virtual, $i > N$) and matrix elements of $u(\boldsymbol{r}_1, \boldsymbol{r}_2)$, $v_{\rm esx}(\boldsymbol{r})$, $v_{c/j}(\boldsymbol{r})$ in the basis of these orbitals. Details may be found in Ref. [13]; here we give a summary and an example.

For the application to the equations satisfied by the DMs (which were discussed in Sec. 3), we are interested in DMs generataed by means of Eqs. (6)–(10) from Ψ^α. The perturbation-theory expansion Eq. (72) for Ψ^α immediately leads to an analogous expansion for the DMs, namely

$$\gamma_p^\alpha(\mathbf{1}\ldots\boldsymbol{p};\mathbf{1}'\ldots\boldsymbol{p}') = \sum_{j=0}^{\infty} \alpha^j \, \gamma_{p/j}(\mathbf{1}\ldots\boldsymbol{p};\mathbf{1}'\ldots\boldsymbol{p}') . \tag{74}$$

The 0th-perturbational-order DMs are just the noninteracting KS matrices, $\gamma_{p/0} \equiv \gamma_p^{\rm KS}$, see Eqs. (32)–(34). The 1st-perturbational-order DMs $\gamma_{p/1}$ are expressed in terms of $v_{\rm x}$ (but not $v_{c/j}$). As an example, we quote [13]

$$\rho_{1/1}(\boldsymbol{r}_1;\boldsymbol{r}_2) = \sum_{k=N+1}^{\infty} \sum_{i=1}^{N} \frac{-1}{\epsilon_k - \epsilon_i} \Big\{ < k|\widehat{v}_{\rm x}^{\rm F} - v_{\rm x}|i> \sum_{s_1} \phi_k(\boldsymbol{r}_1 s_1)\, \phi_i^*(\boldsymbol{r}_1' s_1)$$
$$+ < i|\widehat{v}_{\rm x}^{\rm F} - v_{\rm x}|k> \sum_{s_1} \phi_i(\boldsymbol{r}_1 s_1)\, \phi_k^*(\boldsymbol{r}_1' s_1) \Big\} \tag{75}$$

(see also [18] in the case of real orbitals), where $\widehat{v}_{\rm x}^{\rm F}$ denotes the Fock exchange integral operator [constructed out of the occupied KS orbitals and $u(12)$]

$$(\widehat{v}_{\rm x}^{\rm F}\, \phi)(\mathbf{2}) = \int d\mathbf{3}\, \widetilde{v}_{\rm x}^{\rm F}(\mathbf{23})\, \phi(\mathbf{3})\, ; \qquad \widetilde{v}_{\rm x}^{\rm F}(\mathbf{23}) = -u(23)\, \gamma_1^{\rm KS}(\mathbf{23}) . \tag{76}$$

In general, the DMs $\gamma_{p/j}$ are expressed in terms of KS orbitals and energies, and potentials: u, $v_{{\rm int}/j}$, $v_{{\rm int}/j-1}$, $\ldots$, $v_{{\rm int}/1}$ [see Eqs. (64)–(67)].

7 Perturbation-theory approach to exchange and correlation

7.1 Application to the DVE

Eq. (19) can be easily adapted to the intermediate system at coupling strength α, viz.

$$\boldsymbol{z}(\boldsymbol{r}_1;[\rho_1^\alpha]) - \tfrac{1}{4}\boldsymbol{\nabla}_1 \nabla_1^2 n(\boldsymbol{r}_1) + n(\boldsymbol{r}_1)\boldsymbol{\nabla}_1 v^\alpha(\boldsymbol{r}_1) + 2\!\int\! d^3r_2\, n_2^\alpha(\boldsymbol{r}_1,\boldsymbol{r}_2)\boldsymbol{\nabla}_1 \alpha\, u(\boldsymbol{r}_1,\boldsymbol{r}_2) = 0 . \tag{77}$$

After substituting Eq. (74) for the DMs ρ_1^α and n_2^α, and Eq. (64) for $v_{\rm int}^\alpha$ of the potential v^α, Eq. (61), we obtain from Eq. (77) a set of equations (for each power of α). Because $\gamma_{p/0} \equiv \gamma_p^{\text{KS}}$, the 0th equation is identical with Eq. (41). The 1st equation is

$$z(\boldsymbol{r}_1;[\rho_{1/1}]) - n(\boldsymbol{r}_1)\nabla_1 v_{\text{esx}}(\boldsymbol{r}_1) + 2\int d^3r_2\, n_2^{\text{KS}}(\boldsymbol{r}_1,\boldsymbol{r}_2)\nabla_1 u(\boldsymbol{r}_1,\boldsymbol{r}_2) = 0\,. \quad (78)$$

After subtracting v_{es} from v_{esx}, it can be rewritten as (see Levy and March [18])

$$\nabla_1 v_{\text{x}}(\boldsymbol{r}_1) = -\boldsymbol{f}_{\text{x}}(\boldsymbol{r}_1) = -\{\widetilde{\boldsymbol{f}}_{\text{x}}^{\text{DVE}}(\boldsymbol{r}_1) - z(\boldsymbol{r}_1;[\rho_{1/1}])/n(\boldsymbol{r}_1)\}\,. \quad (79)$$

Thus the second term, depending on $\rho_{1/1}$, represents a correction to the approximate force, Eq. (53), while their sum being an exact exchange force [18], leads to the exact potential

$$v_{\text{x}}(\boldsymbol{r}) = \int_{\boldsymbol{r}}^{\infty} d\boldsymbol{r}_1 \cdot \boldsymbol{f}_{\text{x}}(\boldsymbol{r}_1;[u,n,n_2^{\text{KS}},\rho_{1/1}])\,, \quad (80)$$

the evaluation of which therefore must be path independent.

The jth equation leads to [18]

$$v_{\text{c}/j}(\boldsymbol{r}) = \int_{\boldsymbol{r}}^{\infty} d\boldsymbol{r}_1 \cdot \boldsymbol{f}_{\text{c}/j}(\boldsymbol{r}_1;[u,n,n_{2/j-1},\rho_{1/j}])\,, \quad (81)$$

where

$$\boldsymbol{f}_{\text{c}/j}(\boldsymbol{r}_1) = -\{n(\boldsymbol{r}_1)\}^{-1}\left\{z(\boldsymbol{r}_1;[\rho_{1/j}]) + 2\int d^3r_2\, n_{2/j-1}(\boldsymbol{r}_1,\boldsymbol{r}_2)\nabla_1 u(\boldsymbol{r}_1,\boldsymbol{r}_2)\right\}. \quad (82)$$

Being an exact expression for the jth-order correlation potential, the line integral Eq. (81) is path independent.

7.2 Application to the IDE

A similar procedure applied to the IDE Eq. (47) (adapted to the α-system) gives at 0th order, obviously, the KS form Eq. (46) of IDE. At 1st order, the integral equation for the exchange potential is obtained (see Holas and Levy Ref. [19]), i.e.

$$\begin{aligned} v_{\text{x}}(\boldsymbol{r}_1) = &-\int d^3r_2\, \rho_{\text{x}}(\boldsymbol{r}_1,\boldsymbol{r}_2)\, v_{\text{x}}(\boldsymbol{r}_2) + \{n(\boldsymbol{r}_1)\}^{-1} \\ &\{\widetilde{w}_{\text{x}}^{\text{IDE}}(\boldsymbol{r}_1;[u,n,n_2^{\text{KS}},n_3^{\text{KS}}]) + w_{\text{x}}^{\text{cor}}(\boldsymbol{r}_1;[n,\rho_{1/1},\rho_{2/1}])\} \end{aligned} \quad (83)$$

[cf. Eqs. (54), (57)] where

$$\begin{aligned} w_{\text{x}}^{\text{cor}}(\boldsymbol{r}_1) = &\,\widehat{t}(\boldsymbol{r}_1)\,\rho_{1/1}(\boldsymbol{r}_1;\boldsymbol{r}_1')|_{\boldsymbol{r}_1'=\boldsymbol{r}_1} \\ &+\int d^3r_2\,\{2\,\widehat{h}_{\text{KS}}(\boldsymbol{r}_2)\,\rho_{2/1}(\boldsymbol{r}_1,\boldsymbol{r}_2;\boldsymbol{r}_1,\boldsymbol{r}_2') - n(\boldsymbol{r}_1)\,\widehat{t}(\boldsymbol{r}_2)\,\rho_{1/1}(\boldsymbol{r}_2;\boldsymbol{r}_2')\}|_{\boldsymbol{r}_2'=\boldsymbol{r}_2}\,. \end{aligned} \quad (84)$$

The "free" term $w_{\mathrm{x}}^{\mathrm{cor}}$ represents such a correction to the approximate term $\widetilde{w}_{\mathrm{x}}^{\mathrm{IDE}}$, that a solution of the integral equation Eq. (83) is an exact $v_{\mathrm{x}}(\boldsymbol{r}_1)$.

The equation found at jth order, $j \geq 2$, has a similar structure [19]

$$v_{\mathrm{c}/j}(\boldsymbol{r}_1) = -\int d^3r_2\, \rho_{\mathrm{x}}(\boldsymbol{r}_1, \boldsymbol{r}_2)\, v_{\mathrm{c}/j}(\boldsymbol{r}_2)$$
$$+I_{\mathrm{c}/j}(\boldsymbol{r}_1;[u, v_{\mathrm{KS}}, v_{\mathrm{int}/1}, \ldots, v_{\mathrm{int}/j-1}, n, n_{2/1}, \ldots, n_{2/j-1}, n_{3/j-1}, \rho_{1/j}, \rho_{2/j}]) \,. \tag{85}$$

The "free" term $I_{\mathrm{c}/j}$ of this integral equation is just a combination of integrals involving the listed functions (see Ref. [19] for details).

7.3 Application to the EOM

The spinless version of the EOM Eq. (18), adapted to the intermediate system,

$$\begin{aligned}&\{\,[\hat{t}(1) + v^{\alpha}(1)] - [\hat{t}(1') + v^{\alpha}(1')]\,\}\, \rho_1^{\alpha}(1;1') \\ &+2\int d2\, \alpha\, \{u(12) - u(1'2)\}\, \rho_2^{\alpha}(12;1'2) = 0\,,\end{aligned} \tag{86}$$

leads at 0th order to Eq. (40) — the KS EOM, and at jth order ($j \geq 1$) (see Holas and March [13]) to

$$\begin{aligned}&\{\,\hat{h}_{\mathrm{KS}}(1) - \hat{h}_{\mathrm{KS}}(1')\,\}\, \rho_{1/j}(1;1') - \sum_{\ell=1}^{j} \{v_{\mathrm{int}/\ell}(1) - v_{\mathrm{int}/\ell}(1')\}\, \rho_{1/j-\ell}(1;1') \\ &\qquad + 2\int d2\, \{u(12) - u(1'2)\}\, \rho_{2/j-1}(12;1'2) = 0\,.\end{aligned} \tag{87}$$

The last equation can be solved with respect to $v_{\mathrm{int}/j}$ to give [13] for $j = 1$

$$\begin{aligned}&v_{\mathrm{x}}(1) = v_{\mathrm{x}}(1') + \{\widetilde{v}_{\mathrm{x}}^{\mathrm{EOM}}(1) - \widetilde{v}_{\mathrm{x}}^{\mathrm{EOM}}(1')\} + \\ &\{\rho_1^{\mathrm{KS}}(1;1')\}^{-1}\, [\,\hat{h}_{\mathrm{KS}}(1) - \hat{h}_{\mathrm{KS}}(1')\,]\, \rho_{1/1}(1;1')\,,\end{aligned} \tag{88}$$

and for $j \geq 2$

$$\begin{aligned}v_{\mathrm{c}/j}(1) &= v_{\mathrm{c}/j}(1') + \{\rho_1^{\mathrm{KS}}(1;1')\}^{-1} \Big\{[\,\hat{h}_{\mathrm{KS}}(1) - \hat{h}_{\mathrm{KS}}(1')\,]\, \rho_{1/j}(1;1') \\ &+ 2\int d2\, [u(12) - u(1'2)]\, \rho_{2/j-1}(12;1'2) \\ &- \sum_{\ell=1}^{j-1} [v_{\mathrm{int}/\ell}(1) - v_{\mathrm{int}/\ell}(1')]\, \rho_{1/j-\ell}(1;1')\Big\}\,.\end{aligned} \tag{89}$$

Eq. (88) demonstrates explicitly in the term depending on $\rho_{1/1}$ the correction to the approximate $\widetilde{v}_{\mathrm{x}}^{\mathrm{EOM}}$, Eq. (56), with their sum being the exact v_{x}.

8 Discussion and conclusions

The constrained search method allowed Görling and Levy to define [17] the functional of the density $n(\boldsymbol{r})$

$$F^{\alpha}[n] = \min_{\Psi \to n} \langle \Psi | \widehat{T} + \alpha \widehat{U} | \Psi \rangle = \langle \Psi^{\alpha}[n] | \widehat{T} + \alpha \widehat{U} | \Psi^{\alpha}[n] \rangle = T^{\alpha}[n] + E^{\alpha}_{\text{ee}}[n] \quad (90)$$

— a generalization of the Hohenberg-Kohn (HK) functional for the intermediate system. At the GS density $n = n_{\text{G}}$ of the system discussed in Sec. 6, the minimizing $\Psi^{\alpha}[n]$ coincides with the GS eigenfunction Ψ^{α} of Eq. (68). For $\alpha = 0$, Eq. (90) defines the kinetic energy functional of the noninteracting system

$$F^{0}[n] \equiv T_{\text{s}}[n] \,. \quad (91)$$

Their difference represents the effective interaction energy functional of the α-system

$$E^{\alpha}_{\text{int}}[n] = F^{\alpha}[n] - F^{0}[n] \,, \quad (92)$$

having the following expansion

$$E^{\alpha}_{\text{int}}[n] = \alpha \left(E_{\text{es}}[n] + E_{\text{x}}[n] \right) + E^{\alpha}_{\text{c}}[n] \,, \quad (93)$$

with

$$E^{\alpha}_{\text{c}}[n] = \sum_{j=2}^{\infty} \alpha^{j} \, E_{\text{c}/j}[n] \,, \quad (94)$$

while their functional derivatives

$$v^{\alpha}_{\text{int}}(\boldsymbol{r};[n]) = \frac{\delta E^{\alpha}_{\text{int}}[n]}{\delta n(\boldsymbol{r})} \,, \qquad v_{\text{c}/j}(\boldsymbol{r};[n]) = \frac{\delta E_{\text{c}/j}[n]}{\delta n(\boldsymbol{r})} \quad (95)$$

coincide at $n = n_{\text{G}}$ with the potentials introduced in Sec. 6. As a result, the perturbation theory approach of Görling and Levy [17] remains within the scope of the DFT. However, it must be supplemented by a prescription for the calculation of the exchange and correlation potentials. In their subsequent paper [20], Görling and Levy proposed to evaluate these potentials by performing the functional differentiation in Eq. (95) via a variation of the KS orbitals. The derived expressions involve the inverse of the integral operator of the linear density response, which makes numerical implementation of this procedure difficult and complicated. Three alternative routes ([18], [19], [13]), discussed in Sec. 7, are based on the equations satisfied by the DMs. The potentials, Eq. (95), are expressed in these approaches in terms of expanded DMs — objects directly calculable by means of perturbation theory [17]. Although all four approaches give expressions in terms of the KS orbitals and eigenenergies, they remain within the DFT, because these objects, according to the HK theorem applied to the KS system, are implicit functionals of the density.

The total energy functional of the system at $\alpha = 1$, in terms of the HK functional, Eq. (90), is

$$E[n] = \langle \Psi^{1}[n] | \widehat{T} + \widehat{U} + \widehat{V} | \Psi^{1}[n] \rangle = F^{1}[n] + V[n] \quad (96)$$

with

$$V[n] = \int d1\, n(1)\, v(1)\,. \tag{97}$$

Using the definitions of Eqs. (91)–(94), it can be rewritten as

$$E[n] = T_s[n] + V[n] + E_{es}[n] + E_x[n] + \sum_{j=2}^{\infty} E_{c/j}[n]\,, \tag{98}$$

with the obvious $T_s[n]$ and $E_{es}[n]$, $E_x[n]$ given in Eq. (36), and $E_{c/j}[n]$ terms of the expansion, Eq. (94), for $E_c = E_c^1$, Eq. (37), in the form [17]

$$E_{c/j} = T_{/j} + E_{ee/j} = \tfrac{-1}{j-1}\, T_{/j} = \tfrac{1}{j}\, E_{ee/j} \tag{99}$$

with

$$T_{/j} = \int d1 \sum_{\alpha} t_{\alpha\alpha}(1;[\rho_{1/j}])\,; \qquad E_{ee/j} = \int d1\, d2\, u(12)\, n_{2/j-1}(12)\,. \tag{100}$$

The third form of $E_{c/j}$ in Eq. (99) is of particular interest, because it involves the DM of the $(j-1)$th order only.

The exact GS solution for a given system corresponds to the minimization of the functional Eq. (96), i.e.

$$E_G = \min_{n \to N} E[n] = E[n_G]\,. \tag{101}$$

But, because its correlation energy is available within perturbation theory, Eq. (98), only a limited number of its terms can be taken into account in practice. So, instead of Eq. (101), we are forced to look for a solution to the approximate GS problem:

$$E_G{}^{\{k\}} = \min_{n \to N} E^{\{k\}}[n] = E^{\{k\}}[n_G{}^{\{k\}}] \tag{102}$$

with

$$E^{\{k\}}[n] = T_s[n] + V[n] + E_{es}[n] + E_x[n] + \sum_{j=2}^{k} E_{c/j}[n]\,, \tag{103}$$

corresponding to E_c truncated after the kth term. Assuming good convergence of the perturbation expansion, we expect that the solutions $\{E_G{}^{\{k\}}, n_G{}^{\{k\}}\}$ approach $\{E_G, n_G\}$ with increasing k. The error of this approximation can be estimated to be

$$E_G - E_G{}^{\{k\}} = O([n_G - n_G{}^{\{k\}}]^2) + E_{c/k+1}[n_G] + E_{c/k+2}[n_G] + \dots\,, \tag{104}$$

because, for $\delta n = n_G - n_G{}^{\{k\}}$, we have

$$E^{\{k\}}[n_G{}^{\{k\}} + \delta n] = E^{\{k\}}[n_G{}^{\{k\}}] + \int \left[\frac{\delta E^{\{k\}}[n_G{}^{\{k\}}]}{\delta n}\, \delta n\right] + O([\delta n]^2)\,, \tag{105}$$

with vanishing term linear in δn since $n_G{}^{\{k\}}$ minimizes Eq. (102). It is interesting, that an estimate of the leading term of the error, $E_{c/k+1}[n_G]$ can be evaluated as $E_{c/k+1}[n_G^{\{k\}}]$ within the kth approximation, because it involves the DM $n_{2/k}$ [see Eqs. (99) and (100)], available in this approximation.

It should be noted that the truncated GS problem at $k = 1$, i.e. the exchange-only approximation, is equivalent to the so-called optimized potential method, being thus an alternative to this method.

The computational scheme, leading to the solution Eq. (102), involves iterations, terminated after achieving selfconsistency. For the initial step, one solves the KS equation Eq. (31) using $v_{KS}(\boldsymbol{r})$ constructed with the help of any available approximate $v_{xc}(\boldsymbol{r})$, while for each next step — with $v_{xc}(\boldsymbol{r})$ constructed in the preceding step. A large number of solutions, exceeding significantly the N lowest-energy solutions, is necessary for further calculations. (The unavoidable truncation to a finite number of solutions, say M, introduces some error, which can be always diminished by increasing M.) This allows for a calculation of the KS DMs according to Eqs. (32)–(34). Next, using one of Eqs. (80), (83) or (88), v_x is determined in the following iterative process. Initially contributions of the DMs $\gamma_{p/1}$ are neglected. The approximate v_x obtained in this way is used to calculate $\gamma_{p/1}$ [see, e.g., Eq. (75)], to be used in the next iteration to improve v_x, up to convergence for it. Having v_x and $\gamma_{p/1}$, one determines $v_{c/2}$ in a similar way using one of Eqs. (81), (85) or (89), at $j = 2$. Initially contributions due to the DMs $\gamma_{p/2}$ are neglected, then the approximate $\gamma_{p/2}$ are calculated using the approximate $v_{c/2}$, and so on up to convergence for $v_{c/2}$. Similarly $v_{c/j}$ is to be obtained for $j = 3, \ldots, k$. Having all these potentials determined, their sum is used to construct v_{KS} for the next step of iterations. All of the above-described evaluations are to be performed again. Steps are repeated until selfconsistency (in the density or KS potential).

The calculational schemes, discussed above, are still only propositions awaiting implementation.

References

1. R.G. Parr and W. Yang, *Density-Functional Theory of Atoms and Molecules.* Oxford University Press, New York (1989).
2. R.M. Dreizler and E.K.U. Gross, *Density Functional Theory.* Springer, Berlin (1990).
3. E.S. Kryachko and E.V. Ludeña, *Energy Density Functional Theory of Many-Electron Systems.* Kluwer, Dordrecht (1990).
4. S.B. Trickey, (ed) *Density Functional Theory of Many Fermion Systems.* Academic Press, London (1990).
5. J.K. Labanowski and J.W. Andzelm, (eds) *Density Functional Methods in Chemistry.* Springer, New York (1991).
6. N.H. March, *Electron Density Theory of Atoms and Molecules.* Academic Press, London (1992).
7. E.K.U. Gross and R.M. Dreizler, (eds) *Density Functional Theory.* Plenum, New York (1995).
8. R.F. Nalewajski, (ed) *Topics in Current Chemistry: Density Functional Theory.* Springer, Heidelberg (in press) (1996).

9. H. Nakatsuji, Phys. Rev. **A 14**, 41 (1976).
10. K.A. Dawson and N.H. March, J. Chem. Phys. **81**, 5850 (1984).
11. A. Holas and N.H. March, Int. J. Quantum Chem. (in press) (1996).
12. P. Ziesche, Phys. Lett. **A 195**, 213 (1994).
13. A. Holas and N.H. March, to be published (1996).
14. A. Holas and N.H. March, Phys. Rev. **A 51**, 2040 (1995).
15. W. Kohn and L.J. Sham, Phys. Rev. **140**, A1133 (1965).
16. M.K. Harbola and V. Sahni, Phys. Rev. Lett. **62**, 489 (1989).
17. A. Görling and M. Levy, Phys. Rev. **B 47**, 13105 (1993).
18. M. Levy and N.H. March, Phys. Rev. **A** (in press) (1996).
19. A. Holas and M. Levy, to be published (1996).
20. A. Görling and M. Levy, Phys. Rev. **A 50**, 196 (1994).

Dissipative Quantum Mechanics. Metriplectic Dynamics in Action

Lukasz A. Turski

Center for Theoretical Physics
Polish Academy of Sciences and School of Science
Al. Lotników 32/46. 02-668 Warszawa, Poland

Abstract: The inherent linearity of quantum mechanics is one of the difficulties in developing a fully quantum theory of dissipative processes. Several microscopic and more or less phenomenological descriptions of quantum dissipative dynamics have been proposed in the past. Following the successful development of classical metriplectic dynamics – a systematic description of dissipative systems using a natural extension of symplectic dynamics – we discuss the possibility of a similar formulation for quantum dissipative systems. Particular attention is paid to the Madelung representation of quantum mechanics.

1 Introduction

Quantum mechanics is an intrinsically linear theory. Careful analysis of its foundation has led Gisin to the conclusion that *all deterministic nonlinear Schrödinger equations are irrelevant* [1]. On the other hand several authors, including Gisin himself [2], have attempted to describe the irreversible evolution of quantum systems by means of nonlinear generalizations of the Schrödinger equation; these equations have to be considered as a restricted (simplified) versions of more general quantum Master equations. This point of view, presented in Ref. [1], is complimentary to that discussed by Razavy and Pimpale [3] who have shown, following the method of Caldeira and Leggett [4], how the Gisin equation follows from the restricted description of the "central particle" coupled to a heat bath. The Caldeira and Leggett, or Razavy and Pimpale, description of a quantum system coupled to a heat bath suffers from the same difficulties as all truly quantum many body models. It is therefore tempting to accept the dissipative quantum mechanical equation, for example the one proposed by Gisin, as a phenomenological equations, akin to Navier-Stokes equations of hydrodynamics. The framework within which one can try to use these phenomenological equations is the so called metriplectic dynamics, [1] developed in classical statistical

[1] The name metriplectic dynamics was proposed by Phil Morrison. The method was previously called mixed canonical-dissipative formulation by Charles Enz. Early at-

mechanics and which works remarkably well for many applications ranging from the dynamics of low dimensional magnetic systems to relativistic plasma physics [5].

The plan of this paper is as follows. In Section 2, we shall give a short introduction to the metriplectic dynamics of classical systems. In Section 3, we will introduce the Madelung representation of the Schrödinger equation, which we find a convenient tool for the interpretation of the Gisin damping. In Section 4, the quantum mechanical extension of metriplectic dynamics will be given and it will be shown how the Gisin equation emerges out of that formulation. Section 5 will be devoted to a discussion of the dissipative Schrödinger equation in the Madelung representation. Finally, in Section 6 we shall provide some final comments and conclusions.

2 Classical Metriplectic Dynamics

In classical dynamics of complex systems one often follows the method developed for classical particles in Hamiltonian dynamics and describes the system dynamics in terms of properly chosen (generalized) positions and momenta spanning the even dimensional phase space Γ. Denoting the collection of these coordinates and momenta as $z^A = (q^1, q^2, \ldots, p_1, p_2, \ldots)$ and making the further assumption that the dynamics of the system is governed by Hamilton-like equations of motion we can write them as

$$\partial_t z^a = \{z^A, \mathcal{H}\}, \tag{1}$$

where $\mathcal{H}$ is the system hamiltonian and $\{f, h\}$ denotes the Lie-Poisson bracket, an operation which satisfies three requirements, linearity in both arguments, the Leibnitz rule and the Jacobi identity.

Without going into mathematical details, if the Lie-Poisson bracket is defined over the space of functions on an even dimensional phase space, then (at least locally)

$$\{f, h\} = \gamma_c^{AB} \partial_A f \partial_B h \,, \tag{2}$$

where

$$\gamma_c^{AB} = \begin{pmatrix} 0 & \boldsymbol{I} \\ -\boldsymbol{I} & 0 \end{pmatrix} \tag{3}$$

and $\boldsymbol{I}$ is $n \times n$ unit matrix. The matrix γ_c is obviously antisymmetric and this permits us to generalize the Lie-Poisson brackets to the case of odd dimensional spaces.

tempts to use it can be found in phase transformation literature.

We define the Lie-Poisson bracket for an arbitrary dimension space as

$$\{f,g\} = \gamma^{AB}\partial_A f \partial_B g \quad A = 1\ldots,N, \tag{4}$$

where γ is the antisymmetric tensor. As previously, the equation of motion for any *observable* F is

$$\partial_t F = \{F, \mathcal{H}\}\,, \tag{5}$$

where $\mathcal{H}$ is the system Hamiltonian.

If in some region of the (generalized) phase space $S \subset \Gamma$, the matrix γ has $N-2M$ null eigenvectors $\gamma^{AB} e_B^{(\ell)} = 0, \ell = N-2M$ and $e_B^{(\ell)} = \partial_B C^{(\ell)}$, then, by construction, all of the functions $C^{(\ell)}$ are constants of motion independently of the detailed form of the system Hamiltonian. These functions are called *Casimirs* of the given Lie-Poisson structure. Casimirs play an important rôle in the formal structure of generalized symplectic dynamics. The system evolution is restricted to the leafs in phase space which correspond to a given value of the Casimirs. As we shall see, the Casimirs are also important for a description of dissipative processes. Roughly speaking there are two classes of dissipative processes, those whose dynamics stay on the (initially assigned) Casimir leaf and those which lead to an interleafs transition.

As is well known, not all interesting physical systems can be described by symplectic dynamics. There are wide classes of physically interesting processes which are described by so-called *metric* dynamics. In metric dynamics (a formal generalization of the Langevin equation), the system evolution in phase space is described by

$$\begin{aligned} \dot{z}^A &= -g^{AB}\partial_B S \\ g^{AB} &= g^{BA}\,, \end{aligned} \tag{6}$$

where S is the *potential*, and the symmetric matrix g^{AB} is thought to be non-singular.

A well known example of metric dynamics is the van der Waals, Ginzburg-Landau, Cahn-Hilliard-Wilson equation for the evolution of the order parameter ψ

$$\partial_t \psi(x,t) = -\lambda \nabla^{2a} \frac{\delta \mathcal{G}}{\delta \psi(x,t)}\,, \tag{7}$$

where $a = 0, 1$ for a nonconserved and conserved order parameter, respectively. $\mathcal{G}$ is the appropriate thermodynamic potential, usually the Helmholtz or Gibbs free energy. Here $g^{AB} \to g(x-y) = \lambda \nabla^{2a} \delta(x-y)$.

With the help of a symmetric matrix g, we define a metric bracket as

$$\{\{f,h\}\} = g^{AB}\partial_A f \partial_B h\,. \tag{8}$$

Using both the symplectic, Lie-Poisson and metric bracket, we can now define the *metriplectic* bracket as

$$[f,h] = \{f,h\} + \{\{f,h\}\} \equiv \left(\gamma^{AB} + g^{AB}\right)\partial_A f \partial_B h \equiv D^{AB}\partial_A f \partial_B h\,. \quad (9)$$

The equation of motion for a system described by mixed dynamics is

$$\partial_t A = [A, \mathcal{F}]\,, \quad (10)$$

where

$$\mathcal{F} = \mathcal{H} - \mathcal{S}.$$

We do not claim that *all* dissipative systems can be described by means of the above outlined method. In Ref. [5], however, a rather lengthy list containing important examples from many branches of physics is compiled. In this section we should like to present just two examples important for our application. These are the symplectic structure of hydrodynamics and of Heisenberg magnets.

Recall that in hydrodynamics (Euler representation), the state of the system is described by means of the fluid density $\varrho(\boldsymbol{x},t)$ and velocity $\boldsymbol{u}(\boldsymbol{x},t)$ (or particle current $\boldsymbol{J} = \varrho\boldsymbol{u}$). These two fields span the system phase space. The Lie-Poisson brackets for these variables are well known [6] and read

$$\begin{aligned}
\{\varrho(\boldsymbol{x},t), \varrho(\boldsymbol{y},t)\} &= 0\,,\\
\{\varrho(\boldsymbol{x},t), u^a(\boldsymbol{y},t)\} &= -\frac{\partial}{\partial x^a}\delta(\boldsymbol{x}-\boldsymbol{y})\,,\\
\{u^a(\boldsymbol{x},t), u^b(\boldsymbol{y},t)\} &= \frac{1}{\varrho(\boldsymbol{x},t)}\epsilon^{abc}\left(\nabla\times\boldsymbol{u}(\boldsymbol{x},t)\right)_c\delta(\boldsymbol{x}-\boldsymbol{y})\,.
\end{aligned} \quad (11)$$

The conventional Hamiltonian $\mathcal{H}$ for a fluid consists of the kinetic energy and potential energy expressed in terms of the fields ϱ and $\boldsymbol{u}$. The equations of motion obtained from

$$\begin{aligned}
\partial_t\varrho &= \{\varrho, \mathcal{H}\}\,,\\
\partial_t\boldsymbol{u} &= \{\boldsymbol{u}, \mathcal{H}\}
\end{aligned} \quad (12)$$

are the usual Euler equations for inviscid fluid.

The metric brackets for fluid dynamics leading to the Navier-Stokes equation were first given in Ref. [6]. The metric bracket for the density vanishes, as do those for the density and velocity field. The only nonvanishing bracket is for the velocity field, viz.

$$\{\{u^a(\boldsymbol{x}), u^b(\boldsymbol{y})\}\} \equiv \nu^{ab} = -\left(\eta\delta^{ab}\nabla^2 + (\zeta+\eta/3)\nabla^a\nabla^b\right)\delta(\boldsymbol{x}-\boldsymbol{y})\,. \quad (13)$$

The other interesting example of metriplectic dynamics is the Gilbert-Landau-Lifshitz description of magnetic systems. The Lie-Poisson brackets for the spin variable S_i^a occupying the lattice site i are

$$\left\{S_i^a, S_j^b\right\} = \delta_{ij}\epsilon^{abc}S_j^c\,. \quad (14)$$

The Casimir for this particular bracket is the spin length at a given lattice site $|\boldsymbol{S}_i|^2$. Thus this length is conserved *independently* of the choice of the magnetic model Hamiltonian.

The metric spin bracket [7] is

$$\{\{S_i^a, S_j^b\}\} = -\lambda\delta_{ij}|\boldsymbol{S}_i|^2 \left(\delta^{ab} - \frac{S_i^a S_j^b}{|\boldsymbol{S}_i|^2}\right) . \tag{15}$$

For an arbitrary magnetic Hamiltonian $\mathcal{H}(\boldsymbol{S})$ using the metriplectic union of brackets Eqs. (14) and (15), one obtains the Gilbert-Landau-Lifshitz equation of motion for the spin, namely

$$\partial_t \boldsymbol{S}_i = \boldsymbol{S}_i \times \boldsymbol{B}_i^{eff} - \lambda \dot{\boldsymbol{S}}_i \times \boldsymbol{S}_i \times \boldsymbol{B}_i^{eff}/|\boldsymbol{S}_i| \tag{16}$$

where $\boldsymbol{B}_i^{eff} \equiv -\delta\mathcal{H}/\delta\boldsymbol{S}_i$.

Since the spin length is conserved automatically in our approach, it can be used to construct a fast solver for the molecular dynamics of the spin system. Elsewhere such a solver and its implementations were discussed [14].

The structure of the metric bracket for a spin variable can easily be generalized to the case of an arbitrary Lie-Poisson bracket associated with the Lie algebra, the structure constants of which will be denoted as c^i_{jk}. Let w_i denote the elements of the Lie algebra, then the Lie-Poisson bracket is

$$\{w_i, w_j\} = c^k_{ij} w_k \equiv \gamma_{ij} . \tag{17}$$

Following the standard Lie algebra procedure [9], we can construct from the structure constants the metric tensor (Cartan–Killing tensor) with which we define the scalar product, and, since it is non singular, its inverse with which to raise and lower tensor indices. Explicitly, it is

$$\begin{aligned} G_{ij} &= -c^m_{in} c^n_{mj} , \\ G_{ij} G^{jm} &= \delta^m_i . \end{aligned} \tag{18}$$

It is easy to check that the length of the w_i defined by means of this metric tensor is the Casimir of the bracket Eq. (17).

With the use of the Cartan-Killing tensor, we can define the symmetric tensor

$$b_{ij} = G^{km} \gamma_{ki} \gamma_{mj} . \tag{19}$$

With these definitions, we generalize the spin metric bracket to

$$g_{ij} = \lambda(Casimirs) b_{ij} \equiv \lambda G^{km} \gamma_{ki} \gamma_{mj} , \tag{20}$$

where the damping coefficient λ is an arbitrary function of the Casimirs of the problem.

The metric bracket Eq. (20) describes the most general damping consistent with the algebraic structure of symplectic dynamics. It preserves the Casimirs, that is the damped motion stays on the (hyper) surfaces in the phase space defined by Casimirs, i.e. on the Casimir leafs.

It is very easy to check that the above proposed metriplectic construction is identical to that for spin variables. In the next section we shall show how it works in quantum mechanics.

3 The Schrödinger Equation and Madelung Representation.

We restrict ourselves to the nonrelativistic quantum mechanics of a spinless particle moving in an external force field defined by the potential $V(\boldsymbol{x})$. The fundamental equation of the dynamics, which governs the time evolution of the wave function $\psi(\boldsymbol{x},t)$, is the Schrödinger equation, i.e.

$$i\hbar\partial_t\psi(\boldsymbol{x},t) = \left(-\frac{\hbar^2}{2m}\nabla^2 + V(\boldsymbol{x})\right)\psi(\boldsymbol{x},t) \equiv \hat{H}\psi(\boldsymbol{x},t) . \tag{21}$$

The Schrödinger equation can be derived within symplectic dynamics by defining the Lie-Poisson bracket for the state variables ψ and ψ^* as

$$\{\psi(\boldsymbol{x}),\psi^*(\boldsymbol{y})\} = \frac{1}{i\hbar}\delta(\boldsymbol{x}-\boldsymbol{y}) . \tag{22}$$

Defining the system Hamiltonian as

$$\mathcal{H}(\psi,\psi^*) = \int d^dx\left(\frac{\hbar^2}{2m}|\nabla\psi|^2 + V(\boldsymbol{x})|\psi|^2\right) , \tag{23}$$

we can easily check that the "classical" equation

$$\partial_t\psi = \{\psi,\mathcal{H}\} \tag{24}$$

is actually identical to the Schrödinger equation Eq. (21).

In 1926 Madelung showed that the Schrödinger equation can be cast in a form which closely resembles the hydrodynamics of a non-dissipative fluid. The Madelung representation turned out to be a quite convenient tool for some applications, for example it serves as a natural bridge between quantum mechanics and quantum hydrodynamics [15].

Following Madelung we substitute $\psi(\boldsymbol{r},t) \to \sqrt{\rho(\boldsymbol{r},t)}\exp(i\phi(\boldsymbol{r},t))$ and then separate Eq. (21) into its real and imaginary part. The resulting equations are

$$\begin{aligned}\partial_t\rho(\boldsymbol{r},t) &= -\nabla\cdot\rho(\boldsymbol{r},t)\boldsymbol{u}(\boldsymbol{r},t) ,\\ \partial_t\boldsymbol{u}(\boldsymbol{r},t) + \boldsymbol{u}(\boldsymbol{r},t)\cdot\nabla\boldsymbol{u}(\boldsymbol{r},t) &= -\nabla\mu_Q(\rho,\nabla\rho) - \frac{1}{m}\nabla V(\boldsymbol{r},t) ,\end{aligned} \tag{25}$$

where $u(r,t) = \hbar/m\nabla\phi(r,t)$ is the (potential) velocity field of the "quantum" fluid and μ_Q is the quantum chemical potential. It is this last term which distinguishes the Madelung fluid dynamic equations from the usual Euler equations of hydrodynamics. Indeed,

$$\mu_Q = -\frac{\hbar^2}{2m\varrho}(\nabla^2\rho - 1/2\rho(\nabla\rho)^2)$$

is the only quantity in Eq. (25) containing the Planck constant, and it measures the "internal quantum pressure" responsible for wave-packet spreading. Alternatively the gradient of the quantum chemical potential can be written as the derivative of the quantum stress tensor $\sigma_{ij}^Q = \hbar^2/4m\partial_i\partial_j \ln(\rho)$. Note that the dependence of the quantum chemical potential μ_Q on the density and its gradients is essentially different from that in the generalization of conventional hydrodynamics often used in the theory of phase transitions, where the inhomogeneity of the order parameter (density in van der Waals like theories) is of importance [6, 16, 17]. The other important difference between Madelung and Euler hydrodynamics is that the value of circulation

$$\Gamma = \oint u \cdot dr = n\frac{\hbar}{m} \tag{26}$$

is quantized.

The benefits and/or shortcomings of the Madelung formulation of quantum mechanics are discussed in Ref. [15] and [18].

The Madelung representation of wave mechanics does not require anything beyond the Lie-Poisson brackets Eq. (11) and the Hamiltonian, which differs from that of a classical fluid by the presence of a quantum pressure term.

Indeed, using the Hamiltonian

$$\mathcal{H}(\varrho, u) = \int d^dx \left(\frac{m}{2}\varrho u^2 + V\varrho + \frac{\hbar^2}{2m}(\nabla\sqrt{\varrho})^2\right) \tag{27}$$

and the Poisson bracket relations Eq. (11), we obtain the Madelung equations Eq. (25) as in fluid dynamics, i.e. from Eq. (12).

4 Dissipative Quantum Mechanics - The Gisin Equation

As discussed in Section I and in Ref. [1], the nonlinear generalizations of that equation should be considered as phenomenological equations describing, for example, dissipative systems. Recently Enz [10] has given a comprehensive review of the possible quantization procedures for dissipative systems. A particularly interesting proposal for the description of quantum dissipative systems was given by Gisin [2]. In order to account for the evolution of the quantum state $|\psi\rangle$ of a *dissipative* system, Gisin introduced the phenomenological equation

$$i\hbar\frac{\partial|\psi\rangle}{\partial t} = \hat{H}|\psi\rangle + i\lambda\left(\frac{\langle\psi|\hat{H}|\psi\rangle}{\langle\psi|\psi\rangle} - \hat{H}\right)|\psi\rangle \equiv (1 - i\lambda\hat{Q}_\psi)\hat{H}\psi, \tag{28}$$

in which $\hat{H}$ is the system Hamiltonian and $\lambda \geq 0$ is a dimensionless damping constant. The operator $\hat{Q}_\psi$ is the projection operator $\hat{Q}_\psi = \hat{1} - |\psi\rangle\langle\psi|/\langle\psi|\psi\rangle \equiv 1 - \hat{P}_\psi$.

The structure of the bracketed term on the right-hand-side of the Gisin equation ensures that the norm of the state vector is preserved during the system evolution. This property distinguishes the Gisin equation from several other dissipative quantum mechanical "generalizations" of the Schrödinger equation [11] and permits the retention of most of the conventional interpretation of the quantum mechanics. Another interesting property of the equation, Eq. (28), is that the time evolution of the original Hamiltonian eigenstates is conservative (i.e. no damping). When the initial wave packet, consisting of several eigenstates, evolves in time it will eventually reach a final state which will be the lowest eigenstate present in the initial wave packet. This last property has an interesting implication for the model in which the wave function described by the Gisin equation represents a coherent state of a many boson or spin system [12, 13, 14]. The system *always* proceeds towards the ground state.

The Gisin equation is the only sensible candidate for a dissipative Schrödinger equation. It has been shown in Ref. [3] that this equation can be derived from the general Caldeira-Leggett formulation by assuming a proper form of the coupling between the so-called central particle (in this case a harmonic oscillator) and a heat bath consisting of bosonic oscillators with a given distribution of eigenstates and eigenenergies.

The Gisin equation has been tested on some applications from quantum optics [2]. In Ref. [7] it was shown that the Gisin equation for a Heisenberg spin system is actually equivalent to the quantum Gilbert-Landau-Lifshitz equation describing the time evolution of a dissipative spin system for which the magnetization is conserved. Here the conservation of magnetization is retained by virtue of a single lattice site "conservation law", i.e. by the fact that the spin length is the Casimir for the symplectic spin bracket.

The metric bracket for the wave function, which leads to the Gisin equations, has the form

$$\{\{\psi(\boldsymbol{r}), \psi^*(\boldsymbol{r}')\}\} = -\frac{\lambda}{\hbar}\left(\delta(\boldsymbol{r}-\boldsymbol{r}') - \frac{\psi(\boldsymbol{r})\psi^*(\boldsymbol{r}')}{||\psi||^2}\right) . \tag{29}$$

It is easy to check that the Gisin equation is obtained using the metriplectic bracket for the wave functions $[\psi, \psi^*] = \{\psi, \psi^*\} + \{\{\psi, \psi^*\}\}$ and the Hamiltonian Eq. (23).

5 Dissipative Quantum Hydrodynamics

Having introduced the dissipative Schrödinger equation, we can now proceed with its interpretation via the Madelung, or hydrodynamic, representation. As in Section 3, we decompose the wave function ψ as $\psi(\boldsymbol{x}) = \sqrt{\varrho(\boldsymbol{x})}\exp(i\phi(\boldsymbol{x}))$

and substitute it into the Gisin equation Eq. (28). We obtain a set of equations describing a damped (it is tempting to say viscous) quantum fluid. The momentum equation is the more or less obvious generalization of Eq. (25b), viz.

$$\partial_t \boldsymbol{u} + \boldsymbol{u} \cdot \nabla \boldsymbol{u} = -\frac{1}{m} \nabla V - \nabla \cdot \sigma_Q + \eta_Q \nabla^2 \boldsymbol{u} + \eta_Q \nabla (\boldsymbol{u} \cdot \nabla \log(\varrho)) \tag{30}$$

where $\eta_Q = \hbar\lambda/2m$ is the quantum "kinematic" viscosity coefficient. Note that Eq. (30) is a Gallilean invariant ($\boldsymbol{u} \to \boldsymbol{u} + \boldsymbol{U}$) in contrast to other suggested modifications of the Schrödinger equation for dissipative processes [11] and, further, that the dissipative terms on the right-hand-side of Eq. (30) are different from the Navier-Stokes theory. The Laplacian of the velocity term is present, as expected, but the "compressible" part of the Navier-Stokes equation $\propto \nabla\nabla \cdot \boldsymbol{u}$ is replaced by an essentially different term reflecting coupling between the velocity gradient and the density field which is necessary to account for the Heisenberg uncertain relation.

More intriguing is the continuity equation (conservation of probability) that follows from the Gisin equation, i.e.

$$\begin{aligned} &\partial \varrho + \nabla \cdot \varrho \boldsymbol{u} - \eta_Q \nabla^2 \varrho = \\ &+ \frac{2\lambda}{\hbar} \left[\varrho \langle \hat{H} \rangle - \left\{ \frac{m}{2} \varrho \boldsymbol{u}^2 + \varrho V + \frac{\hbar^2}{2m} (\nabla \sqrt{\varrho})^2 \right\} \right] \equiv \zeta_Q \, . \end{aligned} \tag{31}$$

The left-hand-side of this equation is the usual diffusion equation for the density field. The right-hand-side is more complicated. The "source" term conserves the probability, in the sense that the integral of ζ_Q vanishes identically for $\forall \psi$. Since the wave function can always be normalized, we can assume, without loss of essential physics, that

$$\int d^d \boldsymbol{x} \varrho(\boldsymbol{x}) = 1 \, . \tag{32}$$

Similarly, writing

$$\langle H \rangle \equiv \int d^d \boldsymbol{x} \mathcal{H}(\boldsymbol{x}) = E \, , \tag{33}$$

we can rewrite the source ζ_Q as

$$\zeta_Q = \frac{\eta_Q}{\hbar^2/2m} \left[E \varrho(\boldsymbol{x}) - \mathcal{H}(\boldsymbol{x}) \right] \, . \tag{34}$$

The bracketed term here is the difference between the energy density evaluated by means of the "equipartition" method, i.e. according to the probability density at a given point, and the true energy density. Obviously these two energy densities lead to the same global energy content, *i.e.* $\int d^d \boldsymbol{x} \zeta_Q = 0$.

The dissipative Madelung equations can be derived using the metriplectic brackets from the same Hamiltonian as the usual Madelung equation, provided the metric brackets between the density and velocity field are properly chosen.

These metric brackets are

$$\{\{\varrho(\boldsymbol{x}),\varrho(\boldsymbol{x}')\}\} = -\frac{\eta_Q}{\hbar^2/2m}\left[\varrho(\boldsymbol{x})\delta(\boldsymbol{x}-\boldsymbol{x}') - \frac{\varrho(\boldsymbol{x})\varrho(\boldsymbol{x}')}{\int d^d z\,\varrho(\boldsymbol{z})}\right]$$

$$\{\{u^a(\boldsymbol{x}),u^b(\boldsymbol{x}')\}\} = \frac{\eta_Q}{m}\nabla^a_{\boldsymbol{x}}\nabla^b_{\boldsymbol{x}'}\left(\frac{1}{\varrho(\boldsymbol{x})}\delta(\boldsymbol{x}-\boldsymbol{x}')\right) . \tag{35}$$

Note the difference between the velocity-velocity bracket in Eq. (35) and the one in the metric bracket for classical liquid, Eq. (13). The density dependence of the right-hand-side of the metric bracket Eq. (35b) is again a consequence of the uncertainty principle.

6 Final Comments

As we have seen from the above presentation the metriplectic formalism can equally well be applied to classical physics as to quantum mechanics. The important point is that by following the simple-minded idea of combining together the symplectic and metric brackets one can describe a variety of different dissipative systems using the same *algebraic* procedure. First, the symplectic structure of non-dissipative dynamics is analyzed, the proper Casimirs are found, and then, using the general method, the most general metric brackets consistent with the Casimirs are constructed.

How is one to proceed with those systems and/or models in which dissipation changes the value of the Casimirs, for example in magnetic systems in which the magnetization is not conserved? In such cases the geometrical methods of constructing most general metric bracket consistent with Casimirs, Eq. (19), are of great help. One defines the projection operator acting vertically to the hypersurface of constant Casimirs and then constructs another bracket proportional to that projector. The new damping constant appearing in this bracket corresponds, for example, to the longitudinal damping in NMR theory.

We should stress again that metriplectic dynamics is a convenient, easy and sometimes very powerful description of dissipative systems. It encompasses many interesting applications [5] but there are also important dissipative processes which have not been formulated within this type of an approach. The most attractive property of the metriplectic approach is its close connection with the underlying group theoretical structure of the theory. Once the symplectic brackets, that is the structure constants C^i_{jk} for given Lie algebra, are known, we can construct the symmetric bracket and subsequently the full metriplectic one. This procedure works in the same fashion both in classical and quantum theory and, indeed, is based on the geometical properties of phase space and Casimir leafs. One of the most interesting applications of metriplectic dynamic is the one which generalizes previous work on Boltzmann-Vlasov plasma [19] to that for the coloured plasmas (nonabelian gauge groups) encountered in quark-gluon plasma theory. Work along this line is in progress.

Acknowledgments

I should like to acknowledge discussions on classical and quantum dynamics with Richard Bausch, Andrzej Łusakowski, Rudi Schmitz and Magda Załuska-Kotur. Special thanks for Charles Enz with whom we worked on classical metriplectic dynamics.

References

1. N. Gisin and M. Rigo, J. Phys. **A28** (Math. Gen.), 7375 (1995).
2. N. Gisin, J. Phys. **A14** (Math. Gen.) 2259 (1981); Physica **A111**, 364 (1982).
3. M. Razavy and A. Pimpale, Phys. Rep. **168**, 307 (1988).
4. A.O. Caldeira and A.J. Leggett, Ann. Phys. (NY) **149**, 374 (1983).
5. Ł.A. Turski, Metriplectic Dynamics of Complex Systems. in: *Continuum Models and Discrete Systems*, (Dijon, 1989). vol. 1 G.A. Maugin, ed. Longman, London (1990).
6. C.P. Enz and Ł.A. Turski, Physica **96A**, 369 (1979).
7. J.A. Hołyst and Ł.A. Turski, Phys. Rev. **A45**, 4123 (1992), cf. also: Quantum Dissipative Dynamics, Proc. of the 7-th Symposium on Continuous Models and Discrete Systems, edited by K.H. Anthony and H.-J. Wagner, Trans Tech Publications, Vermannsdorf (1993).
8. M. Olko and Ł.A. Turski, Physica **166A**, 574 (1990); cf. also in *Nonlinear Phenomena in Solids, Liquids and Plasmas*, J.A. Tuszynski and W. Rozmus eds. World Scientific, Singapore (1990). M. Olko, PhD thesis (unpublished) Warsaw University (1996).
9. L.S. Pontriagin, *Continuous Groups*, Moscow (1954) (in Russian); *Grupy Topologiczne*, PWN, Warszawa (1961) (in polish).
10. C.P. Enz, Foundations of Physics **24**, 1281 (1994).
11. M.D. Kostin, J. Chem. Phys. **57**, 3589 (1972).
12. A. Perelomov, *Generalized Coherent States and Their Applications*, Springer, Berlin (1985).
13. R. Balakrishnan and A.R. Bishop, Phys. Rev. **B40**, 9194 (1989); R. Balakrishnan, J.H. Hołyst, and A.R. Bishop, J. Phys. (Condensed Matter) **C2**, 1869 (1990).
14. M.A. Olko and Ł.A. Turski, Physica, **A166**, 575 (1990).
15. Ł.A. Turski, Acta. Phys. Polon. **A26**, 1311 (1995).
16. Ł.A. Turski and J.S. Langer, Phys. Rev. **A46**, 53230 (1973); Phys. Rev. **A22**, 2189 (1980).
17. B. Kim, and G.F. Mazenko, J. Stat. Phys. **64**, 631 (1991).
18. E. Madelung, Z. Phys. **40**, 322 (1926). For recent introduction to the Madelung formulation of quantum mechanics, cf. I. Bialynicki-Birula, M. Cieplak, and J. Kamiński, *Theory of Quanta*, Oxford University Press, Oxford (1992).
19. A.N. Kaufman, Ł.A. Turski Phys. Lett. **A120**, 331 (1987).

Acknowledgements

I should like to acknowledge discussions on classical and quantum dynamics with Richard Bausch, Andreas [illegible], Rolf Schilling and [illegible] Kemer. Special thanks for Charles Enz with whom I worked on classical and quantum dynamics.

References

1. N. Gisin and M. Rigo, J. Phys. A28 (Math. Gen.) 7375 (1995).
2. N. Gisin, J. Phys. A14 (Math. Gen.) 2259 (1981); [illegible] (1984).
3. M. Razavy and A. Pimpale, Phys. Rep. 168, 305 (1988).
4. A.O. Caldeira and A.J. Leggett, Ann. Phys. (N.Y.) 149, 374 (1983).
5. R. Graham, "Stochastic Dynamics of Complex Systems: [illegible] Models and Dissipative [illegible]", [illegible] (1990).
6. C.P. Enz and L.A. Turski, Physica 96A, 369 (1979).
7. [illegible] and L.A. Turski, [illegible] (1984); [illegible] Dissipative Dynamics [illegible] systems, [illegible], [illegible] 1985.
8. N. Gisin and I.C. Percival, J. Phys. A25, 5677 (1992); [illegible] World Scientific, Singapore 1990; [illegible] (unpublished) [illegible] University (1989).
9. L.S. [illegible] (1989); [illegible] (unpubl.).
10. C.P. Enz, Found. of Physics [illegible] (1994).
11. [illegible]
12. [illegible]
13. [illegible]
14. J.H. [illegible] and A.R. [illegible], [illegible] 152, [illegible] (1990).
15. [illegible] and L.A. Turski, [illegible] A168, [illegible].
16. L.A. Turski, Acta Phys. Polon. A[illegible], [illegible] (1991).
17. L.A. Turski and [illegible], Phys. Rev. A44, 6238 (1991); Phys. Rev. A[illegible], [illegible] (1993).
18. [illegible] and [illegible], J. Stat. Phys. 64, 921 (1991).
19. E. Madelung, Z. Phys. 40, 322 (1926). For recent [illegible] to the [illegible] quantum mechanics [illegible], The Quantum Theory of Motion, Oxford University Press [illegible] (1993).
20. A.N. Kaufman, [illegible] Phys. Lett. A[illegible] (1984).

Quantum Analysis and Exponential Product Formulas

Masuo Suzuki

Department of Physics, University of Tokyo
Bunkyo-ku, Tokyo 113

1 Definition of quantum derivative

This paper explains the new concept of quantum analysis,[1-4] namely the non-commutative differential and integral calculus with respect to the relevant operator itself. The derivative $df(A)/dA$ with respect to the relevant operator A is defined by

$$\frac{df(A)}{dA} = \frac{\delta_{f(A)}}{\delta_A}. \tag{1}$$

Here δ_A denotes the inner derivation defined by

$$\delta_A Q = [A, Q] = AQ - QA \tag{2}$$

for any operator Q. Since $\delta_{f(A)}$ is proportional to δ_A for any analytic function $f(x)$, the ratio $\delta_{f(A)}/\delta_A$ in Eq. (1.1) is well defined. The derivative $df(A)/dA$ is a hyperoperator mapping an arbitrary operator dA to the derivation $df(A)$.

2 Some useful formulae

Using the definition Eq. (1.1), we obtain the following formulae:[1-4]

Formula 1:

$$\frac{df(A)}{dA} = \int_0^1 dt f^{(1)}(A - t\delta_A), \tag{1}$$

where $f^{(n)}(x)$ denotes the n th derivative of $f(x)$.

Formula 2:

$$\frac{d}{dA}(f(A) + g(A)) = \frac{df(A)}{dA} + \frac{dg(A)}{dA}. \tag{2}$$

Formula 3:

$$\frac{d}{dt}f(A(t)) = \frac{df(A(t))}{dA(t)} \cdot \frac{dA(t)}{dt}. \tag{3}$$

Formula 4 :

$$\frac{d}{dA}f(g(A)) = \frac{df(g)}{dg} \cdot \frac{dg(A)}{dA}. \tag{4}$$

Formula 5:

$$\frac{d^n f(A)}{dA^n} = \frac{n!}{2\pi i}\oint_C \frac{f(z)}{(z-A)(z-A+\delta_1)\cdots(z-A+\delta_1+\cdots+\delta_n)}dz, \tag{5}$$

where $\oint_C$ denotes an anti-clockwise integration around the contour C and the $\{\delta_j\}$ are defined by

$$\delta_j : (dA)^n = (dA)^{j-1}(\delta_A dA)(dA)^{n-j}. \tag{6}$$

Formula 6: Using Formula 5, we can obtain the following formula:

$$\frac{d^n f(A)}{dA^n} = n!\sum_{k=0}^{n} a_{n,k}(\{\delta_j\})f(A-\delta_1-\cdots-\delta_k). \tag{7}$$

Here the coefficients $\{a_{n,k}\}$ are defined by

$$a_{n,0} = \{\delta_1(\delta_1+\delta_2)(\delta_1+\delta_2+\delta_3)\cdots(\delta_1+\cdots+\delta_n)\}^{-1}, \tag{8}$$

$$a_{n,k} = \frac{(-1)^k}{(\delta_1+\cdots+\delta_k)(\delta_2+\cdots+\delta_k)\cdots\delta_k\delta_{k+1}(\delta_{k+1}+\delta_{k+2})\cdots(\delta_{k+1}+\cdots+\delta_n)} \tag{9}$$

for $1 \le k \le n-1$, and

$$a_{n,n} = \frac{(-1)^n}{(\delta_1+\cdots+\delta_n)(\delta_2+\cdots+\delta_n)\cdots(\delta_{n-1}+\delta_n)\delta_n}. \tag{10}$$

Formula 7 (Operator Taylor Expansion):

$$f(A+xB) = f(A) + \textstyle\sum_{n=1}^{\infty} x^n \int_0^1 dt_1 \int_0^{t_1} dt_2 \cdots \int_0^{t_{n-1}} dt_n f^{(n)}(A - t_1\delta_1 - \cdots - t_n\delta_n) \; : \; B^n. \tag{11}$$

This gives the following Feynman expansion formula:

$$e^{t(A+xB)} = e^{tA} + \sum_{n=1}^{\infty} x^n e^{tA}\int_0^t dt_1 \int_0^{t_1} dt_2 \cdots \int_0^{t_{n-1}} dt_n B(t_1)\cdots B(t_n), \tag{12}$$

where

$$B(t) = e^{-tA}Be^{tA} = e^{-t\delta_A}\cdot B. \tag{13}$$

3 Partial differentiation

The differential of $f(\{A_j\})$ is defined by

$$df(\{A_j\}) = \lim_{h\to 0}\frac{1}{h}[f(\{A_j + hdA_j\}) - f(\{A_j\})]. \tag{1}$$

The partial derivative $\partial f(\{A_j\})/\partial A_k$ is defined in

$$df(\{A_j\}) = \sum_k \frac{\partial f(\{A_j\})}{\partial A_k}\cdot dA_k. \tag{2}$$

The partial derivative $\partial f/\partial A_k$ is a hyperoperator which maps the operator dA_k to the derivation $df(\{A_j\})$ for $dA_j \equiv 0$ $(j \neq k)$, and $\partial f/\partial A_k$ is expressed in terms of $\{A_j\}$ and $\{\delta_{A_j}\}$. We have the following.

Formula 8:

$$\frac{d}{dt}f(\{A_j(t)\}) = \sum_k \frac{\partial f(\{A_j(t)\})}{\partial A_k(t)}\cdot\frac{dA_k(t)}{dt}. \tag{3}$$

Formula 9:

$$\delta_{f(\{A_j\})} = \sum_k \frac{\partial f(\{A_j\})}{\partial A_k}\delta_{A_k}. \tag{4}$$

Formula 10: When $f(\{A_j\})$ is expressed as a convergent non-commutative power series of $\{A_j\}$, namely

$$f(\{A_j\}) = \sum_{\{t_{jk}\}} a(\{t_{jk}\})A_1^{t_{11}}A_2^{t_{12}}\cdots A_n^{t_{1n}}A_1^{t_{21}}\cdots A_k^{t_{jk}}\cdots, \tag{5}$$

we have

$$\frac{\partial f}{\partial A_k} = \sum_{\{t_{jk}\}}\sum_j a(\{t_{jk}\})A_1^{t_{11}}\cdots A_k^{t_{jk-1}}(A_{k+1}^{t_{jk+1}}\cdots)^{\sim}(\frac{d}{dA_k}A_k^{t_{jk}}). \tag{6}$$

Here, we have used the tilde operator $\tilde{f}$ satisfies the following:

$$A_j^{\sim} \equiv \tilde{A}_j = A_j - \delta_{A_j},\ (fg)^{\sim} = \tilde{g}\tilde{f},\ (cf)^{\sim} = c\tilde{f},\ (f+g)^{\sim} = \tilde{f}+\tilde{g} \tag{7}$$

for any number c and any operators f and g.

We define a partial inner derivation $\delta_{Q;A_k} \equiv \delta_{Q;k}$ by such an inner derivation as takes the commutation relation only with respect to the operator A_k. For example,

$$\delta_{Q;k}\cdot(A_kA_jA_k^2) = [Q,A_k]A_jA_k^2 + A_kA_j[Q,A_k^2] \tag{8}$$

for $j \neq k$. Using this partial inner derivation, we find

Formula 11:

$$\frac{\partial f(\{A_j\})}{\partial A_k} = \delta_{f(\{A_j\});k}\delta_{A_k}^{-1}. \tag{9}$$

4 Higher partial derivatives

Similarly we define higher partial derivatives(1–4) in order to obtain an operator Taylor expansion formula(1–4) for the operator $f(\{A_j + xB_j\})$ with respect to the parameter x, viz.

$$f(\{A_j + xB_j\}) = \sum_{n=0}^{\infty} x^n \sum_{j_1,\cdots,j_n} f^{(n)}_{j_1,\cdots,j_n} : B_{j_1} \cdot\cdots\cdot B_{j_n}. \tag{1}$$

The hyperoperator $f^{(n)}_{j_1,\cdots,j_n}$ is too complicated(1–4) to write down here explicitly.

5 Application to exponential product formulae

Quantum analysis is useful in constructing exponential product formulae of higher order:

$$e^{x(A+B)} = e^{xt_1A}e^{xt_2B}e^{xt_3A}e^{xt_4B}\cdots e^{xt_MA} + O(x^{m+1}). \tag{1}$$

It has been shown(5–12) that there exist real parameters $\{t_j\}$ for any positive integer m. In order to determine these values explicitly, we have to solve the following inverse problem, namely to find the operator $\Phi(A, B; \{t_j\})$ satisfying the relation

$$e^{xt_1A}e^{xt_2B}e^{xt_3A}e^{xt_4B}\cdots e^{xt_MA} = e^{\Phi(A,B;\{t_j\})}. \tag{2}$$

The condition on the parameters $\{t_j\}$ is given by the requirement that $\Phi(A, B; \{t_j\})$ should agree with $x(A + B)$ up to the order of x^m. In order to calculate explicitly up to the required order, we derive an operator differential equation for $\Phi(A, B; \{t_j\}) \equiv \Phi(x)$ using the quantum analysis and we solve it iteratively.(2)

More generally we consider the product formula

$$e^{x(A+B)} = e^{C_1(x)}e^{C_2(x)}\cdots e^{C_r(x)} + O(x^{m+1}) \tag{3}$$

for some positive integer m. Here, we have the trivial conditions that $C_1(0) = \cdots = C_r(0) = 0$ and that

$$C_1'(0) + C_2'(0) + \cdots + C_r'(0) = A + B \tag{4}$$

for some appropriate set of basis operators $\{C_j(x)\}$. In order to determine these operators, we express the product on the right-hand side of Eq. (5.3) as the simple exponential operator

$$e^{C_1(x)}e^{C_2(x)}\cdots e^{C_r(x)} = e^{x(A+B)+R(x)}. \tag{5}$$

The correction term $R(x)$ can be obtained by solving the equation

$$\frac{dR(x)}{dx} + (A+B) = \Delta^{-1}(x(A+B)+R(x))\sum_{j=1}^{r} exp(\delta_{C_1(x)})\cdots exp(\delta_{C_{j-1}(x)})\Delta(C_j(x))\frac{dC_j(x)}{dx}. \tag{6}$$

Here, the hyperoperator $\Delta(A)$ is defined by

$$\Delta(A) = \frac{e^{\delta_A}-1}{\delta_A} \tag{7}$$

with the inner derivation δ_A. The above operator differential equation is very convenient for obtaining perturbatively with respect to x explicit conditions on the basis operators $\{C_j(x)\}$ satisfying Eq. (5.3) for some given positive integer m.

If we apply the above general formula Eq. (5.6) to the exponential product formula Eq. (5.1), then we obtain the relation

$$\sum_{j=1}^{M} e^{t_1 x\delta_A}e^{t_2 x\delta_B}\cdots e^{t_{j-1}x\delta_{j-1}}t_jC_j = A+B \tag{8}$$

up to the order of x^{m-1}. Here, we have put $C_j(x) = t_jxC_j$ in Eq. (5.3), where $C_{2j-1} = A$ and $C_{2j} = B$. The inner derivation δ_j is defined by $\delta_j \equiv \delta_{C_j}$ (namely $\delta_{2j-1} = \delta_A$ and $\delta_{2j} = \delta_B$). This new formulation is much more convenient than the previous direct procedures.[5–12]

6 Concluding remarks

Quantum analysis has been applied[2,3] in the derivations of several basic equations for the non-equilibrium density matrix $\rho(t)$ and the entropy operator defined by $\eta(t) = -\log\rho(t)$. It can be also used[2,3] to derive fundamental operator-differential equations on exponential product formulae, from which the Baker-Campbell-Hausdorff formula is easily derived and hybrid exponential product formulae are also systematically obtained.[2,3] There are many other applications in quantum mechanics and statistical physics.

Acknowledgments

The present work is partially supported by the Society of Non-Traditional Technology.

References

1. M. Suzuki, submitted to Commun. Math. Phys.
2. M. Suzuki, submitted to J. Stat. Phys.
3. M. Suzuki, Int. J. Mod. Phys. B, Vol.10, Nos.13 & 14 (June 15 & 30, 1996) Umezawa's Memorial Issue.
4. M. Suzuki, Proceedings of ICPAM in Bahrain, Nov.19-22, 1995.
5. M. Suzuki, Phys. Lett. **A146**, 319 (1990); *ibid.* **A165**, 387 (1992).
6. M. Suzuki, J. Math. Phys. **32**, 400 (1991).
7. M. Suzuki, J. Phys. Soc. Jpn. **61**, 3015 (1992).
8. M. Suzuki, in *Fractals and Disorder,* edited by A. Bunde (North-Holland, 1992), i.e., Physica **A191**, 501 (1992).
9. M. Suzuki, Proc. Japan Acad. **69**, Ser. B, 161 (1993).
10. M. Suzuki, Physica **A205**, 65 (1994), and references cited therein.
11. M. Suzuki, Commun. Math. Phys. **163**, 491 (1994).
12. M. Suzuki, Rev. of Math. Phys. (World Scientific) (1996) (in press), and references cited therein. See also K. Aomoto, *On a Unitary Version of Suzuki's Exponential Product Formula*, Jour. of Math. Soc. Japan (in press).

Coherent Anomaly Method and Its Applications to Critical Phenomena

Masuo Suzuki

Department of Physics, University of Tokyo, Bunkyo-ku, Tokyo 113

1 Basic scheme of the CAM

The present lecture gives a review of the coherent-anomaly method[1-3] (or CAM) and its applications to critical phenomena. The basic idea of this method is to construct a systematic (or coherent) series of mean-field approximations and to evaluate each approximate critical point T_c and the mean-field critical coefficient $\bar{Q}$ at T_c. Then, these data $\{T_c, \bar{Q}\}$ give a coherent anomaly, namely the mean-field critical coefficient $\bar{Q}$ diverges as T_c approaches the true critical point T_c^*, i.e.

$$\bar{Q}(T_c) \to \infty \quad as \quad T_c \to T_c^*. \tag{1}$$

More explicitly, we may put

$$\bar{Q}(T_c) \simeq \frac{C}{(T_c - T_c^*)^{\psi}} \tag{2}$$

with some constant C. The coherent-anomaly exponent ψ can be estimated from the coherent-anomaly data $\{T_c, \bar{Q}\}$. If the relevant physical quantity Q shows a mean-field singularity of the form

$$Q_{mf}(T) \simeq \frac{\bar{Q}(T_c)}{(T - T_c)^{\phi}}, \tag{3}$$

then the true singularity of Q will be given by

$$Q(T) \sim \frac{1}{(T - T_c^*)^{\phi+\psi}}. \tag{4}$$

Therefore, we have only to construct a systematic (or coherent) series of mean-field approximations in order to study non-classical critical phenomena.

2 Envelope theory of the CAM

The simplest explanation of the CAM is to make use of the envelope theory. According to the mathematical theory of envelopes, the tangential points of general solutions with a parameter constitute a special solution not included in general solutions of the relevant problem. In our problem, the general approximate solutions are given by

$$Q_{mf}(T) \simeq \frac{C}{(T_c - T_c^*)^\psi} \frac{1}{(T - T_c)^\phi} \tag{1}$$

with some constant C. The tangential points $\{T\}$ are given by differentiating Eq. (2.1) with respect to T_c and by eliminating T_c both from Eq. (2.1) and from the derived equation

$$T = T_c^* + \frac{\psi + \phi}{\psi}(T_c - T_c^*). \tag{2}$$

Thus, the envelope function is given by

$$Q_{env}(T) = C\frac{(\phi + \psi)^{\phi+\psi}}{\phi^\phi \psi^\psi} \cdot \frac{1}{(T - T_c^*)^{\psi+\phi}}. \tag{3}$$

This gives a fractional singularity of the relevant physical quantity $Q(T)$ in the form

$$Q(T) \sim Q_{env}(T) \sim \frac{1}{(T - T_c^*)^\omega} \tag{4}$$

with

$$\omega = \psi + \phi. \tag{5}$$

This non-classical value of the critical exponent ω is composed of two parts, namely the classical part ϕ and the coherent-anomaly part ψ. The second part can be easily estimated from the coherent-anomaly data $\{T_c, \bar{Q}\}$ for several systematic mean-field approximations.

In particular, the susceptibility $\chi(T)$ is given in the form

$$\chi(T) \sim \frac{1}{(T - T_c^*)^\gamma} \tag{6}$$

with

$$\gamma = \psi + 1, \tag{7}$$

because we have $\phi = 1$ for the mean-field susceptibility. The coherent-anomaly exponent ψ is defined by

$$\bar{\chi}(T_c) \sim \frac{1}{(T_c - T_c^*)^\psi}. \tag{8}$$

3 CAM scaling

The physical basis of the CAM is the scaling of correlation functions or Fisher's finite-size scaling.[4] The mean-field coefficient $\bar{Q}(T_c)$ in Eq. (1.3) is expressed[1-3] in terms of the correlation functions inside of the relevant cluster which is used for the construction of each mean-field approximation. Then, the scaling form of correlation functions gives the expression

$$\bar{Q}(T_c) \sim \frac{1}{(T_c - T_c^*)^{\omega-\phi}} \tag{1}$$

near the true critical point. That is, we obtain

$$\psi = \omega - \phi, \tag{2}$$

namely

$$\omega = \psi + \phi. \tag{3}$$

This is called the CAM scaling relation. Many other CAM scaling relations have been obtained.[1-3]

4 Applications

The first application of the CAM was made to the two-dimensional Ising model by Katori and the present author[5] to confirm the validity of the CAM. In fact, the value $\gamma = 7/4$ for the susceptibility exponent has been confirmed with very high accuracy (four or five-digit accuracy!). Since then, many applications have been reported:

- the Ising-model with long-range interaction by Monroe et al., (App. 57, 58)
- the Blume-Emery-Griffiths model by Chakraborty and by Kolesik, (App. 2 and App. 40)
- the Ising model with four-spin interactions which shows non-universal critical behaviour Minami et al. and by Kolesik and Samaj (App. 52-55 and App. 38, 39)
- the Heisenberg model by Ito and Suzuki, by Oguchi and Kitatani, by Mano and by Tanaka and Kimura, (App. 21, App. 65, App. 48 and App.102)
- the zero-temperature phase transition by Nonomura and Suzuki, (App. 59-64)
- the KT-transition by Suzuki, by Hu and by Fujiki, (App. 93, App. 17, 18 and App. 7)
- spin glasses by Hatano and Suzuki, by Fujiki and by Kawashima et al., (App. 8,9, App. 5 and App. 32)
- the self-avoiding walk by Hu and Suzuki and by Ishinabe et al., (App. 14 and App. 20)

- the percolation problem by Suzuki and by Takayasu et al. and by Lipowski and Suzuki
  (App. 84, App. 100, 101 and App. 45)
- the kinetic Ising model for evaluating the critical slowing-down exponent by Katori and Suzuki,
  (App. 25)
- contact processes by Konno and Katori, and
  (App. 41)
- other miscellaneous problems
  (App. 10, 15, 19, 27, 29, 30, 34, 50, 51, 67, 69, 70, 71, 72, 85, 87, 88, 89, 93, 96, 98, 108, 113)

5 Summary

The basic idea of the CAM theory has been briefly explained. In the present lecture, many applications have been demonstrated explicitly to show how useful the CAM is.

The present work is partially supported by the Society of Non-Traditional Technology.

References

1. M. Suzuki, J. Phys. Soc. Jpn. **55** (1986) 4205.
2. M. Suzuki, M. Katori and X. Hu, J. Phys. Soc. Jpn. **56** (1987) 3092.
3. M. Suzuki, X. Hu, M. Katori, A. Lipowski, N. Hatano, K. Minami, and Y. Nonomura, *Coherent Anomaly Method — Mean-field, Fluctuations and Systematics* (World Scientific, Singapore, 1995).
4. M.E. Fisher and M.N. Barber, Phys. Rev. Lett. **28** (1972) 1516.
5. M. Katori and M. Suzuki, J. Phys. Soc. Jpn. **25** (1987) 3113.

Appendix

List of Papers Concerning the CAM Theory

1. Cenedese, P., J. M. Sanchez and R. Kikuchi: Continuous sequence of mean-field approximations and critical phenomena, Physica **A209** (1994) 257-267.
2. Chakraborty, K. G.: Coherent anomaly method for Blume-Capel model, Physica **A189** (1992) 271-281.
3. Ferreira, A. L. C. and S. K. Mendiratta: Mean-field approximation with coherent anomaly method for a non-equilibrium model, J. Phys. A: Math. Gen. **26** (1993) L145-L150.
4. Frischat, S. D. and R. Kühn: Finite-size scaling analysis of generalized mean-field theories, J. Phys. A: Math. Gen. **28** (1995) 2771.

5. Fujiki, S.: Application of the coherent anomaly method to d-dimensional Ising spin glasses, in *Cooperative Dynamics in Complex Physical Systems*, ed. H. Takayama (Springer-Verlag, Berlin, 1988) pp. 179-180.
6. Fujiki, S., M. Katori and M. Suzuki: Study of coherent anomalies and critical exponents based on high-level cluster-variation approximations, J. Phys. Soc. Jpn. **59** (1990) 2681-2687.
7. Fujiki, S.: Application of the cluster variation method to the ferromagnetic six-state clock model on the triangular lattice and its CAM analysis, J. Phys. Soc. Jpn. **62** (1993) 556-562.
8. Hatano, N. and M. Suzuki: Effective-field theory of spin glasses and the coherent-anomaly method. I, J. Stat. Phys. **63** (1991) 25-46.
9. Hatano, N. and M. Suzuki: Effective-field theory of spin glasses and the coherent-anomaly method. II. Double-cluster approximation, J. Stat. Phys. **66** (1992) 897-911.
10. Hirata, Y. S.: The critical point within the metastable region in $D = 4$ $Z(2)$ lattice gauge theory, Prog. Theor. Phys. **82** (1989) 34-39 (L).
11. Horiguchi, T., K. Tanaka and T. Morita: Coherent-anomaly method for the wave-number dependence of the susceptibility, J. Phys. Soc. Jpn. **59** (1990) 4196-4197 (Comment).
12. Hu, X., M. M. Katori and M. Suzuki: Coherent-anomaly method in critical phenomena. III. Mean-field transfer-matrix method in the 2D Ising model, J. Phys. Soc. Jpn. **56** (1987) 3865-3880.
13. Hu, X. and M. Suzuki: Coherent-anomaly method in critical phenomena. IV. Study of the wave-number-dependent susceptibility in the 2D Ising model, J. Phys. Soc. Jpn. **57** (1988) 791-806.
14. Hu, X. and M. Suzuki: Coherent-anomaly method in self-avoiding walk problems, Physica **A150** (1988) 310-323.
15. Hu, X. and M. Suzuki: Coherent-anomaly method in cooperative phenomena — Some applications to polymer physics, in *Space-Time Organization in Macromolecular Fluids*, eds. F. Tanaka, M. Doi and T. Ohta (Springer-Verlag, Berlin, 1988) pp. 188-197.
16. Hu, X. and M. Suzuki: An approach to critical phenomena from one subsystem based on the CAM analysis, J. Phys. A: Math. Gen. **23** (1990) 3051-3060.
17. Hu, X.: A new possible approach in Kosterlitz-Thouless transitions, Prog. Theor. Phys. **89** (1993) 545-549 (L).
18. Hu, X.: Series of effective-field approximations and coherent anomaly in Kosterlitz-Thouless transitions, J. Phys. A: Math. Gen. **27** (1994) 2313-2323.
19. Inui, N.: Application of the coherent anomaly method to the branching annihilation random walk, Phys. Lett. **A184** (1993) 79-82.
20. Ishinabe, T., J. F. Douglas, A. M. Nemirovsky and K. F. Freed: Examination of the $1/d$ expansion method from exact enumeration for a self-interacting self-avoiding walk, J. Phys. A: Math. Gen. **27** (1994) 1099-1109.
21. Ito, N. and M. Suzuki: Coherent-anomaly method for quantum spin systems, Int. J. Mod. Phys. **B2** (1988) 1-11.

22. Ito, N. and M. Suzuki: Size dependence of coherent anomalies in self-consistent cluster approximations, Phys. Rev. B**43** (1991) 3483-3492.
23. Katori, M. and M. Suzuki: Coherent anomaly method in critical phenomena. II. Applications to the two- and three-dimensional Ising models, J. Phys. Soc. Jpn. **56** (1987) 3113-3125.
24. Katori, M. and M. Suzuki: Coherent-anomaly method in critical phenomena, in *Progress in Statistical Mechanics*, ed. C.-K. Hu (World Scientific, Singapore, 1988) pp. 273-287.
25. Katori, M. and M. Suzuki: Coherent-anomaly method in critical phenomena. V. Estimation of the dynamical critical exponent Δ of the two-dimensional kinetic Ising model, J. Phys. Soc. Jpn. **57** (1988) 807-817.
26. Katori, M. and M. Suzuki: Study of coherent anomalies and critical exponents based on the cluster-variation method, J. Phys. Soc. Jpn. **57** (1988) 3753-3761.
27. Katori, M.: A second-order phase transition as a limit of the first-order phase transitions — Coherent anomalies and critical phenomena in the Potts models —, J. Phys. Soc. Jpn. **57** (1988) 4114-4125.
28. Katori, M. and M. Suzuki: On the CAM canonicality of the cluster-variation approximations, Prog. Theor. Phys. Suppl. **115** (1994) 83-93.
29. Kawasaki, K., K. Tanaka, C. Hamamura and R. A. Tahir-Kheli: Magnetic phase transitions in randomly diluted FCC spin system with competing interactions: The case with ferromagnetic J_1 and arbitrary J_2, Phys. Rev. B**45** (1992) 5321-5327.
30. Kawashima, N. and M. Suzuki: Chiral phase transition of planer antiferromagnets analyzed by the super-effective-field theory, J. Phys. Soc. Jpn. **58** (1989) 3123-3130.
31. Kawashima, N., M. Katori, C. Tsallis and M. Suzuki: Systematic approach to critical phenomena by the extended variational method and coherent-anomaly method, Int. J. Mod. Phys. B**4** (1990) 1409-1422.
32. Kawashima, N., N. Hatano and M. Suzuki: Critical behaviour of the two-dimensional EA model with a Gaussian bond distribution, J. Phys. A: Math. Gen. **25** (1992) 4985-5003.
33. Kinosita, Y., N. Kawashima and M. Suzuki: Coherent-anomaly analysis of series expansions and its application to the Ising model, J. Phys. Soc. Jpn. **61** (1992) 3887-3901.
34. Kitatani, H. and T. Oguchi: Ferromagnetic-nonferromagnetic phase boundary on the two-dimensional $\pm J$ Ising model, J. Phys. Soc. Jpn. **59** (1990) 3823-3826 (L).
35. Kitatani, H., T. Kakiuchi, N. Ito and M. Suzuki: CAM analysis of dimensionality crossover, J. Phys. Soc. Jpn. **63** (1994) 2511-2513 (L).
36. Kobayashi, H. and M. Suzuki: Theory of correlation and susceptibility based on the confluent transfer matrix and its associated transfer matrix, Physica A**199** (1993) 619-639.
37. Kolesík, M. and L. Šamaj: New variational series expansions for lattice models, J. Phys. (France) **I 3** (1993) 93-106.

38. Kolesík, M. and L. Šamaj: Evidence for the nonuniversality of a 3D vertex model, Phys. Lett. **A177** (1993) 87-92.
39. Kolesík, M. and L. Šamaj: Series expansion and CAM study of the nonuniversal behavior of the symmetric 16-vertex model, J. Stat. Phys. **72** (1993) 1203-1226.
40. Kolesík, M.: Coherent-anomaly approach to Blume-Emery-Griffiths model, Int. J. Mod. Phys. **B8** (1994) 113-126.
41. Konno, N. and M. Katori: Applications of the CAM based on a new decoupling procedure of correlation functions in the one-dimensional contact process, J. Phys. Soc. Jpn. **59** (1990) 1581-1592.
42. Lipowski, A.: Coherent anomaly method with modified Bethe approximation, J. Magn. Magn. Mater. **96** (1991) 267-274.
43. Lipowski, A.: Coherent anomaly method with a new mean-field type approximation, Physica **A173** (1991) 293-301.
44. Lipowski, A. and M. Suzuki: Study on phase transitions of antiferromagnetic Ising models using the coherent-anomaly method, J. Phys. Soc. Jpn. **61** (1992) 2484-2490.
45. Lipowski, A. and M. Suzuki: Convergence of mean-field approximations in site percolation and application of CAM to $d = 1$ further-neighbors percolation problem, J. Stat. Phys. **69** (1992) 1-16.
46. Lipowski, A. and M. Suzuki: Exact critical temperature by mean-field approximation, J. Phys. Soc. Jpn. **61** (1992) 4356-4366.
47. Malakis, A. and S. S. Martinos: Variational approximations and mean-field scaling theory, J. Phys. A: Math. Gen. **27** (1994) 7283-7299.
48. Mano, H.: Study of critical phenomena of quantum spin systems by spin-cluster approximation series, J. Magn. Magn. Mater. **90** & **91** (1990) 281-283.
49. Mano, H. and K. Nakano: Study of critical phenomena by spin-cluster approximation series: Ising spin system, J. Phys. Soc. Jpn. **60** (1991) 548-561.
50. Mano, H.: Critical properties of diluted Ising ferromagnets, J. Magn. Magn. Mater. **104–107** (1992) 259-260.
51. Marques, M. C. and A. L. Ferreira: Critical behaviour of a long-range nonequilibrium system, J. Phys. A: Math. Gen. **27** (1994) 3389-3395.
52. Minami, K., Y. Nonomura, M. Katori and M. Suzuki: Multi-effective-field theory: Applications to the CAM analysis of the two-dimensional Ising model, Physica **A174** (1991) 479-503.
53. Minami, K. and M. Suzuki: Coherent-anomaly method applied to the eight-vertex model, Physica **A187** (1992) 282-307.
54. Minami, K. and M. Suzuki: Non-universal critical behaviour of the two-dimensional Ising model with crossing bonds, Physica **A192** (1993) 152-166.
55. Minami, K. and M. Suzuki: A two-dimensional Ising model with non-universal critical behaviour, Physica **A195** (1993) 457-473.
56. Miyazima, S., K. Maruyama and K. Okumura: Critical exponent of bulk conductivity in Swiss cheese model, J. Phys. Soc. Jpn. **60** (1991) 2805-2807 (L).

57. Monroe, J. L.: The coherent anomaly method applied to a variety of Ising models, Phys. Lett. **A131** (1988) 427-429.
58. Monroe, J. L., R. Lucente and J. P. Hourlland: The coherent anomaly method and long-range one-dimensional Ising models, J. Phys. A: Math. Gen. **23** (1990) 2555-2562.
59. Nonomura, Y. and M. Suzuki: Coherent-anomaly method in zero-temperature phase transitions in quantum spin systems, J. Phys. A: Math. Gen. **25** (1992) 85-100.
60. Nonomura, Y. and M. Suzuki: CAM approach to ground-state phase transition in the two-dimensional transverse Ising model, J. Phys. A: Math. Gen. **25** (1992) 5463-5473.
61. Nonomura, Y. and M. Suzuki: CAM approach to correlation function of quantum spin chain in the ground state, J. Phys. Soc. Jpn. **61** (1992) 3960-3965.
62. Nonomura, Y. and M. Suzuki: Coherent anomaly in Kosterlitz-Thouless-type transition in the $S = 1/2$ XXZ chain at the ground state, J. Phys. Soc. Jpn. **62** (1993) 3774-3777 (L).
63. Nonomura, Y. and M. Suzuki: Cluster-effective-field approximations in frustrated quantum spin systems: CAM analysis of Néel-dimer transition in the $S = 1/2$ frustrated XXZ chain at the ground state, J. Phys. A: Math. Gen. **27** (1994) 1127-1138.
64. Nonomura, Y.: New finite-size-scaling analysis in ground-state phase transitions, J. Magn. Magn. Mater. **140-144** (1995) 1495-1496.
65. Oguchi, T. and H. Kitatani: Coherent-anomaly method applied to the quantum Heisenberg model, J. Phys. Soc. Jpn. **57** (1988) 3973-3978.
66. Oguchi, T. and H. Kitatani: Coherent-anomaly method applied to the lattice obtained by the dual transformation in the Ising model, J. Phys. Soc. Jpn. **58** (1989) 3033-3036 (L).
67. Patrykiejew, A. and P. Borowski: Application of the Monte Carlo coherent-anomaly method to two-dimensional lattice-gas systems with further-neighbor interactions, Phys. Rev. B**42** (1990) 4670-4676.
68. Pelizzola, A. and A. Stella: Mean-field renormalization group for the boundary magnetization of strip clusters, J. Phys. A: Math. Gen. **26** (1993) 6747-6756.
69. Pelizzola, A.: Boundary magnetization of a two-dimensional Ising model with inhomogeneous nearest-neighbor interactions, Phys. Rev. Lett. **73** (1994) 2643-2645.
70. Šamaj, L. and M. Kolesík: Self-duality of the $O(2)$ gauge transformation and the phase structure of vertex models, Physica **A193** (1993) 157-168.
71. Sardar, S. and K. G. Chakraborty: Coherent anomalies and critical singularities of the generalized Ising model, Physica **A199** (1993) 154-164.
72. Schmittmann, B and R. K. P. Zia: in Statistical Mechanics of Driven Diffusion Systems *Phase Transitions and Critical Phenomena*, (eds. C. Domb and J. L. Lebowitz) vol.17, p.16 (1995).
73. Suzuki, M.: New methods to study critical phenomena — Systematic clus-

ter mean-field approach, and thermo field dynamics of interacting quantum systems, in *Quantum Field Theory*, ed. F. Mancini (North-Holland, Amsterdam, 1986) pp. 505-531.
74. Suzuki, M. and M. Katori: New method to study critical phenomena — Mean-field finite-size scaling theory, J. Phys. Soc. Jpn. **55** (1986) 1-4 (L).
75. Suzuki, M.: Coherent anomalies and an asymptotic method in cooperative phenomena, Phys. Lett. **A116** (1986) 375-381.
76. Suzuki, M.: Skeletonization, fluctuating mean-field approximations and coherent anomalies in critical phenomena, Prog. Theor. Phys. Suppl. **87** (1986) 1-22.
77. Suzuki, M.: Statistical mechanical theory of cooperative phenomena. I. General theory of fluctuations, coherent anomalies and scaling exponents with simple applications to critical phenomena, J. Phys. Soc. Jpn. **55** (1986) 4205-4230.
78. Suzuki, M.: Nonlinear fluctuations, separation of procedures, and linearization of processes, J. Stat. Phys. **49** (1987) 977-992.
79. Suzuki, M., M. Katori and X. Hu: Coherent anomaly method in critical phenomena. I., J. Phys. Soc. Jpn. **56** (1987) 3092-3112.
80. Suzuki, M.: Power-series CAM theory, J. Phys. Soc. Jpn. **56** (1987) 4221-4224 (L).
81. Suzuki, M.: Continued-fraction CAM theory, J. Phys. Soc. Jpn. **57** (1988) 1-4 (L).
82. Suzuki, M.: Asymptotics, self-similarity and coherent-anomaly method in cooperative phenomena, Sci. Form. **3** (1988) 43-56.
83. Suzuki, M.: Scaling and CAM theory in far-from-equilibrium systems, in *Dynamics of Ordering Processes in Condensed Matter*, eds. S. Komura and H. Furukawa (Plenum Press, New York, 1988) pp. 23-28.
84. Suzuki, M.: CAM estimates of critical exponents of spin glasses and percolation, Phys. Lett. **A127** (1988) 410-412.
85. Suzuki, M.: Super-effective-field theory, general criterion of order parameters and exotic phase transitions, J. Phys. Soc. Jpn. **57** (1988) 683-686 (L).
86. Suzuki, M.: Statistical mechanical theory of cooperative phenomena. II. Super-effective-field theory with applications to exotic phase transitions, J. Phys. Soc. Jpn. **57** (1988) 2310-2330.
87. Suzuki, M.: Super-effective-field theory and coherent-anomaly method in cooperative phenomena, J. Stat. Phys. **53** (1988) 483-497.
88. Suzuki, M.: Super-effective field theory and exotic phase transitions in spin systems, J. Phys. (France) Colloque **C8** (1988) 1519-1524.
89. Suzuki, M.: Super-effective-field CAM theory of dynamical complexity, in *Cooperative Dynamics in Complex Physical Systems*, ed. H. Takayama (Springer-Verlag, Berlin, 1988) pp. 9-16.
90. Suzuki, M.: Canonical series of exactly solvable models for the CAM — Spontaneous broken symmetry of effective Hamiltonians based on the confluent transfer-matrix method —, J. Phys. Soc. Jpn. **58** (1989) 3642-3650.
91. Suzuki, M.: Super-effective-field CAM theory of strongly correlated electron

and spin systems, in *Recent Progress in Many-Body Theories, Vol. 2*, ed. Y. Avishai (Plenum Press, New York, 1990) pp. 277-289.

92. Suzuki, M.: Coherent-anomaly method and super-effective-field theory in magnetism, in *New Trends in Magnetism*, eds. M. D. Coutinho-Filho and S. M. Rezende (World Scientific, Singapore, 1990) pp. 304-318.
93. Suzuki, M.: New trends in the physics of phase transitions, in *Evolutionary Trends in the Physical Sciences*, eds. M. Suzuki and R. Kubo (Springer-Verlag, Berlin, 1991) pp. 141-162.
94. Suzuki, M., N. Hatano and Y. Nonomura: Canonicality of the double-cluster approximation in the CAM theory, J. Phys. Soc. Jpn. **60** (1991) 3990-3992 (L).
95. Suzuki, M.: Possible CAM canonical series in quantum and classical Monte Carlo simulations, Phys. Lett. **A158** (1991) 465-468.
96. Suzuki, M.: Emerging broken symmetry in space and time, in *Thermal Field Theories and Their Applications*, eds. H. Ezawa, T. Arimitsu and Y. Hashimoto (Elsevier, Amsterdam, 1991), pp. 5-15.
97. Suzuki, M., N. Hatano and Y. Nonomura: Numerical CAM analysis of critical phenomena in spin systems, in *Computational Approaches in Condensed-Matter Physics*, eds. S. Miyashita, M. Imada and H. Takayama (Springer-Verlag, Berlin, 1992) pp. 187–192.
98. Suzuki, M.: Coherent-anomaly approach to critical phenomena – proposal of quantum effective-field theory, Trends in Stat. Phys. **1** (1994) 225-234.
99. Suzuki, M., K. Minami and Y. Nonomura: Coherent-anomaly method — Recent development, Physica **A205** (1994) 80-100.
100. Takayasu, M. and H. Takayasu: Application of the coherent anomaly method to percolation, Phys. Lett. **A128** (1988) 45-48.
101. Takayasu, H., M. Takayasu and T. Nakamura: A new approach to generalized diffusion limited aggregation models. The coherent anomaly method, Phys. Lett. **A132** (1988) 429-431.
102. Tanaka, G. and M. Kimura: Critical temperatures and mean-field critical coefficients for the Heisenberg model based on CVM, Int. J. Mod. Phys. **B6** (1992) 2363-2374.
103. Tanaka, K., T. Horiguchi and T. Morita: Coherent-anomaly analysis with cluster variation method for spin-pair correlation function of Ising model on square lattice, J. Phys. Soc. Jpn. **60** (1991) 2576-2587.
104. Tanaka, K., T. Horiguchi and T. Morita: Critical indices for the two-dimensional Ising model with nearest-neighbor and next-nearest-neighbor interactions, Phys. Lett. **A165** (1992) 266-270.
105. Tanaka, K., T. Horiguchi and T. Morita: Critical indices for the two-dimensional Ising model with nearest-neighbor and next-nearest-neighbor interactions. II. Strip cluster approximation, Physica **A192** (1993) 647-664.
106. Tanaka, K., T. Horiguchi and T. Morita: Coherent-anomaly analysis with cluster variation method for two-dimensional Ising model with nearest-neighbor and next-nearest-neighbor interactions, Prog. Theor. Phys. Suppl. **115** (1994) 221-235.

107. Toda, M., R. Kubo and N. Saitô: Statistical physics I — Equilibrium statistical mechanics, 2nd ed. (Springer-Verlag, Berlin, 1992) pp. 167-168.
108. Todo, S. and M. Suzuki: Effective exclusive volume for hard-core systems in the liquid region, J. Phys. Soc. Jpn. **63** (1994) 3552-3555 (L).
109. Uzunov, D. I.: Introduction to the theory of critical phenomena — Mean field, fluctuations and renormalization (World Scientific, Singapore, 1993) p. 362.
110. Wada, K. and N. Watanabe: Coherent-anomaly method calculation on the cluster variation method. I, J. Phys. Soc. Jpn. **58** (1989) 4358-4366.
111. Wada, K. and N. Watanabe: The CAM calculation of critical exponent ν by the cluster variation method, J. Phys. Soc. Jpn. **59** (1990) 2610-2613 (L).
112. Wada, K., N. Watanabe and T. Uchida: Coherent anomaly method calculation on the cluster variation method. II. Critical exponents of bond percolation model, J. Phys. Soc. Jpn. **60** (1991) 3289-3297.
113. Zheng, W. M.: Retrieval of the dimension for Feigenbaum's limiting set from low periods, Phys. Lett. **A143** (1990) 362-364.

Phil Allen

For the Karpacz Winter School of Physics
24 February 1996

Note: *The 5 stanza verse known as a "limerick" is a popular art-form in English speaking countries. The rude language of a real limerick is not repeatable in polite company nor likely to be understood except by native speakers. Therefore this verse has been "edited". "Limerick" also happens to be the name of a county in Ireland. The connection between Limerick and limerick is lost in history*

To the Silesian resort of Karpacz
Jerzy Przystawa brings lots
    of theoretical thinkers
    both sober and drinkers
For tourists and Kindern to watch!

We have Petru and Jedrzejewski to thank
For a program of physics first rank;
    Enz, Tutis, and Makhlin
    Kept the Fermions hopping
Spins spinning, Bosons following Planck.

But Małgorzata Kopczyńska-Dakowska
Gave the program a wonderful glow-ska
    In the wood church of Wang
    Carried here by Max Planck
(*Oder war er Friedrich - Wilhelm des IV-ska?*

So to Jerzy the giant of Wrocław
No praise, blame, or thanks is enough,
    He brought us together
    By kicks, tricks, or whatever,...

At this point the "poet" ran out of rude words and rhymes, and offered a toast to a great friend, and host, and a giant among professors, Jurek Przystawa.

LIST OF KARPACZ SCHOOLS
OF THEORETICAL PHYSICS
Bibliography: 1964 – 1995

I – 1964: R.S. Ingarden (unpubl.), *Particles, Fields and Superconductivity,*

II – 1965: W. Ziętek (ed.), *Symmetries in Particle and Condensed Matter Physics,*Mimeographed Lecture Notes of the University of Wrocław

III – 1966: Z. Galasiewicz (ed.), *Statistical Physics of Condensed Matter,*Acta Universitatis Wratislaviensis, No. 80

IV – 1967: J. Rzewuski (ed.), *Functional Methods in Quantum Field Theory and Statistical Mechanics,*Acta Universitatis Wratislaviensis, No. 88, 89, 90

V – 1968: J. Łopuszański (ed.), *Axiomatic Approach to Quantum Field Theory and Many Body Problem,*Acta Universitatis Wratislaviensis, No. 98, 99, 113

VI – 1969: J. Przystawa (ed.), *Group Theory and Statistical Physics of Condensed Matter,*Mimeographed Lecture Notes of the University of Wrocław

VII – 1970: Z. Galasiewicz (ed.), *Liquid Helium and Many Body Problems,*Acta Universitatis Wratislaviensis, No. 141

VIII – 1971: J. Lukierski (ed.), *New Developments in Relativistic Quantum Field Theory and its Applications,*Acta Universitatis Wratislaviensis, No. 164, Vol. 1 and 2

IX – 1972: K. Wojciechowski (ed.), *Theory of Metals and the Many Body Problem,*Acta Universitatis Wratislaviensis, No. 181, Vol. 1 and 2

X – 1973: J. Lukierski (ed.), *New Developments in Relativistic Quantum Field Theory,*Acta Universitatis Wratislaviensis, No. 207

XI – 1974: J. Łopuszański, A. Pękalski, J. Przystawa (eds.), *Magnetism in Metals and Metallic Compounds,*Plenum Press, London

XII – 1975: W. Garczyński (ed.), *Functional and Probabilistic Methods in Quantum Field Theory,*Acta Universitatis Wratislaviensis, No. 368

XIII – 1976: J. Lukierski (ed.), *Recent Development in Relativistic Quantum Field Theory and Its Application,*Acta Universitatis Wratislaviensis, No. 389

XIV – 1977: Z. Galasiewicz (ed.), *Collective Effects in Condensed Media,*Acta Universitatis Wratislaviensis, No. 436

XV – 1978: W. Karwowski (ed.), *Mathematical Aspects of Quantum Field Theory,*Acta Universitatis Wratislaviensis, No. 519

XVI – 1979: A. Pękalski, J. Przystawa (eds.), *Modern Trends in the Theory of Condensed Matter,*Lecture Notes in Physics No. 115, Springer–Verlag, Berlin

XVII – 1980: L. Turko, A. Pękalski (eds.), *Developments in the Theory of Fundamental Interactions,*Studies in High Energy Physics, Vol. 3, Harwood Acad. Publ., Chur

XVIII – 1981: W. Garczyński (ed.), *Gauge Field Theories: Theoretical Studies and Computer Simulations,*Studies in High Energy Physics, Vol. 4, Harwood Acad. Publ., Chur

1982: *The School did not take place because of the Martial Law in Poland,*

XIX – 1983: B. Milewski (ed.), *Supersymmetry and Supergravity,*World Scientific, Singapore

XX – 1984: A. Pękalski, J. Sznajd (eds.), *Static Critical Phenomena in Inhomogeneous Systems,*Lecture Notes in Physics No. 206, Springer–Verlag, Berlin

XXI – 1985: L. Michel, J. Mozrzymas, A. Pękalski (eds.), *Spontaneous Symmetry Breakdown and Related Subjects,*World Scientific, Singapore

XXII – 1986: A. Jadczyk (ed.), *Fields and Geometry,*World Scientific, Singapore

XXIII – 1987: T. Paszkiewicz (ed.), *Physics of Phonons,*Lecture Notes in Physics No. 285, Springer–Verlag, Berlin

XXIV – 1988: R. Gielerak, W. Karwowski (eds.), *Stochastic Methods in Mathematical Physics,*World Scientific, Singapore

XXV – 1989: Z. Haba, J. Sobczyk (eds.), *Functional Integration, Geometry and Strings,*Birkhaeuser Verlag, Basel

XXVI – 1990: Z. Galasiewicz, A. Pękalski (eds.), *Ordering Phenomena in Condensed Matter Physics,*World Scientific, Singapore

XXVII – 1991: P. Garbaczewski, Z. Popowicz (eds.), *Nonlinear Fields: Classical, Random, Semicassical,*World Scientific, Singapore

XXVIII – 1992: R. Gielerak (ed.), *Infinite Dimensional Geometry in Physics,*Elsevier–North Holland, Amsterdam

XXIX – 1993: T. Paszkiewicz, K. Rapcewicz (eds.), *Die Kunst of Phonons,*Plenum Press, London

XXX – 1994: J. Lukierski, Z. Popowicz, J. Sobczyk (eds.), *Quantum Groups. Formalism and Applications,*Polish Scientific Publishers PWN, Warsaw

XXXI – 1995: (P. Garbaczewski, M. Wolf, A. Weron (eds.)) *CHAOS – The Interplay Between Stochastics and Deterministic,*Lecture Notes in Phys. vol. 457 Springer–Verlag, Berlin